Constructing the Expanding Universe

Constructing the Expanding Universe

First Edition

Dr. Uwe Trittmann

Otterbein University

Bassim Hamadeh, CEO and Publisher
Kassie Graves, Vice President of Editorial
Jamie Giganti, Director of Academic Publishing
Emely Villavicencio, Senior Graphic Designer
Jennifer Codner, Senior Field Acquisitions Editor
Michelle Piehl, Project Editor
Stephanie Kohl, Licensing Coordinator
Casey Hands, Associate Production Editor

Copyright © 2019 by Cognella, Inc. All rights reserved. No part of this publication may be reprinted, reproduced, transmitted, or utilized in any form or by any electronic, mechanical, or other means, now known or hereafter invented, including photocopying, microfilming, and recording, or in any information retrieval system without the written permission of Cognella, Inc. For inquiries regarding permissions, translations, foreign rights, audio rights, and any other forms of reproduction, please contact the Cognella Licensing Department at rights@cognella.com.

Trademark Notice: Product or corporate names may be trademarks or registered trademarks, and are used only for identification and explanation without intent to infringe.

Cover image: Copyright © 2016 iStockphoto LP/den-belitsky.

Printed in the United States of America

ISBN: 978-1-5165-0847-1 (pbk) / 978-1-5165-0848-8 (br)

To the next generation:
ARTHUR, SOPHIE & ANNA

Contents

Preface 1

1 **Observing and Naming: Naked-Eye Astronomy** 3
   1.1 Introduction and Invitation: Patterns in the Sky 4
      1.1.1 Trivial Observations 4
      1.1.2 The Earth Shapes Our Observing Experience 7
      1.1.3 Finer Observations: Constellations as a Frame of Reference 12
      1.1.4 Invitation 15
   1.2 Astronomy as a Science 16
      1.2.1 The Scientific Method 17
      1.2.2 Science Speak 20
      1.2.3 Models and Theories 22
      1.2.4 Measurement and Uncertainties 23
      1.2.5 Evidence and Proof in the Sciences 25
      1.2.6 Mathematics and Its Relation to the Sciences 25
      1.2.7 Plots and Diagrams 29
   1.3 The Motion of the Sun and the Stars with Respect to the Observer 32
      1.3.1 Daytime Astronomy 32
      1.3.2 Finer Observations with a Gnomon 33
      1.3.3 Nighttime Astronomy 36
   1.4 The Celestial Sphere as a Practical Tool to Organize Observations 38
      1.4.1 Seasonal Motion of the Sun Is a Path on the Celestial Sphere Called Ecliptic 39
      1.4.2 Seasonal Motion of the Stars 41
      1.4.3 A Different Observer Location Leads to a Different View of the Sky . 42
      1.4.4 The Celestial Sphere as a Concept 44
   1.5 Observing the Moon 46
      1.5.1 The Moon Moves with Respect to the Stars while Rising and Setting . 46
      1.5.2 The Phases of the Moon 48
      1.5.3 Eclipses Involve the Moon 51
      1.5.4 Interpreting Lunar Observations 52
   1.6 Observing the Planets 54
      1.6.1 Planets Slowly Wander among the Stars while Rising and Setting . . . 54
      1.6.2 A New Pattern: Retrograde Motion 54
      1.6.3 Trying to Interpret Planetary Observations 57

1.7 Summary and Application 58
   1.7.1 Digest of Observations Foundational for the Expanding Universe . . . 58
   1.7.2 Concept Application 59
   1.7.3 Activity: The Sun's Shadow and Position 60
   1.7.4 Activity: Angular Sizes and Distances 60
   1.7.5 Activity: Scaling 61
   1.7.6 Activity: Seasonal Motion 62
   1.7.7 Activity: The Seasons 63
   1.7.8 Activity: The Phases of the Moon 65
   1.7.9 Activity: Retrograde Motion of the Planets 66

2 Inferring & Theorizing: Planetary Astronomy Begets Modern Science   69
2.1 A Brief Summary of the Early History of Astronomy 70
2.2 The Ancient Greeks Discover That the Universe Can Be Discovered 73
   2.2.1 Archaic Greece Was an Ideal Place to Ponder Methods to Understand the Universe 73
   2.2.2 Plato Declares Uniform Circular Motion to Be Ideal 76
   2.2.3 Eudoxus Constructs a Cosmological Model of Homocentric Spheres . . 77
   2.2.4 Aristotle Devises an All-Encompassing Physical Theory of Nature . . 78
2.3 The Hellenistic Greeks Start to Measure the Size of the Universe 81
   2.3.1 Measuring the Skies: The Parallax 82
   2.3.2 Aristarchus Measures the Relative Sizes of the Earth, Moon, and Sun 83
   2.3.3 Eratosthenes Determines the Diameter of the Earth 84
   2.3.4 Hipparchus Discovers the Precession of the Equinoxes by Compiling an Accurate Star Catalog 86
2.4 The Most Enduring Framework: Ptolemaic Astronomy 89
   2.4.1 Ptolemy the Mathematician versus Aristotle the Physicist 89
   2.4.2 Ptolemaic Astronomy Is Based on Epicycles, Eccentric Circles, and Equants 91
   2.4.3 Ptolemy Explains Retrograde Motion with Major Epicycles 94
   2.4.4 The 1,400 Years after Ptolemy Saw Only Subtle Scientific Progress . . 97
2.5 Light at Last: Copernicus's Heliocentric Universe 99
   2.5.1 What Made Copernicus Possible in the Sixteenth, but not the Thirteenth Century? 99
   2.5.2 Copernicus' *De Revolutionibus* Is a Conservative Book That Started a Revolution 101
   2.5.3 The Reception of and Reaction to Copernicus' Book Was Very Slow . 105
2.6 Observing the Universe: Tycho Brahe 108
   2.6.1 Tycho Sets New Standards for Astronomical Observation 108
   2.6.2 Tycho Falsifies Aristotle 111
   2.6.3 Tycho's Copernico-Ptolemaic Model of the Universe 113
   2.6.4 Tycho and Kepler 114
2.7 Describing the Universe: Kepler 114
   2.7.1 Medieval Kepler: Chasing Mathematical Harmonies 115
   2.7.2 Modern Kepler: Using Data to Solve the Problem of the Planets . . . 118

	2.7.3	Bellwether Kepler: From Geometry to Algebra, from Mathematics to Physics	122
2.8	New Methods: Galileo		124
	2.8.1	The Beginning of the Telescopic Era	124
	2.8.2	Galileo's Telescopic Discoveries	128
	2.8.3	Galileo as the Founder of Experimental Physics	131
	2.8.4	Galileo and the Church	133
2.9	The Scientific Revolution		134
	2.9.1	How Revolutionary Was the Revolution?	134
	2.9.2	The Scientific Method: Theory and Practice	135
	2.9.3	Planetary Science between Kepler and Newton	136
	2.9.4	The Vacuum Exists, Falsifying Aristotle and Starting Thermodynamics	141
	2.9.5	Observational Astronomy and a Theory of Light	144
2.10	The Universal Explanation: Newton		149
	2.10.1	Newton's Three Laws of Motion	152
	2.10.2	Newton's Law of Universal Gravitation	155
	2.10.3	Successes and Limitations of Newton's Theory	164
2.11	Summary and Application		166
	2.11.1	Timetable of the Emerging Expanding Universe	166
	2.11.2	Activity: Introducing the Parallax	169
	2.11.3	Activity: Copernicus' versus Ptolemy's Explanation of Retrograde Motion	170
	2.11.4	Activity: Kepler's Laws	172
	2.11.5	Activity: Telescopes	174
	2.11.6	Activity: Venus Phases	175
	2.11.7	Activity: Newton's Three Axioms	175
	2.11.8	Activity: Newton's Law of Universal Gravitation	176

3 Deducing & Checking: Between Newton and Einstein   179

3.1	The Enlightened Century after Newton		180
	3.1.1	Enlightenment and the Age of Reason	180
	3.1.2	Newton Is Finally Accepted	182
3.2	Physical Astronomy		186
	3.2.1	Stellar Aberration: *Eppur Si Muove!*	187
	3.2.2	The Earth as a Spinning Top	190
3.3	The Scale of the Cosmos Is Determined		192
	3.3.1	Measuring the Astronomical Unit	192
	3.3.2	Weighing the World: Quantifying Gravity	194
3.4	The Birth of Chemistry Dissolves Aristotle's Elements		197
	3.4.1	Heat Is Not Temperature. Is Chemistry Physics?	197
	3.4.2	The Delayed Scientific Revolution	199
3.5	The Universe beyond Saturn, before the Earth, and below the Equator		202
	3.5.1	The Deep Sky	202
	3.5.2	The Formation of the "Universe"	205
	3.5.3	W. Herschel as the Father of Modern Astronomy	207
	3.5.4	The Southern Sky and Spiral Nebulae	210
3.6	The Industrialization of Science		212

		3.6.1	From the Eighteenth to the Nineteenth Century	212
		3.6.2	Telescopes and Instrumentation	214
		3.6.3	The Stellar Parallax and Precision Astronomy	218
	3.7	Light Is Electromagnetic Radiation		221
		3.7.1	Light Is a Wave (Again)	221
		3.7.2	Observing Light: Brightness and Doppler Effect	225
		3.7.3	Of Rainbows Dark and Bright Lines: Radiation	228
	3.8	The Rise of Thermodynamics		231
		3.8.1	A First Stab at the Energy Source and Age of the Sun	234
		3.8.2	The Second Law, Heat Death and Other Scary Victorian Stories	236
		3.8.3	Blackbody Radiation	237
	3.9	The Emergence of Astrophysics		242
		3.9.1	Photometry Variable and Binary Stars	242
		3.9.2	Astrophysics, Spectroscopy, and Astrophotography	243
		3.9.3	An Earthly Explanation of Stars: The First Stellar Models	246
		3.9.4	The Classification of Stars and Their Properties	251
	3.10	Summary and Application		253
		3.10.1	Timetable of the Emerging Expanding Universe (II)	253
		3.10.2	Concept Application	255
		3.10.3	Activity: Electromagnetic Waves	256
		3.10.4	Activity: Spectra and Blackbody Radiation	257
4	**Interlude: Discovering the Solar System**			**261**
	4.1	Introducing the Solar System		261
		4.1.1	Timeline of Discoveries	261
		4.1.2	Investigating the Solar System	263
		4.1.3	The Asteroids as Clues to Solar System Formation	266
	4.2	The Earth-Moon System		269
		4.2.1	The Earth	269
		4.2.2	The Moon	276
		4.2.3	The Formation of the Earth-Moon System	278
	4.3	The Terrestrial Planets		280
		4.3.1	Mercury and Venus	280
		4.3.2	Mars	281
		4.3.3	Atmospheres and the Greenhouse Effect	284
	4.4	The Jovian Giants		289
		4.4.1	Jupiter	289
		4.4.2	Saturn, Rings and Outer Planets	291
	4.5	Explaining the Formation of the Solar System		293
		4.5.1	Two Groups of Planets Emerge Due to Temperature Differences in the Early Solar System	293
		4.5.2	The Nebular Hypothesis Is Challenged by Finding Exoplanets	296
	4.6	Observing the Nearest Star		298
	4.7	Summary and Application		303
		4.7.1	Content of the Solar System	303
		4.7.2	Activity: The Greenhouse Effect	304

CONTENTS

    4.7.3 Activity: Formation of the Solar System 305
    4.7.4 Activity: The Sun's Properties 306

5 Extrapolating & Synthesizing: Understanding the Stars 309
5.1 Death by Extrapolation: The Birth of Modern Physics 310
    5.1.1 Relativity 310
    5.1.2 Quantum Mechanics 312
    5.1.3 The Elementary Building Blocks of the Universe 314
5.2 Making Sense of Stellar Classifications 316
    5.2.1 The HR Diagram Is Constructed to Make Sense of Stellar Properties . 316
    5.2.2 Summary of Observations Encoded in HR Diagrams 318
    5.2.3 Applying New Knowledge: The Spectroscopic Parallax 320
5.3 Synthesizing: Putting Radiation into Star Models 322
    5.3.1 Eddington's Standard Model of Stars 322
    5.3.2 Vindication: Stars Consist Mostly of Hydrogen Gas and Vary in Size . 326
    5.3.3 Standard Model 2.0: The Peach Star 328
5.4 Stars Produce Energy by Thermonuclear Fusion 330
    5.4.1 Necessary Preparatory Work 330
    5.4.2 The Microscopic Universe: Subatomic Physics 333
    5.4.3 Putting It All Together: Nuclear Fusion in Stars 334
5.5 Fusion Determines Stellar Structure and Life Cycle 338
    5.5.1 A Universal Theory of Stars 339
    5.5.2 Exceptions to the Rule: White Dwarfs and Neutron Stars 340
    5.5.3 The Construction of the Stellar Life cycle 343
    5.5.4 Predicting: Variable Stars 349
5.6 Nucleosynthesis Fills the Periodic Table 352
    5.6.1 Fusion Leads to Nucleosynthesis in Massive Stars 352
    5.6.2 Supernovae: The Fertile Death of Massive Stars 356
    5.6.3 Extrapolating: The Elementary Evolution of the Universe 365
5.7 Summary and Application 366
    5.7.1 Timetable of the Emerging Expanding Universe (III) 366
    5.7.2 Concept Application 368
    5.7.3 Activity: Hertzsprung-Russell Diagrams 369
    5.7.4 Activity: Stellar Models 370
    5.7.5 Activity: Early Stellar Life 370
    5.7.6 Activity: Cosmic Yardsticks 371
    5.7.7 Activity: Late Stellar Life 372

6 Refining to Infinity: Cosmology 375
6.1 The Milky Way Contains Virtually Everything 375
    6.1.1 The Shape and Size of Our Galaxy 376
    6.1.2 Galactic Structure: The Milky Way Has Three Distinct Parts 379
    6.1.3 Galactic Rotation as Evidence of Dark Matter 386
6.2 Other Galaxies: The Milky Way Is Not the Universe 390
    6.2.1 Galaxies Are Island Universes 390
    6.2.2 Galaxies Come in Different Shapes and Sizes 391

		6.2.3	Galactic Evolution and Collisions	392
		6.2.4	Galaxy Clusters and Superclusters	394
	6.3	Cosmology: The Cosmos as a Single Object		396
		6.3.1	Climbing the Cosmic Distance Ladder	396
		6.3.2	The Large-Scale Structure of the Universe	398
		6.3.3	Cosmological Questions	398
	6.4	The Expanding Universe		400
		6.4.1	Observing the Expanding Universe	400
		6.4.2	Modeling the Expanding Universe: Standard Cosmology	404
		6.4.3	Steady State—Rise and Fall of the Alternative Cosmology	410
	6.5	The Failure of Standard Cosmology at the End of the Millennium		412
		6.5.1	History of the Universe	412
		6.5.2	The Golden Age of Precision Cosmology	415
		6.5.3	The Big Shock: The Universe Is Accelerating	416
	6.6	Summary and Application		419
		6.6.1	Timetable of the Emerging Expanding Universe (IV)	419
		6.6.2	Concept Application	420
		6.6.3	Activity: The Milky Way	420
		6.6.4	Activity: Hubble's Law	421
		6.6.5	Activity: Determining Hubble's Constant	422
		6.6.6	Activity: The Expanding Universe	422
		6.6.7	Activity: The Expanding Ballooniverse	423
		6.6.8	Activity: Space-Time Curvature	424
A	**Useful Data**			**426**
	A.1	Constants		426
	A.2	Properties of Astronomical Objects		428
Glossary				**431**

Preface

This book is based on a semester-long general-education introductory astronomy course the author has taught many times at Otterbein University. The foundational idea of the approach taken is that the expansion of the universe is *twofold*. The universe is *physically expanding*, but this is by no means obvious for humans confined to the Earth and relegated to observing the universe as projected onto the night sky. To realize that they live in an expanding cosmos, humans had to get to know the universe from the ground up. They had to figure out what the different objects in the universe *are* and how they *work*. In fact, humans had to figure out *how* to figure things out and that things *can* be figured out at all. They thus had to expand their knowledge of the universe, and we might as well say with BURKE [11] that *they* expanded *their* universe. The first strand of this double expansion is therefore the standard, *science account* of what is known about astronomy and the universe today. The second is the *historic account* of how this was achieved. The two strands are integrated to give the reader a sense of both what science is and how it is made.

Make no mistake: This is not your standard "intro astro" book! In this book you will *not* find the latest on the planets, black holes, or gravitational waves. These details, however exciting they may be, can confuse and hamper learning. Indeed, difficulties to understand modern scientific results can regularly be traced back to a lack of mastery of basic science and math concepts. The focus of the book therefore is on underlying, fundamental concepts of science that can often be experienced in everyday life and that speak to and nurture the scientist in all of us. *Operational knowledge* as expressed in questions like "How do we know ...?" and "How would you measure...?" is foregrounded. The book puts much emphasis on the *development* of scientific ideas. It aims at elucidating how different descriptions and theories were validated or discarded. The historic development of astronomy and related sciences is presented to show how we got to our modern view of the natural world *and* to motivate and elucidate the scientific results and insights. The hope is to outline a cultural history of science using astronomy as an example—as the oldest and most far-reaching. This is much in the spirit of JAMES B. CONANT's foreword to THOMAS S. KUHN's 1957 book "The Copernican Revolution." CONANT there envisions a common ground between the literary and the scientific traditions which might be realized "in the study of the history of science, particularly if combined with an analysis of the various methods by which science has progressed."[26] Indeed, astronomy and (Western) culture and thought are, without a doubt, deeply entwined. As much as astronomy and science have shaped and defined our modern world, scientific thought and progress were molded by the societies that brought them forth. The story of the construction of the expanding universe is therefore best told in a linear, chronological fashion. This is the guideline I tried to follow in this book.

I have not shied away from presenting mathematical arguments where they are useful.

After all, "the book of the universe is written in the mathematical language." (GALILEO) *This* book is based on a *college-level* course, and I am confident that the reader will have no trouble following the math—which typically does not go beyond eighth-grade proficiency. On the other hand, I have tried to limit or explain **science jargon** as much as possible, while providing ***etymological, epistemological***, and ***historical connections***[1] as part of the main story line and in the form of footnotes.

There are two necessary ingredients for meaningful learning to occur: motivation and grit. Astronomy is an exciting subject, so willingness to learn should not be an issue. But science can be a thorny subject, so perseverance and true engagement with the concepts is key. To help matters, I have included three layers of ancillary materials. At the end of each section, you'll find several *Concept Practice* questions. These are *exercises* in the sense that you should find it straightforward to answer them after you thoroughly read the section's text. At the end of each chapter, there is a *summary* which highlights the important concepts, and displays them in a complementary way, often in the form of a timeline of the evolving universe. Lastly, there are *Concept Applications*. These are problems—not exercises—so they will be more challenging as you have to *apply* the new knowledge and insight gained, which is, of course, harder than just regurgitating it.

It is crucial to practice what you've learned. Practice questions are one way, but don't forget that in astronomy everything is linked to the night sky. So go out, observe, and think about what you are seeing. If you enjoy the universe, learning about it is an unavoidable by-product of the most natural and inspiring flow experience there is: To embrace the beauty of the cosmos!

I am grateful for a sabbatical leave from Otterbein University which allowed me to start this book. The bulk of the text was composed while visiting Siegen University. The generous support of the Department of Physics at Siegen is gratefully acknowledged. The travel was supported in part by a grant from Otterbein's Faculty Scholarship Development Committee. Much of the finishing work was done at the Ohio State University as a *visiting scholar*, and I am grateful for this support. I am indebted to PROF. O. SCHWARZ for help with the literature, stimulating discussions, and a hard copy of his book [34]. I thank DR. S. FRANK for a careful reading of the manuscript. Discussions with PROFS. ROBERTSON, TAGG, REINHARD (Otterbein), BELL, DAHMEN, GRUPEN, FELDMANN, HUBER, KHODJAMIRIAN, KILIAN, MANNEL (Siegen), HECKLER, HEINZ, KOVCHEGOV, PERRY, SHIGEMITSU (OSU) are gratefully acknowledged. Thanks to my project editor M. PIEHL and the folks at Cognella. I thank my parents for hospitality and support during my stay in Siegen. Last but not least, I am indebted to DR. JENNIFER TRITTMANN and our family for love and support while I was busy completing this book.

U.T.

Clintonville, Ohio
July 2018

---

[1] Here and in the remainder of the book, ***boldfaced italics*** are used to emphasize technical or unfamiliar terms that the reader may wish to get accustomed to. SMALL CAPS are used for names throughout, followed by the birth and death dates on first occurrence.

Chapter 1

Observing and Naming: Naked-Eye Astronomy

This chapter sets the stage for the rest of the book by describing the fundamental naked-eye observations on which everything else is based. The reader will undoubtedly feel that she is familiar with many if not all of these observations and their interpretation. So much so, that it seems hardly worth mentioning these patterns in nature. However, it is well documented in educational research that this familiarity is treacherous. For instance, most people are unable to correctly explain the seasons, the moon phases, or how we know that the Earth orbits the sun. Incidentally, most of what we learn in school is **declarative knowledge** (remembering facts), not **operative knowledge** from functions we have performed. Therefore, most of us will be unable to *apply* this knowledge, and not be able, say, to find north or to determine when noon occurs without using technology.

In this chapter, we list the fundamental observations but not their interpretation. For instance, when describing the motion of the moon, we will distinguish its motion with respect to the observer on the ground from its motion with respect to the stars, but refrain from saying that it orbits the Earth. This is done for three reasons. First, to make it clear how few things we can experience immediately with our senses and how much of our world view is *constructed* by agreeing on names and by *theorizing* how the world functions. Second, to uncover the path of science and the struggle to make sense of the patterns we encounter in nature. Only by observations and their careful interpretation do we arrive at an accurate description of nature, mistakes and fallacies notwithstanding. Finally, we live in a world where information and advanced measuring devices are so readily available that it is hard to fathom how people could doubt such "obvious" things like the rotation of the Earth. All these facts are, nonetheless, hard-won insights distilled from myriads of observations. To truly appreciate where our knowledge of an expanding universe comes from, we have to go back to its roots. Every new discovery in science is based on an older one all the way back to the **fundamental observations** described in this chapter that started this amazing and continuing increase in human knowledge.

Guiding Questions of this Chapter

1.1 What is a pattern in nature? What can we see with the unaided eye? What does its appearance tell us about a celestial object? How are simple concepts like *horizon*, *zenith*, and *north* defined operationally? How does our observing location determine our view of the sky? How do the observing date and time determine our view of the sky?

1.2 Which mathematical tools do we need to describe the universe? How should we conduct and record our observations? What is scientific and what is not? Is there a way to ensure progress in science? How should we talk about our observations?

1.3 What is the difference between daily and seasonal motion? What is the difference between a solar and a sidereal day? How does the sun move in the sky? How do stars move relative to the ground?

1.4 What is the celestial sphere? Why is this concept useful? How does the sun move with respect to the stars? What is seasonal motion? What is the reason for the seasons?

1.5 How does the moon move with respect to the stars? What is the reason for the moon's phases? What happens during eclipses? How does the moon rotate?

1.6 How do the planets move with respect to the stars? In which way is planetary motion different from lunar and solar motion?

1.1 Introduction and Invitation: Patterns in the Sky

1.1.1 Trivial Observations

In the beginning humans realized that there are recurring patterns in nature. Well before centers of civilization formed they made many **trivial**[1] observations. They saw that it is dark for about as much time as there is light. The huge, seemingly empty space above the ground they stood on appeared like a half dome touching the ground far in the distance in a circular line that went all around the human observer—no matter where they stood. When there was light, it seemed to come from a bright disklike object in the sky that also made them feel warm when they stood in its light. Even when the bright disk was hidden behind puffy objects far above their heads that sometimes dispersed water, the half dome appeared still fairly bright. They also saw a different, paler object of the same size, sometimes visible when it was light, sometimes only visible when it was dark. This object changed its shape: Sometimes it appeared as a crescent, sometimes as a full disk, and at most times as something in between. Both objects changed their positions with respect to the ground as the hours went by. Both were close to the ground in a special direction, then rose higher in the sky up to a point, before sinking lower and eventually disappearing in a direction roughly opposite to the direction in which they appeared. These early humans also saw tiny specks of light in the sky. These specks were only visible when it was dark, but they were moving in the same fashion as the bright and pale disklike objects: Appearing in a special direction, reaching a highest point, then disappearing in roughly the opposite direction. Some changes happened over much longer periods. It was noted, for instance, that weather would get warmer at the

---
[1] *Trivial* here in the liberal arts sense: **foundational**.

same time that the bright disk was highest in the sky, which was also the time when the sky was bright for much longer than it was dark. Many other patterns are so familiar that we do not list them here, like the tendency of unsupported objects to fall to the ground.

Since these patterns continued with seemingly endless repetition, people gave them names to refer to them when talking to their fellow humans. They called the dark period *nighttime* and the light period *daytime*, and both together a *day*. The bright disk was called *sun*, the pale disk *moon*, the specks of light were labeled *stars*. For orientation, the line where the sky meets the Earth was called *horizon*, and the special direction in which the celestial objects appeared above the horizon came to be known as *east*, its opposite as *west*, and their appearance and disappearance as *rising* above and *setting* below the horizon. The period of warm weather and long daylight was labeled *summer*, and its opposite *winter*. The puffy objects that hide the celestial objects are, of course, *clouds*.

In science *naming* is a vital task. The trick is to to find and label the important concepts and not to get fooled into creating redundant labels by nonessential differences. We often give a name to abstract concepts like *day*. It is the careful choice of concepts or quantities that propels the sciences forward. Note, though, that the name or label itself is arbitrary, and usually does not convey anything about the objects themselves. For instance, a *planetary nebula* has nothing to do with planets at all. Worse, names are often mistaken for *explanations*. As an example, consider the label *gravity*. If you ask someone why unsupported objects fall down, they'll likely say: "Because of gravity," as if it was known what gravity *is* [1]. In fact, nobody knows as of yet, even though both NEWTON's and EINSTEIN's gravitational theories grapple with it. ISAAC NEWTON (1643–1727), for instance, just *postulated* that there is a force of gravity, without explaining where it comes from. Incidentally, most people would be hard pressed to explain what they mean by the related concept we give the name *down*, especially if standing on a sloping hillside, where simple statements like "Down is perpendicular to the ground" do not work.

Therefore, we will use names only for concepts that we have *defined* properly. How do we define a concept? By a description of how to observe it, known as an *operational definition*. So we could say: "*Down* is the direction of motion of falling objects." Note that this definition works on the sloped hillside as well. In fact, a more practical tool to find the *down direction* is a *plumb line*, consisting of a heavy, tipped weight suspended from a string serving as a vertical reference line. We see that the definition of *down* implies the operation (measurement) you have to perform. Now we can continue with "*Up* is the opposite of the *down* direction," and so forth. Let us define another useful concept this way. *Zenith* is the name we give to the point in the sky right above the observer's head.[2] This implies the instruction: "Look straight up to find the zenith." Mathematically, it is the direction that is perpendicular to the plane of the *horizon*. The former definition is more practical if equivalent.

Our account of fundamental observations would not be complete without mention of the patterns in the sky formed by the *stars*. When observing the night sky our brain immediately groups neighboring stars into *asterisms* forming shapes like triangles and rectangles. Over time, we realize that these patterns retain their shapes, though their position in the sky changes. Since they constitute a stable and recognizable pattern, it isuseful to name them; they are *constellations* like *Orion* or *Ursa Major*, the *Great Bear*; see Figure 1.1. We find

---

[2] *Zenith* comes from an Arabic expression meaning "way over the head," which is precisely its operational definition.

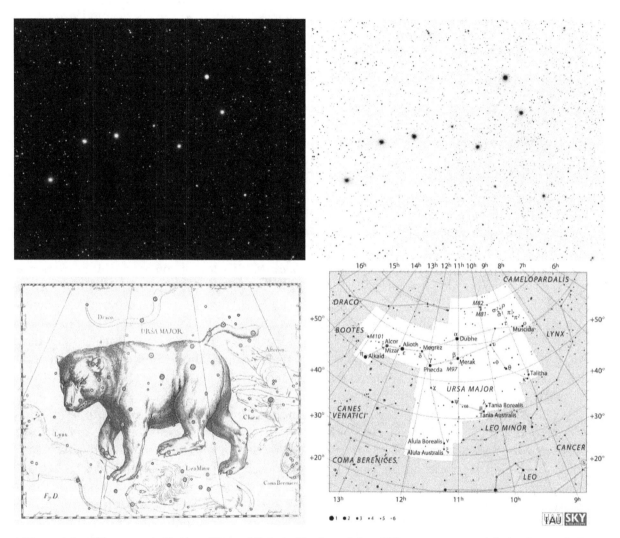

Figure 1.1: **The constellation Ursa Major displayed in different ways.** (a) A photo of the night sky showing an area of about 45° ×25°. Note the different brightnesses and colors of the stars. (b) Inverted photo for better contrast affording easier identification of dimmer objects. (c) A historic star map (HEVELIUS, mid-1600s) showing stars and the mythological figures. Maps like these were in use well into the modern era. Note that the map is reversed, as if looking at a "sky globe." (d) Modern star map showing the stars and official boundaries of the constellation *Ursa Major* in an *equatorial coordinate system*. Different symbols are used for double stars, variable stars, stars of different apparent brightness, and **deep sky objects** such as galaxies and star clusters. Names of stars, constellations, and **deep sky objects** are displayed. Thus, a lot of information is conveyed.

1.1. INTRODUCTION AND INVITATION: PATTERNS IN THE SKY

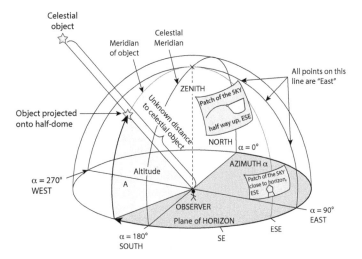

Figure 1.2: **The observer's sky and the horizontal coordinate system.** Shown are the horizon and the zenith with the celestial meridian, that is, the line that runs from north to south via the zenith. The **altitude** A is the angle between the celestial object and the horizon, the **azimuth** is the angle along the horizon between north and the meridian of the object. The distance to the object is irrelevant when determining its position in the sky.

also diffuse objects in the sky. The faintly glowing band that arches across the night sky is known as the **Milky Way** (from Greek *galaxias kyclos* = *milky circle*). Smaller diffuse patches were collectively known as nebulae before their true nature as **stellar clusters**, **galaxies**, or **gas clouds** was discovered.

At this point you may wonder why we haven't mentioned planets. The reason is that planets look, to the unaided eye, like stars. That means it takes a closer look to realize that there are two different categories of **specks of light**. As of yet, there is no reason to introduce another name, because stars and planets, for now, are indistinguishable. Both fit the operational definition of a rising and setting speck of light, visible at night.

1.1.2 The Earth Shapes Our Observing Experience

Until very recently, all astronomical observations were **ground based**, that is, had to be made from some location on Earth. The shape, motion, and atmosphere of Earth influence the way in which we perceive the universe. Take, for instance, the daily pattern of rising and setting of celestial objects. Standing on Earth, we cannot see the part of the sky that is below the horizon.[3] Due to the rotation of the Earth in space, objects like the sun rise higher in the observer's sky during the morning hours, and sink toward the horizon in the afternoon. Hence, there must be a point of highest **altitude**, that is, largest **angle** between the sun and the horizon. The direction in which we have to look to see the sun at its highest point during the day is called **south**,[4] and the time when this happens is called **local noon**. *Local* implies that for a different observer noon may occur at a different time. While we are

---

[3]It is a trivial fact that we cannot see through **opaque** things like rocks, but even the atmosphere is opaque for some forms of light, as we will see later.

[4]Strictly this is true only in the northern hemisphere of the Earth. In the southern hemisphere the sun culminates in the north.

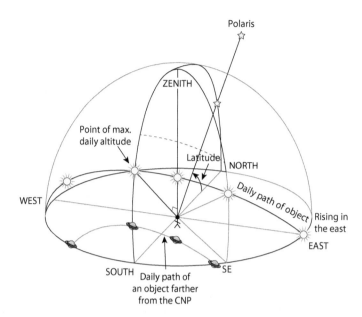

Figure 1.3: **The daily path of celestial objects.** Rising at the eastern horizon, and reaching maximal altitude (that is, culminating) at the meridian, objects are setting at the western horizon. Position at rising and setting, as well as maximal altitude angle depend on the angular distance of the object from the Celestial Pole signified by *Polaris* and different from the zenith. The closer the object is to the pole, the longer it will be above the horizon.

at it, we might as well define the direction opposite of **south** as **north**, and **east(west)** as "direction 90 degrees to the left(right) of south," see Figure 1.2. The latter are more precise definitions of east and west, superseding the rough definitions introduced above. We will see later why this is necessary.

The great circle running from the southern direction in the plane of the horizon through the zenith until it reaches the northern horizon is called the **celestial meridian**, see Figure 1.2. The term is familiar from our division of the day into morning or **a.m.** and afternoon or **p.m.** hours, short for **ante** and **post meridiem**.[5] The meaning is also apparent in the English word *afternoon*, that is, after noon or after midday (sun is highest in the sky), as opposed to mid-night (sun is lowest).

Regardless of the position of a celestial object, regardless of the position of the observer on Earth, and regardless of the object—be it sun, moon or stars—the object always culminates on the meridian. This generality makes the meridian such a *useful* concept; it simplifies descriptions of motion in the sky, and makes our statements more precise. Note, however, that you cannot see or touch the meridian. Albeit operationally defined, it is an abstract concept. From now on, we will say that an object **culminates**.[6] when it crosses the meridian on its daily path in the sky. It is easy to convince yourself that these paths must be concentric circles sharing a common center. This center is a special point in the sky that remains constant as everything else circles around it. It is called the **Celestial North Pole (CNP)** or the **Celestial South Pole (CSP)**, depending on the hemisphere of Earth from which you

---

[5] From Latin: *ante=before* and *post=after*, *medius=middle* and *dies=day*.

[6] This is the **upper culmination** For stars close to the Celestial Poles there is also a lower culmination described later.

1.1. INTRODUCTION AND INVITATION: PATTERNS IN THE SKY

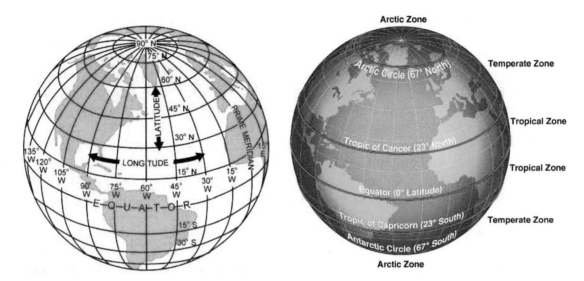

Figure 1.4: **The coordinate system of the Earth.** *Latitude* is counted from the *equator* and reaches 90° at the poles. The origin of *longitude* is the *prime meridian* at Greenwich, UK. (b) The *climate zones* of the Earth: Polar or arctic regions above $66\frac{1}{2}°$, tropical zone between the $23\frac{1}{2}°$ parallels, and the *temperate zones* in between.

observe. For a given position on Earth, only one Celestial Pole will be above the horizon.[7] The Celestial Poles are on the meridian. Their altitude angle depends on the latitude[8] of the observer, as we will see.

When we are standing somewhere on the Earth to observe the sky, we have to keep in mind that the observer sky, as depicted in Figure 1.2, is an *idealization* with a so-called *geometrical horizon*. In practice, we are not observing at sea level with a perfectly flat ocean, but the line of the horizon may be obstructed by trees and buildings, we might be observing from a higher elevation, and the air bends the starlight, so that a star near the horizon appears to be about half a degree higher than it would without the Earth's atmosphere.[9] The view of the sky is different for observers at different locations on Earth. We will discuss all of this in due time. Here we note that it is easier to study the sky than it is to study the gross features of the Earth. The reason is that the overall shape of an object is much easier to assess from a bird's-eye view. For example, an aerial view or map of a city makes it easier to find your way than the "street view" out of your car's windows. Nonetheless, over two thousand years ago, Greek astronomers were able, with clever geometrical reasoning, to determine the shape and size of the Earth. Furthermore, the obscure effects of the atmosphere on light rays were modeled correctly by Kepler more than four hundred years ago, long before the composition of air[10] was known. However, for most of history, things like the shape of continents, the distribution of the landmass, the internal structure of the Earth, or the spatial extent of the atmosphere were unknown, while the heavens were charted with enviable precision.

In summary, for the longest time the role of the Earth in astronomy was simply to provide

---

[7]Technically, both poles are visible if you observe from somewhere on the Earth's equator.
[8]For the definition of *latitude* and *longitude*, see Figure 1.4(a).
[9]This is known as *atmospheric refraction*; see Figure 2.19(a).
[10]Discovery of nitrogen (78% of air) and oxygen (21%) in 1772/1774, of argon (1%) not until 1895.

a horizon, that is, quite literally to stand in the way of observing the sky. We can divide observing locations on the Earth into three zones, as shown in Figure 1.4(b): the **polar regions** or **frigid zones** within 23.5° of the geographic poles, the **tropical zone** within 23.5° of the equator, and the **temperate zones** between the two others. Most people live in the **temperate zone**, from which the view of the sky is in some sense *typical*. In particular, the sun is never overhead, most stars will follow the standard rise-culminate-set pattern, and part of the stars of the opposite celestial hemisphere are visible.[11] The significance of 23.5° boundaries corresponds to the inclination of the Earth's axis with respect to its orbit, as we will see later. Due to the Earth's shape[12]—idealized as a sphere—the horizon will be different for observers at a different geographical longitude as well as latitude. For now, we will just state the **empirical facts**, which you can check yourself when taking a road trip.

- For every degree in latitude an observer in the northern hemisphere is further north, the altitude angle of the CNP will increase by one degree. Of course, the analogous statement is true in the southern hemisphere, just replace "north(ern)" by "south(ern)." This means that the altitude angle of the Celestial Pole *is* the latitude of the observer! For instance, if you see *Polaris* 40° above the northern horizon, you know you are at 40° northern latitude (and you've established the cardinal direction **north** without a compass.)

- For every degree of geographical longitude the observer is further west, the sky will look rotated clockwise by one degree about the Celestial Pole. Incidentally, you will have the same view of the sky as a person one degree further east if you simply wait four minutes, or observe at the same time but one day later.

These observations are summarized in Figure 1.6. Now, the second observation poses a problem for timekeeping. If the sky appears rotated, then local noon occurs at a later time further west. Therefore, it is necessary to agree on a way to record timing of astronomical events *independently* of the location of the observer. In fact, we might go a little further and ponder time and timekeeping itself. Imagine, then, the world about four hundred years ago, a world without mechanical clocks, stopwatches, smartphones, or other timekeeping devices. How would you measure time? This is an important question because all of today's timekeepers have been developed from rudimentary time-measuring methods by constant refinement.

We find *natural* timekeepers in the regular astronomical patterns. For long-term timekeeping, we can use seasonal changes like the varying length of shadows at noon. On a shorter timescale we might use the daily motion of the sun and the stars. The base unit is the **solar day**, defined as the time between subsequent culminations of the sun, that is, the time from noon to noon. It is divided into 24 hours which are conventionally[13] divided into 60 minutes of 60 seconds each. These two timescales are linked, because the time between two instances when the noon shadows are shortest is 365 times longer than the time from noon to noon.

---

[11] In other words, large parts of the southern stars are visible from northern latitudes, and vice versa.

[12] The precise form of the Earth was subject to debate until modern times, Section 3.1.2, but its approximately spherical shape was established very early on. It's a myth that any of Columbus's contemporaries still believed the Earth was flat. Columbus might have misrepresented its size to land royal funds for his expedition, but that's about it.

[13] This convention goes back to the Babylonians who had a number system based on 60, because 60 is divisible by many integers: 1, 2, 3, 4, 5, 6, 10, 12, 15, 20, 30.

1.1. INTRODUCTION AND INVITATION: PATTERNS IN THE SKY   11

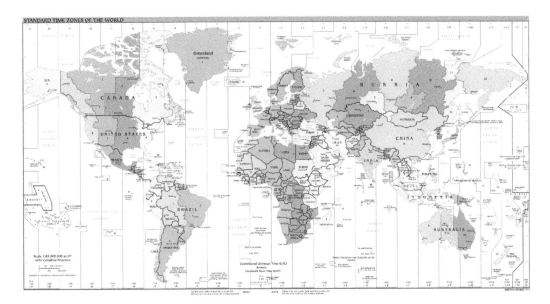

Figure 1.5: **The time zones of the Earth.** The Earth's surface is divided into strips of 15 degrees called *time zones* so that local noon occurs approximately at 12:00 p.m. For political reasons (boundaries between countries and such), the time zones prescribed by law are jagged and distorted. Note that western Spain and eastern Poland are in the same time zone; the sun sets early in Poland and late in Spain. Also the map is distorted, because it is the projection of the curved surface of the Earth onto a flat piece of paper. Note the different size of Greenland compared to Figure 1.4(a)! Star charts suffer an analogous distortion.

In other words, a *year* is 365 solar days[14]. The modern base time unit is the *second*, and a day is

$$T_{SolarDay} = 24\,\text{hr} = 24\,\text{hr} \times \frac{60\,\text{min}}{1\,\text{hr}} \times \frac{60\,\text{sec}}{1\,\text{min}} = 86\,400\,\text{sec},$$

where we have used the occasion to introduce the *conversion factors* between the various time units. For example, since there are 60 minutes in one hour, $\frac{60\,\text{min}}{1\,\text{hr}}$ is simply one!

This solves the timekeeping problem for the local observer, because she can tell time by recording the position of a shadow cast by the sun or the position of stars at night; both move virtually at the same rate of 15 degrees per hour which is, of course, the same as 360 degrees per day. However, communication between local observers poses a problem. On the one hand, the local observer wants the sun to culminate at noon, but local noon occurs later in the west and earlier in the east. On the other hand, two observers want to compare notes about astronomical events, which occur at a specific "absolute" time, regardless of where the sun is in an observer's sky. Two provisions are made to solve these problems. First, the Earth is divided into 24 strips of 15° longitude, so that noon occurs at 12:00 p.m.;[15] see Figure 1.5. Second, there is a special strip at the prime meridian (longitude 0°). Its local time is regarded as the "world time," called *Coordinated Universal Time* (UTC). So if you live in the eastern United States and observe an event at 3:45 a.m. local time, that is Eastern Standard Time (EST), you would inform your fellow astronomers that this event has occurred

---

[14]Technically, the time between shortest noon shadows is the *tropical year* (the time between two *vernal equinoxes*), which is 365 d 5 h 48 m 45 s, or roughly $365\frac{1}{4}$ days long.

[15]Plus (minus) thirty minutes if you live at the extreme western (eastern) boundary of your zone.

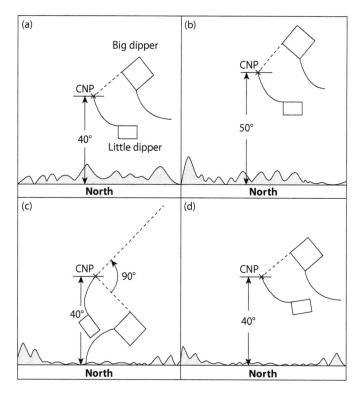

Figure 1.6: **Horizon views from different observing locations.** (a) View from 40° northern latitude. The altitude of the CNP is 40° above the (northern) horizon. (b) View from 50° northern latitude, same time and date. The altitude of the CNP is 50° at this location. (c) Same as (a) but 90° longitude further west, same time and date. The view of the sky is rotated clockwise by 90°. This makes sense, since six hours is a quarter day and 90° is a quarter of a full circle. (d) Same as (c) but six hours later, same date or three months later and same time. The sky view is the same as in (a), although the observer is at a different location. In six hours the sky has rotated 90° counterclockwise. It makes sense that the same view is reproduced three months later, because this is a quarter year later.

at 8:45 UTC, because EST is lagging 5 hours behind UTC. This makes sense, because the eastern United States is around 75° W longitude, and $\frac{75°}{15°/h} = 5\,\text{h}$. So the sun reaches the highest altitude in New York 5 hours later than in London, because the Earth has to rotate into position.

1.1.3 Finer Observations: Constellations as a Frame of Reference

At first glance the night sky is rather confusing. Several thousand specks of light of different brightnesses and colors seem to be randomly distributed. Different cultures in the past have seen different patterns in this distribution, and have grouped the stars into **asterisms**, which they named, usually after mythological figures. An asterism of seven stars always visible at mid-northern latitudes is called "Big Dipper" in English because it looks like a big spoon, but "Großer Wagen" in German because it looks like a great wagon; see Figure 1.1. Technically, it is part of one of the largest **constellations** in the sky, the *Great Bear* or *Ursa Major* From Figure 1.1 it is clear that there are several ways of displaying the same **data**. We see a photo of the night sky. Then there is an old star map showing the positions of the stars and the associated mythological figure. In the modern star map there is additional information about the brightness of these stars, and symbols indicate names of stars and whether they are binary or variable stars. Often, the stars are shown connected with lines to create shapes. In general, these outlines do not resemble the figure for which they are named. Of course, none of this additional information is visible in the night sky.

As the name **asterism**[16] suggests, stars *appear* close to each other from our vantage point. Actually, we have no idea how close they are to us nor to other stars.[17] Astronomers have **agreed** on a set of eighty-eight constellations with **well-defined** boundaries. Note that the boundaries are just convention, and therefore arbitrary. They could have been chosen in a different way—much like the shape of counties in Ohio, of which there are coincidentally also eighty-eight. This convention is **useful** because it makes it easy to describe the position of celestial objects relative to the constellations. We can now say, for instance, that Jupiter is in *Leo* this month, which makes it easier to spot. Indeed, much of science is a quest for convenient, useful concepts and quantities to describe nature's phenomena. The eighty-eight constellations of stars are known by their **Latin names**, often abbreviated by a three-letter combination, such as *CMa= Canis major*. The individual stars are labeled by the letters of the Greek alphabet. Traditionally,[18] the α star is the brightest, β the second brightest, and so forth. The Greek letter is then followed by the *genitive case* of the Latin constellation name, for example α *Canis majoris*. Several hundreds of stars also carry individual names, so often stars carry more than one label. For instance, the brightest **fixed star** α *Canis majoris* is also known as *Sirius*.

By recognizing at least some constellations in the night sky, we can get a sense of **orientation**.[19] The constellations establish a **frame of reference**. For instance, in spring the Big Dipper is high in the sky and easily recognized. You can take it as the starting point of an exploration of the night sky. Extending a straight line connecting its brightest stars, *Dubhe* and *Merak*, about five times leads you to *Polaris* (α Ursae Minoris). *Polaris* is a moderately bright star that happens to be close to the **CNP**. It therefore conveniently shows us where north is. Following the handle of the Big Dipper instead, you end up locating the bright orange star *Arcturus* in the constellation *Boötes*, the Herdsman. Continuing along this arc lies the bluish star *Spica* in *Virgo*. Some people remember this path by saying "Follow the arc to Arcturus and drive a spike to Spica," see Figure 1.7(a). Similar techniques exist to locate the constellations in other areas of the night sky. In the winter, you probably want to start with the familiar figure of *Orion*, which leads you to *Sirius* in *Canis Major*, and so forth. Note that this way of discovering the night sky is mirroring the approach taken in science: Starting with the familiar, we approach new phenomena. After mastering the main constellations, you may find fun in hunting down the dim and obscure ones like *Delphinus (Dolphin)*, *Musca (Fly)*, or *Lacerta (Lizard)*. Otherwise, rest assured that initially only twelve out of the eighty-eight constellations will be important for our exploration of the solar system and even beyond.

Having explored the night sky for a while, you will find that you can make perfect sense

---

[16] From Latin: *aster=star*.

[17] Astronomers tried in vain for thousands of years to figure out the distances to the stars. This most important astronomical problem was solved in 1838 by Bessel [4].

[18] This tradition was started by JOHANN BAYER (1572–1625) in his star atlas *Uranometria* of 1603.

[19] From the Latin *oriens* meaning east; early Christians *oriented* themselves when praying, that is, they faced east.

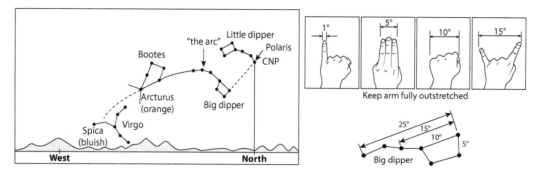

Figure 1.7: **Finding your way around the sky.** (a) *Star hopping*: Starting from a familiar constellation, others can be found by following a string of stars. Shown here is the "arc to Arcturus" from *Ursa Major* to *Boötes*, followed by the "spike to Spica" in *Virgo*. (b) A simple way of estimating angles with your hands on an outstretched arm.

of the initially random-looking distribution of specks of light. You will then be able to "read" the night sky, finding north, and telling what time or season it is.[20] In other words, the constellations provide a ***frame of reference*** that makes it possible to understand the cosmos a little better. This knowledge of the constellations will give rise to new discoveries. You will recognize, as the early humans did, that there sometimes is an "extra star" in one of the constellations. A few weeks later, this "star" has moved on, and the constellation is back to its previous shape. In this way, a new category of objects was discovered, the "wandering stars" or **planets**.[21] Note that by creating the new label **planet**, we have not gained any new knowledge, we are just making it a little easier to refer to this new class of objects. Operationally, we define planets as specks of light that move with respect to the ground and also with respect to the (fixed) stars of the constellations. Careful record keeping reveals that there are five different ones, so we bestow *individual* names on them. The reddish-looking planet is called *Mars*, the brightest one *Venus*, the second brightest, whitish one *Jupiter*, the one that can only be seen shortly after sunset or before sunrise *Mercury*, and a yellowish one *Saturn*. In our modern minds, we associate, say, the name Jupiter with a striped, colorful sphere of gas with a great red spot on it. Remember, though, that this association is due to our access to telescopes and the fine imagery produced by space probes. For a human before 1609 CE, Jupiter literally *is* the second-brightest speck of light moving with respect to the constellations. On the other hand, this human might have attached other properties and astrological influences with the planet associated with the Roman god Jupiter, which have been purged since.

A ***frame of reference*** allows us to make **quantitative** statements about positions of celestial objects. To describe to a friend where we saw Jupiter last night, we could say "halfway up in the eastern sky," but this is not helpful, because Jupiter will change its position during the night at a rate of 15 degrees per hour. Instead of using the ground as a reference frame (the so-called ***alt-azimuth system*** or ***horizontal system*** depicted in Figure 1.2), astronomers use a time-independent system based on the fact that the constellations consist of ***fixed stars***. Before we look at this ***equatorial coordinate system*** in detail, we describe its use by taking the constellations as proxies for the equatorial coordinates. For instance, refining

---

[20] For instance, Orion is on the meridian in the south at midnight in midwinter.

[21] αστηρ πλανητης (*aster planetes*) is *wandering star* in Greek.

our celestial exploration from above, we could say: *Polaris* is 28° past *Dubhe* in a straight line from *Merak*. Our friend knows how to find the two front stars of the Big Dipper, but how is he to measure 28°? Clearly, 28° is an **angle**, that is, the opening enclosed by two rays that meet at a point, Figure 1.8(b). So if you take two sticks, point one at *Dubhe*, and the other at *Polaris*, there would be an angle of 28° between them. This is the **angular distance** between the two stars.[22] The usefulness of the concept of angular distance in astronomy is that it can be measured even if we are clueless about the **actual distance** to these stars; and clueless we have been for most of history. It is the crux[23] of astronomy that directions and positions, that is, angular distances, are easy to measure, but actual distances in miles or kilometers are decisively *not*. Much if not all progress in astronomy has come from devising better methods to measure **actual** distances. In practice, our friend would use her outstretched arms and hands as in Figure 1.7(b) to **estimate**[24] the angular distance between objects. So a description of, say, Jupiter's position as 20 degrees off the constellation *Leo* toward *Virgo*; see Figure 1.27, would make sense. Regardless of the time at night, whether *Leo* is in the east or south, she would be able to find Jupiter two outstretched hand widths left of *Leo*.

A couple of caveats: The sky to us looks like the *inside* of a half dome. Therefore, east is *left*, while we find east on the *right* side of a map, since we are looking down onto it. Similarly, it is fair to use *above* (*below*) and *toward the zenith* (*toward the horizon*) within the alt-azimuth system system. Using the constellations as a reference system, it is better to say *toward Polaris* instead of *above*. For instance, in the spring sky in mid-northern latitudes, *Polaris* is toward the horizon from— and that is, below—the Big Dipper.

In summary, using the constellations as a frame of reference we have found a practical way to refer to positions in the sky. It is quantitative, easy to communicate, and independent of the observation time or location. The stars provide convenient landmarks (or rather "skymarks"). Just like cities on a geographical map, they do not move with respect to each other.[25]

1.1.4 Invitation

Nature exemplified by the star-spangled sky above us is awe inspiring! It invites us to feel as part of the universe, motivating us to think about its deep mysteries. People have been inspired by the heavens for a long time; pretty much until artificial street lighting came about. Since then most people have become more and more detached from nature. In the city you can see and do a lot of things, but observing the night sky is usually not one of them. Instead, we read books exhibiting in great detail and color the marvels of the universe photographed through the lens of the **Hubble Space Telescope**. We look at amazing simulations of the sky theater as an app on our smartphones. We go to science museums and interact with the astonishing exhibits giving us a glimpse of how nature works. We "chat" with "friends" on Facebook about many things, maybe even astronomy.

All of this is great, but no substitute for the *real* thing! I still remember vividly the first time I *really* saw Saturn. Of course, I had "seen" Saturn many times, even in false color to bring out the details in the *Voyager I* and *II* space probe photos. But not until the brutally

---

[22]Technically, *angle* and *angular distance* are synonymous.
[23]The vital or pivotal point; not the constellation *Crux*.
[24]An estimate is the result of a quick and rough but **rational** method to measure or calculate a quantity.
[25]In fact, stars do move with respect to each other, but this so-called **proper motion** is very slow, and was not discovered until the eighteenth century.

cold winter night, when I set my alarm for 4:00 a.m., sneaked out of my parent's house, and aimed my uncle's tiny, wobbly telescope at the ringed planet, did I really see it. There it was, like a caricature of itself in yellow, at magnification 60x, slightly unfocused due to the shortcomings of the telescope, but definitely Saturn, exhibiting the tiniest ring around a barely perceptible disk. Wow! Judging from reactions when I show Saturn (or the moon's craters for that matter) at rooftop observing sessions, everyone has the same sensation: That this is *it*, that this is the *real* thing! Certainly not better than textbook photos—but more real.

I want you to keep this in mind, and to look up—with your unaided eye, with binoculars, with telescopes—and marvel at the wonders of the sky. Talking and learning about astronomy is one thing, experiencing it is another. The two ways are complementary: One shouldn't go without the other. Be inspired, but also study to find out what it truly *is* that you are observing. Read and study, but do not forget to link what you seem to understand better with your experience standing under the star-spangled sky! You can memorize star maps all you want, but you will only learn to find the Big Dipper by going out into the open on a clear night.

There is another point I want to make with the tale of my winterly observing session: It is not always easy nor convenient to observe or "do science." One has to be intentional, motivated, and gritty. I had to *plan* this nightly adventure *by myself*: Borrow my uncle's telescope, set the alarm, and so forth. You might rightly say: Anyone could have done this. Correct, but I *did* it; I think my uncle has never *really* seen Saturn.

These days, we have amazing opportunities to do great things, but most people do not do them. I invite you to not just be amazed and motivated, but to follow through with your plans and intentions. That's what really counts, sets you apart, and results in fulfilling experiences. It also makes you a better, more educated, self-motivated, and enlightened[26] person, able to live and experience your life to the fullest—and to choose the direction it should take.

Concept Practice

1. The fixed stars are *not* moving
    (a) ... in the alt-azimuth (horizon) system.
    (b) ... in the equatorial system.
    (c) ... with respect to the constellations.
    (d) Two of the above
    (e) None of the above

2. What is an operational definition? Explain and give an example.
3. What is the difference between a star and a planet as seen with the unaided eye?
4. What is the difference between an angular distance and an actual distance?

1.2 Astronomy as a Science

Astronomy is without a question the oldest of the sciences. It has been an **exact science** for thousands of years; humans have accurately made and recorded observations of celestial objects for a long time. Often this was driven by practical needs, like navigation or to

---

[26]"Enlightenment is man's emergence from his self-imposed immaturity. Immaturity is the inability to use one's understanding (reason, intellect) without guidance from another." (Kant, 1784) [23]

1.2. ASTRONOMY AS A SCIENCE

establish a reliable and workable calendar to determine optimal planting times in agriculture. Predicting the dates for movable religious feasts (like the modern *Easter Day, Yom Kippur,* or *Eid al-Fitr*) and other religious reasons were probably important early on but became less so later on. All of these motivations imply a fundamental need for accuracy. For this reason astronomy has been driving the human quest for a **quantitative** description of nature. In this section we will look at general features of science, and how astronomy fits into this way of knowing.

1.2.1 The Scientific Method

It is often stated that the right way to do science is collect data, build a theory based on that data, and then test it by further observations. While this is not wrong, it is much too crude and stiff of an algorithm to adequately describe what scientists do. Provocatively put, *the **scientific method*** does not exist. In the words of Nobel laureate PERCY BRIDGMAN (1882–1961) [9]: "Science is what scientists do, and there are as many scientific methods as there are individual scientists .... No working scientist, when he plans an experiment in the laboratory, asks himself whether he is being properly scientific, nor is he interested in whatever method he may be using as method .... I think that the objectives of all scientists have this in common—that they are all trying to get the correct answer to the particular problem in hand." When we say that due to the scientific method there has been much progress in our understanding of nature, we mean the following: A more and more detailed description or explanation of the universe has been achieved by hard work and with many individuals contributing.

In fact, we all act like scientists when we solve problems of everyday life. When your car won't start or the printer doesn't work, you will be trying a lot of (rational) things to get it working again, and often be successful. You will check the fuel level, electric plugs, and so forth, thereby ruling out one by one the different possible explanations of why this machine doesn't work. Also other aspects of science have parallels in everyday life. If someone tries to sell you a five-year-old car with 10,000 miles on it for $1,000, you will be skeptical. This car has been driven 2,000 miles per year. The average mileage per year people put on their cars is almost an order of magnitude (ten times) higher! The price is suspiciously low. You will ask for further information: Was the car in an accident? Is it stolen? Does the buyer need to sell it fast because she is moving? Is there a typo in the ad, maybe it is 200,000 miles? All are plausible reasons to explain this unusual situation. Similarly in science: If a pattern in nature defies expectations, scientists will investigate. Maybe there is a unknown wrinkle of nature that waits to be discovered!

Astronomy is largely an **observational**, not an **experimental** science. It is a profound truth in astronomy that when we observe the sky, we see the celestial objects as they *appear*— not how they *are*. For instance, the sun appears very bright to us. However, unless we know our distance from it, we have no idea how bright it *is*. We will therefore need be careful to distinguish between the former, **apparent brightness** B and the latter, **absolute** brightness called **luminosity** L. When we are closer to an object, it *appears* brighter by a factor that grows quadratically with the diminishing distance, as we will see later in more detail.[27] The object itself, of course, does not get brighter as we approach it; it continues to radiate the

---

[27]This relation between B, L, and distance d is a crucial concept in astronomy, because it allows us to determine the distance to celestial objects.

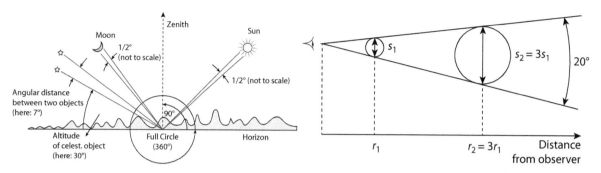

Figure 1.8: **Angles and angular distances.** (a) Illustration of angles and angular distances in the sky. The zenith is a right angle or 90° from the horizon, that is, one quarter of a full circle of 360°. Moon and sun both have a 0.5° angular diameter. The angular distance between celestial objects is measured in degrees (°), just as their altitude above the horizon. No *actual* distances or sizes are involved. The angular measures are used to describe how the sky *appears* to the observer. (b) The *apparent* size of an object depends on both its actual size and its distance from the observer. Here, a smaller, closer object of size s_1 and distance r_1 appears under the same angle (20°) as a larger, more distant object of size s_2 and distance r_2. The ratio s/r is the same for both objects: $s_2/r_2 = 3s_1/3r_1 = s_1/r_1$.

same amount of energy per unit time. Its luminosity remains constant while its apparent brightness grows as a result of the shrinking distance between the observer and the object.

Likewise, the **apparent size** of an object is not its actual size, but its **angular size**. The angle under which an object appears is determined by the distance to the observer, as well as the actual size of the object; see Figure 1.8(b). For instance, the moon appears as big is the sun—about 1/2 degree in angular diameter—yet because the moon is much smaller than the sun, it has to be much closer to appear to be the same size. Indeed, the reason that sun and moon appear to have the same (angular) size is that the ratio of their actual size s to their distance r from us is identical

$$\frac{s_{\text{Moon}}}{r_{\text{Moon}}} = \frac{3480 \text{ km}}{384\,000 \text{ km}} = 0.0091 = \left(\frac{\pi}{180°}\right) 0.52° \approx \frac{1\,392\,000 \text{ km}}{149\,600\,000 \text{ km}} = \frac{s_{\text{Sun}}}{r_{\text{Sun}}} \approx \frac{390\,s_{\text{Moon}}}{390\,r_{\text{Moon}}},$$

where we have used the actual diameters of the sun and the moon as their sizes, and the radii of their orbits as their actual distances. We've also used the fact that the angular size α in radians is the ratio of the diameter to the distance of the object[28]

$$\text{(angular size in radians)} \approx \frac{\text{actual size}}{\text{distance}} = \frac{s}{r}, \tag{1.1}$$

see Figure 1.8. The **conversion factor**[29] from radians to degrees is $\left(\frac{180°}{\pi}\right)$. With the modern values plugged in it is clear that the moon is 390× smaller *and* closer than the sun. In antiquity this factor was vastly underestimated to be 19×, and this false result was used until at least COPERNICUS's times.

Much in astronomy depends thus on our perspective, that is, on our observing location on Earth. We said that the sky looks differently from different places on Earth: How much more so if we'd observed from Mars (sun appears smaller and dimmer), or from α *Centauri*

---

[28]Technically, this is only true for small angles. The exact formula reads $\tan \alpha = \frac{s}{r}$. Since astronomical objects are always very distant, we can use the **approximation** $\tan \alpha \approx \alpha$, where α is necessarily in **radians**.

[29]As all conversion factors, it is equal to 1, since $180° = \pi \text{rad}$; compare to $1\,inch = 2.54\,cm$.

(sun looks like α *Centauri* from Earth and vice versa), or from a galaxy far, far away ... ? This is unfortunate, because we usually do not know the actual distances to celestial objects, and therefore we do not know how much our observation location influences our perception. Is *Sirius* the brightest fixed star because it is so luminous or because it is so close to us? Is the *Andromeda nebula*, which we perceive as a milky blob in the sky, really a diffuse cloud of interstellar gas that is fairly close to us, or a gigantic disk of hundreds of billions of stars very far away from us? As you can see, the ***interpretation*** of what we *see* in the sky depends crucially on our ability to measure the actual distances to these celestial objects. Even the appearance itself might be deceiving. As we remarked in Section 1.1.2, we observe the stars through a layer of air and other material (interstellar gas and dust) that happens to lie along the line of sight from us to the distant object; see Figure 3.25. Care is therefore necessary to ***reduce*** the data, that is, free it from all (known) effects that can influence our measurement of position, brightness, color, and so forth.

In astronomy, we are *observing* how objects appear to us, and draw conclusions by careful interpretation of these measurements. In the *experimental* sciences, on the other hand, we can set up experiments in our labs. We perform specific procedures under well-defined and controlled conditions. For instance, we might measure the velocity of a ball or cart rolling down an inclined track to learn about the acceleration due to gravity. We control all aspects of this experiment, from the inclination of the track to the time we want to start our stopwatch or the experiment. This type of experiment is impossible in astronomy. We cannot observe the motion of the moon around the Earth while omitting the influence of the sun and the other planets. Nor can we decide to collide two black holes to see what would happen. In astronomy, we have to be content with passively observing the patterns in nature that "happen to happen." Also, the idea to carefully control all conditions that can potentially influence the outcome of an experiment is a ***modern*** concept. In laboratory experiments we create an artificial environment, as if we'd taken a piece out of nature and put in a cage. Carefully masking out interfering features of nature allows us to focus on one specific aspect of physical reality. This type of experiment was championed by GALILEO GALILEI (1564–1642), and thus is only about four hundred years old. The extreme focus, and therefore, if you will, a "chopping-up" of reality into manageable little bites, this ***analytic*** approach of separating things and phenomena into their constituent elements, is thoroughly modern. It would have seemed suspicious if not ridiculous to ancient thinkers like ARISTOTLE (384–322 BCE), who had a holistic approach to understanding the cosmos. Yet, it is at the heart of modern science. By breaking up a complex problem into its simpler parts, we can solve many small problems instead of tackling one big question. This has proven so effective, that the sociologist MAX WEBER (1864–1920) remarked in *Science as a Vocation* [43]: "Und wer also nicht die Fähigkeit besitzt, sich einmal sozusagen Scheuklappen anzuziehen und sich hineinzusteigern in die Vorstellung, daß das Schicksal seiner Seele davon abhängt: ob er diese, gerade diese Konjektur an dieser Stelle dieser Handschrift richtig macht, der bleibe der Wissenschaft nur ja fern. Niemals wird er in sich das durchmachen, was man das *Erlebnis* der [modernen] Wissenschaft nennen kann."[30]

Scientists have to have sharp focus, and, often tacitly, will omit anything that is not

---

[30] "And anyone who does not have the ability to put on blinkers, as it were, and to enter into the idea that the destiny of of his soul depends on his being right about [say] this particular conjectural emendation in this manuscript, should stay well away from science. He will never have what may be called the *experience* of [modern] science."

relevant for the problem at hand. This can make it hard to understand scientific writing, since for the outsider it is often *not* clear what the assumptions are or why certain aspects can or have to be omitted to make progress. Here is therefore a brief introduction into the ways that scientists communicate and common misunderstandings thereof.

1.2.2 Science Speak

One could argue that scientists use their own language, ***science speak***. Confusingly enough, scientists often use words of common language, such as ***force, work***, or ***energy*** in a different, much restricted, precise way. Therefore, one must learn to correctly interpret a scientific text, much like it would take effort to understand a text written in, say, Latin. Since the words of science are the same that are used in everyday language, it is only the context that can reveal the true meaning of a scientific statement. As an example, take the term ***energy conservation***. You might know that energy may change its form, for example from kinetic to potential energy, but its quantity never changes; energy is conserved. Therefore, a statement like "We have to conserve energy!" makes no sense if interpreted strictly in the physics sense. What is meant is that we should strive to convert as little electric energy to heat as possible.

In science it is paramount that every statement be precise. This means there should be no way to misunderstand the statement in any way. Neither should there be any relevant information left out, nor should there be superfluous facts. Often, the uninitiated casually parses a "sciency" paragraph and therefore tends to miss the point completely. On the other hand, it is often impractical to explicitly state all information, so the seasoned scientist often omits "obvious" facts. For instance, in the energy conservation example above, the statement "energy is conserved" is only correct if the energy in question is the total energy of an ***isolated system***. This is clear to a scientist, so she doesn't state it explicitly, but probably not to you, which is why you'd have to work hard to understand the concept of energy. In astronomy, the description of motion is crucial and has to be precise, because all motion is relative; relative to a frame of reference, that is. So a statement like "the stars move in the sky" is not acceptable because it is ambiguous: Do we mean the stars are moving with respect to the ground as time goes by (***daily motion***), as days go by (***seasonal motion***), or with respect to each other (***proper motion***)? As another example, consider the following two exam questions.

1. Star A sends out light with a peak wavelength that is half that of Star B's light. Which star has the larger surface temperature?
2. Star A sends out light with a peak frequency that is half that of Star B's light. Which star has the larger surface temperature?

The questions are often answered incorrectly, because they look identical to a casual reader. Only one word is different, but this is the point here. The source of error is confusion about the logical relation between quantities. Wavelength and frequency are related, and therefore often falsely taken to be equivalent. Indeed, they are ***inversely*** related, that is, when one doubles the other halves. We'll learn later that higher frequency means higher energy, and that a star which puts out more energetic radiation is hotter. Hence, in the first question star A is hotter, whereas in the second question it is star B. The moral is to read science texts very carefully and to be familiar with the precise meaning of words in science. Here is a list of commonly used expressions in science and their meaning.

1.2. ASTRONOMY AS A SCIENCE

Approximation An important concept along our way of describing the cosmos, since we have to work *iteratively*. In quite the literal sense, the way (Latin: *iter*) is the goal: By refining the way we describe the universe, that is, by taking more effects into account, we arrive at a better *approximation* to it. As an example, a good first description of planetary orbits is to approximate them as circles. This works because the eccentricities of their *elliptical orbits* are so small. Whenever there is a small quantity involved, we can initially neglect the associated effect. That is the essence of an approximation: It is a fair description of a pattern in nature; not entirely correct, but much simpler than the more precise description. Often, this simplicity is more important on the way to understand nature than to get it "righter."

Assumption Assumptions are things that are taken to be true without proof. If a description of nature is given in the mathematical language, then the result of a calculation is the necessary conclusion of the assumptions. This means that if the results are not agreeing with the observations, the assumptions were wrong and must be replaced.

Behavior In science, often used in describing the dependence of quantities or curves in a plot on parameters, as in "The force as a function of distance *behaves* as follows."

Deduction, induction The two opposite methods to gain knowledge in the sciences. *Deductions* is the extraction of special results from a general rule, *induction* the postulation of a general rule based on isolated data points.

Equivalent Two statements A and B are equivalent if A follows from B *and* B follows from A.

Error In science often used for the unavoidable *uncertainty* inherent in all measurements. Sometimes also used to refer to a *mistake*, for example in a calculation.

Extrapolation When we observe a relationship between two or more quantities over a limited range of their values, we can estimate how this relationship would look like outside this range. The result of this estimate is the *extrapolation*. For instance, we observe that objects attract due to gravity in the solar system, and we extrapolate this to galactic distances and beyond; we *assume* that the inverse square relationship between gravitational force and the distance between the gravitating objects holds even for very large (or very small) objects. Extrapolation frequently goes wrong, but leads to interesting insights, like the breakdown of classical mechanics at large velocities, which gave rise to a better description of nature called *special relativity*. Incidentally, in the limit of small velocities, classical mechanics is an excellent *approximation* to *special relativity*.

Goes up, falls off like/with In science, the description of the behavior of a quantity as a parameter changes: "The force of gravity falls off quadratically with the distance." This means that as the distance grows, the force gets smaller "fast," that is, like the square of the distance: If the distance doubles, the force is down to one quarter (25%) of its value at the original distance.

Implication Scientists often use this term to mean that the result is already contained in the description or formula given. The equation $x + 7 = 2$ implies that $x = 5$, and the label *horse* implies many things, including *mammal* and *animal*.

Omitting/neglecting effects If a pattern in nature resists explanation efforts, scientists *omit* certain parts to simplify the problem. The hope is to come to a crude and sometimes only *qualitative* understanding before tackling the full problem. If the omitted effect is known to be small, we often speak instead of *neglecting* the effect, because there is a stronger mechanism that will explain most of the pattern. Note that it is hard to know what is important and what is not, so one needs to *estimate* the size of the effect, that is, compare it to other relevant quantities.

Postulate To postulate means to tentatively assume the existence of something not observed before to construct a theory or to explain a pattern in nature. NEWTON, for instance, postulated that there is a force at a distance between massive objects (gravity), which explains planetary motion.

Qualitative, quantitative Scientists use *qualitative* to describe a rough understanding of a pattern in nature, whereas *quantitative* means that we can come up with some hard numbers that quantify exactly how this pattern plays out: Exactly how fast it goes, or exactly how large it is, and so forth.

Theory While in everyday language often used synonymously with "unsubstantiated hypothesis," in science a *theory* is a well-established and wide-reaching description of nature, such as *Newton's theory of gravity*.

x **axis, y axis** The former is frequently used incorrectly to label the *horizontal axis* or *abscissa* in a two-dimensional plot, where the *dependent variable* is plotted on the *vertical axis* or *ordinate*, often loosely called y axis.

Words are labels, and we define their meaning; scientists change these definitions as necessary. Sometimes redefinitions are driven by new data; sometimes it is convenient to do so. Science is in flux: Its concepts and results are subject to later revision. For instance, the object labeled *Pluto* used to "be" a planet, and now it "is" a dwarf planet. Of course, *Pluto* has not changed at all. We have merely reclassified it for convenience. Since we discovered, starting in the 1990s, many objects that looked just like Pluto, it became inconvenient to call all of them planets (Who wants to learn all those names?!), so the *International Astronomical Union* decided at a conference in 2006 to call only "big" objects orbiting the sun *planets*. In fact, it is not just *common* in science to redefine objects and concepts—it is the way science works. We stumble upon a useful way or concept to think about a commonly shared observation of a pattern in nature and give it a name. As new data comes in, we might have to refine the concept or redefine what we mean by the name. As we said: Things are in flux in the sciences. By incessant observing, remeasuring, and reinterpreting measurements, we sharpen our language to obtain a better description of nature.

1.2.3 Models and Theories

The concept *model* is central to science. Sloppily speaking, it is a particular way to describe patterns in nature "in one swoop." The term *model* is often used synonymously with *theory*. For example, the *standard solar model* is a well-defined theory of the sun. It describes in detail how the sun produces energy at its core, transports it to the surface, and so forth. Whether the sun indeed *does* it this way is another question. **Standard solar model** is thus just a

1.2. ASTRONOMY AS A SCIENCE

label for an elaborate algorithm to predict properties of the sun like its energy output. If these predictions do not agree with measurements, then the model has to be modified, or discarded.

Humans constantly model things in their environment, not just in science. In our brains we have a model of how the world "works," formed by repeated experiences. You'd be surprised to see people driving their cars on the left side of the road, or pedestrians walking on the street. Neither is impossible, but the normal[31] behavior is so common that a description "Cars are driven on the right side of the road and pedestrians use the sidewalk" is a good if simple *model* for this pattern of human behavior. Note that we would have to modify the model to incorporate contradicting observations made in Great Britain.

In sum, we organize our observations into models collecting and connecting several observations and serving as a platform or reference. It is an attempt to explain the world. Only from this platform can we judge future observations to see if they fit in. The idea of the platform brings home the point that we have to start with things and concepts that are familiar and close, and agree on names and definitions before describing or trying to understand more complicated phenomena. Without a frame of reference we cannot make sense of immediate sensory information. We spend the first years of our lives establishing by trial and error a frame of reference for this world: We learn recognize faces, to use and understand words of our mother tongue, how to grab objects in a three-dimensional world, and so forth. Or think of abstract art: Without knowing something about the artist and his techniques, JACKSON POLLOCK's (1912–1956) famous paintings are mere drips of paint on a canvas that any kindergartner can produce. Only background knowledge allows us to appreciate his works more fully. Without knowledge of the Greek alphabet and vocabulary, we will not be able to read PLATO, and even if we read PLATO, we might not be able to get anything out of it, unless we delve into the concepts of philosophy.

While there is no clear distinction between a model, a theory, a *scientific law* and a *hypothesis*, scientists tend to use the label *theory* for a complex model that has been tested thoroughly, with its predictions having been found to agree with observations many times over. *Law* (examples: *Kepler's laws, Newton's laws*) is a bit of a dated expression, since we came to realize that our description of nature is always subject to revision, while the label *law* suggests an eternal truth which doesn't exist. Often, *hypothesis* is used for a new way to describe a specific pattern in nature. We hypothesize that, say, light is an electromagnetic wave, devise experiments to validate this claim, and then tentatively accept the hypothesis once it passes this *test*. Later, the hypothesis might become part of a larger *theory* dealing with a host of patterns of nature at once. In our example, *Maxwell's theory of electromagnetism* describes not just light as an electromagnetic wave, but virtually any other electric and magnetic phenomenon.

1.2.4 Measurement and Uncertainties

In empirical sciences such as astronomy, the prime source of knowledge is the observation of the outside world, namely the *measurement* of relevant quantities like position, temperature, and brightness. A measurement is, in essence, a comparison with a standard. For instance, if we measure a person's height to be 6 feet, we are comparing the distance from his feet to his head with the length of a measuring device. If the measuring device is a yardstick (1 yard =

---

[31]In fact, it is the *norm*, and, indeed, the law.

3 feet), then our result is equivalent to saying this person is as tall as two yard sticks. The yardstick itself was produced by comparing its length with a standard supplied, for example by the *National Institute of Standards and Technology*. Standards have been around for a long time. For instance, archaeologists have identified a copper-alloy object dating from the third millennium BCE as a Mesopotamian *ell* or *measuring rod*.

No measurement is exact. The result of a measurement depends on the measuring device, the units used, the skill level of the experimenter, the environment, and other factors. Therefore, we associate with every measurement result an **uncertainty** that is, a **probabilistic measure** of the expected spread of the quoted value. Sometimes the uncertainty is called the **error**, but it mustn't be interpreted as a mistake. The uncertainty is an integral part of the measuring process, and needs to be quoted to inform us about the **precision** of the result. Consider an example. A university wants to buy a large number of tables for its classrooms. Company A produces a table of width $w_A = (31.5 \pm 0.2)$ inches $= (80.0 \pm 0.51)$ cm, which means that the company is confident that its tables are between 31.3 and 31.7 inches wide. Here, 0.2 inches is the estimated uncertainty of the quoted result, that is, the width measurement. Company B sells a table of the same width for less money, but admits a larger uncertainty to keep production costs down, $w_B = (31.5 \pm 0.6)$ inches. This means that a table of company A is at most 31.7 inches wide[32] and would fit through a standard 32-inch door, whereas a table of Company B could be slightly wider than the door and therefore will not be a good option. Note that the (average) width of the table is the same in both cases. The uncertainty is thus the most important information because it determines the buyer's decision.

No numerical value is interesting in itself, only when compared to other relevant values. This is also true for the uncertainty. Therefore, we compare the uncertainty to the magnitude of the measured value to find out whether the uncertainty is large or small. The ratio is known as the **relative uncertainty** and determined as follows:

$$(\text{relative uncertainty}) = \frac{(\text{estimated uncertainty})}{(\text{measured value})}. \quad (1.2)$$

Often, the relative uncertainty is small, so we quote it as a percentage

$$(\text{percent uncertainty}) = \frac{(\text{estimated uncertainty})}{(\text{measured value})} \times 100\%. \quad (1.3)$$

In the example above we have a **relative uncertainty** of $\frac{0.2\,\text{in}}{31.5\,\text{in}} = \frac{0.51\,\text{cm}}{80\,\text{cm}} = 0.0063 = 0.63\%$.

As a rule of thumb, we consider an experiment **precise** if its **percent uncertainty** is 10% or less. Note that this doesn't necessarily mean that the result is **accurate**. Accuracy implies agreement with an **accepted value**, typically a world average of results and agreed upon by an international committee of scientists. A measurement of the acceleration due to gravity on Earth quoted as $g = (9.78 \pm 0.01)\,\text{m/s}^2$ is precise because of its estimated uncertainty of only 0.1%, but not accurate, because it does not agree with the standard result of $9.806\,65\,\text{m/s}^2$. The gap between a result and the known or accepted value is often quoted as the **relative discrepancy** to give the reader a sense of the severity of the discrepancy. Using the formula

$$(\text{relative discrepancy}) = \frac{|(\text{measured value}) - (\text{accepted value})|}{(\text{accepted value})} \times 100\% \quad (1.4)$$

---

[32]Since uncertainty is probabilistically defined, there is a very small chance that the table is wider than 31.7 inches.

1.2. ASTRONOMY AS A SCIENCE

in our case yields $\frac{|9.78\,\text{m/s}^2 - 9.806\,65\,\text{m/s}^2|}{9.806\,65\,\text{m/s}^2} \times 100\% = \frac{0.026\,65\,\text{m/s}^2}{9.806\,65\,\text{m/s}^2} \times 100\% = 0.27\%$, which is a small discrepancy indeed.

1.2.5 Evidence and Proof in the Sciences

Scientists tend to be modest when describing research results and their implications, because every measurement is prone to uncertainties. "In science, the burden of proof falls upon the claimant; and the more extraordinary a claim, the heavier is the burden of proof demanded (MARCELLO TRUZZI [39])." **Evidence** comes in the form of a corroborating empirical result, that is, a series of measurements that agree with a hypothesis.

It may come as a surprise that there is **no proof** in the empirical sciences. Proof only exists in an idealized and abstract formal system like mathematics. When measurement is involved, there can be no proof. The reason is that a measurement is necessarily discrete, isolated, and limited. We are measuring a particular quantity at a particular time and a particular location under particular experimental conditions. It is likely but not guaranteed that a measurement at a later time, a different location, and so forth, will yield the same result. We can amass evidence for a theory by making many measurements, but we will never be able to do all logically possible experiments, so we cannot prove that it is correct. With every new experimental agreement it becomes less likely that the theory is wrong, but that still is not a proof. On the other hand, a model can be ruled out or **falsified** easily by making one (credible) measurement that contradicts predictions of the model. The alternative explanation will be, upon falsification of its competitor, **tentatively accepted**, until an even better description of nature is found[33].

In the next chapter, we will look in depth at the prime example for this epic battle for truth in the sciences, namely the **Copernican revolution**. Evidence for COPERNICUS's heliocentric model of the solar system was *not* persuasive until GALILEO observed Venus's phases and falsified the PTOLEMY's geocentric explanation of planetary motion. Often things are not as dramatic. Many times a better description of nature is found that explains all known patterns and then some. In this case, the old model may still be useful as an approximation, as is the case of the **Newtonian gravity theory** replaced by Einstein's general relativity, which is much more complicated. On the other hand, the geocentric model of the solar system lost all usefulness once it was realized that the Earth orbits the sun.

1.2.6 Mathematics and Its Relation to the Sciences

According to GALILEO, the book of the universe is written in the mathematical language. Indeed, it is amazing how useful math is to describe the patterns we find in the cosmos. Math is an abstract language that allows us to talk effectively about the universe. In this sense, mathematics is an auxiliary science, providing valuable **tools** for the quest to understand the universe.

It is wise to keep in mind how rudimentary math was throughout most of history. Up to the early seventeenth century, only geometry, algebra, and arithmetic were available. In a curious case in point, the Italian government, being suspicious of Arabic numbers, outlawed the use of **zero** in the 1500s. Of course, **zero** is a very useful concept, so people continued to

---

[33]Note that even the interpretation of the evidence may depend on the background convictions of scientists, which doesn't make it easier to decide which theory is a better description of nature.

use it illegally and secretively. This illustrates that none of the math underpinning modern science was in use until well after COPERNICUS's death in 1543. Some of them, like calculus and trigonometry, hadn't even been invented. HIPPARCHUS and PTOLEMY used chords as substitutes for sines, and NAPIER's **logarithmic tables** were just coming into use when KEPLER did his monumental calculation concerning planetary orbits. You need to know that before the pocket calculator, logarithmic tables were a great help in doing calculations. You will know that by using logarithms, multiplication can be replaced by addition, and division by subtraction. Since a tremendous amount of tedious calculations have to be done—and done correctly—to find the mathematical laws of nature, logarithmic tables were an indispensable tool to accelerate progress. It is no wonder that science progressed rather slowly before the mathematical tools were available. Imagine building cabinetry without power tools or a skyscraper without a crane. Tools make construction possible, be it of cabinetry or an understanding of the expanding universe.

We will use mathematics to figure out the rough **scales** (sizes, parameters) of astronomical objects and processes. The **dynamic range** of these phenomena is huge, and not just in the literal sense are the numbers involved *astronomical*. We are well advised to use **scientific notation** for numbers. This allows us to suppress leading or trailing zeros, and thus to write very small or large numbers in a compact way. All of this should be familiar, but here are some examples. The mass of the sun is

$$M_{\text{Sun}} = 1.99 \times 10^{30}\,\text{kg} = 199,000,000,000,000,000,000,000,000,000,000\,\text{kg}.$$

The number before the × is known as the **coefficient** (also as **significant** or **mantissa**), here given to three **significant figures**, suggesting an accuracy of this number in the permille range. After the × sign we have 10 as the **base** and 30 is the **exponent**. The resulting number has 30 digits. Likewise, consider the wavelength of green light:

$$\lambda = 500\,\text{nm} = 500 \times 10^{-9}\,\text{m} = 5 \times 10^2 \times 10^{-9}\,\text{m} = 5 \times 10^{2-9}\,\text{m} = 5 \times 10^{-7}\,\text{m} = 0.000\,000\,5\,\text{m}.$$

Here, we have used the Greek letter λ (lambda) as a variable representing the physical quantity **wavelength**. We also demonstrated how easy it is to multiply numbers in scientific notation. Lastly, we introduced the use of **prefixes**, like nm for nanometer, that represent powers of ten. Note that 10^{-7} has seven leading zeros, and that $10^0 = 1$.

Determining a numerical value for an astronomical quantity is hard. Often it is sufficient to know *roughly* what this value is. This is what an **order of magnitude** estimate can do. It produces an estimate for a quantity that is off by less than a factor of ten. The order of magnitude of a number written in scientific notation is easy to determine: It is its exponent. So the order of magnitude of the mass of the sun in kilograms is 30.

In more general cases, we can estimate the order of magnitude of a quantity by rounding the coefficients of the numbers involved in its calculation to one significant figure and adding the exponents. As an example, we estimate the number of atoms in the sun. We assume that the sun is made up (mostly) of hydrogen with an atomic weight $1\,u$, that is, there is 1 gram of mass per **mole** of hydrogen. Each mole of a substance contains 6.022×10^{23} atoms (**Avogadro's constant**), so the number of atoms in the sun is roughly

$$N = \frac{(2 \times 10^{30}\,\text{kg}) \times (6 \times 10^{23}\,\text{mol}^{-1})}{10^{-3}\,\text{kg}\cdot\text{mol}^{-1}} = (2 \times 6) \times 10^{30+23+3}\frac{\text{kg}\cdot\text{mol}}{\text{kg}\cdot\text{mol}} = 12 \times 10^{56}.$$

1.2. ASTRONOMY AS A SCIENCE

So the order of magnitude of the number of atoms in the sun is 57.

Even if we are not interested in getting accurate results in this book, it would be foolish to assume that we can understand the great book of the universe without any math. Therefore, we need to know some of the powerful concepts that the giants of science have developed over the last four centuries. Some are surprisingly simple to use and somewhat like a toolless assembly of a piece of furniture: If the designer has done her job well, the consumer has the benefit of ease of use of the end product.

One such powerful concept is ***scaling***. It is the idea that quantities in nature and math must behave in a special way as a function of their overall size. For instance, the circumference C of a circle grows like its radius r. This means much more than the well-known formula $C = 2\pi r$. In fact, we want to *understand* this formula. Even more, we want to forget about all irrelevant details, like the constant 2π. Remember the sharp focus of the scientist. She mustn't be distracted by details; she needs to grasp the *essence* of the problem at hand. Here, the essential insight is this. The circumference doubles if the radius doubles, and halves if the radius halves. The circumference behaves in the same way the radius does. The radius really is a measure of the size of the object, that is, the circle. We say the circumference *scales* like the radius. Where is the power of this concept? First note that the argument works for any shape, whether it'd be a square, rectangle, star, or even an irregularly shaped object. For a square the circumference is $C_{\text{square}} = 4s$, where s is the length of the side or size of the square, and so on. The power of the concept is therefore that we can forget about the shape of the object, and conclude that any circumference always scales with the same ***scaling factor f*** as the size of the object. If the size goes up by a factor f, the circumference goes up by the same factor f, even if f is as irrational as $\sqrt{2}$. The irrelevant constant turns out to distinguish between the shapes: 2π for circle, 4 for a square, and so forth. It is called, with somewhat derogatory intent, a ***geometrical factor***. Note that even if we don't know the shape of the object, we can still rest assured that if its size goes down by, say, $\sqrt{3} \approx 1.73$, its circumference goes down by the same factor 1.73. Incidentally, this is how scale models work, for example of a model airplane. Now, the manufacturer of such a model[34] will want to know how much plastic material is necessary to produce it. He will therefore not care how the circumference scales with the size, but how the **volume** of the airplane scales with its size. Even assuming the whole airplane is filled with plastic means we are dealing with a complicated shape. However, we learned that the exact shape is an irrelevant detail in the quest to understand the essence of the problem. It turns out that *any* volume scales like the third power of the size of the object. The simple examples are the volume of a cube ($V_{cube} = s^3$) and of the sphere ($V_{sphere} = \frac{4\pi}{3}r^3$), but this works even for an irregularly shaped object like an airplane, a marble statue, or an ***asteroid***. We can write

$$V_{\text{object}} = C_{\text{object}} \times s^3, \tag{1.5}$$

where C_{object} is the geometric constant that is different for differently shaped objects, and s its size, as before. As an example, we'll scale up a sphere by a factor 4. As a motivation for this exercise, imagine a new exosolar planet has been found which is four times as large as the Earth. We want to know the mass of this so-called ***super-Earth***. To simplify we'll assume that the super-Earth has a uniform density equal to the density of the Earth. Then

---

[34]Note the use of "model" in a similar sense as a scientific model. The model is not the same as the "real thing," but it shares many of its features.

the ratio of the mass M of the super-Earth and the mass of the Earth will be the same as the ratio of their volumes V

$$\frac{M_{Super}}{M_{Earth}} = \frac{V_{Super}}{V_{Earth}}.$$

Since $r_{Super} = 4r_{Earth}$, we know that the scaling factor is $f = 4$. Because the volume scales like the third power of the size, we have

$$V_{Super} = f^3 V_{Earth} = 4^3 V_{Earth} = 64 V_{Earth}.$$

So we can fit 64 Earths into one super-Earth. This may seem surprising, because the super-Earth is only four times bigger, but this factor 4 in **_linear_** size translates via the strong power behavior $V = Cs^3$ into an impressive factor 64. So the super-Earth contain 64 times as much mass as the Earth, since we assumed equal densities. Let's do this slowly one more time so you see that there is no black magic involved. Computing the volume of the super-Earth the old-fashioned way by looking up the "volume formula," we have

$$V_{Super} = \frac{4\pi}{3} r_{Super}^3 = \frac{4\pi}{3} (4r_{Earth})^3 = \frac{4\pi}{3} 4^3 r_{Earth}^3 = 4^3 \frac{4\pi}{3} r_{Earth}^3 = 64 V_{Earth}.$$

The added convenience of this method is that we didn't even have to use the calculator or to look up the actual radius of the Earth. It is good enough to know that the super-Earth is four times bigger than the Earth. Note that we do not know what the volume of the super Earth is! We just know it's 64 times larger than the volume of the Earth—but this is the point here! We are not interested in some (astronomically large) number per se, we are interested in the relation of this number to familiar numbers. In other words, if I would quote the volume of the super-Earth in cubic miles, that wouldn't lead to much insight, but if I say 64 times as voluminous as the Earth, then you get this picture in your mind of 64 Earths all fitting into this super-Earth, and you understand just how gigantic this **_exoplanet_** must be!

Probably the most important scaling behavior in astronomy is that of area scaling. The reason is that signals spreading out from a point source cover the surface area of a sphere, see Figure 2.50. If the source of the signal is 2 miles away, the radius of this **_sphere of influence_** it twice as large as at a 1 mile distance, and so forth. Everybody knows that the signal gets weaker the longer it has to travel to reach the observer. Why is that? It's simply because the signal spreads out and therefore dilutes over a larger surface as the radius of the sphere of influence grows. So the signal strength scales like the inverse of the surface of the sphere of influence: As the size grows, the signal fades fast. How does an area A scale with the size? As you can glean from the many "area = length × width" calculations you've done in school, it scales like "size × size," so $A = C's^2$, where C' is a different geometric factor that depends on the shape. For a disk it's $C' = \pi$, as in $A_{disk} = \pi r^2$. As an example, take a 100 W lightbulb. From a distance of 2 meters it looks $(\frac{1}{2})^2 = \frac{1}{4}$ as bright as from 1 meter away. In other words, from 2 meters it looks as bright as a 25 W ($=$100 W/4) lightbulb from 1 meter. The analogous argument applies to stars, so we are back at the **_astronomy dilemma_**: Observing how something _appears_ does not inform you what it _is_. If someone shows you a lightbulb in an otherwise black room, you cannot tell whether it is the 100 W bulb at a distance 2 meters or the 25 W bulb at 1 meter. A smart observer would say: "I can see the size of the bulb, and therefore judge that it is only a meter away. Therefore, I conclude that it must be the 25 W light bulb." Indeed, this kind of _independent_ information is what helps astronomers determine the distance and luminosities of celestial objects.

1.2. ASTRONOMY AS A SCIENCE

Apart from brightness, there are other quantities that scale inversely with the distance squared or area. The force of gravity between two massive objects, and the electrostatic force between two charged objects also fall off this way. Jupiter is about five times farther from the sun than Earth, and therefore feels a gravitational pull that is $5^2 = 25$ times weaker. Telescopes are light-collecting devices, and the amount of light they collect is **proportional** to their area, not their linear size. So an $8''$ telescope collects $4^2 = 16$ times as much light as a $2''$ telescope, much like an upside-down $32''$ umbrella collects $16\times$ the rainwater of an $8''$ umbrella. In that sense, an $8''$ telescope is $16\times$ more useful than a $2''$ telescope. This quantitative difference often makes or breaks the possibility of discovery. Say you want to find a faint object and take a long-exposure photo. If a one-hour exposure shows the object if the camera is mounted to the $8''$ telescope, then you'd have to expose 16 hours with the $2''$ telescope. Since the night is typically less than 12 hours long, there goes your hope of discovery; the $2''$ tool just won't cut it.

As you can see, scaling is a useful concept. Scaling behaviors are everywhere in astronomy! There are some ugly ones: The luminosity L of a star scales like its temperature to the fourth power (**Stefan-Boltzmann's law**, Section 3.8.3), but like its mass M to the 3.5 power ($L = C''M^{3.5}$); the period P of a planet's orbit scales like the three-halves power of its distance a to the sun (Kepler's third law, $P = C'''a^{3/2}$). Sometimes it is obvious where the scaling comes from, but often it is hard to find an explanation. Science progresses by first recording a (scaling) pattern in nature, and then looking for an explanation of that pattern.

1.2.7 Plots and Diagrams

Another useful math tool is the graphical representation of relations between quantities, usually in a two-dimensional (Cartesian) coordinate system like in Figure 1.9. This way of visualizing the observational or experimental data is complementary to the representation of the relation as a mathematical function. Both are helpful, and they inform each other. The graphical representation, also referred to as a **graph** or a **plot**, can be used in two ways. If we represent data by a graph, we might be able to infer the mathematical function that describes the data. We thus reduce the myriad of isolated data points to one equation. This is very efficient. On the other hand, if we already have a formula expressing a relation between measured quantities in mathematical form, plotting it can give us an intuition about the interplay between them. Also, we can read off value pairs instead of computing them from the formula. Furthermore, we will get a sense of the **range of validity** of the formula, and can decide whether it is safe to **interpolate** between data points or **extrapolate** into regions where we have not gathered data.

We establish the terminology by looking at the graph of Figure 1.9(a). In this two-dimensional coordinate graph, we plot the data points with the help of two rectangular axes. The axes are subdivided into fractions of the numerical values x/a and y/b, where x and y are quantities like distance, time, or temperature, and a and b are the appropriate units, like meters, seconds, or Kelvin. Note that we can choose different units like light-years, days, and degrees Fahrenheit if convenient; **conversion factors** will take us from one set of units to another. The numerical value x/a depends on this choice. Therefore, it is crucial to state the choice of units. For instance, $1 \neq 2.54$, but 1 inch = 2.54 cm.

The function displayed in Fiure 1.9(a) goes smoothly through all the data points. We say that it *describes* the data well. The curve $y = f(x)$ has a **minimum** at x_{min} with value

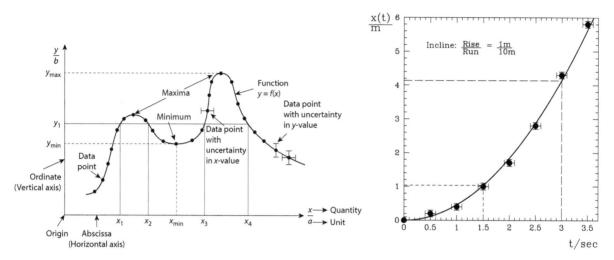

Figure 1.9: **Graphical representation of data.** (a) Data points and a curve smoothly connecting them in a Cartesian coordinate system. (b) Plot of position versus time for the data from the inclined plane experiment.

y_{min}, and two **local maxima**. We read off the value of the function for a specific value of the independent variable by drawing a straight line parallel to the vertical axis through the value, followed by a straight horizontal line once we hit the curve. For instance, following the dashed lines connecting x_{min} to the minimum of the curve, we reach y_{min} on the vertical axis. In this way we associate *one* value of y with each value of x uniquely. On the other hand, the value y_1 corresponds to four different values of the independent variable x_1, x_2, x_3, x_4, as the line drawn through y_1 parallel to the horizontal axis indicates. The data points often carry **error bars**, which indicate the **uncertainty** of the measurements. There may be an uncertainty associated with both quantities, x and y. **Error bars** give us a sense of the quality of the data.

Let's take a look at an example relevant for our quest to understand the universe. Around the year 1600 GALILEO GALILEI sought to describe the motion of free falling objects to gain an understanding of the influence of gravity. In particular, he tried to answer the question how the object speeds up or accelerates. Objects fall awfully fast, and stopwatches were not available. GALILEO's genius let him devise an elegant solution to this practical problem. He slowed the motion down by having the ball roll down an inclined plane. Arguing that free fall corresponds to a fully vertical plane, he *assumed* that the ball would speed up in the same way down the incline as it would in free fall.

In a reenactment of GALILEO's experiment to determine the value of the acceleration due to gravity on Earth g, we put a ball on an inclined plane and measure its position x relative to the starting point at intervals of one second. Say we measure time with a stopwatch and the distance down the incline with a yardstick and estimate their uncertainty to be 1/10 of a second and 0.1 m,[35] respectively. This results in a list of time-position pairs (t_n, x_n) as in Table 1.1. The graphical representation of the **data** in Figure 1.9(b) displays the **units** on the axes, and the error bars afford a sense of the uncertainty involved in the experiment. The data points are *isolated*. Since we observe the motion of the ball to be **continuous**, we may

---

[35] Both are very conservative estimates; we can probably do much better.

1.2. ASTRONOMY AS A SCIENCE

t/s	0.0	0.5	1.0	1.5	2.0	2.5	3.0	3.5
x/m	0	0.115	0.460	1.035	1.840	2.875	4.140	5.635

Table 1.1: **Data of an inclined plane experiment**: The distance x (in meters) of a ball on the plane counted from the starting point is measured as a function of time t (in seconds).

wish to describe the data by one continuous curve. This curve is represented by a function in the mathematical language yielding a position x for every time t, even when we did not make a measurement. It is clear that the ball has to get from, say, $x(t = 1.5\,\text{s}) = 1.035\,\text{m}$ to $x(t = 3\,\text{s}) = 4.140\,\text{m}$, and the curve *interpolates* between these values. Note that the curve does not go through all data points. The curve is a *fit* to the data, and there are sophisticated methods to choose the *best fit*. However, our brain automatically settles on one that is close to correct for simple cases. We will revisit this experiment in the section on GALILEO. Here, we note that the distance traveled in one time interval is increasing such that the difference of subsequent distances is an odd number. This leads to a simple quadratic relation between total distance traveled and time elapsed, namely $x = Ct^2$, where C is a constant proportional to g, the acceleration due to gravity on Earth. This insight helped ISAAC NEWTON formulate his *law of universal gravitation*.

Often it is useful to display the same data in different ways to get a better idea of what is going on. Consider Figure 1.10. As we will discuss in more detail later, starlight sent through a prism decomposes into its color components known as a **spectrum**. The spectrum exhibits black lines on a background of the colors of the rainbow. These so-called **Fraunhofer lines** are indicative of the chemical elements present in the star. It is often useful to display the brightness of the spectrum as a function of the color or wavelength; see Figure 1.10(b). Both graphical representations of the data are useful. The first displays more directly the result of the observation, the second makes it easier to read off the precise position of the Fraunhofer lines.

In sum, it takes effort to understand the information displayed in graphical representations. For instance, perspective drawings like of the observer's sky in Figures 1.2 and 1.3 can be confusing because an aspect of the three-dimensional world is rendered on a two-dimensional book page. To make sense of these drawings like Figure 1.2 it helps to compare them to the *real* sky outdoors.

Concept Practice

1. A scientific *theory*

    (a) ... cannot be falsified.
    (b) ... can be proven.
    (c) ... is a thoroughly tested, widely applicable hypothesis.
    (d) Two of the above
    (e) None of the above

2. What is an *approximation*? Explain and give an example.

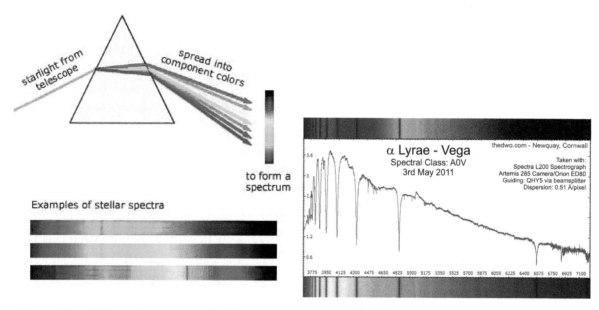

Figure 1.10: **Different ways to display the same data.** (a) Generation of a spectrum by a prism and color photos of the continuous spectrum with superimposed dark Fraunhofer or absorption lines. (b) The spectrum of a star as a brightness versus wavelength plot that reveals finer structures.

1.3 The Motion of the Sun and the Stars with Respect to the Observer

1.3.1 Daytime Astronomy

We described the basic pattern of solar motion with respect to the observer on Earth earlier: Rising in the east, the sun drifts higher in the sky until it crosses the meridian at noon, reaching its maximal daily altitude above the horizon. It continues to move west (increasing its azimuth angle)[36] while sinking toward the horizon; see Figure 1.3. It disappears at the western horizon about twelve hours after it had risen. This daily or **diurnal** motion of the sun repeats after about twelve hours of darkness.

If we consider this description of motion of the sun in the sky as a model, it is immediately clear that this **approximation** to the actual solar motion is flawed. The model can be **falsified** by any of the following observations:

- The sun is above the horizon for much longer (shorter) in the summer (winter).
- The sun at noon is higher in the sky in the summer than in the winter.
- The sun rises in the northeast in the summer and in the southeast in the winter.
- The sun sets in the northwest in the summer and in the southwest in the winter.

As a consequence, we have to *modify* our description of the solar motion to accommodate the new contradicting data. As a first step, we record the exact position of the sun as time goes by with the help of a simple measuring device, called a ***gnomon***.

---

[36]The azimuth angle is counted from north, so of north it is 0°, of east 90°, of south 180°, of west 270°; see Figure 1.2(b).

1.3. THE MOTION OF THE SUN AND THE STARS

1.3.2 Finer Observations with a Gnomon

The **gnomon** is essentially a stick driven perpendicularly into the level ground. Let the length of the stick protruding out of the ground (that is, its height) be H. The sun will cast a shadow of length L in a direction β that we will measure in degrees from north. We note that a larger gnomon affords a more accurate measurements of the position of the sun in the sky because the *relative uncertainty* will become smaller according to Equation (1.2). It follows that accurate celestial measurements could be made well before the advent of the telescope by using very large gnomons.

To measure the position of the sun during the day all we have to do is to measure the length and the direction of the shadow cast by the gnomon in regular time intervals. This type of an observing setup depicted in Figure 1.11(a). The most important measurement is that of the shortest shadow, which we anticipate to happen at local noon. Several factors influence when exactly that'll be, so we better make more measurements around the time when our wristwatch shows 12:00 p.m., lest we miss the time of the shortest shadow. The outcome of such a set of measurements is shown in Table 1.2.

	Shadow		Sun	
Time	Length (m)	Azimuth (°)	Azimuth (°)	Altitude(°)
6:40 a.m.	143.24	269.9	89.9	0.4
7:00 a.m.	13.62	270.1	90.1	4.2
8:00 a.m.	3.58	283.1	103.1	15.6
9:00 a.m.	2.01	294.1	114.1	26.5
10:00 a.m.	1.36	307.4	127.4	36.4
11:00 a.m.	1.02	324.1	144.1	44.4
12:00 p.m.	0.86	344.9	164.9	49.4
12:20 p.m.	0.83	352.5	172.5	50.2
12:30 p.m.	0.83	356.4	176.4	50.4
12:40 p.m.	0.83	0.4	180.4	50.4
12:50 p.m.	0.83	4.3	184.3	50.3
01:00 p.m.	0.84	8.2	188.2	50.1
02:00 p.m.	0.95	30.1	210.1	46.4
04:00 p.m.	1.75	62.2	242.2	29.7
06:00 p.m.	7.30	84.1	264.1	7.8
06:40 p.m.	572.96	90.6	270.6	0.1

Table 1.2: **Measuring the sun's position in the sky.** The table displays typical data taken with the help of a gnomon of height $H = 1\,\text{m}$ at 40° N latitude and 83° W longitude.

The sun's position in the two-dimensional sky is given by the <u>two</u> coordinates **altitude** A and **azimuth** α. Since the shadow points in the direction opposite of the sun, we can determine A by subtracting 180° from our values for the direction of the shadow β; see Table 1.2. To determine altitude, we have to do a bit of trigonometry. A look at Figure 1.11(b) will convince you that the tangent of the altitude angle is the ratio of the shadow and gnomon height, so we have

$$\tan \alpha = \frac{H}{L} = \frac{\text{(gnomon height)}}{\text{(shadow length)}}.$$

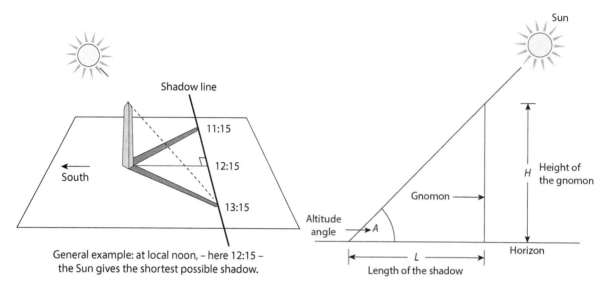

Figure 1.11: **The shadow cast by the sun and a gnomon.** (a) Sketch of the gnomon casting a shadow at different times around noon. Precisely at noon the shadow is shortest and pointing north. Local noon typically does not occur at 12:00 p.m., since a wristwatch shows official (timezone) time, not local (astronomical) time. (b) The geometry of the situation: The ratio of the shadow length and the gnomon height is the tangent of the sun's altitude angle A.

We should always cross-check our math, so consider three special cases in which we can do the calculation in our head.

- Case 1: Shadow length and gnomon height are the same $\frac{H}{L} = 1$: In this case we have an isosceles triangle, which has two 45 degree angles, so $A = 45°$. This makes sense, since $\tan 45° = 1$.

- Case 2: There is no shadow, $L = 0$. The math here is simple: We attempt to divide by zero, so the result tends to infinity, as does the tangent for a right angle, so $A = 90°$. This makes sense, because this means that the sun is at the **zenith** or directly overhead. Note that this *never* occurs in the contiguous United States! For the sun to be directly overhead, you have to observe from the **tropical zone** of the Earth between the tropic of *Cancer* and *Capricorn*; see Section 1.1.2.

- Case 3: The shadow is exceedingly long, $L \to \infty$. Clearly, $\tan A = 0$, hence $A = 0$, so the sun is at the horizon. This makes sense, since at sunrise and sunset we observe long shadows.

So our math and logic are consistent. In the field, use the following rule to check if you typed the numbers correctly into your calculator: If the shadow is shorter (longer) than the gnomon height, the angle is larger (smaller) than 45°.

If we determine the minimal shadow and therefore the maximal altitude of the sun for many days of the year, we find that it oscillates, see Figure 1.12(a). The same thing is true if we plot the azimuth (direction) of sunrise or sunset for many days in the year in Figure 1.12(b). Taking these observations together, we conclude that the path of the sun in the sky, while always following the approximate behavior of rising, culminating, and setting, is different on different dates, and that the extremes occur on June 21 and December 21. This

1.3. THE MOTION OF THE SUN AND THE STARS

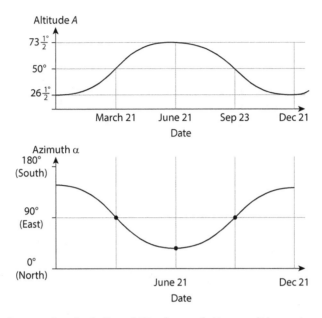

Figure 1.12: **The sun's maximal daily altitude and its position at sunrise.** (a) The sun's maximal daily altitude as a function of the date for an observer at 40° northern latitude. Maximal altitude at noon is reached on the **solstice** on June 21. The altitude at noon $A_{Equinox}$ on the **equinoxes** (March 21 and September 23) is indicative of the observer's **latitude** $= 90° - A_{Equinox}$. (b) Approximate position of sunrise as a function of the date. The position is given as an azimuth angle. Recall that east is 90°.

behavior is visualized in Figure 1.13, which makes it obvious that the sun is on a daily circular path around the Celestial Pole. This path is circular because its angular distance from the Celestial Poles is constant during a day. Furthermore, it makes sense that on the date of longest daylight most of the sun's daily circular path is above the horizon. This implies that the sun is closest to the Celestial Pole, and casts the shortest shadow, that is, has maximal altitude at noon. That is, its maximal daily altitude is larger than on any other day of the year.

It follows that the sun is not a **fixed star**; its position with respect to the **Celestial Pole** changes! Of course, we cannot directly observe the sun's position among the stars since its brightness makes it impossible to see stars during daytime hours. However, it is not hard to see that the motion of the sun as observed with a gnomon means that it moves with respect to the stars. This is related to the *seasonal* motion of the stars. Since the sun by definition always culminates at noon, the stars' culmination time will shift as the sun moves relative to the stars.

As the name suggests, this motion is the **reason for the seasons**. Since the sun is highest in the sky in summer, its light rays hit the Earth at a larger angle. Therefore, they spread out less, and more energy per area is deposited, which heats up the Earth's surface, see Figure 1.13. Also, extended hours of daylight mean that in the summer the sun's rays can heat up the surface for a longer time during a 24-hour period than in the winter. Indeed, we *observe* that during the summer the average temperatures are *much* higher than in the winter. This is a large effect. We can get a sense of just how great the seasonal shift is by comparing the length of the shortest daily shadow on June and December 21. For a typical

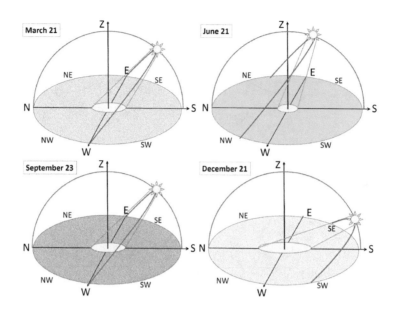

Figure 1.13: **The sun's path in the sky on four different dates from mid-northern latitudes.** Shown are March 21, June 21, September 23, and December 21. The altitude of the sun determines how much its rays are spread out. Less altitude means more spread, resulting in less energy deposited per surface area. This is the situation in winter.

United States city like Columbus, Ohio (40° N latitude), and a gnomon of height $L = 1.0$ m, the shadow lengths are $S_{June} = 0.3$ m and $S_{Dec} = 2.0$ m, that is almost an order of magnitude (factor 6.7) difference! For now these are are just empirical facts, that is, borne out of our observations. Later we will see—you may know this already—that all of these patterns can be explained by the inclination of the Earth's axis with respect to its orbit around the sun.

We have thus established that the sun moves with respect to the observer *and* with respect to the stars. It oscillates between being closer and farther from the **Celestial Pole** during the year. We say it changes its **celestial latitude**. Is the sun also moving in **celestial longitude**? This is impossible to tell with the data we've collected thus far. We need to map the motion of the sun with respect to the stars. Since the sun's brightness wipes out all hope of seeing stars during the daytime, we must continue our quest to understand the cosmos by a thorough study of the night sky.

1.3.3 Nighttime Astronomy

Earlier, we identified the constellations as groups of **fixed stars**. They move with respect to the ground, that is, the observer, but retain their shapes and position relative to the other constellations, establishing a frame of reference. The notion of a moving fixed star seems paradoxical, but we saw above that motion is always motion with respect to a frame of reference. With two frames of reference—the observer's horizon and the constellations—the term **motion** is ambiguous. Specifying the reference frame removes the paradox: An object can be at rest (fixed) in one frame while moving in another.

With respect to the ground, most stars move in virtually the same fashion as the sun: They rise at the eastern horizon, move higher in the sky until they cross the meridian at their

1.3. THE MOTION OF THE SUN AND THE STARS

maximal altitude, and then sink toward the western horizon. However, some stars are so close to the Celestial Pole that they will always be above the horizon in their circular motion around the Celestial Pole. These **circumpolar** stars never rise or set; see Figure 1.18(d).

Note that north, east, south and west are *not* directions in astronomy. They are families of directions. South, for instance, is anywhere on the meridian between the southern horizon and the zenith, see Figure 1.2. As in an actual family, you have to specify the family name and the first name for complete identification. As an example, the position of *Polaris* in the sky from 45° northern latitude is north *and* halfway up in the sky.

If we take a closer look at the motion of the stars with respect to the observer, we will notice after a couple of days that the view of the night sky is changing even if we observe exactly at the same time. The stars that were on the meridian at midnight last week are already past the meridian at midnight this week. Since we divide our days into twenty-four hours to make sure that the sun always culminates at noon, it must be that the stars circle the Celestial Pole *faster* than the sun. This is a small but **cumulative** effect, twice as large after two days, ten times as large after ten. We can ignore it for a day or two, but not for weeks.

To establish the size of this effect, we measure the **rotation period** of the stars. This means practically to go out at night with a good timekeeping device (smartphone will do), and take note when a bright star goes behind a building due to its motion relative to the ground. The time it takes a certain star to complete a full 360° rotation around the Celestial Pole is known to astronomers as the **sidereal**[37] **day** to distinguish it from the standard 24-hour, noon-to-noon **solar day**. The next night you do exactly the same thing (same star, same spot, same building) and then compare the two readings of your timekeeping device. You will find that *less* than twenty-four hours have passed: The star vanishes four minutes before a full **solar day** is completed. This means the star rotates faster around the Celestial Pole than the sun if by only 4 min/24 hr = 4/1440 = 0.3%. It also means that the sun is slow compared to the stars. Note that it doesn't make sense to say that the sun is slow—slow compared to what? Precise language providing a reference helps avoid paradoxical or ambiguous statements here.

This small daily effect is *accumulative* and adds up to two hours after a month, and twelve hours after six months. This means that the stars that culminated at midnight six months ago now culminate at noon (practically invisible, but oh well!). This in turn means that after half a year we see the other half of the sky at night; the one we couldn't see earlier because the sun was above the horizon! In other words, this slow drift of the sun with respect to the stars (or the other way around—motion is relative), gives us the chance to see all stars at night although the Earth blocks half of them at any instance. Hence, we can measure the position of these stars and draw a star map of the entire sky, even though at any moment in time we can only see half of the sky, due to the physical horizon beneath our feet courtesy of the Earth.

In our quest to understand the motion of the sun with respect to the stars, we now need to connect daytime and nighttime observations. With our star map in hand, this is not hard but tedious. However, we can simplify our task greatly by introducing a new concept. This tool to discuss positions and motions in the sky is called the **celestial sphere**.

---

[37]From Latin: *sidus* = star.

Concept Practice

1. The shadows cast by a sun
    (a) ... are longest at noon
    (b) ... are in the west in the morning
    (c) ... are shorter in winter
    (d) ... vanish at noon
    (e) None of the above

1.4 The Celestial Sphere as a Practical Tool to Organize Observations

The sky above the observer looks like a giant dome. To ancient observers it was not clear whether it actually *is* a giant sphere. It is not, but nonetheless many of the diverse observations that we have made so far can be described in one fell swoop by **modeling** the sky as a gigantic, hollow, transparent sphere with stars affixed to its inner surface. The Earth with the observer is at the center of the sphere (which is assumed to be so huge that it doesn't matter whether the observer or the center of the Earth is at its center), Figure 1.15. The sphere rotates once a day around an axis that goes through the Earth's geographic poles and extends into the sky where it pierces the celestial sphere at the CNP and CSP. The Earth's equator is projected onto the sky, where it becomes the **Celestial Equator**, separating the sky's northern from the southern hemisphere. This way we can speak about northern and southern stars. Note that many southern stars can be seen from the northern hemisphere of Earth and vice versa. Is the **celestial sphere** a model or a description of the universe? Is it a practical tool? We will see that its interpretation has changed over the millennia. Note that we cannot observe the sphere itself. In that sense the sphere is certainly not real. Only the stars and their positions (changing as a whole due to the rotation of the sphere) are observable. Also, we cannot tell whether the sphere or the Earth is rotating. Motion is relative, and we would have the same sensation, that is, make the same observations, whether the celestial sphere rotates from east to west, or the Earth rotates from west to east. The latter as the former interpretation gives us the **impression** that the stars circle westward. Here then are two different explanations of our observational data. Both theories explain the observations and **save the appearances**, as we say. Hence, we need further, independent observations to falsify one of them.

The celestial sphere is in line with our observations of the **daily motion** of the sun and stars. Since its rotational axis is the same as Earth's, we can describe the position of an object on the celestial sphere in complete analogy to the latitude/longitude system we use to specify a point on the surface of the Earth, Figure 1.4(a). This is the **equatorial coordinate system** we alluded to earlier. The celestial sphere and the Earth are idealized as mathematical **spheres**, that is, curved two-dimensional surfaces. Therefore, we need two **celestial coordinates**. They are called Right Ascension (RA, or **celestial longitude**) and declination (Dec., or **celestial latitude**). **Declination** is defined just like latitude on Earth. For instance, the star Regulus (α *Leonis*) has declination $+10.5°$, which means that it sits $10.5°$ north of the celestial equator. **Right Ascension** is expressed differently. Instead of using degrees, the RA angle is quoted in hours, where 24 hours is equivalent to 360 degrees. It may seem strange to measure an angle in hours, but it is convenient in astronomy, since positions of stars can be

1.4. THE CELESTIAL SPHERE AS A PRACTICAL TOOL

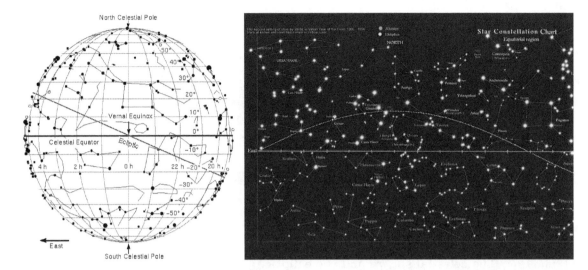

Figure 1.14: **Sky maps.** (a) The coordinate system of the celestial sphere. ***Declination (celestial latitude)*** is counted from the ***Celestial Equator***, ***Right Ascension (celestial longitude)*** is counted from the ***vernal equinox***. (b) Corresponding star chart of the eastern hemisphere of the celestial sphere. Note that this map is distorted, because it is the projection of the curved surface of the sky onto a flat piece of paper. Also visible is the ***ecliptic***, the apparent path of the sun as it moves with respect to the stars. East and west are switched because we are looking at the *inside* of a sphere.

measured with ease. For instance, observing the bright star Betelgeuse (α *Orionis*) culminating, that is, transiting the meridian 4 hours and 13 minutes[38] before Regulus, we immediately know that Regulus's Right Ascension is 4 hours and 13 min or $\frac{4\frac{13}{60}\,\text{hr}}{24\,\text{hr}} \times 360° = 63\frac{1}{4}°$ greater than Betelgeuse's. To establish an ***origin*** or zero point of ***Right Ascension***, we agree on a ***primary direction*** from which we measure the RA angle. This is analogous to the longitude angle on Earth, which is measured from the ***prime meridian*** running through Greenwich, UK.[39] Astronomers agree to measure the RA angle from the celestial position of the sun on March 21 toward the east. This makes sense, since the sun on this day, called the ***vernal equinox*** or ***First Point of Aries***, is exactly on the ***Celestial Equator***, that is, has zero declination, as we have seen in Section 1.3.1.

For most purposes it is easier to use stars as proxies for coordinates to describe positions in the sky, but you should at least be aware of the concept of ***celestial coordinates*** as opposed to coordinates on Earth and alt-azimuth system or observer coordinates.

1.4.1 Seasonal Motion of the Sun Is a Path on the Celestial Sphere Called Ecliptic

We saw earlier that the stars are moving faster than the sun in the observer's sky. Based on our observations, we had to distinguish the ***solar day*** defined as a 360° complete rotation of the sun with respect to the observer ("noon to noon") from the shorter ***sidereal day*** defined as a 360° complete rotation of the celestial sphere of fixed stars. To describe the daily rising and setting of stars we therefore have to conclude that the celestial sphere rotates once every

---

[38]These are ***sidereal hours*** and minutes, often displayed as 4^h13^m.

[39]Greenwich is a completely arbitrary choice. There is nothing special about Greenwich, other than that it was the location of the Royal Observatory of the most powerful nation on Earth at the time.

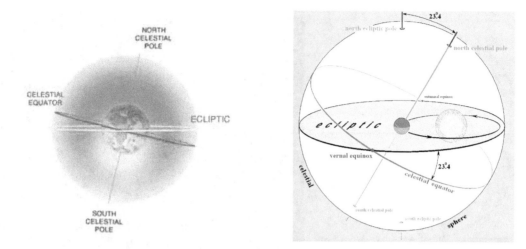

Figure 1.15: **The celestial sphere and the ecliptic.** (a) Shown as a crystalline sphere with affixed stars rotating around an axis through the poles. (b) The *ecliptic* as the plane of the Earth-sun motion. Projected onto the celestial sphere, the ecliptic is a **great circle** inclined 23.5° with respect to the Earth's and the **Celestial Equator**.

23 hours and 56 minutes, that is, once every **sidereal day**. This in turn means that the sun cannot be fixed to the sphere. While barely noticeable over a day or two, the sun invariably moves among the fixed stars and traces out a path on the celestial sphere. The shape and orientation of this path can be determined from earlier observations. Namely, determining the sun's position at noon, Section 1.3.1, and the positions of the stars at night, Section 1.3.3, allows us to draw the apparent path of the sun, called the **ecliptic** into our star maps. Exhibited on the celestial sphere, it appears as a circle inclined 23.5° with respect to the celestial equator, Figure 1.15(b). This angle means that the Earth's equator is not parallel to the **ecliptic plane**, today identified with the plane of the orbit of the Earth around the sun. There are four special or cardinal points along the ecliptic. The points of maximal and minimal celestial latitude (closest to the north and south Celestial Pole, respectively) are labeled **solstices** and happen on June and December 21, respectively. The two points where the ecliptic intersects the celestial equator, that is, the two points when the sun is *on* the celestial equator, are known as **vernal equinox** (Latin for "equal night") and **autumnal equinox**. The sun is at the equinoxes on March 21 and September 23, respectively. The angle between the ecliptic and the equator can be determined by measuring the noon altitude of the sun at the December solstice from the noon altitude half a year later (June solstice). Dividing the difference between the extreme noon altitudes and dividing by two yields the **obliquity of the ecliptic**, that is: 23.5°.

The apparent path of the sun among the stars leads through twelve special constellations known as the **zodiacal constellations**: *Aries, Taurus, Gemini, Cancer, Leo, Virgo, Libra, Scorpio, Sagittarius, Capricorn, Aquarius,* and *Pisces*. Due to the modern definition of constellation boundaries, they have an unequal share of the sun's path. Traditionally, the zodiac is subdivided into twelve exactly 30° long intervals known as the **signs of the zodiac**, since $12 \times 30° = 360°$. When this system was set up about two thousand years ago, the intersection between ecliptic and celestial equator, that is, the origin of celestial longitude, was in *Aries*. Due to the **precession of the equinoxes**, see Section 2.3.4, this point has drifted into *Pisces*,

1.4. THE CELESTIAL SPHERE AS A PRACTICAL TOOL

but is still called the **Aries point** (♈). During the June solstice the sun is in *Gemini*, which means that the constellation *Gemini* is unobservable, because it culminates at noon, that is, is above the horizon during daytime.

Note that the sun's daily motion is slower than the stars'. It moves slowly **eastward** with respect to them, while the whole sky turns **westward** fast. That means that in the twelve hours during which the stars go from the eastern to the western horizon, the sun has moved a tiny bit east, so that it takes a little bit longer before the sun reaches the western horizon. To illustrate the point, consider a schoolbus moving westward at 25 mph, Figure 5.1(a). A school student moves from the front to the back of the bus, that is, eastward, at 5 mph. With respect to the ground or sidewalk, he is thus moving slower at 20 mph, but he is still moving westward.

Let us explore the consequences of the inclined path of the sun on the celestial sphere. Since the ecliptic is closer to the CNP at the summer solstice than at other times, it appears higher in the sky at noon. We have already seen that stars close to the CNP never set, so it is reasonable to expect that the closer an object on the celestial sphere is to the CNP, the longer it will stay above the horizon. This is, of course, exactly the case when the sun is at the June solstice, that is, in the summer in the northern hemisphere: The days are long, the nights are short, the sun is high in the sky, and the temperatures are up. The opposite is apparently true at the December solstice. In sum, we have established that the inclination of the ecliptic with respect to the celestial equator is responsible for the seasons on Earth!

By recognizing the path the sun traces out in one year among the stars of the the celestial sphere, we have successfully modified our description of the phenomena in the sky. We are able to predict when and explain why the sun is high in the sky and the days are long, and explain why the stars and the sun move at slightly different rates with respect to the ground. We will see shortly that we can also completely understand the daily motion of stars even from different locations on Earth. In other words, the model or concept of a celestial sphere will turn out to be very useful indeed.

1.4.2 Seasonal Motion of the Stars

There is no seasonal motion of the stars. Before you start to argue that this cannot be true because, say, *Orion* is only visible in winter months, let me point out that this (intentionally provocative) sentence is ambiguous. It is useless, because it does not specify a frame of reference. Motion is always relative motion, and, of course, the fixed stars are fixed; stars don't move with respect to the stars. The positions of the stars on the celestial sphere are, according to our observations thus far, unchanging.

What is meant by the **seasonal motion** of the stars is their changing **visibility at night**. Orion is considered a **winter constellation** in the northern hemisphere of Earth because it is culminating at midnight in the winter. This means it is culminating at noon in the northern summer,[40] that is, when the sun is culminating. Hence, Orion is above the horizon for twelve hours *all year*, but due to the glare of the sun not visible in the summer months. To make the point even clearer, imagine a total solar eclipse at noon in the summer. For a few minutes you would be able to see all the **winter constellations** in the summer! Also consider that

---

[40]Remember that stars gain about 4 minutes per day on the sun, that is, $4\,\frac{\min}{d} \times \frac{365\,d}{2} \times \frac{1\,hr}{60\,\min} \approx 12$ hours in six months.

due to the smooth shift of the sun along the path of the ecliptic, you will be able to see the winter constellations just after sunset in spring, and just before sunrise in the fall.

Why all of this happens is a different question. For now, we are just recording our naked eye observations, not speculating as to why stars and sun should behave this way. Nonetheless, we can already conclude that the relative motion of the sun with respect to the stars, this yearly motion of the sun along the ecliptic which is inclined with respect to the celestial equator, is responsible for the fact that, among other things, Orion is not visible in the (northern) summer months. This in turn means that we can consider the motion of the sun along the ecliptic as the *reason* for the seasonal visibility of the constellations. We already saw that the sun's motion with respect to the stars is the reason for the longer daylight and hotter temperatures in the summer. In other words, the motion of the sun along the ecliptic explains the seasons.

We conclude that this new and improved description of the universe as a celestial sphere of fixed stars with a moving sun on it summarizes *all* observations we have made thus far. It is pretty amazing that so many different patterns in nature can be described so efficiently: The rising and setting of stars and sun, the changing altitude and rising and setting position of the sun, the seasons and the seasonal visibility of stars. Collecting our many individual observations so efficiently allows us to use the celestial sphere as a reference or platform from which we can make refined observations and ponder additional questions. An obvious one: *Why does the sun move along the ecliptic?*, has been answered differently over the course of history. Today we "know" that the Earth's motion around the sun makes it look like the sun is moving with respect to the stars. This is the so-called **heliocentric model**. But how exactly do we know that this is true? There must be an observation we can make that ***falsifies*** the alternative hypothesis, namely that the Earth is standing still and the sun *appears* to be moving because it *is* moving around the Earth (**geocentric model**). Surprisingly, there is no naked-eye observation which distinguishes between these two explanations, which is the reason why the **geocentric model** was regarded a viable hypothesis for such a long time.

1.4.3 A Different Observer Location Leads to a Different View of the Sky

We have now a consistent description of the sky, and with it can understand why a different observing location on Earth leads to a different view of the sky.

We start with an observer at the Geographic North Pole, Figure 1.16(a), where the situation is simplest, because the zenith is in same direction as the CNP. Incidentally, this means that the altitude of the latter is 90°, and hence the same as the geographical latitude. This relation is true for any latitude. Therefore, the celestial equator is at the horizon, and the sky evolves around the CNP, that is, the zenith. Therefore, stars never get closer to the horizon, and therefore never rise or set, Figure 1.16(b). In other words, the observer at the pole always sees the same stars, namely, the northern stars, that is, the stars north of the celestial equator. If you think about it, there is not even a notion of east or west. At the pole, you will see exactly half of the stars in the sky. You will never see the other half. If you are at the Geographic North (South) Pole you will see all of the northern (southern) stars. This suggests a label for the demarcation line: Northern and southern sky are separated by the **Celestial Equator**. Indeed, we have defined these names above for similar reasons. The pole is a strange observing location: No rising nor setting of stars occurs, and the sun the sun changes its altitude only due to its seasonal motion! As a consequence there is daylight

1.4. THE CELESTIAL SPHERE AS A PRACTICAL TOOL

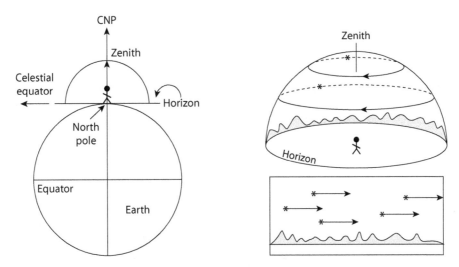

Figure 1.16: **Observing from the Geographic North Pole.** (a) The sky with zenith and horizon as defined by the observer's location on Earth. (b) A view of the horizon: The stars circle around the observer without rising higher or sinking lower in the sky.

for half a year at the poles, because the sun on the ecliptic is north of the celestial equator, and therefore above the horizon for six months out of the year.

The other extreme is the view of the sky from the equator, Figure 1.17(a). Here, the altitude of the Celestial Pole(s) is zero because we are observing from zero geographical latitude. In other words, while the zenith is by definition 90° from the horizon, the Celestial Poles are at the horizon. Therefore, stars rotate around the observer at the equator as if she were in a great barrel: They rise and set perpendicular to the eastern and western horizon, Figure 1.17(b). From the equator, an observer can see *all* the stars of the sky: all the northern *and* all the southern stars. A telescope close to the equator is thus highly desirable, and indeed, several of the largest modern telescopes are near the equator (Hawaii: 19° N, Arecibo 18° N, Paranal 25° S).

The generic situation at mid-latitudes is what most people on Earth are experiencing, Figure 1.18(a). The Celestial Pole for these observers is between the zenith and the horizon, with an altitude equal to the geographic latitude. Since stars rotate around the Celestial Pole, most of them will rise on the eastern horizon and set on the western horizon, Figure 1.18(b). Around the **meridian** stars will appear to move parallel to the ground, Figure 1.18(c). Around the pole, stars will get closer to the horizon, but not sink below it as they rotate, Figure 1.18(d). These are called **circumpolar stars**. If you think about it, you'll realize that the part of the sky with circumpolar stars is as big as the part of the sky that is completely unobservable.

As an example, consider an observer at 40° northern latitude, that is, set the latitude equal to 40° in Figure 1.18(a). Then the celestial equator is 90° − 40° = 50° from the southern horizon, which means the observer can see stars down to declination −50°, that is −50° celestial latitude. In other words, the region of the celestial sphere which is unobservable for this observer is 40° around the south Celestial Pole. On the other hand, stars closer than 40° to the north Celestial Pole are circumpolar for this observer. Finally, stars with declinations between −50° and +50° will show the usual rising and setting behavior. Note that stars close to

44 CHAPTER 1. OBSERVING AND NAMING: NAKED-EYE ASTRONOMY

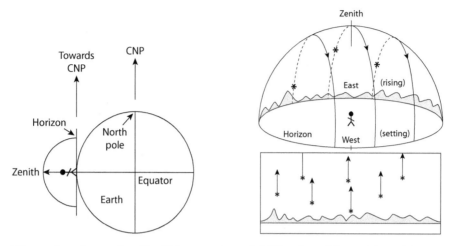

Figure 1.17: **Observing from the Earth's equator.** (a) The sky with zenith and horizon as defined by the observer's location on Earth. (b) A view of the eastern horizon as seen from a location on the Earth's equator: The stars rise vertically.

declination +50° will be above the horizon almost twenty-four hours, stars close to declination −50° will be visible only for a very short time around their culmination.

So far, we have focused on location differences in **latitude**. What about observers at different **longitudes**? Well, we have implicitly covered this earlier. This question leads to the constitution of time zones, because local noon is different at different longitudes. Since the sun represents the orientation of the celestial sphere, we conclude that the celestial sphere appears to rotate if you are moving in longitude. Recall from Section 1.1.2 that for "every degree of geographical longitude the observer is further west, the sky will look rotated clockwise by one degree." Loosely speaking, for the observer further west it is earlier, and therefore the sky has "not turned as much" as for an observer further east.[41]

In sum, we have come to an understanding of how the observer's location on Earth affects her view of the sky.

1.4.4 The Celestial Sphere as a Concept

Conceptually speaking, the celestial sphere is a way to make sense of the fundamental patterns in the sky. We might work in our insights about the observing location by constructing a **two-sphere model** [42] of the cosmos. By *assuming* that we observe the (inside of the) celestial sphere by standing on the surface of a spherical Earth, and that the two spheres are rotating relative to each other, we can describe basically every pattern in nature we have talked about so far. Thereby we have achieved the following. Instead of having to track the individual motions of thousands of stars for observers at many different locations, we have one **unified** description of these millions of position changes. How do we know that these are reasonable assumptions? Because they describe correctly what we see in the sky, and they make predictions about how the view of the sky should change—if I wait an hour, or go north

---

[41]In an absolute sense, you are just looking in a different direction, and the Earth obscures a different part of the celestial sphere.
[42]The name was coined in [26].

1.4. THE CELESTIAL SPHERE AS A PRACTICAL TOOL

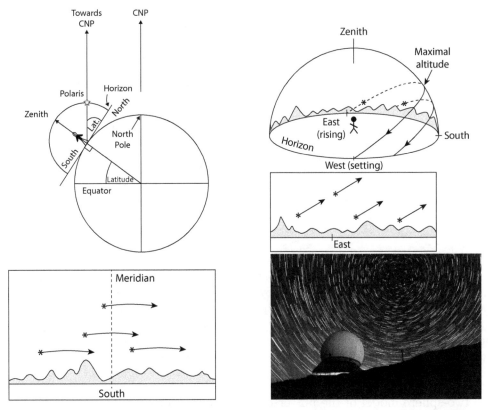

Figure 1.18: **Observing from mid-northern latitudes.** (a) [Top left] The sky with zenith and horizon as defined by the observer's location on Earth. (b) [Top right] A view of the eastern horizon: The stars rise at an angle to the horizon. (c) [Bottom left] A view of the southern horizon: The stars move westward at almost constant altitude that peaks at the meridian. (d) [Bottom right] A view of the northern horizon: The stars circle clockwise around the CNP, with an altitude angle equal to the latitude of the observer.

1,000 miles.[43] If we check these many predictions many times, we will become convinced that this **model** makes sense, that is, that this is a good description of what goes on in nature.

Additionally, we will have a breeze to describe *anything* that is happening in *our* sky and translate it into the view that another observer would have. Almost without noticing, we have made a model of the universe. But have we understood anything? Note that merely by describing in an efficient way *what* is happening, we have acquired predictability, which is comforting and of practical use (orientation, navigation, agriculture). Still, we have not explained *why* the sun is moving along its path with respect to the observer *and* with respect to the stars, but by seeing the **pattern** repeating itself over and over again, we can safely *assume* that this will keep happening in this way. Assumptions like these are seeds of (scientific) models of the world. Although we can't say for sure if the world is still the same tomorrow, we assume it will be, and therefore abstract from the real world to an idealized description of it. It may seem silly to posit that something catastrophic will happen tomorrow, but we don't know *in advance* that everything is going to be as it was yesterday. Therefore, our assertion that nothing bad is going to happen is reasonable, but nonetheless an **assumption**.

---

[43]Note that in order to quantify the latter prediction, we would need to know the size of the sphere.

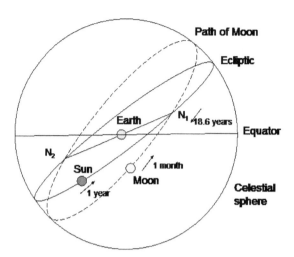

Figure 1.19: **The moon's path on the celestial sphere.** The path is inclined with respect to both the celestial equator and the ecliptic, here shown as the path of the sun with respect to the fixed stars. The moon takes one (sidereal) month to completely traverse its path, whereas the sun takes one year to go once around the ecliptic. Both are slow motions compared to the daily rising and setting, which is about 27 and 365 times faster, respectively. The intersection between the moon's path and the ecliptic is known as the **node line**. It travels around the ecliptic in 18.6 years.

The two-sphere model of the universe is, of course, nothing but a start in understanding the universe. In may ways we still have to fill it with real things. All we have done so far is to understand how the **specks of light** and the *bright disk* in the sky are moving. We still do not know what they *are* nor why they move in this way. In the next chapter we will see that humans have struggled for millennia to answer these questions.[44] First, we need to complete our discussion of celestial motion by considering the movements of the moon and the planets.

Concept Practice

1. The celestial sphere
    (a) ... is based on the relative rotation of Earth and stars.
    (b) ... is a crystalline sphere on which the fixed stars are situated.
    (c) ... cannot be used to describe the position of sun and moon.
    (d) Two of the above
    (e) None of the above

1.5 Observing the Moon

1.5.1 The Moon Moves with Respect to the Stars while Rising and Setting

The moon moves in a fashion very similar to that of the sun. Its daily motion takes it from east to west, reaching the highest position at the meridian—but usually not at noon. It also

---

[44] As late as 1859 even scientists believed that "what the stars are we do not know and will never know" [48].

1.5. OBSERVING THE MOON

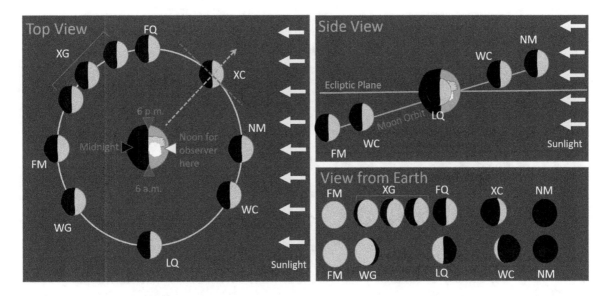

Figure 1.20: **Lunar phases.** The relative position of sun, moon and observer on Earth produces the phases of the moon. (a) *Bird's-eye view.* the moon's hemisphere facing the sun is always illuminated. An observer on Earth sees different fractions of the illuminated side depending on the relative position of Earth, moon and sun. The observer can see the part of the moon closer than the (red) line perpendicular to the line of sight (green arrow at XC). This creates the **lunar phases**: new moon (NM), waxing crescent (WC), first quarter (FQ), waxing gibbous moon (XG), full moon (FM), waning gibbous moon (WG), last quarter moon (LQ) and waning crescent (WC). (b) *Side view.* the moon's orbit is inclined by about 5° with respect to the **ecliptic plane**. (c) *View from Earth.* The extreme positions are **new moon**—when the moon stands between Earth and sun, and we see only the dark side—and **full moon**—when the moon is opposite of the sun, that is, the Earth stands between moon and sun, and we see only the illuminated side.

moves with respect to the stars, and therefore traces out a path on the celestial sphere. We can easily determine this path by sketching the position of the moon relative to the constellations. This is possible because the moon is not nearly as bright as the sun, and stars are visible around the moon at night. We observe that the moon moves eastward with respect to the stars at about 13° per day. As depicted in Figure 1.19, the moon's path is inclined with respect to the celestial equator by an angle similar to the inclination of the ecliptic. However, there is also a 5.2° inclination of the moon's path with respect to the ecliptic. This inclined path of the moon implies that the moon's angular distance to the Celestial Poles oscillates. To complete a full circle on the celestial sphere, the moon takes approximately twenty-seven days, which makes sense since $27d \times 13°/d \approx 360°$. This is known as the **sidereal month** $T_{sid} = 27.322$ days. It is *the* rotation period of the moon, since the moon rotates a full 360° in an absolute sense, that is, with respect to space itself. In the next paragraph we will see that the moon's motion gives rise to another pattern—the **moon phases**—with a slightly different time period.

Recall that labels have to have precise meanings, and since the label **month** is ambiguous, we use the qualifier **sidereal** to make it clear that we mean rotation period with respect to the stars.

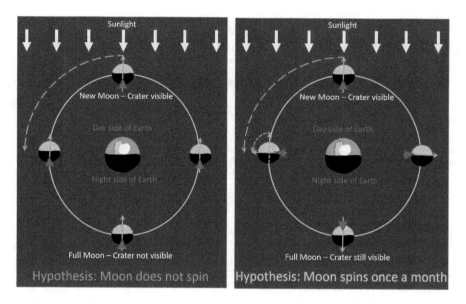

Figure 1.21: **Synchronous rotation of the moon.** Shown is the moon in its orbit as seen from far above the Earth's North Pole. A crater (triangle-shaped for illustration) is located on one side of the moon. Two hypotheses are considered to explain the synchronous rotation of the moon. (a) Hypothesis A: the moon does not spin, that is, does not rotate around itself. (b) Hypothesis B: the moon spins once while rotating once around the Earth. Only B predicts that the moon keeps the same face toward Earth in accordance with observations. Hypothesis A is falsified.

1.5.2 The Phases of the Moon

The sun always appears as a full, bright disk in the daytime sky. The moon on the other hand changes its appearance while moving on the celestial sphere. Often, we only see part of the moon's disk lit up. This pattern of lunar[45] illumination is known as the **phases of the moon**, see Figure 1.20(c). It follows that the moon does not shine by its own light—otherwise the lunar disk would always be fully lit up. Rather than emitting light like the sun, the stars or a light bulb, the moon **reflects**[46] light. What we see here on Earth is sunlight that hit the moon's surface, bounced off, and found its way into the observer's eyes on Earth. In other words, this light has traveled through space, made contact with the lunar surface, and traversed the Earth's atmosphere before being "collected" by the observer. Later we will see that all parts of its journey leave their mark on the light beam.

How the moon appears to us on Earth depends therefore on the relative position of the moon, the sun, and the Earth. It is important to realize that the moon, like all objects illuminated by a distant light source,[47] is bright on the side that faces the sun, and dark on its other side. For the same reason there is nighttime on half of the Earth's surface and daytime on the other half. How come, then, that we don't see see the moon always half lit up, a phase we call *quarter moon*? The reason is that the observer on Earth may see more or less

---

[45]Latin: *luna*, the moon. Incidentally, the Roman moon goddess was *Selena*.

[46]All surfaces reflect light; most of them not nearly as perfectly as a *mirror*, though.

[47]Try it out with a baseball and a flashlight in a dark room. As you hold the baseball in your hand with an outstretched arm, you'll see the dark side of the ball when its in the direction of the flashlight mounted on a shelf. Half of the side of the ball facing you is lit up if you move the ball in a direction perpendicular to the flashlight, and likewise for the other "phases."

1.5. OBSERVING THE MOON

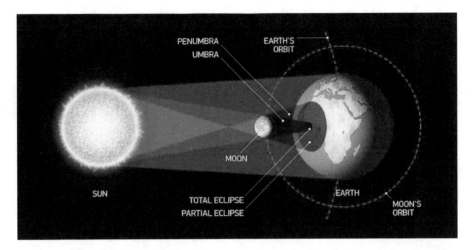

Figure 1.22: **The position of sun, Earth and moon during a solar eclipse.** Shown is the area on Earth which experiences a *total* eclipse, because it is located in the **umbra** cast by the moon.

of the bright half of the moon. In fact, the *quarter moon* phase occurs when the observer can see half of the bright half of the moon, and $\frac{1}{2} \times \frac{1}{2} = \frac{1}{4}$, so the name makes sense. Incidentally, the *quarter moon* phase occurs when the moon has completed one quarter of its monthly cycle, or has one quarter to go. From Figure 1.20 it should be clear how the lunar phases are a result of the changing relative position of moon, sun, and Earth. When the moon is at position NM is is standing roughly in the direction of the sun. This phase is called *new moon*, and is conventionally counted as the first phase of the moon. We are careful here to say *roughly in the direction of the sun*, because the figure leaves out the motion of the moon in the direction perpendicular to the ecliptic. The *new moon* culminates on the meridian at the same time the sun culminates on the meridian. In the observer's sky, the motion of the *new moon* is therefore the same as the motion of the sun: It rises with the sun in the east, it culminates with the sun at noon at the meridian, and it sets with the sun in the west, and is thus only up during *daytime*. Typically, the new moon will be "above" or "below" the sun, that is, higher or lower in the sky than the sun. Sometimes the moon stands *exactly* in the direction of the sun. This is the configuration leading to a **solar eclipse**, Figure 1.22, Section 1.5.3. Generally, the *new moon* is invisible because it stands in the direction of the sun, and we can only see its unilluminated side, as its bright side is facing away from the Earth.

A day after new moon we can perceive a thin sliver—the **crescent moon**—about 12 degrees east of the sun. We can estimate that it will be visible for less than an hour after sunset, since daily motion happens at a rate of 15 degrees per hour or 360° in 24 hours. The crescent waxes thicker over the next days while the moon continues to move east with respect to the sun at the rate of about 12 degrees per day. This means that during a full night it moves 6° or 12 lunar diameters, which a careful observer will notice if she compares the moon's position to the position of bright stars close to it. Note that the moon moves eastward with respect to the stars slowly, whereas the celestial sphere of stars moves westward fast with respect to the ground. So the moon will rise *later* than the stars each night. In fact, each day the moon rises about one hour later, since its 13° movement per day with respect to the *stars* is about the same as the sky's 15° rotation in one hour.

After about a week, half the disk is illuminated at **first quarter moon**. During the

Figure 1.23: **Annular eclipse of the sun.** The moon does not completely cover the sun's disk. A bright ring (Latin: *annulus*) remains around the eclipsed part of the sun's disk.

next week this half disk grows smoothly in its *gibbous moon* phase until it becomes the *full moon* two weeks after being lost in the sun's glare as the *new moon*. Since the illuminated part grows or waxes during these two weeks, these are known as the *waxing* phases of the moon. The moon then goes through two weeks of waning phases (*waning gibbous moon*, *last quarter*, *waning crescent*) before concluding this cycle after 29.53 (solar) days as a *new moon* disappearing in the glare of the sun. This is not the same time it takes the moon to make a full rotation with respect to the stars, which we had called the *sidereal month* of 27.3 days. The reason is that the moon moves slower with respect to the sun (12° per day) than with respect to the stars (13° per day), which in turn is due to the sun's motion with respect to the stars of about 1° per day or 360° per 365 days. As a cross-check we compute that $\frac{360°}{29.53\,\text{d}} \approx 12°/\text{d}$. We call the time between two full moons the *synodic month*. Note that this synodic cycle only concerns the *phases* of the moon, and does not describe a cycle in position: A full moon can be high or low in the sky, can be in *Gemini* or *Virgo*, can be in the east or the west.

Even with the unaided eye, we can make out surface features on the moon. We see dark spots and bright areas. The dark spots are called *maria* or oceans, because that's what they look like—of course, that's not what they *are*. The bright areas are known as *terrae* or *highlands*. Looking at the moon during its cycle of phases, Figure 1.20(c), we realize that the degree of illumination changes dramatically, but the surface features remain in the same spot. In other words, the moon always shows us the same side or face. This is a remarkable pattern in nature that needs to be explained. You might think that the moon must not be rotating if it always shows us the same face, but this is wrong. The orbiting of the moon around the Earth would show us all sides of the moon if it didn't rotate, see Figure 1.21(a). Only if the moon rotates around itself in precisely the same time as it rotates around the Earth, as shown in Figure 1.21(b), does the same hemisphere always face Earth. Loosely speaking, the moon's day is as long as the moon's year. We have *reasoned* this out: The so-called *synchronous rotation* of the moon is responsible for the pattern in nature we observed. We have *inferred* from our observation that the moon rotates synchronously, that is, that its two (sidereal) rotation periods—about its axis and about the Earth—have the same length. This is a nice achievement, but a skeptic might say that we did not explain anything. We just observed and used our reason to figure out *what* is happening. It is much harder to figure out *why* the moon should rotate in this way. After all, the Earth's day is hundreds of times shorter than

1.5. OBSERVING THE MOON

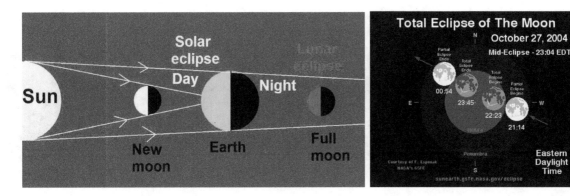

Figure 1.24: **Lunar eclipses.** (a) The moon moves into the shadow cast by the Earth. This can only happen during full moon at night, and is visible from the entire night side of Earth. (b) The total lunar eclipse of October 2004. The moon stayed in the shadow of the Earth for almost 1.5 hours.

its year. As is typical in science, we need to gather more information before we can answer this question in Chapter 4.

Concept Practice

1. Find a strikingly simple reason why the moon, contrary to the sun, cannot be emitting its own light.
2. Formulate an argument or devise an observation that can be made to show that the phases of the moon are *not* due to the shadow of the Earth falling unto the moon.
3. What is the angular separation of sun and moon at: (a) First Quarter moon (b) Full moon (c) New moon.

1.5.3 Eclipses Involve the Moon

The moon can eclipse the sun because both have virtually the same apparent size. The new moon obscures our view of the sun when it stands exactly in the direction of the sun. This is known as a **total solar eclipse**, Figure 1.22. Since this is a delicate configuration of sun, moon, and observer, the totality phase of solar eclipses does not last long, typically five minutes or less. The shadow of the moon is known as the **umbra**.[48] Due to the rotation of the Earth and the motion of the moon along its orbit, the tip of the umbra moves along the surface of the Earth, creating an **eclipse path**. This path is narrow, and to experience a total eclipse, you'd have to observe from this small fraction of the Earth's surface. For most observers, therefore, the alignment isn't perfect and they see a **partial solar eclipse**, during which the moon covers only *part* of the sun's disk. **Partial solar eclipses** can be seen from the much larger region covered by the **penumbra** or half shadow of the moon, Figure 1.22. Whenever the moon moves in and out of a total solar eclipse, we experience a **partial solar eclipse**. There is a third kind of solar eclipse. It happens when sun and moon are perfectly aligned, but the apparent size of the moon is too small to completely cover the sun. Since there is a bright ring left, this is called an **annular eclipse**, from the Latin word for ring, *annulus*, see Figure 1.23.

While at solar eclipses the Earth moves into the shadow cast by the moon, during a **lunar eclipse**, the moon moves into the shadow cast by the Earth, see Figure 1.24(a). Since the

---

[48]Latin: *umbra = shade*, cf. *umbrella = little shade*.

Figure 1.25: **The moon's appearance during a lunar eclipse.** The photos taken with a digital camera using an 8″ telescope show the change of the moon's color during the 2004 lunar eclipse. Several photos were taken with different exposure times and then digitally stitched together. The orange, eclipsed moon on the right in reality is as dark as the unilluminated part of the moon on the leftmost photo.

Earth is much larger than the moon, its shadow is much larger, and lunar eclipses last much longer than solar eclipses, typically several hours; see Figure 1.24(b). Lunar eclipses may not be quite as spectacular as solar eclipses, but they can be seen from anywhere on the night side of Earth. This is is due to the fact that the full moon is opposite of the sun, and therefore visible from the half of the Earth that does not face the sun. As with solar eclipses, there are partial lunar eclipses when the moon does not entirely move into the shadow of the Earth because Earth, sun, and moon do not perfectly line up, see Figure 1.26. In fact, most of the times when the moon is opposite of the sun, that is, at full moon, the moon is either "above" or "below" the shadow of the Earth.

The eclipsed moon appears orange because some of the sunlight from the day side of Earth is scattered by the atmosphere into the shadow region. The atmosphere filters out much of the blue part of the sunlight, so anti-blue or orange remains. The scattered light is only a fraction of the full sunlight, so the eclipsed moon appears much dimmer; see Figure 1.25.

1.5.4 Interpreting Lunar Observations

As we have seen, there is a host of phenomena or patterns in nature associated with the moon: The moon rises and sets at different times during the month, the moon traces out a path among the stars that is similar to the ecliptic, the moon shows phases and generates eclipses.[49]

We started to organize our observations by pointing out a correlation between the visibility of the moon above the horizon and its phase. Since the new moon is so close to the sun in the sky, it has the same daily motion as the sun. It rises at sunrise and sets at sunset. The full moon is opposite of the sun and exhibits the opposite daily motion: The full moon rises when the sun sets and vice versa. While the full moon is above the horizon only at night and the new moon only during the day, the moon is visible partly during the night and partly during the day during most of its monthly cycle. As an example, the first quarter is 90° east of the sun[50] and will therefore rise a quarter of a day (6 hours) later than the sun. Of course, it will

---

[49]It is from *eclipses* that the *ecliptic* gets its name: They happen when the moon is on the ecliptic.

[50]Cross-check: Seven days of approximately 13° motion each day yield approximately 90°, and a quarter of

1.5. OBSERVING THE MOON

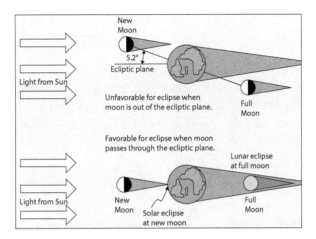

Figure 1.26: **The inclination of the moon's orbit.** The orbit of the moon is inclined with respect to the ecliptic. Only if the moon is on the ecliptic can eclipses occur. (a) Moon above or below the plane: Lunar shadow misses the Earth. (b) Moon in the ecliptic plane: Eclipses occur.

also set 6 hours later than the sun, so it is visible 6 hours at night and 6 hours during the day. It will culminate 6 hours later than the sun, namely at 6:00 p.m. Hence, we generated a good and consistent description of the motion and the phases of the moon as seen from Earth.

What does this tell us about the *actual* position, motion, and distance of the moon? It is immediately clear that the moon must revolve around the Earth. The fact that we see moon phases rules out the possibility of the Earth revolving around the moon. In contrast, we cannot yet rule out that the sun is going around the Earth, since we do not have independent observations analogous to the lunar phases. When we say that the moon rotates around the Earth we have implicitly introduced the notion of an orbit. Note that we can never *observe* an orbit. We can only observe the position of celestial objects and *infer* that there must be a motion and therefore a trajectory of the object. The trajectory or orbit is therefore an *abstract concept*. It is *useful* because it helps us describe how the object moves.

Solar eclipses show us unambiguously that the moon is closer to Earth than the sun. We'll see later that we can get much more out of observing eclipses. For instance, the size of the shadow of the Earth during a lunar eclipse was used by the ancient Greeks to determine the size of the moon and its distance by a clever geometric argument.

We already pointed out that the rarity of eclipses means that the orbit of the moon is inclined with respect to the ecliptic. The inclination causes the new and full moon to be above or below the plane of the ecliptic, and therefore prevents the shadow of the new moon from falling onto the Earth and the shadow of the Earth from obscuring the full moon; see Figure 1.26. We knew that already from observing the path of the moon on the celestial sphere; see Figure 1.19.

Summarizing, we have interpreted our lunar observations and were able to come to an understanding of the patterns of nature associated with the moon. They are caused by the moon orbiting the Earth in 27.3 days and spinning exactly once during that time in an orbit inclined by 5.2° with respect to the ecliptic plane. Interestingly, it will not be so easy to interpret the motion of the planets.

---

the full 360° monthly motion of the moon on the celestial sphere is 90°.

Concept Practice

1. The *lunar month*
   (a) ... is about 27 days.
   (b) ... is about 29 days.
   (c) ... is either 30 or 31 days.
   (d) ... is not properly defined. There are several different ways to define a month.
   (e) Two of the above
   (f) None of the above

2. The moon always shows us the same hemisphere. This means
   (a) ... it does not rotate.
   (b) ... it rotates once a month.
   (c) ... it rotates once a year.
   (d) ... it rotates opposite to the Earth's rotation direction.
   (e) None of the above

1.6 Observing the Planets

1.6.1 Planets Slowly Wander among the Stars while Rising and Setting

Like the sun and the moon, the planets move slowly on the celestial sphere, while following the daily motion of rising, culminating, and setting. We already understand the **daily motion**: Since the celestial sphere turns relative to the Earth, all celestial objects follow suit and seem to rotate around the observer. To learn something about planetary motion, we must therefore *mask out* the daily motion to focus on the motion that is due to the planets themselves. The easiest way to do this is to measure the position of a planet with respect to the stars. We can, for instance, measure the position of Jupiter as 8.5 degrees off of Regulus in *Leo* in the direction toward Spica in *Virgo*; see Figure 1.27. If we do a series of these measurements over several months, we see that Jupiter slowly drifts from *Leo* toward *Virgo*. This is in the general direction of the ecliptic, so very similar to the motion of the sun, only much slower. If we extrapolate the data taken, we see that Jupiter moves about one zodiac constellation or about 30° per year. Since there are twelve zodiac constellations, it will take Jupiter about twelve years to complete a full circle: 30°/yr×12 yr=360°. Planets are specks of light like the stars, but since they move slowly[51] with respect to the fixed stars, they cannot be modeled as being fixed to the celestial sphere. In other words, this second type of specks of light are sufficiently different to deserve their own label (**planets**), as we remarked earlier. Note that only five planets were known in antiquity.

1.6.2 A New Pattern: Retrograde Motion

So far the motion of the planets with respect to the stars looks remarkably similar to that of the sun and the moon. But we are in for a surprise. If we monitor the position of a planet for about a year, we find that the planet at some point slows down and then reverses its direction of motion on the celestial sphere (not with respect to the ground!); see Figure 1.28. This is the so-called **retrograde** motion of the planets. The name suggests a backward motion, and

---
[51] Slow compared to their daily rising/setting motion!

1.6. OBSERVING THE PLANETS

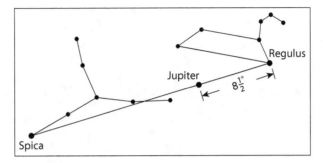

Figure 1.27: **Jupiter in *Leo*.** Its position is specified as an angular distance from Regulus toward Spica.

indeed that's what it is. It is counter to the normal or **prograde** motion of the planet. These labels can be confusing, since it is tacitly understood that planetary motion is with respect to the stars, that is, on the celestial sphere. While the daily motion takes the celestial objects westward, the normal or prograde motion of the planets is actually *eastward*[52]. Retrograde motion is a slow drift of the planets westward **among the stars**. So slow is this drift that it virtually has no effect on the daily rising and setting of the planet. It is emphatically *false* that a retrograde planet rises in the west. The retrograde motion is thus virtually independent of the daily motion.

Figure 1.28(a) is a composite photo of many sightings of Mars during several months. Therefore, it is misleading: In the sky you will see, of course, Mars only once in front of the background of the stars. Nonetheless, this processed photo is closer to reality than schematic drawings like Figure 1.28(b), in that it shows the sightings of Mars as isolated points. We'd be surprised if Mars wouldn't move smoothly between the sightings, but strictly speaking, this is an assumption. We *infer* from our isolated data that there must be a smooth curve that connects the positions of Mars on different nights. This inferring of a general relationship from isolated data points, here of observation time and celestial position, is a trait of the empirical sciences. It illustrates the process of **induction**. Once we have this general relationship, we can use **deduction**, that is, the method of finding special instances from this general relationship. Here, we could **predict** the position of Mars at a date in the future after we inductively found a description of its motion as a function of time.

As you can see from Figure 1.28, Mars slows down (sightings closer together) and turns around in a complicated way that forms a **retrograde loop**. The two turnaround points are known as **stationary points**, because the planet has to come to a standstill when changing its direction of motion. At these points the planet changes from pro- to retrograde and from retro- to prograde motion. The individual photos compiled here into a single representation were taken about a week apart. This gives you a sense of how slow planetary motion really is. While Mars moves a few degrees among the stars, the celestial sphere has done over 100 of its 360° rotations, that is, over 100 days have passed.

Retrograde motion is unique to the planets—we have not seen anything like it for the sun nor the moon. In fact, this is so curious a behavior that it took millennia to correctly explain it. Indeed, the retrograde motion of the planets can rightly be considered the seed for the

---

[52] Recall that also the sun and the moon travel eastward *with respect to the stars*. This is why the **solar day** is longer than the **sidereal day** and why the moon rises an hour later with each passing day.

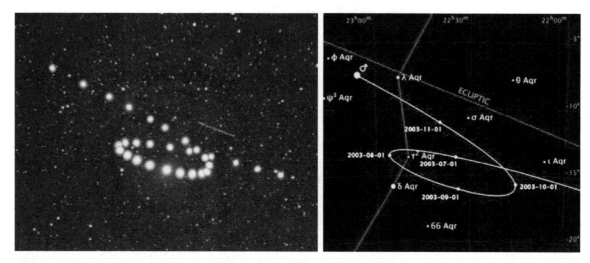

Figure 1.28: **Retrograde loop of Mars (and Uranus) in 2003.** (a) A composite photo by Tunc Tezel showing the position of Mars in Aquarius. Uranus appears as a much dimmer and shorter line of dots right of the center. (b) A schematic rendering of the retrograde loop with dates.

Copernican revolution. We will, for now, not attempt to explain this complicated motion but rather gather more data to make it easier to describe how exactly the planets move. The hope, as always in science, is that eventually the amassment of data leads to a clue as to what causes this pattern of nature. And indeed, planets show an additional pattern. When planets are in retrograde motion, that is, moving westward with respect to the stars, they are *brighter* than when they are in prograde or generally eastward motion. We can safely assume that the planets do not change physically the short time we are observing them, that is, we can assume they do not suddenly get bigger or more reflective. Therefore, the only interpretation of the changing (apparent) brightness of the planets is that the planets are closer to us when they appear brighter, and farther when they appear dimmer. What's more, we find that they they are closest when they are at the midpoint of their retrograde loop.

We also find that planets are brightest when they culminate at midnight. Since this means that they are 180° off the sun in the sky (the sun culminates twelve hours later at noon), they are opposite of the sun. We call this special configuration of Earth, sun, and planet the **opposition** of the planet. In Figure 1.28, Mars reaches opposition on August 28, 2003. However, not all planets culminate at midnight. In fact, Venus and Mercury stay curiously close to the sun in the sky, and so they tend to culminate around noon as the sun. Historically, these two planets have been called **inferior planets** because of this behavior, whereas the other planets are called **superior planets**[53]. Superior planets can appear in the same direction as the sun, that is, culminating at noon. This configuration is called the **conjunction**[54] of the planet. Physically, a superior planet in conjunction is behind the sun, and cannot be seen due to the sun's glare.

The inferior planets Mercury and Venus are never visible at midnight, which means that their cycle of apparent motion plays out differently. The inferior planet can be seen some time after sunset when it is east of the sun, Figure 1.29. The sun is then west of the planet and

---

[53] Since Copernicus, this distinction boils down to the fact that Venus and Mercury orbit the sun inside the Earth's orbit, and the other planets are farther from the sun than the Earth.

[54] From Latin *coniugare* = *join together*, that is, sun and planet are close together.

1.6. OBSERVING THE PLANETS

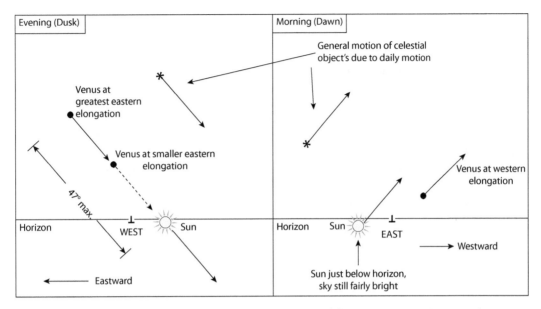

Figure 1.29: **Venus as evening and morning "star."** (a) Venus is visible just after sunset on the western horizon. (b) Venus is visible just before sunrise on the eastern horizon. The greater the elongation, the longer an inner planet can be seen before sunrise or after sunset. If the elongation is too small, the planet disappears in the glare of the sun in an inferior or superior conjunction.

already below the horizon. There is a maximal possible angle between an inferior planet and the sun. This configuration is known as **greatest elongation**, and can happen east (evening visibility of the planet) or west of the sun (morning visibility), see Figure 1.29. You may have heard the *planet* Venus being called the evening or morning *star*. This is not as ridiculous as it sounds. Recall that without a telescope, a planet is a speck of light and in that respect not different from a star. It is also nontrivial to realize that the morning and evening star are two names for the same object, something we want to avoid in science. Since in between its two apparitions Venus vanishes in the glare of the sun, it is hard to establish continuity here. Between the two elongations is a configuration called the **inferior conjunction**. This is when the planet is between us and the sun (and therefore in the direction of the sun). By contrast, the **superior conjunction** happens when the planet is behind the sun (and therefore again in the direction of the sun). The planet is much closer to us at inferior conjunction. Without a telescope we cannot see the associated change in apparent size. However, we will note that the planet appears much brighter at inferior conjunction. Still, these considerations show how hard it is to judge from the *apparent* motion of the planets in the sky how they *actually* move in space.

1.6.3 Trying to Interpret Planetary Observations

Intricate patterns in nature—such as the retrograde loops of planets—are hard to describe or even explain. The first step toward an understanding of planetary motion is to carefully gather as much data as possible. We can then sift through the data looking for similarities, repetitions, and so forth. In this respect, the **synodic period** is a useful concept. It is defined as the time between two successive identical configurations of sun and planet in the sky. For example, the time between two consecutive **inferior conjunctions** of Mercury

Planet	Mercury	Venus	Mars	Jupiter	Saturn
Synodic Period (days)	116	584	780	399	378

Table 1.3: **The synodic periods of the classical planets.**

is 116 days, and the time between two **oppositions** of Mars is 780 days. In fact, we have encountered this concept earlier when discussing the motion of the moon. Namely, we called the 29.5 days between two full moons the **synodic month**. From Table 1.3 we see that there is at least some regularity. The retrograde motion recurs in regular intervals: 116 days, 584 days, and so forth. Furthermore, if this is the correct order of the planets[55] then the synodic periods are shorter the farther the planet is from Earth. The synodic period of Mercury seems strangely short, which doesn't increase the confidence into our ability to interpret planetary observations. As we pointed out, this "problem of the planets" as it has been called, took millennia to figure out. This section should have given you a hunch of just how confusing the observational data is without a theory guiding our search for an explanation.

Concept Practice

1. A planet in *retrograde motion*
   (a) ... moves from east to west with respect to the sun.
   (b) ... moves from east to west with respect to the stars.
   (c) ... moves from east to west with respect to the ground.
   (d) None of the above

2. The *synodic period* of a planet
   (a) ... is its rotation period around the Earth.
   (b) ... is its rotation period around the sun.
   (c) ... is its rotation period with respect to the stars.
   (d) None of the above

1.7 Summary and Application

1.7.1 Digest of Observations Foundational for the Expanding Universe

- We see the universe from Earth as projected onto a two-dimensional half dome, aka the sky.

- The observer's sky is the result of the Earth's blocking of the sky below the **horizon**. Observer coordinates of a celestial object are **altitude** above horizon and **azimuth**, that is, direction from north in the horizontal plane.

- Observers at different locations on Earth have a different view of the sky.

- The position of celestial objects can be described by **equatorial coordinates** on the **celestial sphere**, where the **Celestial Poles** and the **Celestial Equator** are the main features.

---

[55] This was not clear until COPERNICUS, although PTOLEMY guessed the order correctly.

1.7. SUMMARY AND APPLICATION

- For most observers, objects rise in the east, culminate at the meridian, and set in the west. This is the **daily motion**.

- The solar day (noon-to-noon) is exactly 24 hours long. The **sidereal day** is about four minutes shorter, and represents one full revolution of the celestial sphere.

- The sun's apparent path among the stars or on the celestial sphere is the **ecliptic**. It is a great circle inclined 23.5° with respect to the **Celestial Equator**. The sun goes around the ecliptic once per year in its **seasonal motion**, also called the **annual motion**. This in turn causes the **seasons** and determines the **seasonal visibility** of stars and constellations.

- The moon traces out a path among the stars, which it completes in a **sidereal month** of 27.3 days. This path is inclined with respect to the ecliptic by 5.2°.

- The moon shows **phases**. It goes through a complete cycle of phases in a **synodic month** of about 29.5 days.

- The motion of the planets is more complicated than lunar and solar motion. Planets rise, culminate, and set like the sun and moon, but exhibit **retrograde motion** when they reverse the normal or **prograde** direction of their motion to westward with respect to the stars.

1.7.2 Concept Application

1. Sketch the horizon system with the observer in the center. Label the zenith, the meridian, and the altitude angle.

2. Consider two cubes. One has twice the width of the other. How many of the smaller cubes fit into the bigger one?

3. Draw a simplified diagram showing the sky for an observer at (a) 60° N (b) 30° S.

4. Redraw the observer sky, Figure 1.18(a), for an observer at mid-southern latitude.

5. Redraw Figure 1.18(b)[bottom], for an observer at mid-northern latitude centered on *west* instead of *east*.

6. How large is the region of circumpolar stars for an observer at (a) 30° latitude, (b) on the equator, (c) at the South Pole.

7. Compute the angular separation of the sun and the moon three days after new moon. Use the length of the sidereal month and neglect the seasonal motion of the sun.

8. Explain the difference between a partial and an annular solar eclipse.

9. Someone claims to have seen an annular lunar eclipse. Refute the claim.

10. Naively, how many eclipses (lunar and solar) would you expect in a year?

11. Can Mercury ever be seen at midnight? Explain.

12. What distinguishes inferior from superior planets?

13. Venus is at greatest eastern elongation. Find a way of telling, with unaided eyes, whether Venus is headed for inferior or superior conjunction. If you succeed, you have found a way of telling whether Venus orbits clockwise or anti-clockwise.

1.7.3 Activity: The Sun's Shadow and Position

To study the motion of the sun in the sky, we need to measure its position accurately. However, the sun is too bright the aim a telescope at it and then measure the orientation of the telescope. In fact, you shouldn't even look directly into the sun with a naked eye! What to do? There is a simple way to circumvent the problem: Use a stick called a gnomon and measure the height H of the stick above ground, and the length L of the shadow the sun casts. The ratio of the two is the tangent of the altitude angle, see Figure 1.11. Use two meter or yardsticks and measure the position of the sun, that is, its altitude angle A and its azimuth α (counting from north).

1. Write down the height H of the gnomon.
2. Measure the length L of the shadow keeping the gnomon is perpendicular to the ground.
3. Estimate the direction of the shadow (called the azimuth angle), for example northeast.
4. If $H = L$ then $A = 45°$. Use this to estimate the altitude of the sun crudely, that is, decide whether it is close to, bigger or smaller than $A = 45°$?
5. From what you know (latitude of your city, current season, time of the day) does this make sense? Why or why not?
6. Predict how the position of the sun (altitude and azimuth) will be different (if at all):
    (a) In one hour.
    (b) Same time tomorrow.
    (c) Same time in a month.
    (d) Same time in a year.
7. When (if ever) is there no shadow cast by the sun at your observing location?
8. When (if ever) is the shadow cast by the sun at your observing location infinitely (or exceedingly) long?

1.7.4 Activity: Angular Sizes and Distances

When observing the sky, we see how big objects *appear*, but not how big they really *are*. Therefore, we use angles to measure the "size" of objects and the "distance" between objects. The former is called **apparent** or **angular size**, the latter **angular distance**. For example, if the full moon and the sun appear in the east and the west, respectively, we say they are 180° apart, but have no idea how many miles or kilometers there are between them. In this activity, you will explore how angular sizes and distances are measured and what we can or cannot do with this information.

1. Use your hand on an outstretched arm as shown in Figure 1.7(b) to measure the angular distance between two nearby objects or the angular size of a nearby object, for example two close trees, or a window on a nearby building.
2. What, if anything, does this tell you about the actual distance or size (in meters or feet) of the object or between objects?
3. Describe how the angular distance or size changes as you move closer to the object(s).

1.7. SUMMARY AND APPLICATION

4. Repeat 1–3 for far away object(s).

5. Imagine the objects are very far away—like the moon or the stars.

    (a) Are the angles involved big or small?

    (b) Does it make a difference if you are moving a substantial distance on Earth, for example, does the moon look bigger on the equator or from Mt. Everest? Why or why not?

6. Try to think of a method of measuring the size of faraway objects like the moon, or list methods that will work for close but not for faraway objects.

1.7.5 Activity: Scaling

Scaling is a very important concept, because it allows us to predict behavior of physical systems in an almost abstract fashion, that is, without having to deal with unnecessary details. As an example, consider the surface area of a sphere. How does it *scale* with the *size* of the sphere? As we saw in Sec. 1.2.6, areas *always* scale like the *square* of the size. So the surface of a sphere of double radius quadruples ($2^2 = 4$). The amazing thing is that this works for *any* area. Also the surface of a *cube* of double side-length quadruples! Note that we usually are not interested in the precise formula. For instance, the fact that a sphere is not a cube is often an uninteresting detail. If we compare the two area formulae ($A_{square} = 6s^2$, $A_{sphere} = 4\pi r^2$), we see that the quadratic dependence on the size variable (side-length s and radius r) is their most important feature. This reckoning can lead to interesting conclusions. For example, half a gallon of milk is eight times lighter than a gallon of milk, regardless of the shape of its container.

1. Draw a rectangle on a piece of graph paper, then scale it up by a factor of two.

2. Measure the perimeter of the two rectangles. How do they relate?

3. Calculate the area of the two and compare by taking the ratio between the two numbers.

4. Convince yourself that this is correct by counting the number of squares in the small and big rectangles and comparing them.

5. If the linear scaling factor would be 1.5 instead of two, by what factor does the area change?

6. An artist wants to cast a bronze statue, and makes a model that is half the intended size. As he scales up the statue, he needs to know how the model sizes relate to the actual sizes of the bronze version. If the linear dimensions are doubled, by what factor will the size of the following increase?

    (a) The circumference of the arm.

    (b) The cross-sectional area of the arm.

    (c) The total surface of the statue.

    (d) The volume of the required material for the cast.

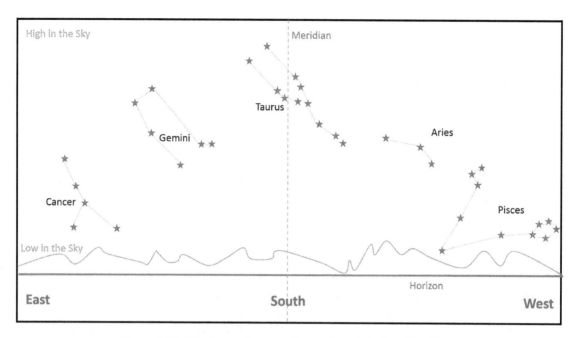

Figure 1.30: **Horizon view on December 1 facing South.**

1.7.6 Activity: Seasonal Motion

Stars, sun, moon and planets move with respect to the observer from east to west and reach their maximal altitude in the sky on the meridian in the south for observers in the northern hemisphere). This is called **diurnal** or **daily motion**. Close inspection shows that the sun moves slightly slower than the stars: a **solar day** of 24 hours is four minutes longer than the **sidereal day** of 23h 56m. In other words, the stars rise four minutes earlier each day, since our clocks and watches follow the 24 hour solar day which guarantees that the sun always culminates at 12 p.m., that is, noon. This slight shift between sun and stars adds up to two hours in a month and is known as **seasonal motion**.

Figure 1.30 is a horizon view showing what an observer sees facing south at mid-northern latitudes at midnight on December 1. High in the south is *Taurus*, medium high in the southeast resides *Gemini*, low in the east is *Cancer*, medium high in the south-west is *Aries*, and low in the west you see *Pisces*.

1. Where is the sun? Describe its position using words like east, south, west, north, and high in the sky, below the horizon, etc., reflecting the fact that the sky is two-dimensional.

2. There are twelve **zodiac constellations** along the 360° path of the sun among the stars, a circle we call the **ecliptic**. On average, how wide (in degrees, that is, angular size) is each of these constellations?

3. The celestial sphere turns once a day. How many degrees does it turn in one hour?

4. How long does it take until the next constellation in the figure will be at the position now taken by *Taurus*?

5. Which constellation will that be?

1.7. SUMMARY AND APPLICATION

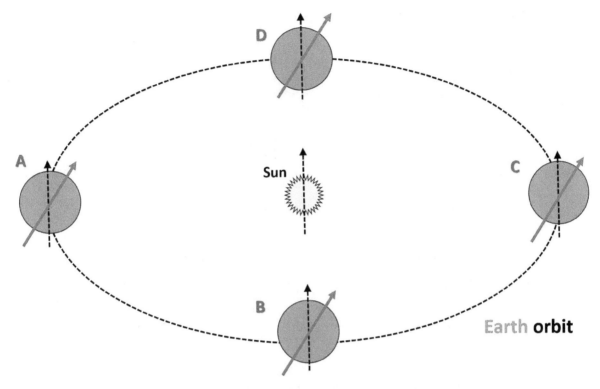

Figure 1.31: **The Earth's axis tilt throughout the year.**

6. The sun goes around the *ecliptic* in one year, that is, 365 days. How much does in move each day

    (a) ... with respect to the ground or observer?
    (b) ... with respect to the celestial sphere or stars?

7. Make a list of four dates and times (for example, December 1 at 12:00am) when the sky will look exactly like in the Figure 1.30.

8. Draw the horizon view at midnight a month earlier.

9. Draw the horizon view at midnight a month later.

1.7.7 Activity: The Seasons

The seasons are produced by the tilt of the Earth's rotational axis with respect to the plane of its orbit. To understand how the orientation of the axis with respect to the sun changes during the year, consider Figure 1.31, showing the Earth orbiting around the sun while rotating around its axis. Note that the Earth's axis always points in the same (absolute) direction, which means it's sometimes pointing somewhat toward the sun, and sometimes away from it. The other relevant axis is dashed in the figure and represents the rotation of the Earth around the sun. It is perpendicular to the Earth's orbital plane. The two axes are tilted 23.5° with respect to one another.

1. Where do the solid arrows point to? What do the represent?

	Sun's position at noon	Length of Day	Sunbeam Spread	Temperature in Area
A				
B				
C				
D				

Table 1.4: **Seasonal changes due to the sun's position in the observer sky.**

2. Where do the dashed arrows point to? What do the represent?

3. Draw in the Earth's equator in positions A–D in the figure or redraw it on a separate sheet.

4. Draw in an observer's horizon and zenith in positions A–D in that figure. Assume the observer is at 45° northern latitude.

5. Where in the sky would an observer see the sun at noon in positions A–D? Rank positions A–D in order of closeness of the sun to the zenith starting from closest.

6. If the sun is lower at noon, will it rise earlier or later than when it's higher at noon?

7. If the sun is lower at noon, will it set earlier or later than when it's higher at noon?

8. If the sun is lower at noon, will it stay up longer above the horizon or a shorter time than when it's higher at noon?

9. When a beam of sunlight hits the Earth, it spreads out over a certain area, depending on how high above the horizon it stands. Obviously, the same is true for the shadows it casts if objects are in its way. When the sun is lower, is the area over which the sunlight spreads out larger or smaller than when the sun is higher in the sky?

10. A sunbeam has a certain amount of energy. What happens if it spreads out over a larger area?

11. Fill in Table 1.4 with words like longest, highest, biggest, their negatives, and neutral terms. Assume the observer is in the northern hemisphere.

12. In which direction is the Earth moving in Figure 1.31? Is it $A \to B \to C \to D$ or $A \to D \to C \to B$? Can you tell?

13. The Earth is moving counterclockwise around the sun when viewed from above the plane of the orbit. Draw the correct direction of motion into the figure.

14. Match the four seasons in the northern hemisphere with the letters $A - D$.

15. Match the four seasons in the southern hemisphere with the letters $A - D$.

16. At the December solstice Earth is farthest from the sun: True   False

1.7. SUMMARY AND APPLICATION

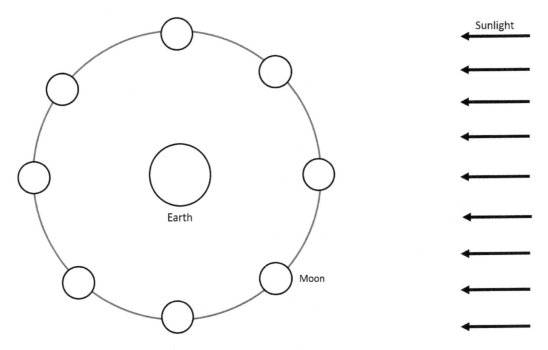

Figure 1.32: **The moon at various points around its orbit.**

1.7.8 Activity: The Phases of the Moon

The moon's phases are a result of the relative position of the three objects sun, moon and Earth (observer). In particular, the moon is always half lit up by the sun (day side) and half in the dark (night side). The observer on Earth sees different fractions of the dark and illuminated sides at different times. The moon orbits the Earth counterclockwise.

1. Color in the dark parts of Earth and moon with a pencil (8 circles).
2. Indicate the time for an observer at different spots on the Earth in 6 hour increments starting from noon (Recall when and where noon occurs!).
3. Identify the phases of the moon in the 8 positions shown.
4. At what time of the day is the waning gibbous moon at the meridian, that is, at its highest daily altitude?
5. Can a person at noon see the first quarter moon? If *yes*: where in the sky? If *no*: why not?
6. Draw the shadows cast by the Earth and the new moon into the figure.
7. During what phase could the shadow of the Earth fall onto the moon?
8. This is called a *lunar eclipse*. Why does it not happen once a month?
9. Say the moon is in its waxing crescent phase when it rises. What will the phase of the moon be when it sets?
10. How much of the entire surface of the moon is illuminated during the crescent phase?
11. When does the new moon rise, when does it set?

Date	RA (Cel. Longitude)	Declination (Cel. Latitude)
1. May 15, 2003	21 h 06 m	−18° 43'
2. June 1, 2003	21 h 43 m	−16° 36'
3. June 15, 2003	22 h 10 m	−15° 00'
4. July 1, 2003	22 h 35 m	−13° 37'
5. July 15, 2003	22 h 49 m	−13° 05'
6. Aug 1, 2003	22 h 56 m	−13° 29'
7. Aug 15, 2003	22 h 50 m	−14° 36'
8. Sep 1, 2003	22 h 34 m	−16° 04'
9. Sep 15, 2003	22 h 21 m	−16° 30'
10. Oct 1, 2003	22 h 16 m	−15° 41'
11. Oct 15, 2003	22 h 21 m	−13° 59'
12. Nov 1, 2003	22 h 39 m	−11° 01'
13. Nov 15, 2003	23 h 01 m	−8° 4'

Table 1.5: **The position of Mars during the 2003 opposition.**

12. Where in the sky does the new moon rise, where does it set?

13. When does a last quarter moon rise, when does it set?

14. Where in the sky does the last quarter moon rise, where does it set?

15. How far apart in the sky are the last quarter moon and the sun?

16. How far apart in the sky are the full moon and the sun?

1.7.9 Activity: Retrograde Motion of the Planets

Usually, planets drift eastwards along the *ecliptic*. In other words, they move just like the sun and moon with respect to the stars. This is known as *prograde motion*. From time to time, a planet will slow down and reverse its normal motion with respect to the stars. This westward drift is called *retrograde motion*. It is important to note that regardless of this slow motion with respect to the stars, the planet will move fast from east to west due to the apparent rotation of the celestial sphere. Planets always rise in the east and set in the west. To get a sense how this works, consider Table 1.5 detailing Mars's changing celestial coordinates in 2003. In this year Mars came historically close to Earth (of course, still being about 60 million km away). The region of the sky between 21h and 0h of right ascension (RA) or celestial latitude just south of the celestial equator (that is, slightly negative declination or celestial latitude) is occupied by the zodiac constellations *Capricornus* and *Aquarius*. So Mars appeared in these constellations which lie along the *ecliptic*, that is, the sun's apparent path among the stars.

1. Plot the position of Mars using Figure 1.33.

2. During which time period was Mars moving prograde?

3. During which time period was Mars moving retrograde?

4. How long was Mars moving retrograde?

1.7. SUMMARY AND APPLICATION

5. How big is Mars's retrograde loop?
6. Where did Mars rise on the day of **opposition**, that is, in the middle of its retrograde motion?
7. Define the term **opposition**—what does it mean for a planet to be in opposition?
8. What does the apparent motion of Mars in the sky tell us about the actual motion of Mars in space?
9. What maximal altitude did Mars reach in the sky of an observer at 40° N on September 1, 2003?
10. Where in the sky is Mars when it reaches maximal daily altitude?
11. At what time does this happen?
12. Is the intersection of the **celestial equator** in Figure 1.33 and the **ecliptic** the **vernal** or **autumnal equinox**?

Image Credits

Fig. 1.1a: Copyright © 2007 by A. Fujii, (CC BY 4.0) at https://www.spacetelescope.org/images/opo0706b/.
Fig. 1.1b: Adapted from Copyright © 2007 by A. Fujii, (CC BY 4.0) at https://www.spacetelescope.org/images/opo0706b/.
Fig. 1.1c: Source: https://commons.wikimedia.org/wiki/File:Ursa_Major_constellation_Hevelius.jpg.
Fig. 1.1d: Copyright © 2011 by Roger Sinnott and Rick Fienberg, (CC BY 3.0) at https://commons.wikimedia.org/wiki/File:Ursa_Major_IAU.svg.
Fig. 1.4a: Source: http://www.dauntless-soft.com/PRODUCTS/Freebies/Library/books/AK/8-2.htm.
Fig. 1.4b: Source: https://dr282zn36sxxg.cloudfront.net/datastreams/
f-d%3Ad4c3daef1332c480f4be8da9e006f5d88002f53cf7ac1816f4350a25%2BIMAGE%2BIMAGE.1.
Fig. 1.5: Copyright © 2011 by TimeZonesBoy, (CC BY-SA 3.0) at https://commons.wikimedia.org/wiki/File:Standard_time_zones_of_the_world.png.
Fig. 1.5: Copyright © 2011 by TimeZonesBoy, (CC BY-SA 3.0) at https://commons.wikimedia.org/wiki/File:Standard_time_zones_of_the_world.png.
Fig. 1.7b: Source: https://dept.astro.lsa.umich.edu/ugactivities/Labs/coords/fingers.jpg.
Fig. 1.10a: Source: https://www.classzone.com/books/earth_science/terc/content/
investigations/es2803/images/es2803_p1_light_rays_b.jpg.
Fig. 1.10b: Source: http://www.frankjohns.uk/stellar—a-stars.html.
Fig. 1.11a: Copyright © by ESO.
Fig. 1.14a: Source: http://www.cv.nrao.edu/ rdickman/images/radecspin.gif.
Fig. 1.14b: Source: https://onthebrinkofdiscovery.files.wordpress.com/2015/10/star-map-equatorial.jpg.
Fig. 1.15: Source: https://commons.wikimedia.org/wiki/File:Axialtilt.png.
Fig. 1.18d: Copyright © 2016 by 0x010C, (CC BY-SA 4.0) at https://commons.wikimedia.org/wiki/File:2016-05_Grand_Ballon_circumpolar_star_trails_01.jpg.
Fig. 1.22: Source: http://cityofdepoebay.org/images/news/2017newsSolarEclipseInfographic.png.
Fig. 1.23: Source: http://www.pulseheadlines.com/annular-eclipse-practice-investigation-scientists/59679/.
Fig. 1.24a: Copyright © 2014 by Tomruen, (CC BY-SA 4.0) at https://commons.wikimedia.org/wiki/File:Solar_lunar_eclipse_diagram.png.
Fig. 1.24b: Source: https://eclipse.gsfc.nasa.gov/LEmono/TLE2004Oct28/image/TLE2004Oct-GMT1.GIF.
Fig. 1.26: Copyright © by C.R. Nave.
Fig. 1.28a: Copyright © 2003 by Tunc Tezel.
Fig. 1.28b: Copyright © 2008 by Eugene Alvin Villar, (CC BY-SA 3.0) at https://commons.wikimedia.org/wiki/File:Apparent_retrograde_motion_of_Mars_in_2003.gif.

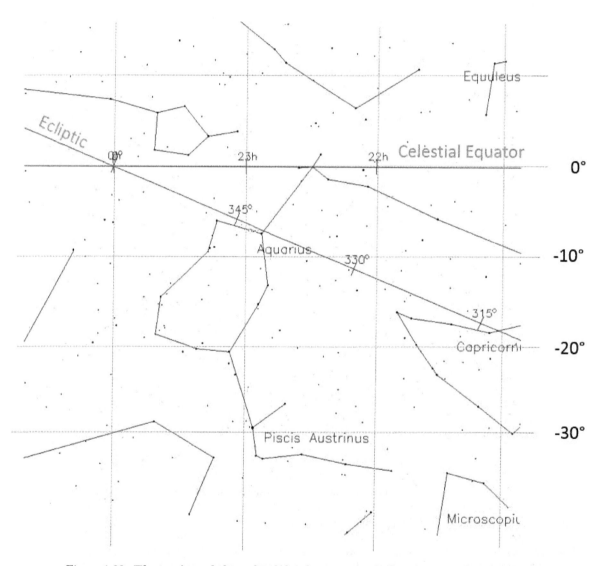

Figure 1.33: **The region of the celestial sphere around *Capricorn* and *Aquarius*.**

Chapter 2

Inferring & Theorizing: Planetary Astronomy Begets Modern Science

"Die Sonne tönt, nach alter Weise,	"The sun sings out, in ancient mode,
In Brudersphären Wettgesang,	His note among his brother-spheres,
Und ihre vorgeschriebne Reise	And ends his pre-determined road,
Vollendet sie mit Donnergang.	With peals of thunder for our ears.
Ihr Anblick gibt den Engeln Stärke,	The sight of him gives Angels power,
Wenn keiner sie ergründen mag;	Though none can understand the way:
die unbegreiflich hohen Werke	The inconceivable work is ours,
Sind herrlich wie am ersten Tag."	As bright as on the primal day."
	(Archangel Raphael in GOETHE*'s Faust)*

After describing the fundamental astronomical observations in Chapter 1, we will now switch to a ***chronological***[1] account of our ever-improving descriptions or explanations of the universe. In the ***liberal arts*** scheme the first chapter is the ***grammar*** or *coming to terms*, that is, the definition of the labels and names used in astronomy. There is thus no ***canonical*** order in presenting these observations. They all are equally fundamental, known since prehistoric times in most cultures. Now we will see how different people and cultures have ***interpreted*** this data set differently in different eras due to different pressures, preferences, and obsessions as well as historical, societal, and religious backgrounds. People over the centuries used the observational results to ***infer***, that is, make up, theories that connect the data points. This resulted in the chronological order of models of the universe we follow from hereon. The development of concepts is often connected to historic persons or events, and the narrative may, at times, seem overly historical. Keep in mind, therefore, that the science results—not the historic details—are essential. Names and dates are meant to be used as ***labels*** to help you keep track of the chronological and logical order of developments and concepts. Much like *Ursa Major* invokes the mental image of a northern constellation and pointer to the pivotal **Celestial North Pole** for orientation, names like COPERNICUS, KEPLER, and NEWTON suggest a chronological line of achievements in the understanding of the expanding universe. While exact dates are not important, every educated person should know that "COPERNICUS was before KEPLER was before NEWTON." A secondary goal of this chronological approach is to show that science is "made" in small steps, with lots of wrong turns and dead ends. Because of limited space, we have to be somewhat judgmental here in the spirit of Weinberg [44]; we are biased toward developments that "worked out," that is, proved useful, or, through their failure, led to other fruitful ideas.

---

[1] Greek: *chronos = time, logos = word, opinion, reason.*

Guiding Questions of this Chapter

2.1 To what extent and for which reasons did ancient people start to make sense of the various patterns in the sky? How do we know they did?

2.2 Why did science as we know it originate in old Greece and not somewhere else? Which are the crucial questions to understand nature?

2.3 How do we measure distances in the universe while confined to Earth? Why is the assumption of a stationary Earth reasonable?

2.4 How did PTOLEMY's geocentric theory become so successful even though it is wrong? How can we explain more than one thousand years of scientific hibernation after all the successes culminating with PTOLEMY?

2.5 What made COPERNICUS possible in 1500 but not in 1200? What was so revolutionary about his ideas?

2.6 Why were TYCHO's observations ten times better than anyone else's even though he did not use a new tool? How does accuracy drive progress?

2.7 How was KEPLER able to decipher the cosmic enigma of planetary motion even when everyone else failed for two thousand years?

2.8 How did GALILEO falsify PTOLEMY's theory?

2.9 Was the *scientific revolution* a revolution?

2.10 How can NEWTON's theory explain everything from cannon shots to planetary motion to the tides?

2.1 A Brief Summary of the Early History of Astronomy

Astronomy is the oldest of the sciences and has been practiced since at least 4000 BCE in many places of the world, notable the Near East. The pyramids in Egypt and structures like Stonehenge in England make it clear that celestial objects were important and fascinating to humans of these eras. The efforts necessary to erect these monuments were immense and therefore give testimony of the strong motivation these early people had to study the heavens. Probably several sources of motivation were at play, among them practical needs like **direction finding** and **timekeeping**. We can infer these goals by the precise alignment of the pyramids with the cardinal points (north, east, south, and west), and by the positioning of pillars at Stonehenge affording sightings of the solar solstices, the northernmost moon set, and more. Indeed, the amount of sophistication evident in these structures is amazing, and speaks to the stability of these societies that must have spent a good fraction of their productivity to construct them, sometimes over many centuries.

Paradoxically, the **timekeeping** aspect of astronomical activities stands a bit outside of the chronological order adopted in this book. The reason being that it took an astonishingly long time to learn how to keep time accurately. Mechanical clocks weren't invented until the late Middle Ages, and accurate pendulum clocks not until the seventeenth century. But even the mundane establishing of a reliable calendar was not achieved until the great Gregorian reform of 1582, which in turn wasn't universally adopted until after the Russian October

2.1. A BRIEF SUMMARY OF THE EARLY HISTORY OF ASTRONOMY

Revolution.[2] Let's take a look at this in detail, since there are many interesting concepts involved. For one, it shows how much astronomy is part our our culture with roots running so deep that we are mostly unaware of their astronomical origin. For instance, there is nothing special about the date of New Year's Day. Nothing astronomically relevant happens on this day. It is merely **convention**, that is, an agreement, that we call January 1 the beginning of the year. The fact that we have weeks of seven days has a weak astronomical significance in that the main lunar phases (new, first quarter, full, last quarter moon) are about seven days apart. The Babylonians had a seven-day, the Romans, much later, an eight-day week. The likely reason for the seven-day week is the fact that seven "planets"[3] were known in antiquity, after which the days are named. Obviously we have Saturn-day, Sun-day, and Moon-day. The others are closer to the planets' names in the Romanesque languages. In French, *Mardi* (Tuesday) is Mars-day, *Mercredi* (Wednesday) is Mercury-day, *Jeudi* (Thursday) is Jupiter-day, and *Vendredi* (Friday) is Venus-day. In the Germanic languages these days are named after Norse gods: Thursday is Thor-day, Friday is Freya-day, and so forth. The biblical account of creation also implies a seven day week with a day of rest (*Sabato* is Saturday in Italian), which the Christians shifted to Sunday, while Muslims rest on Friday.[4]

The astronomically relevant temporal patterns in nature are the length of the day, the year, and the month. Unfortunately, no integer number of months, that is, lunar orbital periods, makes up a year, and the length of the month itself is somewhat variable. Therefore, calendars like the Jewish and Muslim ones that are based on the moon, have various disadvantages. One being the seasonal variability of religious feasts. The holy month of *Ramadan*, when Muslims are to fast from sunrise to sunset, sometimes occurs in winter when days are short, and sometimes in summer, when days are long and temperatures high, which makes fasting more strenuous. From this it is clear that the ideal calendar is one that ensures constancy of the seasons. In other words, the spring equinox should always happen on the same date. The time period from one vernal equinox to the next, known as the **tropical year**, was measured accurately early on. Unfortunately, its length is not an integer number of days: $T_{tropical} = 365$ days 5 hours, 48 minutes and 46 seconds; it is about 11 minutes shy of $365\frac{1}{4}$ days. An extra quarter day is the reason why there is a **leap day** every four years in our calendar. The extra day in February was introduced during the calendar reform initiated by JULIUS CAESAR with the help of the Greek astronomer SOSIGENES in 45 BCE. Incidentally, this was the year before Caesar's *brutal* assassination by BRUTUS. While he was at it, JULIUS renamed the fifth Roman month, *Quintilis*, after himself. His successor AUGUSTUS followed suit and renamed the sixth.[5] CAESAR also decreed January to be the first month—and hence July the seventh. Before CAESAR's reform, the year began in March, which made sense since the vernal equinox happens in March. The **tropical year** is a little shorter than the **Julian year** which is, by construction, exactly $365\frac{1}{4}$ days long. Hence, even after the reform, the seasons slipped by one day every 130 years, so that a new reform became necessary. It was delayed for centuries, and eventually happened under pope GREGORY XIII (1502–1585), advised by the astronomer CHRISTOPHORUS CLAVIUS (1538–1612). The reform aimed at **approximating** the length of

---

[2] Which happened, as it were, in November.

[3] That is, celestial objects moving with respect to the fixed stars.

[4] This means you can always find someone at work on the weekend in Jerusalem, holy city of all three religions.

[5] Other emperor's names didn't stick, so we still call the seventh (*septem*), eighth (*octo*), ninth (*novem*) and tenth (*decem*) month by their pre-Caesarian Roman names.

the tropical year even better. Therefore, it was agreed to drop the leap day every hundred years, but not if the year was divisible by 400. So 1900 was not a leap year, 1904 and 2000 were. This way the discrepancy between the new Gregorian year and the tropical year is only 38 seconds. Hence, millennia will go by before the two are more than a day apart. This struggle for an adequate calendar shows how much thought and observational effort is involved in such *trivial* [6] things as timekeeping. We are fortunate that these fundamental problems have been solved, so we don't have to think about them anymore.

Indeed, many cultures have carefully observed the naked-eye sky, and recorded over long periods the positions of and events related to the celestial objects. For instance, the **heliacal rising**[7] of the bright star Sirius was important for the Egyptians, because it coincided with the flooding of the Nile River. Only the dispersing of fertile soil and water onto the river's banks made agriculture in the Egyptian desert possible. As interesting as this field of **archaeoastronomy** is, we have to focus on the eventual emergence of science from these beginnings. Therefore, we note that given the sophistication of observing and record keeping, it is surprising that no systematic *explanation* of patterns of nature such as the heliacal rising of Sirius was achieved or attempted as far as we know. Early cosmologies, that is, models of the world and how it came to be, were rather rudimentary, and only loosely related to astronomy. The motivation to create cosmologies is grounded in the basic need of humans to have some kind of understanding of the world around them. Indeed, the world must have seemed rather hostile to early humans, and so they tried to model and therefore explain the strange patterns and phenomena experienced in nature by familiar things and processes. For example, the Egyptians thought they lived in a long and narrow world with the goddess of the sky, Nut, her body full of stars, arching over it from horizon to horizon. The sun was thought of as being on a boat that flowed across her. The boat entered the region of the dead when setting in the west below the Earth, thus explaining nightfall. Traveling through the region in a different boat, the sun emerged in the morning back in the east. You can see at once how this world-view was modeled after the narrow Nile River region where the Egyptians lived, and how it accounts for the basic celestial phenomena while using only familiar concepts. While poetic and comforting, this model of the world does not lend itself to quantitative predictions, nor to real insight into the workings of the heavens. Beliefs and attempts at explaining natural phenomena emerged probably in all societies. Eventually, they became the fertile ground from which the progenitors of science emerged. In fact, the very idea that nature itself is a stable pattern which can be explained is something that had to be invented and accepted before anything else could be done. Being subject to a hostile and seemingly unpredictable environment could easily suggest the opposite: that the world is a chaotic, random place and that it makes no sense to attempt to understand it. Indeed, clouds, rain, thunder, and lightning are hard to predict yet also part of the world and even the heavens. In view of this, the early, rudimentary and *qualitative* Greek models of the world to which we owe so much seem like real progress.

---

[6]Again the word is used in its *liberal arts* sense: *Trivial* is something that we have to get right before we can go any further.

[7]The first time a star is visible just before sunrise after a time of being invisible due to being close the sun on the celestial sphere.

Concept Practice

1. In which month did the Russian Revolution take place? Explain why a qualifying statement is necessary.
2. How would our calendar change if the days were twice as long, that is, 48 hours?
3. If the tropical year were $185\frac{1}{8}$ days long, how often would a leap day occur, if at all?

2.2 The Ancient Greeks Discover That the Universe Can Be Discovered

The origins of modern science and much of Western thought lie in ancient Greece. It was here that humans began to learn how to learn and how to ask the right, that is, relevant, questions of nature. Why not somewhere else? After all, astronomy was done—even eclipses predicted—in many places since ancient times. This interesting—and, as of yet, unanswered—question is a historic one. It belongs to a different **way of knowing**, not to science. While the **heuristic method** of history is distinct from the way science progresses, they have much in common. In particular, we gather evidence and use rational thought to solve problems, whether they are historic or scientific in nature—or everyday problems like a car that won't start. To widen the scope and deepen the mystery, keep in mind that at the same time, around 600 BCE, when the Greeks started to think like we do, other cultures produced astonishing thoughts. Buddha in India, Lao Tse and then Confucius in China lived around that same time!

Often, we will use the terms **science** and **philosophy** synonymously. In fact, the term *philosophy*, coined with humble intent by PYTHAGORAS (c. 570–495 BCE), means "friend of wisdom" and encompassed pretty much any rational thought and endeavor. Science comes from the Latin word *scientia* for knowledge, which is also pretty broad, but the modern connotation is often **natural science**, such as astronomy, physics, and chemistry. None of these sub-specialties existed as recently as 400 years ago. They were part of philosophy, that is, **natural philosophy**, the part that deals with nature.

Ancient Greek history is usually divided into four periods: *Archaic Greece* lasting from about 800 to 480 BCE, followed by *Classical Greece* until the death of ALEXANDER the Great in 323 BCE, succeeded by *Hellenistic Greece* ending with the Roman conquest in 146 BCE, and finally *Roman Greece* enduring until 330 CE, when Constantinople replaced Rome as the capital of the Roman Empire.

2.2.1 Archaic Greece Was an Ideal Place to Ponder Methods to Understand the Universe

Ancient Greece was an ideal place for philosophy and scientific thought to start. The Greeks, by the topology and geography of their country, were drawn to the sea. They inhabited the many islands of the Aegean Sea, and colonized parts of Sicily, Italy, and Asia Minor. They traded with the Orient, Egypt, and much of the peoples around the Mediterranean Sea. With the goods also ideas and opinions were exchanged. Fortunately, the Greeks were exposed to those ideas without being forced to adopt them, as is the case so many times when nations are conquered by more powerful ones. While the Persian Empire to the east was a constant threat, the classical Greeks lived in relative stability within their *democratic* city states, where free discourse and discussion were guaranteed and even encouraged. On top of that, the religious or priest caste in Greece never had much political power, and so did not interfere with nascent science, contrary to Egypt, for instance. All this makes it plausible why ancient Greece was the place where the first leap in human thought occurred that eventually led to the development of modern science some two thousand years later.

Indeed, an intriguing idea was proposed by several Greek philosophers around in the seventh century BCE: That the cosmos has a fundamental order, and that the human mind is capable of figuring it out. This does not sound like much, and on top of this, it is merely an *assumption*. Due to the apparent disorder and myriads of patterns in nature this assumption doesn't even seem very plausible. Nonetheless, this belief in order generated many new ideas as to how to describe the motions of celestial objects, because it was so liberating. The thought alone *demystified* the universe, because it allowed *rational* access to its secrets.

Of course, it was not at all clear what the fundamental or important properties of the universe are. Nor was it clear which were the important questions to ponder, nor was it settled which approach one should use to decipher the universe. Nonetheless, while not much was known at the time, the method of *gaining knowledge* was discovered, namely an interplay between observation of nature and of theorizing, that is, rational thinking about reasons for the observed patterns. Furthermore, these reasons had to be understandable, and therefore familiar. To us, some of these explanations of natural phenomena seem strange if not ridiculous, but at the time they could not be ruled out. Furthermore, their real merit was the refusal to invoke gods or other imponderables in explaining patterns in nature. Therefore, the *archaic* Greeks made this important point: The universe is comprehensible for humans!

THALES OF MILETUS in the seventh century BCE, for instance, is thought to have thought that everything is made out of water, and that this fact explains all patterns in nature. Water is very abundant and important on our planet and can exist in solid, gaseous, and liquid form. However, his theory was not completely convincing even at his own time. Yet it was exemplary in its *modeling* of incomprehensible patterns with familiar ones. Comparing with the Egyptian model of the world, the progress becomes evident. Instead of replacing one mystery with another ("the sky is like the body of a goddess"), Thales links the universe to a familiar *substance*. Since everyone has access to the substance, a conversation and investigation can start to determine the merits of this model of the universe. It certainly wasn't clear that water is the right substance to explain the universe. ANAXIMANDER, also of Miletus, held that the right choice is a mysterious substance he called the *unlimited*. Yet another Milesian, ANAXIMENES, had yet another choice (*air*), and HERACLITUS OF EPHESUS (535–475 BCE) thought that the substance must be *fire*.

The general idea that *substance* is the key to unlock the secrets of the cosmos was developed further by EMPEDOCLES OF AGRIGENTUM (495–430 BCE) and DEMOCRITUS around 450 BCE. EMPEDOCLES held that the fundamental *elements* explaining all things were *earth, fire, air*, and *water*. So in some sense, he collected the ideas of his predecessors to develop an *eclectic theory* of the universe. DEMOCRITUS sought to explain the cosmos based on the assumption that matter consists of smallest units he called *atoms*, because they cannot be subdivided (*a-tomon* means "un-cuttable" in Greek). While there is only one kind, the atoms can be arranged in various ways to explain the wealth of substances we can discern. This is an idea that was realized later in modern chemistry, but not very successful or useful in his own time. In short, it didn't lead anywhere: It could not be tested experimentally and, while explaining certain patterns in nature, it did not predict any new ones. Some of the early Greek scientific writings are more like poetry: Pleasurable and uplifting to read, yet of little use to understand nature. This is not meant to degrade the archaic Greeks (nor poetry), but rather to make the point that thoughts and writings can be important even if they do not have an immediate practical impact.

Other Greeks thinkers entertained the idea that the key to the universe is more abstract. This line of thought gave rise to mathematical explanations of the cosmos, and to abstract mathematics itself. The importance of numbers was championed by PYTHAGORAS (c. 570–c. 495 BCE), known for the theorem concerning right triangles attributed to him. He was born around 580 BCE in Samos and was inspired by the discovery that sounds made by strings sound harmonious if the strings' lengths are related to each other by integer ratios. For instance, a vibrating string of double length produces a tone that is an octave below the tone of the original string. Special significance was given to **special numbers** like 10, which is the sum of the first four integers $(1 + 2 + 3 + 4 = 10)$ which were taken to represent a point, a line, a surface, and a volume, respectively. Hence, 10 was taken to represent all forms. PYTHAGORAS founded a spiritual community called the Pythagoreans, who tried to understand the universe in terms of mathematical relations. They were the first to introduce the **sphere** as the perfect geometrical form, and therefore the prejudice that the sphere was fundamental in describing the universe. Indeed, the sphere is special in the sense that a single number, its radius, is enough to completely describe it. Also, it is perfectly symmetric, looking the same from all aspects, and a point on a sphere returns to itself when rotating. A circular motion has, in that sense, no beginning nor end. Furthermore, the sky looks spherical, and there were good reasons already at that time to think that the Earth was a sphere, too. So it all made perfect sense, at least as an abstract idea. Unfortunately, the Pythagoreans were not the last to "adjust reality" if it wouldn't fit their theories. Consider this: In his quest to understand the mighty, presumably eternal, absolutely beautiful universe, a lowly inhabitant of a chaotic planet stumbles upon this perfect, abstractly defined, form—the circle. Clearly, then, the circle has to have something to do with the secret of the universe, right? Well, not just in this case the answer is no. Often, seemingly related concepts are not related. The history of the sciences is full of frustrating stories like this.[8] Here, I want you to share the sense of excitement when a scientist, or any person for that matter, thinks she figured out some problem, or has explained some pattern in nature. The euphoria of this **eureka moment** is amazing! It is probably a large part of the motivation that keeps scientists going, even when in the end it turns out that nature did not realize this beautiful concept but rather chose to function differently. With THOMAS EDISON (1847–1931 CE) we may then say: At least we know now how nature *doesn't* work.

The Pythagoreans were able to construct an astonishingly concrete though not predictive model of the universe with the abstract idea of the perfect sphere. They hypothesized that a series of spheres move the seven "planets" (sun, moon, and naked-eye planets) and the celestial sphere of stars. An additional sphere rotates this set of concentric or **homocentric spheres**[9] once every day around the stationary Earth. To model the motion of sun, moon, and planets with respect to the stars, the inner, "planet" spheres had to rotate a little slower. All in all, a reasonable approximation to the actual motion that we observe in the sky; see Chapter 1. The **problem of the planets**, that is, their retrograde motion, was not explained. For the Pythagoreans there was an even bigger flaw, though. If you count the number of spheres,

---

[8]As a researcher of theoretical particle physics, I have many a tale to tell about my own follies.

[9]From Greek *homos* = *same*, not to be confused with Latin *homo* = *human*. The spheres all have the same center, namely the stationary Earth.

then there are only nine, one shy of the special number ten! The Pythagoreans fixed that by postulating that there is a "counter-Earth" that rotates with the Earth on the tenth sphere around a central fire. Conveniently, the counter-Earth is always on the opposite side of the fire from the Earth, and can therefore not be observed. You can probably figure out that this is an incredibly weak theory, because it cannot be falsified. It is as if someone claimed to be invisible, but only if nobody looks. Furthermore, the theory was weakened by the addition of a counter-Earth which serves only the (highly questionable) purpose of bringing the number of spheres to ten. This makes the theory more complicated than it needs to be. These days, we would say it fails **Occam's razor**, the heuristic rule of the medieval thinker WILLIAM OF OCCAM (c. 1287–1347 CE) that a model should be as simple as possible; all else should be "shaved off." In fact, it is the Pythagorean model's superfluous complexity that makes it less testable, and therefore questionable.

The thrust of the model was a different one, anyway. The rotation of the spheres was linked to the idea of the integer-ratio relations between the lengths of vibrating strings that sound harmonious. Assuming that these "harmony relations" are universal, the Pythagoreans introduced the notion of the **harmony of the spheres.** Namely, they thought that if the speeds of the spheres carrying celestial objects exhibit small integer ratios, then the whole universe should make a harmonious sound. In other words, the explanation of the speeds of the sphere, that is, the workings of the cosmos, was said to be due to the criterion that the cosmos should *sound pleasing*. Today, this may seem far fetched, but at a time when none of the modern principles of science were known, it was a rational and concrete model, worthy of being tried and tested. Incidentally, the harmony of the spheres came back two millennia later to haunt KEPLER, who couldn't resist its *siren call*.

In sum, the thinkers of *archaic Greece* came up with the idea that nature exists independently of the human mind but its inner workings can be discovered and comprehended by humans. Naturally, they did not agree on a correct model, and produced a great variety of them. Some of their ideas reappeared later, almost as if these dead theories decomposed to create a fertile ground on which new theories could grow.

2.2.2 Plato Declares Uniform Circular Motion to Be Ideal

We are now entering *classical Greece*, epitomized by the three great philosophers SOCRATES (ca. 470–399 BCE), PLATO (428–348 BCE), and ARISTOTLE (384–322 BCE), each a pupil of the former. It is PLATO who cemented the idea that the cosmos has to be described in terms of **perfect** geometric shapes like sphere and circle. The associated motion reflecting the same perfection is **uniform circular motion.** It is ideal because the moving object never changes its distance from the circle's center, nor its speed, and the motion has no beginning and no end, as it always returns to where it started from. In this way it is almost featureless, completely described by two numbers: the radius of the circle and the period or time it takes for one revolution of 360°.

PLATO's obsession with perfection comes from his mistrust in our senses. The abstract world of perfect, well-defined shapes or forms is, in his opinion, the true content of the cosmos. Humans cannot experience it, however, because our senses are only able to perceive how these perfect shapes *appear*. PLATO illustrated this in his **cave analogy.** A person chained behind a wall in a cave sees other people move in the cave. Alas, she doesn't see the people themselves, but only their shadows cast by a fire burning in the cave. The shadows therefore are how

2.2. THE UNIVERSE CAN BE DISCOVERED

the people appear to her, and her intellect has to figure out how the people "really" look like. This is the essence of PLATO's criticism of our senses that influenced generations of philosophers. IMMANUEL KANT, for instance, voiced the suspicion that we cannot see how things *are*, only how they *appear*. Incidentally, this is also true for the astronomer, as we saw in the first chapter. Chained to the Earth and observing how celestial objects *appear* in the sky, the astronomer as an **observational scientist** has to figure out with his intellect how the celestial objects "really" look like. Unfortunately, PLATO's criticism introduced a bias against observations. Not in a blunt way, but such that observations were taken lightly, as only one of the factors to help unravel the mysteries of the cosmos. Today, we would say that observations and experiments are the **primary sources** for an understanding of nature that comes from *interpreting* the *empirical data*.

Somewhat cynically, one could summarize PLATO's influence on the course of science as having left astronomers with the following unsolvable homework: Given the naked-eye observations of the celestial objects, construct a model of the cosmos that produces the correct positions of the objects, that is, **saves the appearances**. The only motions of the objects in the model that are permissible are uniform and circular. This is like telling a craftsman to construct kitchen cabinets without using hinges or screws. It is possible, but not very practical.

2.2.3 Eudoxus Constructs a Cosmological Model of Homocentric Spheres

PLATO had a colleague named EUDOXUS (c. 406–350) who is credited with one of the first systematic models of the universe. Owing much to the Pythagoreans and PLATO who devised an eight-sphere model, he modeled the cosmos as a series of 27 spheres, centered on a stationary Earth; see Figure 2.1.

EUDOXUS's scheme was elaborate. Each of the five naked-eye planets was described by four spheres. For the sun and the moon he had three spheres each, and the outermost celestial sphere to which all others were attached rotated once each day to account for daily motion of all celestial objects. The set of four spheres for a planet was hinged in a complex way to make it possible to produce something as complicated as retrograde motion. It is implausible for one uniformly rotating sphere to suddenly change direction. To reproduce the looping paths of the planets, four spheres are necessary as it turns out. Furthermore, their axes of rotation have to be inclined with respect to each other. EUDOXUS' model did not account for the nonuniform motion of the sun,[10] and did not work well for Mars nor Venus. In fact, the discrepancy between predicted and observed positions of the planets, sun, and moon were as large as 5°. This is the apparent width of a hand on an outstretched arm, Figure 1.7(b), and thus a large discrepancy between prediction and observation. On the other hand, the criterion that prediction and data should match as closely as possible is a surprisingly modern notion. You might think that this model would, in turn, be ruled out and discarded swiftly, but scientists were quite content with a *qualitative* description of nature until the modern era. As we will see, sometimes even small discrepancies between prediction and observation make all the difference in figuring out how nature "works,"[11] so nowadays we take discrepancies

---

[10] The sun moves faster on the celestial sphere in January than in July—in the northern hemisphere winter is several days shorter than summer.

[11] KEPLER, for instance, prevailed in figuring out the laws of planetary motion due to a discrepancy of his current (wrong) theory with TYCHO's observations of only 4 minutes of arc. This is less than a seventh of the

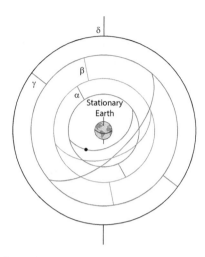

Figure 2.1: EUDOXUS's **model of planetary motion.** The planet is attached to the equator of the innermost sphere which rotates around an axis α inclined with respect to the second sphere rotating around an axis β, and so forth. The two innermost spheres emulate the ***retrograde motion*** of the planet. The third sphere rotating around the axis γ rotates once in a planet's ***sidereal year***. The fourth, outer sphere rotates once a day from east to west around the axis δ to account for the daily motion of the planet. An individual set of four **homocentric spheres**, centered on the stationary Earth, for each of the five naked-eye planets was necessary to describe the motion of the planets in a **Qualitatively** correct way. Together with three spheres each for sun and moon and one sphere for the fixed stars, a total of $5 \times 4 + 2 \times 3 + 1 = 27$ spheres, individually hinged, inclined, and rotating described the cosmos. This was *not* a unified nor a mechanically plausible model.

very seriously. EUDOXUS' model, as far as we can tell, had mostly a psychological function, namely to comfort people to think that they (more or less) understand the universe. Probably his model was never used to make ***quantitative*** predictions. Neither is it clear that EUDOXUS thought that his scheme of spheres was real in the physical sense or how the hinging of the spheres could have been accomplished practically.

2.2.4 Aristotle Devises an All-Encompassing Physical Theory of Nature

The thread spun by PLATO was further developed into a prescription how to understand nature by his pupil ARISTOTLE. Due to his systematic study of nature he is considered by many as the first scientist. He is also the founder of logic and many other fields. His eminent place in world history is due to his tutoring of the son of the Macedonian king Philip II. This son was ALEXANDER the Great (356–323 BCE), who would conquer one of the greatest empires in history. It measured about 5,000 km from the Greek west to the Indian and Afghan east. In the middle of the vast kingdom we find Persia, location of ancient Babylonian astronomical wisdom. In the history of science ARISTOTLE has a unique place, because he influenced the course of science immensely over more than a thousand years! During his lifetime he founded the Lyceum,[12] his writings were embraced by Islam about one thousand years later, and another several hundred years went by before they were merged with Christian doctrine to forge ***scholasticism***, the medieval form of philosophy. His style was different from PLATO's—argumentative and "unmathematical."

Unfortunately, ARISTOTLE also injected some notions and concepts into scientific thinking

---

apparent diameter of the moon!

[12] A school in Athens like PLATO's Academy.

2.2. THE UNIVERSE CAN BE DISCOVERED

that proved wrong yet persistent. They set science on a wrong path that wasn't corrected until the advent of modern science in the 1600s. For instance, he wrongly distinguished between two different ways in which things change. A natural way, like the growth of plants and the vertical fall of an unsupported object, and an artificial or forced way, like the cutting of a flower or the upward motion of a hurled stone. This wrong **dichotomy** had two regrettable consequences. First, it pushed him and his successors away from a focus on specific, well-defined and controlled situations as they are common in modern science experiments. Experiments are **artificial** in Aristotle's thinking, and therefore they are less interesting than the holistic study of motions categorized as natural. The irony is that ARISTOTLE's categories are often artificial, even senseless in hindsight. While ARISTOTLE was a good observer of nature, as his writings on almost all aspects of nature show, he deemphasized systematic and detailed observations and experiment. Second, his definition of what constitutes a natural change or motion led to **Aristotelian physics** which had to be "unlearned" before it could be replaced by the modern, Newtonian way of thinking about physics problems. Many of ARISTOTLE's ideas are strikingly convincing[13]—and convincingly wrong. When applied to the motion of objects, his ideas amount to saying that, on Earth, vertical motion is natural: Flames naturally rise, stones naturally fall down, and so forth. The scheme was extended by attributing natural motion to the four elements: *Fire* and *air* rise, *water* and *earth* fall. This was explained by the claim that these elements wanted to be in their **natural place**. Water's natural place is to surround the Earth, as the oceans do. Steam rises until it reaches its natural place in the clouds, where it comes to rest. Fire rises until it surrounds the atmosphere, farthest from the Earth because it was considered the lightest element. Earth's natural place was at the center of the universe, so everything containing *earth* naturally falls down. By and by, these ideas stuck and ossified, because they make superficial sense, are based on (casual) observation and mutually support each other. It is hard to fathom that this consistent scheme is wrong, especially if it has been in place for hundreds of years. As has been pointed out by historian Thomas Kuhn [26], the consistency of the Aristotelian cosmological system guaranteed its tenacity. It is virtually impossible to take one of its building blocks out without destroying the entire conceptual edifice. In ARISTOTLE's universe, the Earth has to be at rest at the center of the universe, because otherwise stones wouldn't naturally fall down, water wouldn't flow downstream, and so forth. ARISTOTLE had a clever (if wrong) argument for a spherical Earth. Since the sphere is the mathematical shape with most volume per surface area, you can therefore pack the most *earth* into the smallest possible volume. Since the center of the universe or Earth is the natural place of the element Earth, its accumulation must generate a spherical shape. Oddly enough, this is the correct reason for the sphericity of the sun, moon, and planets, if the correct assumption is made that massive objects attract due to gravity. Fortunately, ARISTOTLE did use his observational virtues to produce independent evidence for the **sphericity of Earth**, which was, contrary to common belief, widely accepted from his time on. Namely, he noticed that the shadow of the Earth during lunar eclipse is always round. Since only spheres produce circular shadows when illuminated from different aspects, he concluded correctly that the Earth must be spherical.

So far we have only considered ARISTOTLE's explanations of changes and motion on Earth. What were his thoughts on the heavens? Again he introduces a false dichotomy that stuck around for a millennium. Namely, that there is a fundamental difference of what happens

---

[13]So much so that even in today's introductory physics courses, many Aristotelian views held by the students have to be purged before modern concepts can be learned [1].

on Earth and what happens in the sky. We don't see celestial objects change, and this must mean in ARISTOTLE's opinion that they are always at their natural place. However, the four elements all surround the Earth. As ARISTOTLE would put it, they are confined to the **sublunar sphere**. This means that the **superlunar sphere**, that is, the rest of the universe, must consist of a new, fifth element, called **quintessence**.[14] By construction, things made out of quintessence are unchanging and perfect in ARISTOTLE's universe. It will not have escaped your attention that there *is* change in the heavens: The celestial objects move with respect to the observer and with respect to each other. Well, ARISTOTLE explained this away with the notion that there is a natural way for celestial object to move, namely in the most perfect way possible, as a combination of **uniform circular motions** as his teacher PLATO had taught. Hence, the toolkit for the practicing astronomer to describe the cosmos was reduced to one tool: uniform circular motion.

Over the centuries ARISTOTLE's opinions petrified to an extent that it became almost unthinkable that there is change in the heavens. Indeed, if it were to be found, the whole Aristotelian universe was in danger of imploding. Change would mean that celestial objects are not made out of unchangeable quintessence, which means that they are not at their natural place, and so forth. This "danger of doom" later became reality, but you have to understand that ARISTOTLE's theory was thus testable, falsifiable, and a **bona fide theory** in the modern sense. Its weaknesses were of a different nature. For all its impressive, qualitatively correct descriptions of patterns in nature, it had trouble explaining some of the simplest ones. Why does a ball tossed in the air keep moving even after it left the hand? This question vexed ARISTOTLE and his successors for centuries. Strange explanations were produced, such as the **impetus theory**, which held that the ball keeps moving because it acquires a temporary power to do so (the impetus) from the hand that started this forced or artificial motion. The concept of impetus is completely unnecessary, if one tosses out the wrong Aristotelian belief that continued motion means a continued force. Today we know that not motion, but change in motion (acceleration) needs to be explained by a force. Furthermore, insisting on forced or artificial motion leads naturally to the question of what agent caused the motion. Since this agent must necessarily be moving artificially, it needs to be moved by another agent, and so forth, until we come to a first or **prime mover**. While completely unnecessary in our modern understanding of motion and dynamics, the prime mover was often identified with (a) God. This in turn led to many (false) arguments for the existence of God and mixed the realm of science with the realm of religion, muddying the waters considerably. ARISTOTLE's misleading concept of *natural* places or behaviors is mirrored in his inclination toward **teleological reasoning**. This study of nature is driven by the attempt to see or ascribe a purpose to every thing or phenomenon. In Aristotle's view, for instance, an acorn has the intrinsic goal—*telos* in Greek—of becoming an oak tree, much as water finds its natural place around the Earth.

When ARISTOTLE merged his ideas with EUDOXUS's, he conceived of a workable, almost mechanical universe made of real, crystalline spheres that were physically connected. The **prime mover** of this system was the outermost sphere of stars, which rotated *naturally* once a day, and distributed this motion to the next sphere, which moved the subsequent one, and so forth. All told, he used more than fifty spheres to account for all movements in the sky. The end result is a comforting description of the cosmos. Since by construction, nothing ever

---

[14] From Latin *quintus* = *five* and *essence* = *substance*; sometimes *quintessence* is used synonymously with *aether* or *ether*.

changes in the heavens, close observation of the sky is no longer necessary. This completeness of the Aristotelian cosmological theory is highly suspicious for modern tastes, but it fulfilled a basic psychological need: to be at home in a universe that makes sense, even if chaos and unpredictability reign on Earth.

In sum, the cosmos was a very different place at the end of classical Greece. It was, quite literally, filled with quintessence because ARISTOTLE held that a vacuum cannot exist. It was finite due to the existence of an outermost sphere. In fact, it had to be finite, because only finiteness guarantees a center, and this center is the natural place for Earth.[15] It was basically the solar system swaddled in starry wrapping paper. A small universe indeed! *Classical Greece* came to an abrupt end when, within one year, Alexander the Great and his teacher ARISTOTLE died (323/322 BCE).

Concept Practice

1. Explain the roles of the five elements in Aristotle's cosmology.
2. Why is Aristotle's model of the universe called *qualitative*? When is a model considered *quantitative*?

2.3 The Hellenistic Greeks Start to Measure the Size of the Universe

After the lofty ideas and models of *Archaic* and *Classical Greece*, the practicality of astronomers and thinkers of *Hellenistic* and *Roman Greece* is astonishing. The capital of Hellenistic Greece was Alexandria, in modern Egypt. A bit of historic background helps to explain the progressiveness of the era. Alexandria had been founded in 330 BCE by the Egyptian king PTOLEMY I. He was one of the **diadochs**, that is, generals-turned-kings when Alexander's empire was divided up after his death. In turn, Alexandria at the delta of the Nile, far off the Greek mainland, became the capital of Ptolemaic Egypt. The Museum of Alexandria was founded as a large library dedicated to the nine Muses,[16] one of which was URANIA of astronomy. We would probably call it a first-class research center today. At the end of the Hellenistic period, Alexandria became part of the Roman Empire in 31 BCE, when PTOLEMY XV, son of CLEOPATRA and JULIUS CAESAR, was murdered. The museum, however, survived unharmed and remained for centuries the epicenter of scientific progress.

The astronomical wisdom of Babylonia had been made accessible to the Greeks by ALEXANDER's conquers, and helped further astronomical research. The emphasis shifted from constructing all-encompassing theories of the cosmos to solving specific problems and explaining individual patterns in nature. Due to this focus, many discoveries were made. We will highlight three, associated with ARISTARCHUS OF SAMOS (310–230 BCE), ERATOSTHENES OF CYRENE (276–196 BCE), and HIPPARCHUS OF NICAEA (190–120 BCE). The first two worked at the Museum, HIPPARCHUS in Rhodes.

---

[15] The element and the planet.
[16] Hence the name *muse-um*.

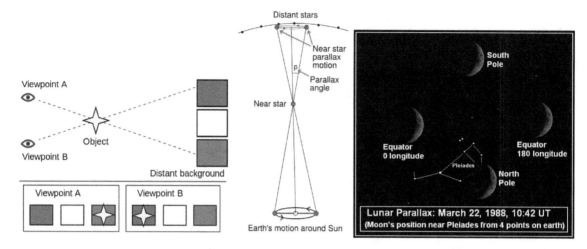

Figure 2.2: **Parallax.** (a) Geometry with baseline and two angles (b) Annual parallax due to the Earth's moment. Note that the parallactic angle p is *half* the angle spanned by the Earth's motion in half a year. (c) *The lunar parallax.* The moon's position with respect to background stars is shifted for observers at different locations.

2.3.1 Measuring the Skies: The Parallax

The guiding question for this section is really a ***dilemma***: How can we measure distances in the universe when we are confined to the Earth? We cannot do a ***direct measurement***, for example by taking a very long yardstick and comparing its length with the distance between celestial objects. In fact, this method already fails on Earth, when we want to measure the distance between cities, or, for that matter, anything that is farther apart than about 100 yards. **Land surveyors**, for instance, use **geometrical methods**, that are based on **geometry** and **trigonometry**. Since an arbitrary triangle is completely determined by one of its three sides and two of its three angles, we can determine the distance to a distant object by looking at it from two different vantage points A and B which are a known distance apart, see Figure 2.2. Surveyors use **theodolites**[17] to accurately determine the angles α and β from the two endpoints of the **baseline** between A and B to calculate the lengths of the other sides of the triangle, and therefore the distance to the object. In astronomy there is another straightforward way of measuring these angles by using the ***parallactic effect***. Namely, the *apparent* position of an object will be different when viewed along two different lines of sight. You can get a sense for the ***parallax***[18] by noticing how objects near to you seem to shift with respect to distant background objects when you observe your vicinity with your left eye versus your right eye closed. The same thing holds true for astronomical observations. If you look at the sky from a different point on Earth, it will look slightly different. Since an observer's vantage point changes naturally due to the Earth's rotation, there is a daily or ***diurnal parallax***. The closer an object is, the bigger the effect, that is, the ***parallactic***

---

[17] *Theodolites* are basically small telescopes with a finely grained scale to read off the direction in which the telescope is pointing. Thus, they were not available before the invention of the telescopes around 1610.

[18] The Greek word *parallaxis* means *alteration*.

2.3. THE HELLENISTIC GREEKS MEASURE THE SIZE OF THE UNIVERSE

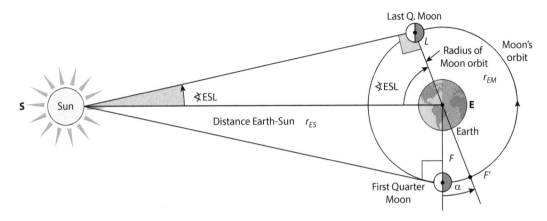

Figure 2.3: **The distance to the sun relative to the distance to the moon as determined by Aristarchus.** At *first* and *last quarter moon*, Earth, sun, and moon form a right triangle with the 90° angle at the moon. Therefore, the positions at first and last quarter moon are not exactly opposite. The deviation α from 180° is very small and hard to measure. Since the angle α is the ratio distance to the moon to the distance Earth-sun, it can be used to determine the distance to the sun relative to the distance to the moon.

angle. Therefore, the parallactic angle p and the distance d are inversely related[19]

$$d = \frac{1}{p}. \qquad (2.1)$$

In short, by measuring the parallactic angle and due to a known baseline, we can determine the distance to the object. This is the **parallactic method**.

2.3.2 Aristarchus Measures the Relative Sizes of the Earth, Moon, and Sun

ARISTARCHUS used a clever geometric argument to figure out the *relative* sizes of the Earth, moon, and sun. This was a huge achievement, because he was able to measure the size of celestial objects, that is, superlunar objects, while being confined to Earth, that is, the sublunar sphere. It must have felt empowering that the human mind is able to devise a method to size up the cosmos from a distance!

ARISTARCHUS measured the distance to the sun by exploiting the geometry of the Earth-moon-sun configuration at first and last quarter moon, see Figure 2.3. At its quarter phase, the moon is at an exact right angle with respect to the sun, because we see exactly half of its illuminated hemisphere. Since the sun is not infinitely far away, the position of the moon at first and last quarter (points F and L in the figure) are not exactly opposite (180°) of each other along the lunar orbit. Unfortunately, the sun is so far away that the deviation from 180° ($\alpha = \sphericalangle ESL$ in Figure 2.3) is only ten arc minutes. That is well beyond the accuracy of antique naked-eye observing techniques.

It seems quite unbelievable that the angle is so small, so let's do a quick cross-check. Since the triangles SEL and FEF' are **similar** or **congruent**, they share the same angles. in particular $\alpha = \sphericalangle ESL$. Therefore, α is the angle under which the radius r_{EM} of the moon's

---
[19]The formula is technically correct when d is the distance in parsecs and p half the angle by which the celestial object moves in half a year.

orbit appears as seen from the sun, a distance r_{ES} away from the Earth. The angle therefore is according to Equation (1.1) (see also Figure 1.8)

$$\alpha = \frac{r_{EM}}{r_{ES}},$$

that is, the ratio of moon distance to sun distance, which we know today is $\frac{0.384\,\text{million km}}{149.6\,\text{million km}} \approx \frac{4}{1500} = 0.0027\text{rad} = 0.15° = 9'$, in fair agreement with our claim of $10'$.

Due to the angle's smallness, ARISTARCHUS had to more or less *guess* its value. Unfortunately, he grossly overestimated the angle to be $3° = 20 \times 0.15°$. Since his angle was $20\times$ too large, his estimate for the distance of the sun was $20\times$ too small. In his reckoning the sun is 20 times farther than the moon and therefore 20 times larger. Actually, it is roughly 400 times farther and therefore 400 times larger.

Still, ARISTARCHUS's result that the sun is much farther and therefore larger than the moon naturally led to the next question: Is the sun bigger than the Earth? To answer it, ARISTARCHUS devised the following method to relate the distance to the moon to the radius of the Earth. By direct observation one can determine the angular or apparent size of sun and moon to be equal and roughly $0.5°$. During a lunar eclipse, Figure 1.24, the moon travels through the shadow of the Earth cast by the sun. Aristarchus figured out that the shadow of the Earth at the orbit of the moon has a size of about $\frac{8}{3} = 2.\overline{3}$ lunar diameters by measuring the time it takes the moon to move a distance equal to its diameter. He then could produce a drawing like Figure 2.4, but to scale, that is, with the sun 20 times farther off the Earth than the moon, and, placing the Earth into the sketch centered at the intersection point E, he could read off the relative sizes of the Earth, sun, and moon. Unfortunately, he did not do the observations himself and took a value for the angular size of the sun and the moon that was a factor four too big ($2°$). Though his result was not accurate, ARISTARCHUS's method was sound, and he figured out correctly that the sun is much bigger than the Earth (about 5 times in his reckoning, 110 times in reality). He thus figured out by scaling (see Sec.1.2.6) that the Earth would fit $5^3 = 125$ times into the volume of the sun. ARISTARCHUS's was impressed enough by the immense size of the sun that he pondered the possibility of the sun, not the Earth, being at the center of the universe instead of moving around a stationary Earth. Unfortunately, his ideas did not catch on and were forgotten.[20] The reason being that it seems so "obvious" that the stars and planets rotate around us, and, more importantly, that we do not *feel* the rotation of the Earth. Only much later was it understood that there is no reason why we *should* feel the effects of the rotation [44]. Until that time it was reasonable to assume that the Earth is stationary. We have already seen that this assumption of a stationary Earth at the center of the universe became the cornerstone on which ARISTOTLE's world system was erected.

2.3.3 Eratosthenes Determines the Diameter of the Earth

The first to make a measurement of *actual* distances in the cosmos was ERATOSTHENES. He was director of the museum in Alexandria and a geographer by trade, which helped him to invent a simple method to determine the size of the Earth. He had heard that the sun in Syene on the Nile River in upper Egypt is directly overhead on June 21, the date of the summer solstice. At noon on that day, the sun's rays would fall into a deep well, illuminating

---

[20]COPERNICUS read about them one and a half millennia later in a book.

2.3. THE HELLENISTIC GREEKS MEASURE THE SIZE OF THE UNIVERSE

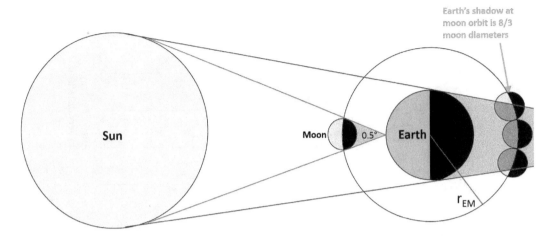

Figure 2.4: **The distance to the moon relative to the Earth diameter as determined by Aristarchus.** ARISTARCHUS knew that the moon and sun appear under the same angle. From his estimate of the distance to the sun, he assumed that the sun in the drawing must be 20× farther from the Earth than the moon. Following a lunar eclipse, he determined the diameter of Earth's shadow at the moon orbit to be 8/3 moon diameters. He could then draw *to scale* a sketch like the figure shown—which is drawn *not* to scale to show the salient features. This geometric construction completely determines all *relative* sizes. In particular, one can read off the distance to the moon r_{EM} in terms of the radius of the Earth r_E. Note that the observer is shown here to be located at the center of the Earth. Strictly, this is nonsense, but a very good approximation since the radius of the Earth is so much smaller than the distance to the sun or the moon: $r_E \ll r_{EM} \ll r_{ES}$.

its waters. In Alexandria, meanwhile, a gnomon would cast a shadow, because the sun was not directly overhead. Since the Nile flows approximately due north, Syene is due south of Alexandria. From Section 1.11 we know that the position of the sun above the horizon can be determined by measuring the length of a shadow cast by a gnomon. ERATOSTHENES found it to be 7.2° or one fiftieth of a full circle off the zenith in Alexandria. Since the sun is very far away, as measured by his predecessor Aristarchus, its light rays hit the Earth nearly parallel. From Figure 2.5 it is then evident that Alexandria and Syene are 7.2° apart in **latitude**. This allows us to set up a simple proportionality. The distance between Alexandria and Syene compared to the circumference of the Earth is the same as the angle between Alexandria and Syene (7.2°) compared to the 360° of a full circle

$$\frac{\text{Distance Alexandria to Syene}}{\text{Circumference of Earth}} = \frac{7.2°}{360°} = \frac{1}{50}.$$

You'll notice that while the distance between the two cities can be measured, the circumference C of the Earth is the only unknown, and determines the radius r_E of the Earth via $C = 2\pi r_E$. It is not quite clear which value for the distance between the cities ERATOSTHENES used, but his value was certainly within 20% of the accurate result ($R_{Earth} = 6371$ km).

The importance of this measurement can hardly be overstated. Once the absolute size of the Earth is known, it can be plugged into ARISTARCHUS's calculations, and yields the absolute sizes of the sun and the moon. This means that ERATOSTHENES was able to bring the heavens to Earth by linking a measurement between two places on Earth to a distance between two objects in outer space!

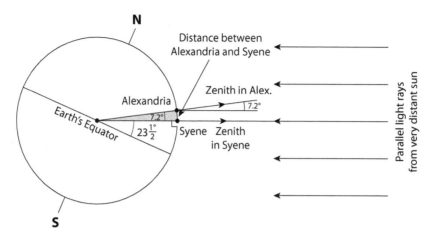

Figure 2.5: **The radius of the Earth as determined by Eratosthenes.** At noon on June 21, the sun is directly overhead, that is, at the zenith in Syene. At the same time a *gnomon* in Alexandria casts a shadow showing that the sun is 7.2° off the zenith; see Figure 1.11. The ratio of the distance between the two cities to the circumference of the Earth must be the same as the ratio of the angular distance (latitude difference) between the two cities to the 360° of a full circle. This allowed ERATOSTHENES to calculate the unknown circumference of the Earth and thereby its radius.

2.3.4 Hipparchus Discovers the Precession of the Equinoxes by Compiling an Accurate Star Catalog

HIPPARCHUS was active around 150 BCE on the island of Rhodes, and his achievements make him the greatest astronomer of antiquity rivaled only by PTOLEMY. His biggest achievement was a star catalog of about 850 stars whose positions he measured with unprecedented accuracy. The catalog was extensively used for more then one and a half millennia, until TYCHO BRAHE's even more accurate star catalog became state-of-the-art in 1602 CE.[21] By dividing the stars into six groups of different apparent brightnesses or **magnitudes** (the first magnitude stars being the brightest), he introduced a classification that is still used today. Furthermore, he developed methods of trigonometry to help him with his extensive calculations, and he introduced a mathematical description of the nonuniform motion of the sun with respect to the stars, the *eccentric circle*.

The **precession of the equinoxes** is an incredibly slow motion of the CNP around the so-called **pole of the ecliptic** as depicted in Figure 2.6. One rotation takes about 26,000 years which means 72 years go by before the position of the Celestial Pole changes by one degree! How was HIPPARCHUS able to detect such a minuscule effect? For one, his measurements were of exquisite accuracy, but more importantly, the effect is **accumulative.** That means, if you compare data 720 years apart, you'd find a discrepancy of 10° which is impossible to ignore. He had an old set of data, probably of TIMOCHARIS (c. 320–280 BCE), which helped him figure out the precessional pattern. HIPPARCHUS saw that the stars' positions had changed in the roughly 150 years in between by about 2°. That is four apparent lunar diameters, and thus much more than can be explained by measurement uncertainties. He had discovered a real change in the heavens!

How did HIPPARCHUS discover the discrepancy between his and the older data? Positions

---

[21]This catalog of 777 stars was part of the book *Astronomiae Instauratae Progymnasmata* (1602); the full *Thousand-Star Catalog* appeared in the *Rudolphine Tables* of KEPLER in 1627.

2.3. THE HELLENISTIC GREEKS MEASURE THE SIZE OF THE UNIVERSE   87

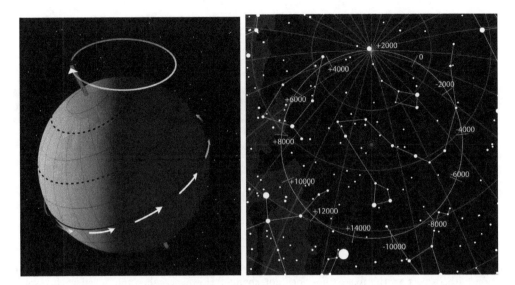

Figure 2.6: **Precession.** (a) The slow motion of the Earth axis around the pole of the ecliptic. (b) The associated slow motion of the Celestial North Pole around the pole of the ecliptic. The motion of about one degree in 72 years means that *Polaris* (α *Ursae Minoris*) will not remain the pole star. For instance, 5,000 years ago, α *Draconis* was the pole star, and in 14,000 years it will be *Vega* (α *Lyrae*). As the pole moves, so do the *equinoxes*, and in 500 years the position of the **vernal equinox** will shift from *Pisces* into *Aquarius*. Therefore, the **tropical year** from equinox to equinox is slightly shorter than the **sidereal year**, the true 360° rotation of the Earth around the sun (or of the sun around the celestial sphere in the geocentric model). The **interpretation** of the **precessional motion** of the equinoxes is different in the helio- and geocentric models.

of stars are listed relative to a **coordinate system** or **frame of reference**. As we pointed out in Section 1.4, there are several that make sense: the coordinate system of the observer with **zenith** and horizon, the **equatorial coordinate system** with the **Celestial Poles** and the **Celestial Equator** based on the rotation of the celestial sphere,[22] and the **ecliptic system** based on the apparent path of the sun among the stars. Likely, ecliptic coordinates were used in the old table. The shift of positions is then simply a shift in ecliptic longitude; see Figure 2.7. This pattern ruled out the possibility of a flaw in his or the older data, because the stellar positions were shifted **systematically**, whereas errors would have been **randomly** distributed. He got even more lucky: One of the stars appearing in the old data set was Spica in *Virgo*, which is very close to the ecliptic, and thus exhibits the maximal effect. Spica had shifted by 2° in the 150 years. HIPPARCHUS concluded that there was a pattern hidden in the shifted positions of the stars. He hypothesized that the two data sets would agree if the direction of the **vernal equinox**[23] had shifted for some reason. Put the other way around, he made the **assumption** that the **equinoxes** shift continuously, which explained the observed pattern of shifts. He could even determine the size of the effect. The equinoxes had shifted by about 2° in 150 years, and setting up a simple proportionality, HIPPARCHUS could figure out the period of a complete rotation:

$$\frac{360°}{2°} = \frac{x}{150 \text{ years}} \implies x = 180 \times 150 \text{ years} = 27{,}000 \text{ years},$$

---

[22] Rotation as any motion is *relative*, so we do not have to commit here as to what rotates; it could be the celestial sphere or the Earth.

[23] The intersection of the ecliptic with the celestial equator.

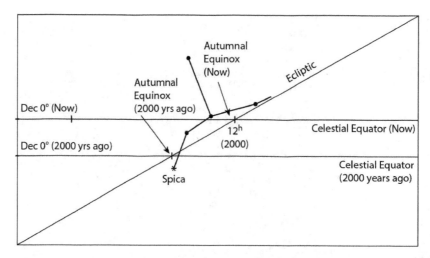

Figure 2.7: **Shift of the equinox.** Spica in *Virgo* is close to the ecliptic and close to the *autumnal equinox*. Its position with respect to the fixed stars does not change. However, when measured with respect to either the *equatorial* or the *ecliptic* coordinate systems, its position slowly changes. The conclusion is that the coordinate systems themselves change, that is, the intersection point of celestial equator and ecliptic shifts as depicted. As you can see, the *coordinates* of *Spica* change: Its declination (*celestial latitude*) gets smaller, as does its **Right Ascension (*celestial longitude*)**. Millennia later, precession was interpreted as a wobbling of the Earth's rotation axis due to the gravitational influence of the other planets; see Section 2.10.

in good agreement with the modern value. This is a simple description of a complex pattern. If the position of the equinox and therefore the Celestial Pole are slowly changing, it results in a change of both the **celestial latitude** and the **celestial longitude** of the stars. A two-dimensional shifting pattern is complicated, but, with enough grit, dedication, and luck, Hipparchus was able to figure it out. Note also that the basis of the discovery was immaculate record keeping and measurement technique, not the least on the part of Timocharis.

Very influential in the quest to find the correct mathematical way to describe planetary motion (though irrelevant today), was Hipparchus's introduction of the **eccentric circle**. It is his version of having a cake and eating it too. The cake here is uniform circular motion, and eating it means to destroy uniformity by a changing pace of the moving object, in his case the sun. His trick is as follows; see Figure 2.8. The sun is moving at constant speed around a circle, as required by Plato, but the observer on Earth sits *off center*, so the solar motion *appears* faster when at B than when it is at A. That is, from the observer's perspective the angular distance traveled with respect to the stars is greater at B. Hipparchus used this trick to model the nonuniform motion of the sun and the moon so well that he could predict lunar eclipse to within one hour. On the other hand, the eccentric circle, already an **oxymoron** as a label, because circles are by definition not eccentric, is a dirty trick and certainly not in the spirit of Plato. Copernicus' disgust for tricks like these led him to propose the heliocentric model of the universe.

Concept Practice

1. What did Eratosthenes measure?

    (a) The distance to the sun
    (b) The distance to the moon
    (c) The size of the moon

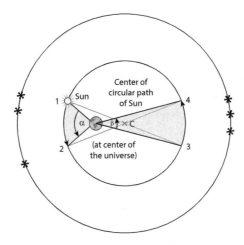

Figure 2.8: **Eccentric circle.** HIPPARCHUS introduced this geometric construction to *model* nonuniform apparent motion, in particular of the sun. The sun moves faster with respect to the stars in January than it does in July. Therefore, the northern winter is several days shorter than the summer. To conform to Plato's ideals, HIPPARCHUS's sun moves uniformly around a circle with center C, which does not coincide with the position of the Earth at the center of the universe. It is thus an *eccentric circle*. The sun moves from $1 \to 2$ in the same time as from $3 \to 4$. Since the sun is closer to Earth during the first time interval, for an observer on Earth it *appears* that it's moving faster. The **angular distance** traveled by the sun with respect to the background stars is bigger from $1 \to 2$ (angle α) than from $3 \to 4$ (angle β): $\alpha > \beta$.

   (d) The radius of the Earth
   (e) Two of the above
   (f) None of the above

2. What did ARISTARCHUS measure?

   (a) The distance to the sun
   (b) The distance to the moon
   (c) The size of the moon
   (d) The radius of the Earth
   (e) Two of the above
   (f) None of the above

2.4 The Most Enduring Framework: Ptolemaic Astronomy

Three hundred years went by before someone did something comparable to HIPPARCHUS's amazing discoveries. It is again in Alexandria that CLAUDIUS PTOLEMY (no relation to the Ptolemaic kings of Egypt) around 150 CE advances the sciences to such a degree that his work can be described as a capstone. For one, he summarized and synthesized the work done before him while injecting a healthy dose of fresh, original ideas. But his was also the last, definitive model of the universe for well over a thousand years.

2.4.1 Ptolemy the Mathematician versus Aristotle the Physicist

PTOLEMY wrote several books that became standards for centuries to come, including one on astrology (sic!) and geography. However, the book that garnered him an eternal front place in the history of science is his compendium of astronomy, the *Almagest*. PTOLEMY named it the Great Composition (Greek: *Megale Synthaxis*), but because it was lost during the Dark Ages and resurfaced as an Arabic translation of the Greek original, it is known by its Arabic name as "Al Majisti" or "The Greatest." Incidentally, quite a few astronomical names and labels are Arabic, because of the great influence of the Muslim scientists in the early Middle Ages, when the torch of wisdom passed into the Islamic world. The Arabic equivalent of "the" is "al," so when you come across a name beginning with "Al-" it probably is Arabic, like Aldebaran (α Tauri), algorithm, or algebra.[24]

Before we delve into the details of the Ptolemaic cosmological system, we have to bust some myths about PTOLEMY and his theory. It is often wrongly said that PTOLEMY "only" compiled and streamlined what others had done before him. This is most certainly not true, as he introduced new and essential features like the **equant** to the geocentric theory that bears his name. He was a highly original thinker as his influential books attest. PTOLEMY's theory is often derided as the epicycle theory, with the word epicycle connoted with "overly complicated or cumbersome" or even "inaccurate." Some even go so far as to identify PTOLEMY's theory and COPERNICUS' theory with "wrong theory" and "right theory." None of this is correct. Surprisingly, the simplest version of Ptolemy's epicycle theory with a central, stationary Earth and the simplest version of COPERNICUS' theory of circular planetary orbits around a central, stationary sun *are equivalent*. Most people are also not aware that Ptolemy's theory is fundamentally different from the system of **homocentric spheres** of the Pythagoreans, PLATO, EUDOXUS and ARISTOTLE—at least in our modern view. In the old days, these two strands of ancient science were strangely interwoven. We saw that ARISTOTLE devised a virtually all-encompassing description of the world. He had a reasonable, **qualitative** explanation of how stars move, how water flows, how steam rises, how plants grow, and of many other patterns in nature. **Quantitatively** however, his explanations did not fare well. To pick one example, even his fifty-plus crystalline spheres could not reproduce the positions of the planets well enough to agree with the crude, naked-eye observational data available back then. For us this agreement between predictions of a theory and observation is a necessary condition to accept a theory. But in ancient science, it was not seen this way. The general lack of attention to detail was not a bug but a feature of the old way of doing science.

Only in that vein it makes sense that the comprehensive, physical explanation of the world of ARISTOTLE merged with the focused, mathematical explanation of the heavens of PTOLEMY. To us it seems strange, since the notion of an epicycle, that is, of a small circle carrying the planet rotating around a point on a rotating second circle (**deferent**); see Figure 2.9, is quite incompatible with the concept of solid, crystalline spheres. It did not seem to bother the contemporaries that you could choose from two contradicting theories, depending on what you'd like to understand about the world. If you'd be interested in quantitative predictions of planetary positions, eclipses, and lunar motion, you'd go with PTOLEMY. If you wanted a holistic, comforting, qualitative reason why things happen the way they do, you'd go with Aristotle. In fact, from the fifth century on, that is, several centuries after PTOLEMY, we find these two theories so intertwined in the heads of the scientists, that it became possible to calculate the size of the cosmos. In this calculation it was assumed that the epicycles with sizes determined by PTOLEMY are contained in thick, solid homocentric sphere. Adding up the widths of these spheres, astronomers determined a (much too small) radius of the outermost sphere of the stars, that is, they determined the size of the cosmos. Still, there was

---

[24] A counter-example is *alma mater*, Latin for *nourishing mother* (university).

2.4. THE MOST ENDURING FRAMEWORK: PTOLEMAIC ASTRONOMY

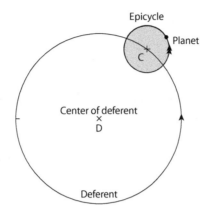

Figure 2.9: **Anatomy of an *epicycle*.**: A circle carrying the planet and called an *epicycle* rotates uniformly around a point C on the circumference of a second circle called the *deferent* which has a center D. The rotational speeds of the planet around C and of the epicycle around D as well as their radii could be chosen by the *model builder*.

a rift between the two ways of doing science, and the chasm between the two groups widened over the centuries. One group was characterized as the "physicists," natural philosophers or Aristotelians, the other as and "mathematicians" or followers of PTOLEMY. The Aristotelians wanted to understand the mechanisms, that is, the physics, underlying the cosmological clockwork, whereas the disciples of PTOLEMY wanted to devise a mathematical scheme that reproduces the positions of the celestial objects, that is, that saves the appearances.

2.4.2 Ptolemaic Astronomy Is Based on Epicycles, Eccentric Circles, and Equants

The *epicycle* is a crucial feature of Ptolemy's system securing fair agreement with observations. The models with **homocentric spheres** had no epicycles, and did not agree well with the patterns observed in the sky.

To see why this is true, we'll describe in more detail the features of Ptolemy's model in the next subsection. Since it was the longest-lasting theory ever devised, that is, hard to *falsify*, it is well worth our time. It is an important milestone on the way to the expanding universe. Recall which *appearances* had to be saved, that is, which observations the theory had to agree with. Here is a list.

1. All celestial objects show daily motion around the Celestial Pole.

2. The sun and the moon move on two circular paths around the celestial sphere; these paths do *not* coincide with the celestial equator.

3. The sun and the moon show a slow eastward motion relative to the stars while moving west; they are "slower" than the stars.

4. Planets exhibit retrograde motion.

5. Planets are brightest, so probably closest, when they are moving in retrograde fashion, that is, westward with respect to the stars.

6. Planets always stay close to the ecliptic, the path of the sun among the stars.

7. Mercury and Venus are always close to the sun in the sky; their angular distance from the sun is never more than 28° and 47°, respectively.

The accuracy of observational data he had to match were as follows.

1. Planetary positions were known with an uncertainty of about one degree (two apparent lunar diameters or a fingers width, Figure 1.7(b)).

2. The lengths of the year and the lunar month were measured by HIPPARCHUS with amazingly small uncertainties (a few minutes and a second, respectively).

3. The absolute sizes of sun and moon and their distances, according to ARISTARCHUS, were known.

4. The Babylonian explanations of the eclipses and mathematical procedures to predict them accurately were available thanks to Alexander's conquest of Persia.

5. Star positions on the celestial sphere were known to better than a degree (two apparent lunar diameters) due to HIPPARCHUS's star catalog.

6. The precession of the equinoxes was discovered by HIPPARCHUS and amounts to about one degree per seventy-two years.

Finally, PTOLEMY's assumptions were based on the obsessions of the day:

1. All celestial objects move in (combinations of) uniform circular motion.
2. The Earth is stationary at the center of the universe.

As we pointed out, describing the universe as a combination of homocentric, crystalline spheres does not work well. The quantitative results of the Aristotelian scheme don't agree with observation. Worse, the brightening of the planets during retrograde motion was completely unaccounted for. What other description could be used? **Epicycles** had been used since before the time of HIPPARCHUS and APPOLONIUS OF PERGA (262–190 BCE) to solve several problems with nonuniform motion. Also, the *eccentric circle* championed by HIPPARCHUS was a useful geometric construction. PTOLEMY himself invented another one, the ***equant***.

Let's understand how these **constructs** work. An *epicycle*, Figure 2.9, is a circle that carries the planet around a point in uniform circular motion. This point, the center of the epicycle, is itself uniformly moving on the circumference of a second, larger circle called the deferent. The result of this "circular motion on circular motion" are quite intricate trajectories, see Figure 2.10. In essence, it is a way to circumvent the Platonic requirement of uniform circular motion. The combination of *two* uniform circular motions is not necessarily uniform nor circular. So the letter but not the spirit of PLATO's "law" are honored. As evident from Figure 2.10, depending on how big the epicycle is, and how fast it is turning, many "orbits" can be dialed in: circular, shifted circle, eccentric circle, and so forth. The ***major epicycle***, Figure 2.10(d), a fast-turning epicycle, can describe retrograde motion: While the center of the epicycle is moving prograde (eastward) slowly, the planet on the epicycle is moving westward fast, which results in a retrograde (westward) motion. Better yet, the planet on a major epicycle is also closest to Earth when moving retrograde, as you can see from the figure. Other, so-called ***minor epicycles*** can be used to describe, for example the nonuniform motion of the sun, accounting for the different lengths of the seasons. I pause

2.4. THE MOST ENDURING FRAMEWORK: PTOLEMAIC ASTRONOMY

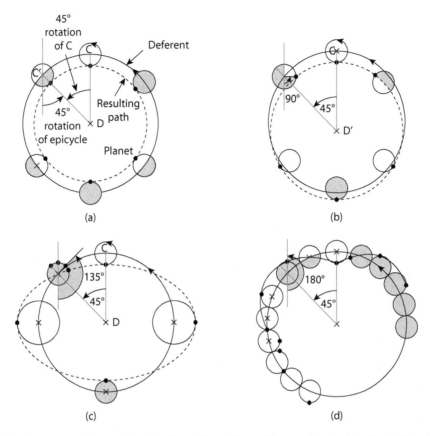

Figure 2.10: **Paths created by epicyclical motion.** Depending on the sizes and rotational speeds of the epicycle and the deferent, the path traced out by the planet will vary. Some examples are as follows. (a) A smaller circle results if the planet rotates around the epicycle at the same rate as epicycle's center rotates around the deferent. (b) A circle with an offset center at D', that is, an **eccentric circle**, results if the epicycle rotates twice as fast as the deferent. (c) An elliptical path results if the epicycle rotates 3× as fast as the deferent. (d) **Retrograde motion** is achieved when the epicycle rotates 4× as fast as the deferent. Since this epicycle rotates so fast, it overcompensates the motion of its center in the opposite direction. Note that in all examples the epicycle and the deferent rotate in the same direction (counterclockwise).

here to give you time to put yourself in the shoes of the person who *invented* the epicycle. How exciting it must have been to discover that this geometric construction you thought of can fulfill PLATO's wishes *and* describe the complex retrograde motion *and* predict that the planets are closest in the middle of the retrograde loop! The puzzle pieces fit so well that it can't be coincidence; it has to be true, it has to be the key to the heavens, right? It is not, as we know now. But maybe you could feel a little bit of the "Eureka!" excitement of a scientific discovery. It's for moments like these that scientists do science. Now, you may say that it wasn't a discovery because planets do not travel on epicycles. However, the epicycle solved a lot of problems that the Aristotelian homocentric spheres couldn't. So back then, the epicycle was progress. And while it led to a dead end eventually, it was an exciting new idea to try.

We already described the **eccentric circle** as a geometric tool in the section on HIPPARCHUS. The final geometric device in the Ptolemaic astronomer's toolbox was invented by the master himself—the **equant**. It is a strange construction to soften the requirement

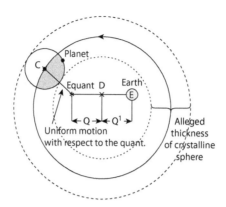

Figure 2.11: **The equant.** The epicycle center C moves on a deferent with center D. It moves **uniformly** with respect to the equant, and consequently nonuniform with respect to D and the Earth. The equant sits opposite the Earth with respect to D, but at equal distance from D, hence $Q = Q'$ and the name. The equant was seen as an arbitrary, **ad hoc** invention, and despised in particular by Copernicus. However, it improved the accuracy of Ptolemy's model. Centuries later, the size of the epicycle was thought to determine the thickness of the **crystalline sphere** of the planet. This (wrong) assumption was used to calculate the size of the Aristotelian universe.

that the deferent had to move uniformly. It improved the accuracy of the model. PTOLEMY decreed that the center of the epicycle would move uniformly with respect to a point that is equally distant from the center of the deferent, but on the opposite side of the Earth; see Figure 2.11. This point around which the deferent turns uniformly is called the **equant**. It is a slap in the face of the true believer of Platonic ideal uniform circular motion. So much so, that COPERNICUS would later rather sacrifice Earth's central position in the universe than use the equant. Here, then, is a versatile geometric tool kit that can accommodate all kinds of heavenly motions: epicycle, eccentric circle, equant. The task for astronomers like PTOLEMY was to combine them, size them, and make them rotate with appropriate speeds to build a mathematical description of planetary motions. Working with these tools within the geocentric framework became known as **Ptolemaic astronomy**. It was the leading astronomical theory from about 300 BCE (well before PTOLEMY!) to about 1450 CE. An enduring theory indeed!

2.4.3 Ptolemy Explains Retrograde Motion with Major Epicycles

PTOLEMY constructed the geocentric world system using epicycles, equants and eccentric circles as follows. By construction, the Earth is at the center. Sun, moon, and planets circle it; each with a period that reflects its **sidereal period**. For instance, the sun's deferent turns once a year, whereas Jupiter's turns in twelve years. Next, each planet is assigned a **major epicycle** to explain retrograde motion. The sun and the moon show no retrograde motion, so there is no solar or lunar major epicycle. PTOLEMY then had to fix the order of the planets. From solar eclipses it is clear that the moon is closer than the sun, and the "orbit"[25] closest to the Earth in Ptolemy's system is that of the moon. The order of the other "planets" is not at all obvious, because their distances were not known. PTOLEMY made the reasonable assumption that the slower a planet moves, the farther out it is. That assumption picks the

---

[25] **Orbit** is used here to mean a mathematical curve traced out in PTOLEMY's description, not the actual path of a planet in three-dimensional space.

order of the planets to be Mercury (next from the moon), Venus, sun, Mars, Jupiter, and Saturn on the farthest sphere; see Figure 2.12. The Ptolemaic system cannot determine the distances to the planets. To reproduce observations, a fixed ratio of the size of the epicycle to the size of the deferent is sufficient; see Figure 2.13. Given the planetary observations in Section 1.6, we can construct the major epicycles of the Ptolemaic model. Since the inferior planets Mercury and Venus are always close to the sun in the sky, their deferent must turn once a year like the sun's. To obtain the correct **synodic period**, Table 1.3, their epicycles—sizes fixed by maximal elongations of 18° and 47°—must turn in 88 and 225 days, respectively. For the **superior planets**, PTOLEMY chose a different route. The sizes of the major epicycles are fixed by the apparent size of their retrograde loops. They ensure the correct number of retrograde loops per sidereal period, he had to stipulate that the epicycle turn once per year, which meant that the radii of the epicycles of Mars, Jupiter, and Saturn had always to be parallel to the Earth-sun line. The deferent of the superior planets turns once every sidereal year (1.88 years, 11.9 years, 29.7 years). This construction reproduces planetary motion fairly well. A weakness of Ptolemy's theory is that the main features are put into the model; they do not follow from the model. It is from *observations* that we know how long the sidereal and synodic periods are, how great the maximal elongations are, and so forth. These are therefore no features to brag about. Furthermore, there is a good deal of **fine-tuning** involved: Why should the deferents of Mercury and Venus turn *exactly* the same and *exactly* in a year, why should the epicycles of the superior planets turn the same and *exactly* in a year? As you can see, even in the Ptolemaic system the sun has a special place. All of these facts would make a modern astronomer rightly suspicious. However, Ptolemy's theory was just meant to save the appearances, which it did fairly well.

Before we conclude, let's bust one more myth. It is often said that COPERNICUS did away with epicycles, because he put the sun at the center of the universe. This is wrong. The central sun eliminates only the need for **major epicycles**. As myths always contain a grain of truth, we will see that the elimination of the major epicycles was the major appeal of the Copernican theory, because of the associated beautiful explanation of retrograde as relative motion. Also, we can now substantiate our claim that the Ptolemaic and Copernican theories in their simplest forms are equivalent. The crucial difference is, of course, that COPERNICUS has the sun at the center, and the observer on Earth moving around it. However, this is just a matter of perspective. By a simple switch of the **reference frame** you could base your coordinate system on the Earth instead of the sun. In this coordinate system, the Earth would be stationary, and the sun would be moving around the Earth. Planetary motions observed from the Earth *look* complicated, as we have seen in Section 1.6, but they can certainly be correctly described by some mathematical model. One can even show that the epicycle theory can describe *any* motion with arbitrary precision if epicycles on epicycles are allowed[26]. For all of these reasons, the Ptolemaic theory is not *wrong*. Rather, it is not useful, because it doesn't lead to any new insight. If you patch up the theory, as Ptolemaic astronomers have done for almost two thousand years, with equant, minor epicycles, and the like, you get a more accurate description of planetary motion, but one that is strangely sterile. What is to be learned from the fact that Jupiter's motion is described by a deferent, a major epicycle of size X, several minor epicycles of size Y, rotating in times T_1, T_2, and so forth? Nothing substantial about the inner workings of the cosmos unfortunately! Before we come down too

---

[26] See https://www.youtube.com/watch?v=QVuU2YCwHjw for an epicycle description of a planet whose "orbit" looks like Homer Simpson.

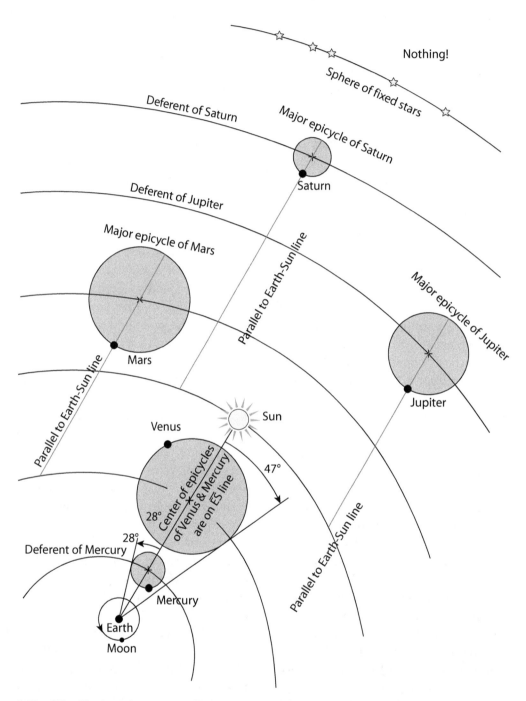

Figure 2.12: **The Ptolemaic system.** Shown is a simplified version of the Ptolemaic model of the cosmos; only *major epicycles* are shown. Several examples of *fine-tuning* are apparent: The centers of the epicycles of Venus and Mercury are situated on the line connecting Earth and sun; the ratio of their epicycle to deferent size is such that they appear under an angle of 47° and 18°, respectively, from the stationary Earth at the center of the universe; the position of the superior planets along their epicycles is such that the line planet-epicycle center is parallel to the Earth-sun line. The cosmos ends at the sphere of the fixed stars.

2.4. THE MOST ENDURING FRAMEWORK: PTOLEMAIC ASTRONOMY

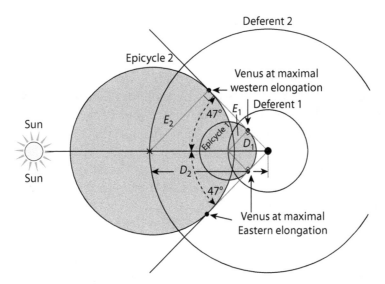

Figure 2.13: **Determination of Venus's major epicycle from its maximal elongation in Ptolemy's model.** The observed maximal elongation angle of 47° can be reproduced without committing to a particular epicycle size. Both the small epicycle of radius E_1 on a deferent of radius D_1 and the large epicycle (radius E_2) on a large deferent (radius D_2) predict the same elongation angle, as long as $\frac{E_1}{D_1} = \frac{E_2}{D_2} = \sin 47°$. Note that these two alternatives predict a different position of Venus at maximal elongation. This is of no consequence, since distances to Venus were not observable before the advent of the telescope. In other words, there was no way to distinguish the "large –" from the "small epicycle model."

hard on the Ptolemaic model, let's reiterate that any model has the benefit of constituting a frame with respect to which observations can be categorized and understood. Only when we interpret data with a theory in mind can we make judgments like "there is a discrepancy between theory and observation," or "we do not understand this pattern in nature yet." In this way, the Ptolemaic theory was fruitful. It allowed mathematical astronomers to try their best at describing nature in a rational **approximation** scheme. When they failed, it led to the eventual insight that two essential features of nature had been missed: the centrality of the sun, and the ellipticity of the planets' orbits.

Note that superficially, the motive for COPERNICUS' and KEPLER's theories of planetary motion was no better than PTOLEMY's. They, like PTOLEMY, set out to find an algorithm or a formula to describe planetary motion. All these theories were **useful** for astronomers because they allowed the prediction of planetary positions. Only later did KEPLER's theory turn out to be fruitful, enabling NEWTON's discovery of the universal law of gravity.

2.4.4 The 1,400 Years after Ptolemy Saw Only Subtle Scientific Progress

Most of the fantastically long time between PTOLEMY's *Almagest* (c. 150 CE) and COPERNICUS' major opus *De Revolutionibus* (1543 CE) is classified as the Middle Ages. Indeed, this era of decline of astronomy in the West is wedged in the *middle* between learned antiquity and the beginnings of the scientific revolution in the early modern period. The label **Dark Ages** is sometimes used to describe the early medieval era, often in a derogatory sense. The original term *saeculum obscurum* is less negative and refers to an "obscure" era, mostly because significantly fewer documents have been written or survived from this than other periods. Indeed, while it can be argued that much has been achieved during in the Middle Ages, for the Early Middle Ages (fifth to tenth century CE), a decline in population, economic wealth, trade, and writings compared to earlier periods in the West is well documented. We cannot do justice to this interesting period in the few paragraphs we are able to dedicate to it. The multitude of medieval developments and discoveries were, though not spectacular, in sum the necessary ammunition in the hands of early modern scientists—used to sweep away false conceptions and assumptions of Aristotelian science and Ptolemaic astronomy.

In broad strokes a short history of the period goes like this. Due to the dramatic political, religious, and economic changes in Western Europe, interest in science and writing in general declined substantially, and many ancient scriptures were lost or forgotten. The rise of Islam around 700 CE saw the establishment of a vast Arab empire reaching from India to Portugal. The supplanting of Roman rulers and their barbaric successors made the scriptures located in these lands available to the Arabic cultures. They were studied, commented on, translated into Arabic, and therefore rescued. Arab scientists made important discoveries, in particular in mathematics. This Eastern wisdom was rediscovered in the Middle Ages by people in the West, and became (again) part of the Western tradition. Many classical works were translated from the Arabic into Latin and found renewed interest in the West. In particular, ARISTOTLE's philosophy was made compatible with Christian doctrine by THOMAS AQUINAS (1125–1174 CE) which gave rise to the powerful medieval method of study, learning, and discourse, namely *scholasticism*.

Starting in the twelfth century, precursors of modern universities developed from *monastic* schools in Italy, France, and elsewhere. There, *scientific ideas* could be openly discussed by scholars hailing from all parts of Europe.

Concept Practice

1. Which are the main "ingredients" of Ptolemaic astronomy?

   (a) Crystalline spheres
   (b) Homocentric spheres
   (c) Epicycles, eccentric circles and equants
   (d) Epicycles, ellipses, and equants
   (e) Major epicycles
   (f) None of the above

2. In which era did PTOLEMY live?

   (a) Classic Greece
   (b) Hellenistic Greece
   (c) Roman Greece
   (d) Muslim Egypt
   (e) None of the above

3. Which pattern of nature does PTOLEMY's but not ARISTOTLE's model describe correctly?

   (a) Retrograde motion
   (b) Changing brightness of planets
   (c) Elliptic orbits
   (d) Two of the above
   (e) None of the above

2.5 Light at Last: Copernicus's Heliocentric Universe

2.5.1 What Made Copernicus Possible in the Sixteenth, but not the Thirteenth Century?

The dark picture we painted of the scant scientific progress between PTOLEMY and COPERNICUS has to be taken with a *grain of salt*. Sure, the landscape of scientific methods and problems changed so little during this long time, that PTOLEMY, time traveling to the sixteenth century, would have been "completely comfortable with the problems addressed by contemporary astronomers and the conceptual framework within which they worked" [7]. Nonetheless, important inventions (plough, gunpowder, mechanical clocks, spectacles) were made. Moreover, the *metaphysical* interpretation of the Aristotelian and Ptolemaic model of the world shifted. Instead of thinking of the cosmos as organic and purposeful, it was increasingly seen as lifeless and mechanical [27].

While ARISTOTLE's was a comprehensive description of nature in virtually all of its facets, cracks started to appear in this Aristotelian-Ptolemaic world view. Scholastic philosophers worked by breaking up texts into small pieces, discussing and commenting on every detail. It soon became clear that Aristotle's homocentric spheres and PTOLEMY's epicycles did not fit all that well together. As we have pointed out earlier, they weren't supposed to. Nonetheless, these inconsistencies in the work of *the philosopher*, as ARISTOTLE was called with awe in the High Middle Ages, startled the scholastic scientists. They planted a seed of doubt, and over the next centuries, important questions were raised as to what was and what wasn't acceptable in ARISTOTLE's teachings. Indeed, there was an increasing collection of patterns in nature for which explanations by ARISTOTLE and his followers did not hold up under scrutiny, like projectile motion or the alleged impossibility of a vacuum. Those discrepancies couldn't be swept under the carpet in an era that used routinely technology not available in antiquity, for example cannons and underground mining.

As an important example that triggered important revisions of ARISTOTLE's teachings, consider simple projectile motion. An object, say a stone, is thrown vertically upward. Nowadays we would say it moves under the influence of Earth's gravity which results, near its surface, in a constant acceleration. However, concepts like *gravity* and *acceleration* were not *invented* until centuries later. So from ARISTOTLE to the late Middle Ages people had a hard time explaining why the stone keeps on rising until it reaches a maximal altitude without being pushed. According to ARISTOTLE, the natural place of the stone (made out of the element *earth*) is the center of the Earth, whither it should fall. Initially, it is forced upward by the thrower's hand, in a motion considered *unnatural* by ARISTOTLE, but why does it rise after it leaves the hand? This is not an *academic question* because it is the prototype for any kind of motion, and because it has important applications, for example to aim a cannon at military targets. ARISTOTLE himself came up with an awkward explanation. He held that the air under the stone would keep pushing it. As JEAN BURIDAN (c. 1295–1363), rector of the University of Paris, pointed out, this makes little sense, since air should *resist* the motion as the stone moves through it. He instead came up with the concept of *impetus*. Basically, *impetus* is the ability of an object to move, and is transferred by the hand to the stone. Eventually, the impetus is used up, which is why the stone slows down its upward motion, stops, and falls down. The concept is a clumsy precursor of the modern *momentum* (*mass* times *velocity*) and therefore an important idea. Although *impetus* itself had to be discarded, it was used for centuries; COPERNICUS around 1500 and even GALILEO around 1600 studied the concept at their universities of Padua and Pisa, respectively. More importantly, it helped to put the idea in people's minds that there exists a dynamic property of an object, a ***quantity of motion*** as NEWTON called it centuries later, that determines how it moves. In modern terms we would say that the ***momentum*** of the object becomes smaller due to a ***force*** acting on it for a time.

We see that scientific progress was made even though the new idea (impetus) is only partially correct. By distilling an idea, that is, focusing on its useful parts or connotations, we can arrive at a better description of nature without being completely right. Indeed, BURIDAN's student NICOLE ORESME (1325–1382) came closer to the modern understanding of motion as follows. BURIDAN had considered the possibility of a rotating Earth, because this would handily explain why all celestial objects display daily motion.[27] A rotating Earth in a stationary universe is more natural than a whole universe rotating around a stationary Earth. However, there were several arguments why the Earth had to be stationary in the Aristotelian cosmos. One of them was disarmed by ORESME and led to a new view of the universe. The argument is as plausible as it is Aristotelian and wrong: If the Earth rotated, then a stone thrown straight up would land west of the thrower, because while it is in flight, the Earth rotates eastward. BURIDAN succumbed to the argument and rejected the idea of a rotating Earth. His student ORESME took BURIDAN's impetus theory and applied it correctly, that is, also to the horizontal direction, as one should for consistency. The stone gets a vertical impetus from the thrower, and a horizontal one from the Earth, hence it and the Earth would travel eastward, and no horizontal relative motion would occur; see Figure 2.14. This is close to the modern view, which simply notes that since no horizontal force acts on the stone, no horizontal velocity change[28] occurs. The stone keeps traveling eastward with the Earth, and therefore appears to move only vertically for the thrower who also travels eastward due to the Earth's rotation. The moral here is that inventing a theory and (correctly) applying it are two things, and sometimes it takes two minds and much time to bring a idea to fruition.

Many small advances like BURIDAN's and ORESME's were necessary to pave the way for COPERNICUS. However, the most important revolution which arguably made all others possible[29] might have been the invention of the ***printing press*** by JOHANNES GUTENBERG (c. 1400–1468) in Mainz, Germany, around 1440 CE. Before this invention, manuscripts[30] had to be reproduced by hand in small numbers, inviting copying mistakes. Mass production of books and pamphlets became possible and was commonplace all over Europe within a few decades. Scientific progress by building on the work of others became much easier. Details in publications could be cited reliably, because it was guaranteed that a copy of a book in Paris was exactly the same as the copy in Rome.

---

[27]Incidentally, this is a nice example of ***Occam's razor***: Isn't it simpler to account for the apparently rotating sky by assuming *one* rotation (of the Earth) rather than *thousands* of rotations (of the stars)? Of course, the Aristotelian would say that only *one* rotating (celestial) sphere is necessary, but the assertion that this sphere exists is an *additional* assumption.

[28]A change in velocity is proportional to the stone's acceleration.

[29]Certainly COPERNICUS's knowledge of ARISTARCHUS's heliocentric ideas and the ***dissemination*** of Protestant ideas through flyers and pamphlets was due to the invention of the printing press.

[30]Latin: *manus = hand, scribere = to write*.

2.5. LIGHT AT LAST: COPERNICUS'S HELIOCENTRIC UNIVERSE

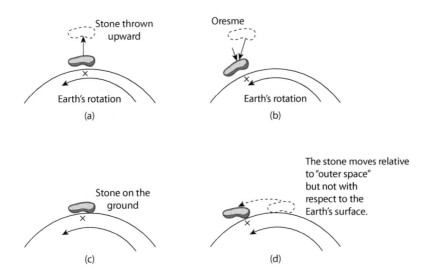

Figure 2.14: **Oresme's argument showing the possibility of a rotating Earth.** ORESME countered an old Aristotelian argument that the rotation of the Earth would lead to predictions in contradiction with observations. A stone is thrown vertically upward and lands at the same place on Earth (X) from where it was thrown. The Aristotelian argument assumed that if the Earth rotated eastward during the time the stone is in the air, it would land west of the launch position. ORESME applied BURIDAN's medieval *impetus* theory to both the vertical and the horizontal motion of the stone. As the Earth, the stone moves eastward—whether moving vertically or not. Thus, relative to the Earth's surface, it does not move horizontally at all.

2.5.2 Copernicus' *De Revolutionibus* Is a Conservative Book That Started a Revolution

NICOLAUS COPERNICUS (1473–1543), was born in Toruń, Poland. Being the protégé of an influential uncle with connection to the Polish king, he followed the academic traditions of the day and studied at Krakow and several Italian universities for over a decade. Studying theology, law, and medicine (the "big three" back then), and also mathematics and astronomy, he received a well-rounded liberal arts education. His diverse contributions to his community included reforming the local currency system, organizing a defense against the Teutonic Knights, and weighing in on the Gregorian calendar reform which eventually happened almost 40 years after his death; see Section 2.1. He lived a rather peaceful life as a lay **church canon** in an era that saw many changes and upheavals. When he was 21 years old CHRISTOPHER COLUMBUS (c. 1451–1503) discovered America, and when he was 38, MARTIN LUTHER (1483–1546) started the Protestant Reformation. Arguably, he himself started the *scientific revolution* on his deathbed, at age 70, after having finally agreed to publish his book *De Revolutionibus Orbium Coelestium* (*On the revolutions of the heavenly spheres*) that he had been working on for several decades. Colloquially speaking, in his lifetime the area of the known world doubled, as did the number of (Western) Christian church denominations, and the center of the universe shifted from the Earth to the sun.

Often COPERNICUS' work is summarized and thus reduced to the content of the last half sentence, that is, decreeing that the universe is **heliocentric**. COPERNICUS is said to have exorcised epicycles, to have laid out an accurate and efficient way to predict planetary positions, to have made the sun a fixed star, and to have proclaimed the infinitude of the heavens. None of this is true, as our close look at COPERNICUS will show. What remains true

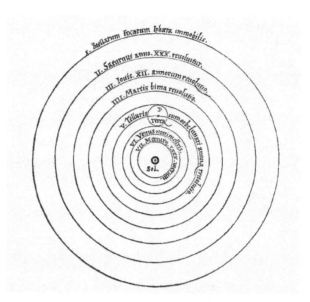

Figure 2.15: **The Copernican model of the cosmos.** The sun is at the center, only the moon orbits Earth, and the sphere of fixed stars is much larger than the outer planetary sphere of Saturn. Note that *Copernicus's model* predicts the relative sizes of the planets' paths, contrary to *Ptolemy*'s. Copernicus was aware of the fact that Jupiter and Saturn are much farther from the sun than the rest of the planets, contrary to what is apparent in the figure. The nonobservation of a parallax of the stars resulted in Copernicus's hypothesis that the stars are much farther away than previously believed.

is that Copernicus played a crucial role in expanding the universe. His major contribution was to enable his followers to revolutionize the way we think about the cosmos. He himself had rather conservative motives when he constructed his model of the universe, as we will see. Changing the way we think about the universe is, in some sense, synonymous with changing the universe. Indeed, the Aristotelian-Ptolemaic cosmos was a finite sphere of quintessence centered on the Earth. In the aftermath of the Copernican Revolution, the universe became sun-centered, vastly bigger, but mostly empty. Even an infinite universe became ponderable, which has, as a consequence, no center.

So let's set the record straight and separate what Copernicus did from what his followers did and his adversaries claimed he did. First off, Copernicus's *De Revolutionibus* is not a revolutionary book, yet it eventually started a revolution. Its title is often misunderstood in two ways. *Revolutionibus*[31] is better translated as *rotation*, although our word *revolution* comes from the connotation of an upside-down turning of old values and beliefs—as in a rotation. Secondly, *orbis* is not to be confused with our modern concept of *orbit*, but appears in the Aristotelian sense as a sphere. Still, Copernicus is assuming the planets are moving on circles around the sun, and thus is tacitly introducing the concept of an orbit.

The paradoxical nature of Copernicus' book stems from the discrepancy between his motivation to reform the monstrous theory of the Ptolemaists, and the reception of the book as a completely different way to think about the universe. In Section 2.4 we have seen that Ptolemy's *Almagest* was an elaborate calculational scheme to predict, with reasonable accuracy, the positions of the planets. Ptolemy used geometrical constructs like epicycles, minor epicycles, and the equant to achieve his goal, namely to **save the appearances**.

---

[31] I left the word in the Latin *ablative* here for clarity.

This branch of **mathematical astronomy** had grown, with additions and modifications of Arabic and medieval scholars, into a tree of different schemes, that were all based on PTOLEMY, but all different. There was no standard, and at the time of COPERNICUS this collection of theories could easily be perceived as a monstrosity. Many scientists had been working diligently for over a millennium, and yet no clear solution of the **planetary problem** had been achieved. COPERNICUS felt emboldened to think outside the box, because the box didn't seem to contain the solution. It was one Ptolemaic construct in particular that earned COPERNICUS' scorn due to its arbitrariness. COPERNICUS was offended enough by the **equant** to start his reformation that was called a revolution. It is tempting to draw a parallel to the Protestant Reformation that happened at the same time. MARTIN LUTHER, COPERNICUS' junior by only ten years, was disgusted enough with the Catholic Church's sale of indulgences to start his revolution that was called a reformation. And just as LUTHER sought to reform some minor church doctrines and divided the entire church, COPERNICUS sought to remedy some technical problems of the Ptolemaic system, and brought down the entire Aristotelian way of doing science and thinking about the cosmos. Though COPERNICUS' book was written in the medieval scientific tradition, it became "the source of a new tradition that ultimately destroys its parent." [26]

Concretely, COPERNICUS based his new theory of the universe on the following assumptions; see also Figure 2.15:

1. All celestial objects except the moon rotate around the sun, which is essentially the center of the universe (heliocentrism); only the moon rotates around the Earth.

2. The daily motion of the celestial sphere (stars) is a result of the rotation of the Earth around its axis; the celestial sphere does not rotate.

3. The distance between the Earth and the sun is very small compared to the distance to the stars.

While these three assumptions explain the main features of planetary motion, it should be noted that others are necessary to elevate the accuracy of the model to an acceptable level. The third assumption is necessary to explain why the yearly **stellar parallax** is too small to be observed at the time.

From the assumptions it follows that the apparent motion of the sun results from the rotation of the Earth around its axis together with its rotation around the sun. This physically explains the discrepancy between **solar** and **sidereal day**. In Section 1.3.3 we had argued that the seasonal motion of the sun among the stars causes it. In *Copernicus's model*, the seasonal motion of the sun is due to the orbital motion of the Earth around the sun. Since the Earth moves about a degree per day along its orbit, it has to rotate around its axis about 361° so that the sun appears at the meridian the next day. Since the Earth rotates 15° per hour, it takes $\frac{60\,\text{min}}{15} = 4$ minutes to rotate one degree. The apparent *retrograde motion* of the planets is the consequence of a **moving observer** on Earth passing the superior planets (Mars, Jupiter, Saturn) or being passed by the inferior planets (Mercury and Venus), Figure 2.16. This beautifully simple explanation is what convinced the likes of KEPLER and GALILEO that heliocentrism is the way to go. Indeed, one of the key differences between the heliocentric and the geocentric models is that COPERNICUS' explains retrograde motion by just one assumption (# 1), whereas PTOLEMY has to introduce a major epicycle for each planet and individually size them up. In some sense, PTOLEMY had to devise a new theory for

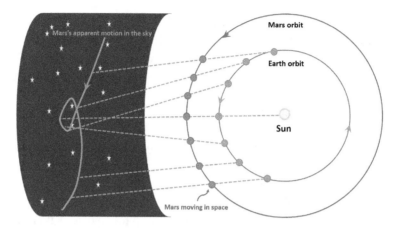

Figure 2.16: **Copernicus's explanation of retrograde motion.** In *Copernicus's heliocentric model* both the Earth with the observer, as well as the observed planet are moving, that is, rotating around the stationary sun. The retrograde motion of the planets is due to the *relative* motion of planet and observer. In the figure, Earth is closer to the sun and moving faster around the sun than Mars. When Mars is overtaken by Earth, it *appears* as if Mars is moving opposite to its usual, eastward motion. The complicated *apparent* motion of Mars is elegantly explained by *two* simple motions of the two planets involved.

each planet. As strange as it may sound in a scientific debate, COPERNICUS wins on *aesthetic* grounds. The fact that *all* retrograde motions follow from *one* assumption is simply *beautiful*. While it is hard to define *beauty* as a criterion, it plays an important role to this day in the construction of scientific theories. Often beauty is synonymous with simplicity and symmetry, and sometimes scientists will call a beautiful theory the most *natural*. Keep in mind, though, that in COPERNICUS' times, it was decidedly *unnatural* to assume a moving Earth. As we have pointed out before, a stationary Earth is what we glean from our senses; we are quite unaware that we rush with 30 km/s through space, and rotate with up to 40,000km/day around the Earths axis. It seems obvious to us now (because it was pounded into us in school) but it is hard to devise an experiment that **positively** shows that the Earth is orbiting the sun (Bessel's parallax 1838) and rotating (Foucault's pendulum 1851). Therefore, COPERNICUS' opponents used the apparent lack of evidence for the motion of Earth to dismiss and ridicule his heliocentric model. At the time, a moving Earth was simply against common sense. COPERNICUS audacity to nonetheless abandon a stationary Earth and to forge a more *pleasing* theory of planetary motion is what made him a revolutionary scientist. If you look at Figure 2.16, it will hardly escape you how elegant and efficient COPERNICUS' explanation of retrograde motion is. As the inner, faster planet overtakes the outer, slower planet, the slower planet seems to move back in front of the background stars for an observer on the faster planet (here: Earth). Also note that the middle of the retrograde loop is precisely when the planets are closest together, which explains the change in brightness of the planets; see Section 1.6. So COPERNICUS' model is a real, physical explanation of retrograde motion. In this sense he is on the side of the physicist ARISTOTLE, not the mathematician PTOLEMY. What we *perceive* in the night sky is a loop of the planet among the fixed stars in the sky. We contrast this with PTOLEMY's philosophy. He devises a mathematical scheme that reproduces the weird loop in the sky. It is unlikely that PTOLEMY or his followers believed that the planet is *indeed* swirling around fixed to a circle on a circle. It is unlikely because Ptolemy's scheme is relative, not absolute, contrary to COPERNICUS'. We had pointed out that PTOLEMY could—and had to—fix the size of his epicycles and deferents by hand. Any size is possible, as long as one gets the **ratio** of the epicycle to deferent radius correct. In principle, one could make Mercury's deferent larger than Saturn's and still **save the appearances**! This arbitrariness is gone in COPERNICUS's model, and therefore a good argument for the latter. COPERNICUS could—and wanted to—determine unambiguously the sizes of the planet's orbits. This was real progress! Consider the simplest determination of Venus's orbit, Figure 2.13. It is merely a matter of basic trigonometry to determine the unknown side of the right triangle formed by the Earth, sun, and Venus at maximal elongation, since the angle is known from observation. For the exterior planets a similar scheme is used, in which the **quadrature** (right angle between sun and planet) is used instead of maximal elongation, and compared to the configuration at opposition (180° between sun and planet), see Exercise 2.17. COPERNICUS was more of a theorist than an observer himself, but he did exceedingly well with the data he could glean: The largest discrepancy between his and the modern value (for Saturn's orbital radius) is only 4% [7]. The following comments should dispel some common myths about COPERNICUS' *magnum opus*.

1. COPERNICUS explains retrograde motion with the relative motion of the observer on Earth and the observed planet. He thus does not need major epicycles in his theory.

2. COPERNICUS needs minor epicycles in his theory to explain things like the different lengths of the seasons and the changing speed of the moon along its orbit.

3. Minor epicycles complicate COPERNICUS's theory to the point where it is only marginally simpler and only slightly more accurate than Ptolemy's theory. The reason for the complications is that COPERNICUS, in classical tradition, held steadfastly that the planets move in circles around the sun. In fact, they move in ellipses. This was not discovered until more than sixty years after COPERNICUS' death.

4. The Copernican and Ptolemaic systems seem to differ much more in theory than they do in practice. Both are rather clumsy calculational schemes to predict the positions of celestial objects. Both do so with only fair accuracy. On the basis of the naked-eye observations made up to COPERNICUS' death, there was **no way to decide** whether the helio- or the geocentric theory is correct. However, the Copernican system *was* a tad easier to apply, and was used by many astronomers (and astrologers) *even* if they were opposed to heliocentrism.

5. The reception of the book was surprisingly slow; it was first revered for its observational data and tables [6].

6. The controversy about the heliocentric system was anticipated by its author but much delayed, and not a science-versus-religion debate. COPERNICUS' *De Revolutionibus* was not put on the Index of forbidden books until 1616, although it remained there for two centuries. As a corollary, the Protestant as well as the Catholic Church were opposed to COPERNICUS' heliocentrism at the height of the conflict.

2.5.3 The Reception of and Reaction to Copernicus' Book Was Very Slow

You might think that a book that changed the universe in such a dramatic way as COPERNICUS' *De Revolutionibus* must have caused an instant sensation or scandal—but it didn't.

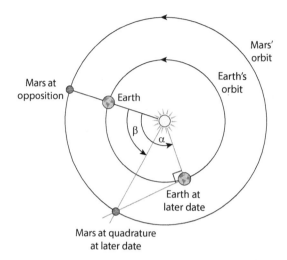

Figure 2.17: **Determining the size of Mars's orbit.** Two separate configuration of Earth and Mars are used. At *opposition*, Mars, Earth and the sun form a straight line, Mars being opposite of the sun in the sky. At a later date, Mars is at *quadrature*, that is, at right angles in the sky with respect to the sun. Between the two dates, Mars has moved an angle β, and the Earth a larger angle α.

As always when something contradicts our predictions or assumptions, we should ask why. There are several reasons. For one, COPERNICUS wrote his book for the experts only. "Mathematics is for mathematicians" is what he says in the preface of his book dedicated to pope PAUL III (1468–1549). While the group of experts in mathematical astronomy included both astronomers and astrologers, imagine how small of a group this must have been! Few people could read at the time, even fewer could understand Latin, and the mathematical expertise necessary to understand this very technical text was immense and mastered by very few indeed. This was quite the intent of the author, and probably a blessing for the book's reception. In the same vein, a controversial anonymous foreword was included in the first printing, which emphasized the mathematical, and therefore nonphysical, nature of COPERNICUS' framework. In other words, it was suggested that COPERNICUS' book was not about the cosmos itself, but merely a mathematical scheme by which one could predict where the planets are to be found in the sky. It was later discovered by KEPLER that this foreword was included without COPERNICUS' consent by ANDREAS OSIANDER (1498–1552). By paying lip service to the foreword, astronomers could then use COPERNICUS's mathematical scheme without having to buy into its dramatic consequences, for example the moving Earth. And using it they did, because COPERNICUS book was *useful*. It was immediately agreed by the experts of the day, who considered Copernicus as the leading astronomer of the first half of the sixteenth century, that it was on par with PTOLEMY's *Almagest*. Written 1,400 years later, it contained 1,400 years' worth of observations, which made its data tables superior to PTOLEMY's.[32] COPERNICUS' calculational scheme was slightly easier and slightly more accurate than PTOLEMY's, so it became somewhat silly to use the older scheme. Indeed, one of the "early adopters," GEORG JOACHIM RHETICUS (1514–1576), produced with his book *Narratio Prima (First Account)* of 1540(!) an "executive summary" of COPERNICUS' algorithm, and ERASMUS REINHOLD (1511–1553) a new standard of **ephemeris tables**.[33] The latter *Prutenic Tables* were much

---

[32] Of course, 1,400 years' worth of errors and copying mistakes also found their way into the book.

[33] Ephemeris tables list daily or monthly positions of celestial objects.

superior to the medieval *Alphonsine Tables* (1272) and widely used even by people in strong opposition to a moving Earth. This use of *De Revolutionibus* by a small group of experts went on for decades, and made it indispensable. Its methods were being taught at the universities, for example by MICHAEL MAESTLIN (1550–1631), KEPLER's teacher, in Tübingen.

If you use someone's algorithm today, you agree to the *terms of use*. The tacit acknowledgment that COPERNICUS' "users" agreed to was that a moving Earth is thinkable. COPERNICUS' universe was so dramatically different from the Aristotelian-Ptolemaic cosmos that it took time to change the universe in the minds of the people. The universe became heliocentric in a battle that lasted over a century. Due to a lack of **positive evidence** for either side before the advent of the telescope, it became a matter of belief which of the two major world systems you subscribed to. Oddly enough, at the same time it was literally a matter of belief which of the two major Christian churches you subscribed to. In the heat of the struggle, often local authorities decided both questions by law.[34]

Here, then, is the next paradox: How did the *academic question* whether the Earth or the sun is at the center of the universe become of interest for kings, rulers, popes, and therefore everyone? For starters, whenever the topic came up among nonexperts, COPERNICUS and his anti-common-sense scheme that includes a moving Earth was ridiculed. This did much to turn the public opinion against COPERNICUS, since these texts of "poets and popularizers" [26] were read much more widely than scientific writings for experts. At the time to appeal to common sense itself, aided by an Aristotelian world view, was enough to make COPERNICUS's moving Earth seem implausible. Put bluntly, people simply said: We do not feel nor see that the Earth is moving, hence it must be stationary. For the churches the Copernican problem was long a minor one. Surprisingly, it was initially the Protestant church that was anti-Copernican. MARTIN LUTHER, and later his collaborator PHILIP MELANCHTON (1497–1560), denounced the Copernican heliocentrism even before *De Revolutionibus* was printed. As everyone else, they were led astray by the contemporary Aristotelian common sense. Furthermore, they sought refuge and arguments in a literal interpretation of the bible. This is understandable. In the newly founded Protestant church, and supported by the **humanist movement**, the bible in its original and exact meaning was seen as the main line of defense against the Catholic Church. Having separated from much of the mother church's inheritance and wisdom of the church fathers and scholars, the Protestants could only rely on the holy scriptures as God's own word for support of their new interpretation of the Christian faith. Quite out of habit, the main argument of these church men became therefore to find passages in the bible that contradict COPERNICUS. The prime example is Joshua 10:12 and 13, where Joshua orders the sun to stand still, and not the Earth. Hence, in their opinion, the Earth must have been stationary already. One can interpret every text literally or metaphorically, and when the Copernican theory was accepted eventually, much work was done to adapt church doctrine accordingly. The Catholic anti-Copernican sanctions were much fiercer, it has to be said, but so delayed that their cause was almost immediately lost. Indeed, at the time when COPERNICUS' book was put on the ***Index*** in 1616, GALILEO had already falsified Ptolemy's geocentrism, and KEPLER's correct treatment of planetary motion with ellipses had already eclipsed COPERNICUS' theory as well, as we shall see.

---

[34] *Cuius regio, eius religio* (As the region, so the religion) was the "compromise." If you didn't like the religion of your country or ruler, you were free to leave.

Concept Practice

1. Which model was simpler?
   (a) Copernicus's was slightly simpler.
   (b) Copernicus's was much simpler.
   (c) Ptolemy's was slightly simpler.
   (d) Ptolemy's was much simpler.
   (e) Both were equally simple.

2. Which model was more accurate?
   (a) Copernicus's model (b) Ptolemy's model (c) Neither

3. Which model used epicycles?
   (a) Copernicus's model (b) Ptolemy's model (c) Neither

4. Which model used major epicycles?
   (a) Copernicus's model (b) Ptolemy's model (c) Neither (d) Both

5. Which model used equants?
   (a) Copernicus's model (b) Ptolemy's model (c) Neither (d) Both

6. Which model explained retrograde motion?
   (a) Copernicus's model (b) Ptolemy's model (c) Neither (d) Both

7. Which universe was bigger?
   (a) Copernicus's model (b) Ptolemy's model (c) Neither

2.6 Observing the Universe: Tycho Brahe

Danish Renaissance man TYCHO BRAHE (1546–1601) is the first of a trio of scholars representing the three pillars of the scientific method: From **data** collected by observation or experimentation, a mathematical description of the data is inferred in the form of a **hypothesis**, which is then tested and refined as necessary to form a *tentatively accepted theory*. TYCHO, the eminent astronomer of his time and most accurate naked-eye observer in history, provided the **raw data** to finally solve the **problem of the planets**. The positions of stars and planets as measured by TYCHO became the indispensable foundation upon which his colleague Johannes KEPLER (1571–1630) erected his **phenomenological** description of the solar system. KEPLER's *description* of planetary motion in turn enabled Isaac NEWTON (1643–1727) to find the very reason why the planets should move in the complicated way KEPLER had prescribed for them. NEWTONs *theory of universal gravitation* was the crowning achievement and solution[35] of the problem of the planets.

2.6.1 Tycho Sets New Standards for Astronomical Observation

To appreciate the novelty of TYCHO's way to do astronomy, we will take a top-down approach. Namely, we will ask, reversing the chronological order, how he became the observational genius who set a high standard for any future astronomer. We start with the sentence that is typically used to characterize TYCHO: The guy that provided accurate positions of the planets for KEPLER. While not entirely wrong, this summary diminishes TYCHO's role in the history of astronomy, and, worse, gives a wrong impression as to how science is done. Indeed, TYCHO did not, with a stroke of a genius, suddenly produce data ten times (a whole order

---

[35]It was tentative in the sense that Einstein's theory of gravity, *general relativity*, superseded it in 1915.

2.6. OBSERVING THE UNIVERSE: TYCHO BRAHE

Figure 2.18: **Tycho's great mural quadrant and the art of measuring.** Tycho is seen standing and point to a hole in the whole through which starlight falls unto the quadrant. An assistant at F is reading the altitude angle off the quadrant's finely graded scale. In the background we see Tycho's other instruments, a conference room, and an underground alchemy laboratory. Note the assistant who is reading off the time of the star's transit on the mechanical clock.

of magnitude!) more accurate than anyone had done before him. Scientific achievements of that caliber are almost always the fruits of a long period of intense labor and effort. Being an exceptionally gifted person is only necessary, but not sufficient to secure success.[36] If you take into consideration that the telescope was not used until almost a decade after TYCHO's death, a perplexing question arises. How *was* TYCHO able to produce, with essentially the same type of instruments that others had used for millennia, such accurate data?

The simple answer is—money! He was able to garner the support of the Danish king FREDERICK II (1534–1588) who generously supported TYCHO's research with what is estimated to be as much as 5% of the Danish gross national product (GNP) at the time. Imagine what scientists could do today if the United States pumped 5% of its GNP into a single science project![37] This is a surprising amount of money, which begs the question: Why would a king hand so much money to an astronomer working on the esoteric *problem of the planets*? The answer to our question yielded thus another question—as is often the case in science. We

---

[36]Unfortunately, as Max Weber points out [43], neither is hard work nor dedication nor talent able to guarantee success; in fact, nothing can.

[37]For comparison, the Manhattan Project that produced the atomic bomb was 0.14% of the United States's GNP, and the Apollo program that put man on the moon was (at its height in 1966) 0.75% of GNP.

hold off answering for now, since a more immediate question arises from the fact that money does not buy data. How did TYCHO turn money into data? He build, on the island *Hven* that FREDERICK II gave him as an estate, a magnificent castle dedicated to astronomy: Uraniborg[38]. He erected observational instruments like his famous **mural quadrant**, Figure 2.18 that were crucially improved, and often much bigger and more stable versions of instruments used in centuries past. TYCHO had gained expertise and observing experience by traveling and visiting other astronomers and observatories. In 1576 he was ready to make the most accurate naked-eye measurements of the heavens. The support of the king allowed him to enlist the help of several able assistants to help him with his observations. Several of them later became eminent astronomers of their own right. He even had a paper mill, a printing office, a farm, and fishponds on his estate [29]! In sum, he established a new, professional way to do astronomy. It was firmly based on *repeated* and *independent* observations made with *different instruments* and a thorough understanding of the *errors* and *uncertainties* of the method and instrumentation employed. This seems common sense to us today, but it was a novel, cutting-edge approach distilled from decades of observing, trial-and-error, collaborations and discussions with other observers, most notable the landgrave of Hesse-Kassel, WILLIAM IV (1532–1592). Indeed, the fundamental role of empirical data came to be realized surprisingly late in the development of modern science. Remember that PLATO was highly suspicious of empirical knowledge. Even half a century after TYCHO, "Mr. Cogito-ergo-sum"[39] René DESCARTES (1594–1650) favored *ratio*, that is, reason, over data. The latter were merely impressions of the fallible senses of man in DESCARTES's view. Be this as it may, producing superb *data* was life's calling for TYCHO, and it changed the *modus operandi* of astronomy irrevocably.

Alas, not just the way astronomy was done had changed, but with the new quality of data came new questions—and more work to be done. If you are measuring the patterns in nature with unprecedented accuracy, you will discover new patterns.[40] Unfortunately, you may also have to take effects into account that could alter your data if aiming at higher accuracy. One such effect that TYCHO was aware of but had no proper way of taking into account was **atmospheric refraction**.[41] Light rays falling through the transparent atmosphere will be bent or refracted, as if the atmosphere were a gigantic lens, Figure 2.19(a).

But back to our question: What *was* the reason that FREDERICK II endowed TYCHO with so much money? It was a perceived importance of planetary astronomy for politics. Namely, its power to predict planetary positions was revered due to its astrological implications. Back then, many astronomers cast **horoscopes** for the nobility to earn money. Many believed, for lack of better knowledge or superstition, that celestial objects exert an influence on humans and their fate. This has since been soundly refuted, but you have to remember that many correlations between patterns in nature were not understood at all back then. The king could simply not afford to lose an astronomical heavyweight like TYCHO by seeing him settle abroad, and decided to keep him in Denmark with his generous offer of an island and loads of money to finance his research.

---

[38]Recall the muse of astronomy, *Urania*, and *borg* as the Germanic word for castle, as in *Pittsburgh*.

[39]I think, therefore I am (sure that I exist).

[40]For instance, using a telescope instead of the unaided eye, you will literally see new patterns—many stars that were invisible, craters on the moon, and so forth.

[41]This was later achieved by KEPLER, who needed to know where the planets "really were" when TYCHO measured their positions as they *appeared* when observing the cosmos through the turbulent atmosphere.

2.6. OBSERVING THE UNIVERSE: TYCHO BRAHE

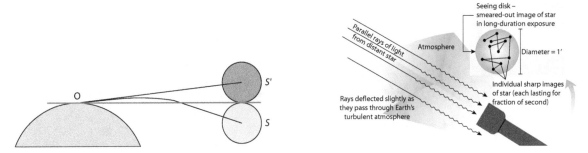

Figure 2.19: **Atmospheric refraction and seeing.** (a) Atmospheric refraction. The atmosphere bends light. As a consequence, celestial objects appear higher in the sky than they are. The effect is largest at the horizon and zero at the zenith. (b) Seeing is a term astronomers use to describe the overall quality of the observing conditions. Good seeing refers to calm, clear skies when minute details are discernible. Bad seeing means that turbulence, haze and aerosols in the atmosphere blur out details. The perfectly parallel starlight gets distorted by the turbulent air which results in a *seeing disk* of about 1 arcsecond, limiting the resolution of *ground-based* observing. This is also the reason why stars seem to *twinkle*: It is not the star which twinkles, but the atmosphere which disturbs its light so it *seems* to twinkle.

2.6.2 Tycho Falsifies Aristotle

How did Tycho become the famous astronomer worthy of the king's support? It didn't hurt that he was born into an influential family, with ties on both sides to the upper echelon of Danish nobility and clergy. Of interest for us are, of course, mainly his astronomical merits. And here it was in particular the discovery of a new star in the constellation Cassiopeia in November 1572 that made him famous. His account of the discovery, *De Nova Stella (On the new star)*, gave rise to the name *nova* for this pattern in nature, that is, a star changing its **brightness**, and, indeed, its **luminosity**, so dramatically that it seems to suddenly appear and then fade away over the course of several months. Now, why would someone become famous just because he discovered a new star? The answer is that not the discovery itself, but the subsequent research and interpretation of the discovery garnered the credit, as it is often the case in science. The more surprising the discovery, and the more dramatic the consequences and explanation, the greater the fame. In Tycho's case the astonishment of the star's appearance came from the fact that in the Aristotelian cosmos was thought to be unchanging. The heavens were made out of unchanging, perfect material, the **quintessence**, and the only thing "quintessential" objects could do is to rotate uniformly. For an Aristotelian, therefore, this new star had to be part of the changing **sublunar sphere**, and could not be part of the eternally constant **superlunar sphere**—problem solved! Tycho, however, was able to show that this new object did not have a **parallax**, that is, its position relative to other fixed stars did not change when observed from different positions on Earth, [42] and therefore it had to be very far away. Indeed, the moon and objects closer than the moon exhibit a small change of position in front of the background of fixed stars that can be observed with the naked eye, see Figure 2.2(c). Since the new star did not have a parallax, its distance must therefore be comparable to that of the distant background stars. After duly checking his reasoning and data, he published *De Nova Stella* the next year, and was not afraid to openly state that Aristotle's teachings had been falsified. Clearly, this discovery and its interpretation against all odds were "big thing." Too, this was not a case of getting lucky and

---

[42] Or at different times, when the observer (or the sky) is "transported" due to rotation to a different position.

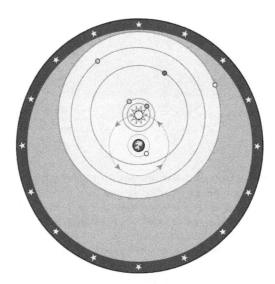

Figure 2.20: **Tycho's model of the cosmos.** Tycho's model was published in 1588 in his book *De mundi aetheri recentioribus phaenomenis (Phenomena of the aetheric world recently observed)*. All planets except the Earth rotate around the sun, which rotates around the Earth. Note that the stars are close to the planets, because a parallax is avoided due to the Earth being stationary at the center of the universe. Also note that the paths of the sun and Mars intersect. Thus, the crystalline spheres cannot be impenetrable—if they exist at all.

looking up as something happens in the sky. To get a sense how hard Tycho worked *after* the discovery, consider that he observed the brightness of the new star over several months until it faded and became unobservable. So detailed were his recordings that centuries later German astronomer Walter Baade (1893–1960) used Tycho's data to reclassify the new star as a **Type I supernova**.[43] As if to prove himself right, Tycho repeated his feat and independently falsified Aristotle's teachings by showing that also the great comet of 1577 did not show a parallax and therefore was beyond the moon: A different type of object—appearing and disappearing—heralded change in the heavens contrary to Aristotle. Two implications of this discovery are important, because they signify the slow, but inevitable path toward modernity in the sciences. First, it was realized how important *independent* evidence is. Not only did two different patterns in nature (new star *and* comet) point to the same conclusion that Aristotle is at least partially wrong, but both patterns were independently verified by other observers, thus ruling out any human factors. Second, the comet was not just appearing but moving with respect to the stars and planets. This convinced Tycho that the crystalline spheres of Aristotle carrying the planets were not real, or at least not hard nor impenetrable. This led eventually to a very modern question: If not the spheres, what carries the planets, what makes them move? It is modern, because it brings physics into play. No longer were astronomers satisfied with a **mathematical** description of the heavenly motions, they also needed to see a **physical** realization, a mechanism by which the cosmos functions. These are the seeds of the mechanical universe constructed by Isaac Newton.

---

[43] We will learn about this violent end of a star's life in Chapter 5.

2.6.3 Tycho's Copernico-Ptolemaic Model of the Universe

As much as his achievements paved the way to modernity, BRAHE was traditional, even conservative in his scientific views. For instance, while he saw that the Ptolemaic model was flawed, he disliked the Copernican heliocentrism, because its successes were bought at the prize of a moving Earth. Others, like ERASMUS REINHOLD, had pointed out that the motion of the Earth in *Copernicus's model* is not essential. All geometric relations between the planets can be preserved with a fixed Earth. Today we would say that the solar system can be described in a frame of reference centered on the Earth, since all motion is relative to a reference frame. Inspired by this, TYCHO came out in 1588 with his own model of the solar system, Figure 2.20. It is quite literally a blend of the best of COPERNICUS and PTOLEMY: The planets move around the sun, which moves around the stationary Earth. All the advantages of heliocentrism in a geocentric model! While more of a historic curiosity, you can immediately see the psychological benefits for a generation of astronomers during the transition from a medieval to a modern view of the universe. Now astronomers could use the Copernican methods in good faith. This was important because the implications of a moving Earth for religion and science were grave and had to be dealt with slowly before Copernicanism could become the world view accepted by most. In this sense—inadvertently and unintentionally—TYCHO's model prepared the way for COPERNICUS's acceptance and astronomy's march toward modernity.

To understand the reasons for the dramatic change of observational astronomy inaugurated by TYCHO, we might finally ask about his motives. Why did he choose to give up a career in law, contrary to his family's wishes, to dedicate his life to the heavens? The initial impulse came when he was a teenager and observed a partial solar eclipse. Not so much the event itself, but its predictability influenced him. It inclined the young mind toward a science that can measure and understand the motions of the heavenly bodies so well as to predict their movements decades into the future. At a time when so little about patterns in nature was known, and chaos and unpredictability seemed to reign the Earth, it must have been amazing to think that man could understand the heavens and prescribe the paths of celestial objects! This inspiration turned into diligence, because TYCHO became convinced that understanding comes from dedicated observation. Monument to his dedication is his star catalog of about one thousand stars. Their positions were measured with unprecedented accuracy of about one minute of arc, or 1/30th of the apparent moon diameter. Note that this was the first observational star catalog since HIPPARCHUS's almost two thousand years earlier.

TYCHO himself ascribed the reason why he switched to astronomy to his dissatisfaction with the faulty prediction of the Jupiter-Saturn *conjunction* of 1563. This close encounter of the two planets had been predicted using the Ptolemaic *Alphonsine tables*, and the Copernican Prutenic astronomical tables. While the former was off by a month, the latter fared slightly better. TYCHO had observed the event with a simple ***cross-staff***,[44] yet his crude method was enough to realize that the problem of the planets was far from being solved, even in *Copernicus's model*. His own model, essentially a change of coordinates from COPERNICUS's didn't fare much better, and the search for a better description of planetary motion occupied TYCHO for the last years of his life.

---

[44] A simple instrument to measure angular separations in the sky; a precursor of the ***sextant***. A cross-bar was movable on a long graduated pole which explains the name.

2.6.4 Tycho and Kepler

Trying to understand how the "torch of astronomical wisdom" was passed on to the next generation in the person of JOHANNES KEPLER gives us an excuse to highlight some of TYCHO's flamboyant personality and life. Likely his less likable character traits ended his funding by the new Danish king Christian IV in 1597. Among them were irascibility and sensitiveness due to his nose, lost in a duel and replaced by a metal prosthesis. After leaving Denmark in 1597 with all his entourage and instruments, TYCHO made station in Hamburg, then part of the *Holy Roman Empire of the German Nation*. There he published his *Astronomiae instauratae mechanica (Instruments for the restoration of astronomy)*, a detailed and illustrated exposition of his instruments and observatories. Wisely he dedicated the book to Emperor RUDOLF II (1576–1612), and sure enough was offered the position of Imperial Mathematician. The offer included a castle near Prague, where TYCHO resumed his observations and, most importantly, enlisted KEPLER as his assistant. TYCHO was struggling to perfect his model of planetary motion, and had made extensive observations of the planet Mars in particular. He rightly considered the Martian motion as the hardest problem, and eventually handed it to his assistant KEPLER to solve it. TYCHO was otherwise carefully watching over his data[45] (and stingy with the money he paid KEPLER), but he couldn't have chosen a better data set for KEPLER to work on. Mars's orbit is the most eccentric of the superior planets, and so is best suited to falsify the assumption that planets move in perfect circles. Before the two eminent astronomers could clash due to their opposing personalities, BRAHE died, and KEPLER slyly secured TYCHO's data before his heirs could get their hands on them. This was very fortunate for progress in science, because KEPLER was arguably the only person able to put the data to use and to finally solve the *problem of the planets*.[46]

Concept Practice

1. What type of model was TYCHO's? (a) Heliocentric (b) Geocentric (c) Neither (d) Both
2. What faction did TYCHO belong to? (a) Copernican (b) Anti-Copernican (c) Neither (d) Both
3. Did TYCHO's model use major epicycles? (a) Yes (b) No

2.7 Describing the Universe: Kepler

JOHANNES KEPLER (1571–1630) was born in Weil der Stadt (Figure 2.21(a) shows his birthplace), a town in the Protestant part of Germany. He lived at a time when one's religion could easily become one's fate. During KEPLER's lifetime, the conflict between the Christian religions reached its climax and open war broke out, devastating Germany for thirty years (1618–1648). His life was at the very crossroads between medieval scholarship and modern science. It can be argued that, at least for astronomy, KEPLER pushed science into modernity. While TYCHO pulled science forward in one field (observational astronomy) and remained conservative otherwise, KEPLER stood by his **Pythagorean ideals**, whether they

---

[45] He was in a lengthy quarrel with previous Imperial Mathematician NICOLAUS REYMERS BAER (REYMARUS URSUS) (1551–1600) about who invented the "Tychonic" model; TYCHO claimed BAER had stolen papers of his.

[46] At least he definitively showed *how* the planets move; NEWTON then finished up the problem by explaining *why* they move in this way.

2.7. DESCRIBING THE UNIVERSE: KEPLER

Figure 2.21: **Kepler museum.** (a) KEPLER's birthplace in Weil der Stadt, Germany, now the *Kepler museum*. (b) Model of the *Mysterium Cosmographicum* at the museum shows KEPLER's construction of the cosmos following a Neoplatonic idea in his 1595 book. By nesting the five platonic solids into the crystalline spheres of the six known planets, he believed to have constructed a unique, mathematically harmonious rule to determine the distances of the planets from the sun. We now know that the distance pattern and the mathematical fact are completely unrelated, but the inferred rule works surprisingly well.

served him well or ill. His belief in simple mathematical truths or *harmonies* in nature and his deep religiosity were strong convictions that made it possible for him to decipher the laws of planetary motion. They saw him through almost a decade and thousands of pages worth of calculations to finally solve the problem of the planets. In other lines of research his *Pythagorean ideals* led him completely astray, and we find in his works most brilliant solutions and trite **numerology**. It has to be said that the latter condemning judgment is due mostly to hindsight; in KEPLER's times, his ideas were entirely reasonable, because it was unknown which facts could be explained or related, and which were unrelated or accidental. Thus, KEPLER had "one foot firmly planted in the medieval past, the other in the scientific future" [7]. We might add that his body and mind were tormented in a terrible present. For instance, religious fervor and fanaticism resulted in **witch trials**, which his mother was subjected to. He himself remained an outcast as a protestant in largely Catholic lands, became displaced during the counter-reformation that recatholicized Austria, and his position as protestant Imperial Mathematician to a Catholic emperor was precarious. He was even cut off from his Protestant roots by being excluded from the Protestant Holy Communion for refusal to accept certain church doctrines. Such was the backlash for a pious man struggling with his faith as much as his science a few years before the Thirty Years' War.

2.7.1 Medieval Kepler: Chasing Mathematical Harmonies

Before describing his lasting and well-known legacy, namely **Kepler's laws** of planetary motion, let's look at his early, seemingly mystic work. While irrelevant today, it illustrates the transition of science into modernity. The erratic, meandering path of science toward a better

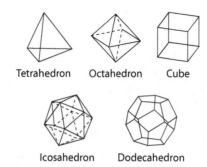

Figure 2.22: **Platonic Solids.** The pyramid or tetrahedron, the octahedron, the cube, the dodecahedron, and the icosahedron are the only three-dimensional solids which have identical faces.

understanding of nature becomes clear in KEPLER's struggle to know the universe and, in his words, to read God's mind. As a devout Copernican, probably due to his teacher MICHAEL MAESTLIN at the University Tübingen, KEPLER asked the question why there are exactly six planets. As a passionate Pythagorean, he found the answer in the mathematical fact that there are exactly five regular three-dimensional solids: the pyramid or tetrahedron, the cube, the octahedron, the dodecahedron, and the icosahedron, Figure 2.22. These Platonic solids are special, because their faces are identical polygons, that is, triangles, squares or pentagons. There is mathematical proof that only five such solids exist in three dimensions. While in lecture in Graz, Austria, KEPLER had an *epiphany*. Six crystalline spheres carrying planets have exactly five surfaces where one touches the next. His Pythagorean mind-set interpreted this as the simple, harmonic truth: That there are six planets because there are five platonic solids which fit in between their spheres, see Figure 2.21(b). Nowadays, it seems absurd to think that a relation could exist between some simple mathematical fact, and a complicated physical configuration such as a solar system. Still, KEPLER's construct works surprisingly well—excluding the "new" planets Uranus and Neptune. By assuming the spheres to be as thick as COPERNICUS's minor(!) epicycles suggest, and inscribing a cube into Saturn's sphere, a pyramid into Jupiter's, and so forth, the solar distances of the planets come out to better than 10%.[47] KEPLER's book based on his "discovery", namely the *Mysterium Cosmographicum (Secret of the Universe)* is often used to show his Pythagorean, and hence alleged anti-modern side. However, looking for simple mathematical harmonies is still en vogue today.[48] Also, the book contains the modern, anti-Pythagorean opinion that there must be a *physical* reason for the motion of the planets. In its first edition, KEPLER invokes the mystical concept of an *anima motrix*, the soul of the mover (here: the sun), which works together with the soul of the moved object (the planet). Taking into account the observed motion of the planets, KEPLER concludes: "We thus have to chose one of these two facts: Either the moving souls [of the planets] are weaker the farther they are from the sun, or there is just one moving soul, at the center of all orbits, that is, in the sun, which drives a body the harder the closer it is to the sun, and which, for the remote planets due to the longer distance and the [associated] weakening of its power becomes forceless."[49] This passage shows very well how concepts in science are invented in a creative process. By toying with several plausible

---

[47] Nobody knows why; maybe this is just coincidence.
[48] In particle physics and other fields it is cherished as the *beauty of symmetry*.
[49] Loosely translated following [14, 46].

2.7. DESCRIBING THE UNIVERSE: KEPLER

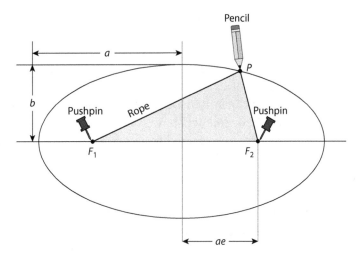

Figure 2.23: **Characteristics of an ellipse.** An ellipse is characterized by two numbers, for example the semi-major axis a (analogous to the radius of a circle) and the eccentricity e. Alternatively, a and the semi-minor axis b can be specified, since $e = \sqrt{1 - \frac{a^2}{b^2}}$. The distance between the two focal points F_1 and F_2 is $d_{F_1 F_2} = 2ae$. An ellipse with zero eccentricity $e = 0$ is a circle because the distance between the focal points vanishes. Ellipses can be constructed easily with a rope, a pencil and two pushpins as shown. The reason is that mathematically an ellipse is defined as the set of points which have a constant distance—the length of the rope $L = 2a(1+e)$—from *two* focal points.

ideas, KEPLER is led to the idea that the sun at the center of the solar system has an active role to play. Also realize his tacit invocation of **Occam's razor**: The second alternative (that only the sun has an *anima motrix*) is simpler and therefore more likely. In the new edition of 1623 KEPLER goes a long way toward NEWTON when he adds the paragraph "If one replaces 'soul' (*anima*) with the word 'force' (*vis*) then one has precisely the principle on which the **physics of the heavens** is based." This is the first time that the concept "force" in its physics meaning appears in an astronomical context [46].

Foremost, the *Mysterium Cosmographicum* is testament to KEPLER's strong Copernican views. The first part of the book is a defense of Copernicus. He then goes on to improve *Copernicus's model* with Copernican methods. Indeed, *Copernicus*'s still had Ptolemaic "left-overs" that KEPLER purged. For instance, in KEPLER's improved model the planes of the planetary orbits cross at the center of the sun, and not the Earth (as COPERNICUS had it). By being consistent and more Copernican than COPERNICUS himself, KEPLER was able to solve several problems plaguing both models. This episode is important for another reason. KEPLER's insistence that the solar system is ruled by the sun, and that the Earth is merely a planet just like the others, mended the gap between the mathematical and physical strand of astronomy, or at least his work had the effect of guiding scientific thinking in this direction. By insisting that there must be some kind of influence or force from the sun that causes the planets to rotate around it, he put out an idea that should change the universe yet again. The **law of universal gravitation** of NEWTON realized this idea and fulfilled the dream of not just mathematically describing but understanding the physical reason for the motion of the celestial bodies; the schism between PTOLEMY and ARISTOTLE was thus swept away together with their cosmologies.

2.7.2 Modern Kepler: Using Tycho's Data to Solve the Problem of the Planets

Mysterium Cosmographicum, published in 1596, garnered the interest of TYCHO BRAHE, who employed him as his assistant in Prague in 1600. There he eventually inherited TYCHO's invaluable data on planetary positions, and began his decades-long research that led to his three laws of planetary motion. In a time when no computer, calculator, not even logarithmic tables were available, Kepler had to do an incredible amount of "number crunching." TYCHO's data had to be **reduced**, that is, freed from errors due to atmospheric refraction and the like, before they could be used. Then a hypothesis, say that the planet's orbits are ovals, had to be formed, tested, and typically falsified. KEPLER had the courage to ponder noncircular orbits, but which was the right choice: oval, egg-shaped, elliptical? Each choice was better than PTOLEMY and Copernicus could ever dream of, but TYCHO's data was so accurate that hypothesis after hypothesis was ruled out, failing to agree with observations. Incidentally, KEPLER's insistence that the predictions of a hypothesis agree *up to the published uncertainty* of the observational data is innovative and thoroughly modern. We have seen before how scientists used to be satisfied with a rough or qualitative agreement of their theories with observations. KEPLER could have chosen the same stance—and would have failed to put his name into the annals of astronomy. Instead, he insisted that his hypothesis agreed to within a minuscule one or two arc minutes (1/30th of 1/360th of a full circle) with TYCHO's data. Once he had to throw out 900 pages of calculations due to his high standards [50] On the bright side, only one possibility remained. Mars and his brethren had to move on ellipses, and they had to move faster when closer to the sun. These two conclusions are now known as KEPLER's first two laws. They were published after nine years of calculations in 1609 in a monumental 650 page book called *Astronomia Nova (New Astronomy)*.[51] The book is written in a very personal style, quite dissimilar to scientific publications today, and gives ample account of KEPLER's struggles to find the correct solution to the problem of the planets. Comprising descriptions of false starts and dead ends entirely absent in today's publications, it is almost a historical account, in the style of "the making of KEPLER's laws". Kepler's first two laws read:

1. The orbits of the planets are **ellipses**, with the sun being at one focus, the other one being empty.

2. The planet moves along the ellipse such that the line connecting sun and planet sweeps out **equal areas** in equal times.

In hindsight, these laws make perfect sense. The ellipse is the **generalization** of a circle. A circle is completely specified by one number, its **radius**, an ellipse by two numbers, typically taken to be its **semi-major axis** a and its eccentricity e, see Figure 2.23. The semi-major

---

[50] "To us, on whom Divine benevolence has bestowed the most diligent of observers, Tycho Brahe, from whose observations this eight-minute error of Ptolemy's in regard to Mars is deduced, it is fitting that we accept with grateful minds this gift from God, and both acknowledge and build upon it. ... Now, because they could not be disregarded, these eight minutes alone will lead us along a path to the reform of the whole of Astronomy, and they are the matter for a great part of this work." [24]

[51] Full title: *Astronomia Nova AITIOΛOΓHTOΣ seu physica coelestis, tradita commentariis de motibus stellae Martis ex observationibus G.V. Tychonis Brahe (New Astronomy, Based upon Causes, or: Celestial Physics, Treated by Means of Commentaries on the Motions of the Star Mars, from the Observations of Tycho Brahe, Gent.)*

2.7. DESCRIBING THE UNIVERSE: KEPLER

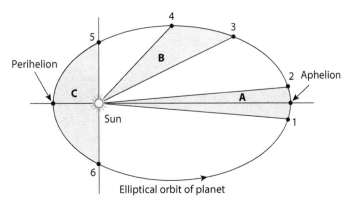

Figure 2.24: KEPLER's **equal area law.** KEPLER's second law states that the planet-sun line sweeps out equal areas in equal time. If the shapes A, B and C have the same area, the planet moves from $1 \longrightarrow 2$, from $3 \longrightarrow 4$ and from $5 \longrightarrow 6$ in the same time. It therefore travels a larger distance when it is closer to the sun. Hence, its speed (distance traveled per time elapsed) is largest when it is at **perihelion** and smallest when at **aphelion**.

axis determines the size of the ellipse and is the analog of the circle's radius. The eccentricity tells us how eccentric the ellipse is, that is, how much it differs from a circle. Incidentally, a circle is an ellipse with zero eccentricity. The eccentricity is a measure of how far apart the two foci of the ellipse are. If they are very close together, the ellipse looks very much like a circle. This is the case for all planetary orbits, and the reason why circular orbits are good **approximations** to the actual, elliptical orbits. On the other hand, this is why astronomers were fooled for millennia into thinking that only circular motions are possible in the heavens. An ellipse can be constructed easily with pencil, rope, and two pushpins; see Figure 2.23. The construction reflects the precise mathematical definition of this **geometric curve**: It is the set of points that have the same distance from the two foci, that is, if you add the distance of a point on the ellipse from one focal point to the distance of that point from the other, you get the same number, a constant determined by the size and eccentricity of the ellipse. There are two special points along the ellipse. Disregarding the empty focus, there is a point closest to the sun in the other focus which is called **perihelion**, and a point farthest from the sun, called **aphelion**.[52] Note that for a circle, the distance to the sun is always the same, so it doesn't make sense to define perihelion and aphelion. The second law is also called the **equal area law**, for obvious reasons. It looks a little complicated, but essentially it says, cf. Figure 2.24, that the planet moves along its orbit faster when it is closer to the sun. Hence, it is fastest at **perihelion** and slowest at **aphelion**. This makes sense for us today, because we "know" that gravity drives the planets along their orbits, and the gravitational force between two objects gets much weaker when the distance between the objects grows. KEPLER did not know this, but as a staunch Copernican, was convinced that the sun in the center rules the planets. It exerts an influence on the planets, and it is very natural to assume that the influence diminishes as the distance between sun and planet grows. He imagined the physical process to be something like the force between magnets. The study of magnetic forces was *en vogue* at the time, and WILLIAM GILBERT (1544–1603) had published a book on magnets in 1600 that KEPLER had read. This nascent way of thinking about the *physics* of the solar system eventually led to NEWTON's assertion that mass attracts mass. Having

---

[52]Greek: *helios = sun, peri = around, apo = off, away.*

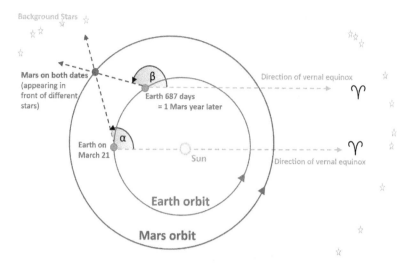

Figure 2.25: **Kepler's construction of Mars's orbit.** Contrary to Copernicus, who assumed a circular orbit and just had to find its radius, KEPLER needed to find the *shape* of Mars's orbit. He constructed it by finding two sightings of Mars exactly one *sidereal* Mars year (687 days) apart. Since Mars is at the same point along its orbit after a Mars year, the two different angles α and β due to Earth changed position (365.25 days $\neq$ 687 days) allow for a *triangulation* of Mars's position. The angles are measured with respect to a reference point, usually the *vernal equinox*, that is, the *apparent* position of the sun among the stars on March 21.

realized that the orbits are elliptical, that is, that the distance planet-sun changes, it was natural for KEPLER to infer that the speed of the planet along its orbit must change. The reason is that he, as a Copernican, knew that Saturn was the farthest and the slowest planet, and that, in general, a planet takes more time to circle the sun the farther it is away from it. What is true for the planets as a group, must be true for the individual planet, and hence the second law is perfectly reasonable.

Yet, every solution is reasonable in hindsight—and confusing when confronted with the unsolved problem. So we need to ask how KEPLER managed to figure out what no one had been able to do before him. What led KEPLER to conclude that the orbit of Mars and the other planets are ellipses? There are really two steps here. First, KEPLER had to accept the fact that no combination of circles would do, very much like COPERNICUS became convinced that a stationary Earth would not solve the problem of the planets. The harder step was to figure out *just which* "non circle" would do the trick. There are a gazillion curves out there, and KEPLER had no good reason to prefer one over the other. A depressing outlook. You may have encountered something similar when facing a difficult homework problem. Faced with a new problem, it is easy to feel confused, and not to know what to do, where to start, or which part of the problem could be related to what you've learned before. Even worse is the situation of the researcher, who doesn't know whether a solution even exists and neither if the question makes sense at all. In situations like this it is always better to try *something* than to do nothing out of fear of trying something that doesn't lead anywhere. It took Kepler a long time to convince himself that circles wouldn't work. After that, it was "only" hard work and trial-and-error that stood between him and the solution. In principle, he knew what to do. COPERNICUS had constructed the size of Venus's and Mars's orbits from special configurations, cf. Figures 2.13 and 2.17. KEPLER devised a similar method to determine single points along Mars's orbit. Essentially, he used the fact that after a *sidereal* year of

2.7. DESCRIBING THE UNIVERSE: KEPLER

Mars (687 days, definition in Section 1.6), Mars is back where it was along its orbit, because it has rotated exactly 360° with respect to the stars, that is, space, that is, the stationary sun. But Earth is not back where it was, because its *year* is much shorter, see Figure 2.25. Since we *know* that Mars is at the same point in space, and we can *observe* its **apparent position** in the sky, we can **triangulate** its **absolute position**. It's simple geometry to determine all sides of the triangle with vertices EME' if you know the angles $\sphericalangle ME'E$ and $\sphericalangle MEE'$ and the angle $\sphericalangle ESE'$. The distance to Mars in units of the Earth-sun distance (the so-called **astronomical unit**) can thus be determined. Next, KEPLER had to repeat this calculation as many times as possible, that is, he had to find pairs of Mars observations in TYCHO's data that were exactly 687 days apart. He then tried out which geometric curve best described this set of points in space, and eventually settled on an ellipse.

Compared to the systems of epicycles and equants of PTOLEMY, and even to Copernicus's model, KEPLER's solution of the problem of the planets is amazingly simple. One law to describe the *shape* of the orbit, the other to prescribe the *varying speed* along the orbit, are enough to calculate the planetary positions at any time, be it in thousands of years in the past, now or next month. Incidentally, KEPLER exhibited both his religiosity and his mathematical genius by subsequently applying his laws to see if the biblical star of Bethlehem could have been a close encounter of two or more planets. He proposed that a **conjunction** of Jupiter and Saturn in the year 7 BCE was interpreted by the magi as the star enunciating the birth of the messiah, Jesus Christ. This is still considered a viable hypothesis.

It will not have escaped your attention that these two KEPLER laws completely describe planetary motion, and that we failed to mention the third law. Indeed, it took KEPLER another decade to find his third or harmonic law which connects the duration of the planet's year, that is, its period P with its distance from the sun, that is, the semi-major axis a of its orbit. This was very important input for NEWTON, because it showed that and how these two naively independent quantities are related. Specifically, it reads:

3. The square of the period P of the planet's orbit is proportional to the cube of the semi-major axis a of its orbit, or $P^2 = Ka^3$, where K is a constant.

If both quantities are expressed in "terrestrial units" ($P_{Earth} = 1$ year, $a_{Earth} = 1$ AU), then the ratio $\frac{P^2}{a^3}$ equals exactly one—for any planet. This makes it possible to compute the size of an orbit from the easily measured length of its **sidereal year**! For instance, since we measured the Mars year to be 687 days = 1.88 years, we can compute from $a^3_{Mars}(\text{AU}) = P^2_{Mars}(years)$ that

$$a_{Mars}(\text{AU}) = P^{2/3}_{Mars}(years) = 1.88^{2/3} = 1.52. \tag{2.2}$$

In other words, Mars is 52% farther from the sun than the Earth. Note that this calculation only yields relative size of an orbit; from KEPLER's third law we cannot determine how far it is to Mars in kilometers or miles. On the other hand, this one-liner replaces the 900 plus pages of calculations KEPLER had to do to determine the orbit form TYCHO's observations! This stark contrast gives us an idea as to how much harder it is to *discover* a new useful fact about nature than to *use* it.

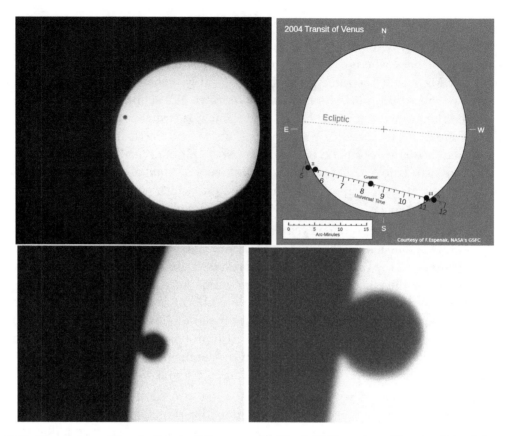

Figure 2.26: **Venus transiting in front of the sun.** (a) [Top left] Venus transiting the disk of the sun. (b) [Top right] A schematic drawing of the rare event. Venus transits occur only about once in a century in pairs 8 years apart. Due to the substantial inclination of Venus's orbit with respect to the ecliptic (3.39° ≈200'), Venus usually passes above or below the sun's disk (0.5° = 30') at *inferior conjunction*. I happened to observe both the 2004 and the June 6, 2012 transits. The next transit will not occur until December 11, 2117, long after my passing. (c) [Bottom left] The dreaded **black drop effect** which makes it hard to decide when exactly Venus reaches the solar limb (so-called *phase III*). (d) [Bottom right] The existence of Venus's atmosphere manifests itself in the dim arc around Venus on this photo. It was discovered by MICHAIL LOMONOSOV during the 1761 transit. The photos (a)(c)(d) were taken with a digital camera through an 8" reflector with a solar filter (exposure time 1/180 second) on June 8, 2004.

2.7.3 Bellwether Kepler: From Geometry to Algebra, from Mathematics to Physics

If COPERNICUS started the *scientific revolution*, KEPLER drove it forward by delivering results necessary for any future research. His *Rudolphine Tables*[53] finally published in 1627 made it possible for the first time to predict Venus and Mercury transits across the disk of the sun; see Figure 2.26. Tragically, KEPLER did not live to see the Mercury transit he predicted for November 7, 1631. Instead, the French priest and natural philosopher PIERRE GASSENDI (1592–1655) became the first person to observe a Mercury transit, confirming KEPLER's prediction gloriously; he had been only five hours off.

---

[53]These tables are basically precise initial positions of the planets and instructions how to compute their positions at any other time. Like the *Alphonsine* and *Prutenic tables*, they were used to predict celestial events, but with an incredible accuracy—many times better than any of its predecessors.

2.7. DESCRIBING THE UNIVERSE: KEPLER

It is no coincidence, that KEPLER's third law, Equation (2.2), is the first algebraic equation we mentioned in the history of astronomy. Before KEPLER, most astronomers used geometric constructions like the circles and equants of PTOLEMY to describe patterns of nature like planetary motion. This changed rapidly after KEPLER. By the time of NEWTON, equations were the main tool to get from a theory, that is, a set of assumptions, in a logically consistent way to its predictions or results. KEPLER published his equation by describing it in words. The use of letters for variables, the equal sign, and so forth, were not commonly used before RENÉ DESCARTES invented *analytic geometry*, the study of geometry with equations, in the decades following KEPLER's death.

KEPLER's influence on the course of the ***scientific revolution*** was far-reaching. He relentlessly searched for a *physical reason* why the planets should move in the way he had prescribed. *Why* are the planets faster when closer to the sun, *why* do the planets orbit the sun in ellipses? These questions, later answered by NEWTON, were new and a by-product of KEPLER's modifications to COPERNICUS's universe. If planets are moved by crystalline spheres, these questions don't arise. But ellipses are incompatible with crystalline spheres, and a "freestanding" elliptical orbit raises new questions. KEPLER did much to lead his successors toward the modern way of thinking about motion in terms of forces. His preliminary thoughts on how the sun's influence on the planet, and the planet's own kinematic state (for example its velocity) work together to form a resulting, elliptic path can be seen as a precursor to NEWTON's explanation involving his first law of motion and universal gravity, cf. Section 2.10.3.

Concept Practice

1. Did *Kepler's model* use major epicycles? (a) Yes (b) No
2. Did *Kepler's model* use minor epicycles? (a) Yes (b) No
3. What does the semi-major axis a of a planet's orbit tell you?

    (a) How far off the ellipse is from a circle.
    (b) How far the planet is from the sun.
    (c) How long it takes the planet to orbit the sun.
    (d) The average distance of the planet from the sun.
    (e) None of the above

4. Which are the two numbers characterizing an ellipse?
5. To which number characterizing an ellipse does the radius of the circle correspond to?

    (a) Eccentricity
    (b) Period
    (c) Semi-major axis
    (d) There is no correspondence.
    (e) None of the above

6. In which "direction" is KEPLER's third law usually used? *(Hint: Think about which quantity (P or a) is easier to measure.)*

    (a) As an equation for P.
    (b) As an equation for a.

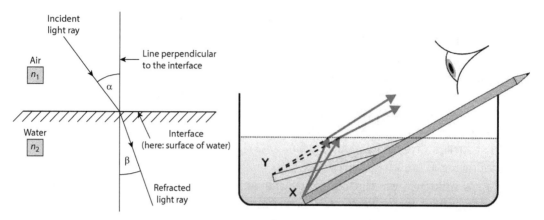

Figure 2.27: **Refraction.** (a) The *law of refraction* relates the angle α of a light ray incident on the interface between two transparent media to the angle β of refraction under which the ray exits the interface. Transitioning into a denser medium $n_1 < n_2$ leads to a smaller angle $\alpha > \beta$; the ray appears to be bent *toward* the line perpendicular to the interface. (b) A straw in a glass of water *appears* to be bent due to refraction when transitioning from the optically thicker water into air.

2.8 New Methods: Galileo

GALILEO GALILEI (1564–1642) had an immense influence on the development of modern science. We focus here on his work in astronomy, but hasten to note that he was an excellent physicist and mathematician, and made major contributions to all of the sciences he engaged in. He set new standards for experimentation and "lived" the scientific method in his research. He is therefore often considered the founder of *experimental physics*.

2.8.1 The Beginning of the Telescopic Era

In astronomy, GALILEO's name is connected with the beginning of telescopic observations. He was one of the first to use the telescope, invented in the Netherlands in 1608. The account of his telescopic discoveries of 1609 in a book called the *Starry Messenger (Siderius Nuncius)* (1610) was very accessible and highly influential. Before we discuss his results, let's recall how a telescope works and why it was invented rather late in science history.

Single lenses and their optical properties were known for centuries before the invention of the telescope. This is surprising, because simple telescope are a combination of merely two lenses. These two-lens telescopes are known as refractors, because they can be understood entirely by the law of refraction

$$n_1 \sin \alpha = n_2 \sin \beta. \tag{2.3}$$

This formula describes how a light ray traveling in a transparent medium characterized by a **refractive index** n_1, and hitting an interface with another medium of index n_2, is deflected, Figure 2.27(a). If the incident angle is α, the formula can be used to compute the angle β at which the ray exits the interface. The law of refraction can be demonstrated by an apparent bending of a straw half submerged in a glass of water, Figure 2.27(b). The light rays coming from the end of the straw X are refracted as they transition into the optically thinner air ($n_{air} < n_{water}$, $\beta > \alpha$). For our eyes it thus looks like the light rays come from point Y, and the straw *appears* to be bent. The "bent" straw in Figure 2.27(b) teaches us another lesson. What we *see* is a bent straw, but the straw *is* perfectly straight, it just *appears* to be bent.

2.8. NEW METHODS: GALILEO

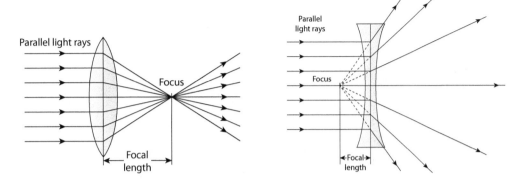

Figure 2.28: **Lenses.** (a) A positive (usually convex) lens focuses light rays, that is, refracts all of them such that they pass through a common *focus* or *focal point*. (b) A negative (usually concave) lens produces beams diverging from a common *focus*. The stronger the lens, the shorter the focal length, that is, the more dramatically the light rays are refracted. Here, the distance between focus and lens is shorter for the diverging lens; it is the stronger lens. Also note that no light rays pass through the focus of the diverging lens!

If we didn't know of the law of refraction, we would have no way of *interpreting* correctly what we see. Likewise, what we *see* in the sky has to be interpreted with the help of other things we know about nature to reveal what we really *observe*. Sometimes we do not have all relevant information, and are led to a faulty interpretation of what we see.

The interface between media in the example of the "bent" straw, Figure 2.27(b), is the flat surface of the water. The law of refraction also holds—and is even more interesting—for curved surfaces, the prime application being the focusing properties of lenses. By grinding a rigid, optically dense medium like glass into an appropriate shape, one can redirect light beams such that they all meet at one point. The **convex lens** in Figure 2.28(a) is an example. The focusing of light has amazing applications: One can start a fire by concentrating sunlight on combustible materials.[54] Most importantly for us, a combination of two lenses will show an object magnified. The two major types of lenses are **convex lenses** focusing light rays, and **concave lenses** producing diverging light rays, as in Figure 2.28. Simple two-lens telescopes produce a magnified *image* of an *object*; see Figure 2.29. The first, large lens or **objective** collects the light rays and forms an image. The second, small lens or **eyepiece** takes this image as its object and produces a second image that the observer sees. Note that the second image is magnified. In a **Galilean telescope**, a combination of a large diverging lens and a small converging lens, an **upright, virtual** image is produced. This means that the orientation of the image is the same as the orientation of the object, and that no actual light rays go through the image which the observer sees. The advantage of the Galilean telescope is that its upright image makes it easier to navigate the skies, its disadvantages are its rather small magnification, and its virtual image which does not allow for projection or photographic[55] detection of the image. The **Keplerian telescope**, invented and explained by KEPLER in his *Dioptrice* of 1611, is the type used by modern astronomers. It produces an **inverted image** that is real, in the sense that it can be photographed or projected onto a screen, as in Figure 2.34. Incidentally, the projection method shown is the *only safe way* to observe the sun with a telescope without a solar filter.

---

[54]The same focusing property of mirrors was allegedly used by ARCHIMEDES (c. 287–212 BCE) in Syracuse to set invading Roman ships on fire 1,800 years earlier.

[55]Of course, photography wasn't invented until about two hundred years after GALILEO's death.

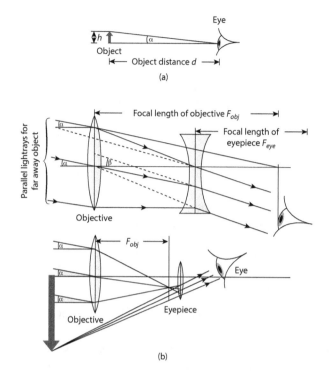

Figure 2.29: **Refracting telescopes.** (a) To the naked eye, an object appears under an angle α, depending on the object's size h and distance D, namely $\alpha = \frac{h}{D}$. Using a refracting telescope means looking at the image produced by a two-lens system: A large, converging lens or **objective** and a smaller eyepiece lens or **ocular** (Latin: *oculus=eye*). A refracting telescope alters—by **refraction**—the direction of the light rays coming from the object such that it appears under a larger angle β. The magnification is simply $M = \frac{\beta}{\alpha} = \frac{f_{obj}}{f_{eye}}$. Since the eyepiece can be a diverging or a converging lens, two types of refractors exist. (b) *Galilean telescope*: The eyepiece is a diverging lens and renders the converging light rays of the objective parallel. It thus sits closer than the objective's focal length. The image is **upright** and **virtual**. (b) *Keplerian telescope:* The eyepiece is a converging lens. The image is **inverted** and **real**.

The magnification M of a refracting telescope can easily be derived[56] from elementary **ray optics**. It is simply the ratio of the focal length f_{obj} of the objective lens and the focal length f_{eye} of the eyepiece lens

$$M = \frac{f_{obj}}{f_{eye}}. \qquad (2.4)$$

For instance, the magnification of a refractor with a 2000 mm objective lens is 50× if used with a 40 mm eyepiece, and 100× if used with a 20 mm eyepiece. In practice, one uses an eyepiece with long focal length to *find* an object, because the field of view is large, and switches to an eyepiece with short focal length to look at details. Objective lenses cannot be switched; they define the telescope's quality, that is, light collecting power. More light results in a better image quality. Incidentally, the light collected is proportional to the *area* of the lens, not its linear diameter.[57]

Upon hearing of the invention of the telescope,[58] GALILEO realized its potential. He

---

[56]A typical euphemism used by scientists and meaning "While nearly impossible for a novice, for an expert it is relatively straightforward to derive the formula."

[57]So is, unfortunately, its sticker price.

[58]The word was coined by the Greek theologian JOHN DEMISIANI at a banquet honoring GALILEO [29], since

2.8. NEW METHODS: GALILEO

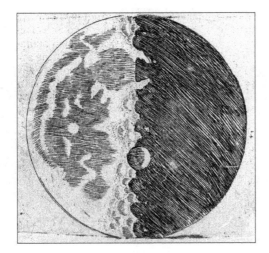

Figure 2.30: **Galileo's eyepiece sketches.** (a) The surface of the moon as published in *Siderius Nuncius* (1610). Note the impressive crater in the middle of the sketch. In reality, it does not exist. (b) Jupiter's four brightest moons as drawn by Jesuits in 1620—an important confirmation of GALILEO's discovery.

started to construct telescopes by grinding his own lenses and making major improvements on the Dutch design. His first telescopes magnified only two or three times, and even his best instrument reached a fairly meager 30× magnification. As a reference, standard binoculars magnify 7×, and small modern telescopes typically start at 50× magnification. On the other hand, a scientist that has an instrument that is 30x better than that of his competitors, can do marvelous things. Imagine how drastic of change the advent of the telescope must have been! Suddenly, things could be seen that nobody had been able to see before. Invisibly faint stars and interstellar gas clouds were appearing in the new telescope. Not surprisingly, skeptical contemporaries doubted that what was visible through a telescope was real at all! Maybe the telescope itself produced these sensations?! Indeed, for the first time it was possible to perceive something that was not accessible to the senses of man himself. Nowadays, we are used to apparatus that show us in great detail what we cannot see, feel or smell: Microscopes reveal a whole microcosm of organisms unheard of before, particle accelerators give us a clear pictures of how particles ten thousand times smaller than atoms interact. In some sense, all this modern technology started with the telescope, which expanded the universe immensely. For the first time, new celestial objects by the thousands and millions were available for scientific study. And with the available magnifying powers of the new instruments, known celestial objects like the moon and the planets could be studied in unprecedented detail, revealing unknown patterns in nature that had to be explained. It is easy to see that the telescope was a sensation that caused a revolution and an explosion in astronomical research yielding unparalleled insights into the nature of the cosmos. GALILEO anticipated this reaction, and wrote his account of his telescopic discoveries in a style that could be understood by nonexperts, inviting others to follow up with independent observations. This was arguably the beginning of amateur astronomy and of science as a hobby [7]. Over time, these **amateurs** have made important contributions to science. A famous example is BENJAMIN FRANKLIN (1706–1790).

---

in Greek *tele=afar* and *skopeo= I look at*, cf. *television*.

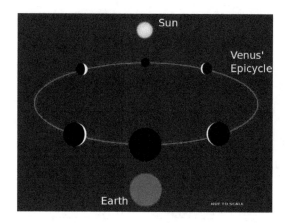
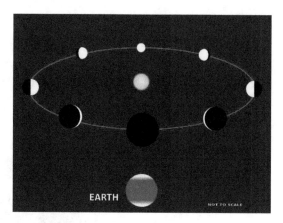

Figure 2.31: **The predicted phases of Venus.** The predictions of the Ptolemaic, geocentric model and the Copernican model differ crucially. (a) In the Ptolemaic model, the sun is always "behind" Venus; the model predicts that Venus shows us mostly its dark hemisphere; only crescent and new phases are possible. (b) The Copernican model predicts that all phases should be observed: from full Venus at superior conjunction, to new Venus at inferior conjunction.

GALILEO squeezed as much scientific insight out of his discoveries as possible by carefully interpreting what he saw. Several of his discoveries cast serious doubts on the Aristotelian system in which he was educated, and his observation of the phases of Venus falsified the geocentric theory of PTOLEMY. It that vein, it is strange that he refused to accept KEPLER's theory of planetary motion. GALILEO remained a Copernican and believed in circular planetary orbits even after KEPLER's successes in predicting planetary positions with unprecedented accuracy. This is a behavioral pattern that is quite typical in science. New ideas of others are often rejected by established scientists.

2.8.2 Galileo's Telescopic Discoveries

1. **The surface of the moon features mountains and valleys.** GALILEO saw that the moon's surface looks surprisingly similar to the Earth's, exhibiting mountains and valleys, and also craters; see Figure 2.30(a). This was contrary to Aristotle, since the moon should belong to the heavenly sphere, and therefore be a uniform, featureless, perfect object. GALILEO was artistically trained, and produced impressive, if not necessarily correct, renderings of the moon's face. His genius was such that he investigated further and estimated the height of the mountaintops on the moon. He found that they are comparable in size to mountains on Earth.

2. **Jupiter has four great moons rotating around it.** Looking through his telescope, GALILEO saw a couple of "stars" close to the disk of Jupiter,[59] sometimes two, sometimes three or four. To establish beyond doubt that these are moons of Jupiter and not background stars, he observed the Jupiter system over many weeks and recorded his observations in a notebook; see Figure 2.30(b). For an Aristotelian scholar this poses several problems. Since the moons keep orbiting Jupiter, which is moving, an old argument that a moving Earth would lose its moon, is null and void. Furthermore, the moons orbiting Jupiter which orbits the sun (or the Earth for Aristotelians) shows that

---

[59] The moons would be visible to the unaided eye save for Jupiter's glare.

2.8. NEW METHODS: GALILEO

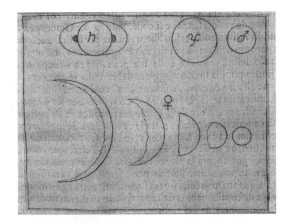 

Figure 2.32: **The observed phases of Venus.** (a) GALILEO's renderings of the phases he saw in his telescope. His observation of half and gibbous moon Venus falsifies the Ptolemaic model. (b) Photos of Venus taken in 2004 show the change in Venus apparent diameter. Note that the June 8 photo of new Venus is from the day of the Venus transit; see Figure 2.26. The "halo" around new Venus shows that Venus possesses an atmosphere that scatters light from the illuminated to the dark hemisphere. *Courtesy: Statis Kalyvas - VT-2004 programme.*

there are at least two centers of rotation, not just the stationary Earth at the center of the universe as claimed.

3. **There are more stars than than the naked eye reveals**. When GALILEO pointed his telescopes to the sky, he saw more stars in the constellations—stars that no one had seen before. His eyepiece sketches of the region around Orion's belt shows tens of stars around the three stars visible with the unaided eye. This meant that the ancients like ARISTOTLE did not know everything, and revealed parts of the universe that were, up to then, completely unknown.

4. **The Milky Way consists of millions of stars**. The human eye is unable to resolve the individual stars and therefore perceives our galaxy as a milky band. The telescope reveals patterns in nature that were unknown to ARISTOTLE and his contemporaries. Worse, an argument could now be made that God had not revealed all his knowledge to the people, and that it takes effort to truly perceive the universe in all its details.

5. **In the telescope, planets appear magnified and disklike, whereas stars continue to appear point-like, regardless of magnification.** This points to a *physical* difference of the two types of celestial objects. Before GALILEO, both stars and planets were specks of light. They were distinguished (and therefore labeled by different names) only on the basis of their motion, not their properties. The fact that stars are not magnified suggests that they are far away, and therefore support Copernicus theory. Recall that Copernicus was forced to *assume* that the stars are very far away, because otherwise they should show ***parallactic motion*** due to the Earth's motion around the sun. In some sense, the exceedingly large distance of the stars is a prediction of the Copernican theory. If the stars had been magnified in GALILEO's telescope, the inference would have been that they are close, which would have falsified Copernicus's theory. Having survived this test, Copernicus's theory was strengthened.

6. **Venus shows phases similar to the moon.** This discovery was arguably GALILEO's most important. It showed directly that the planets are not shining by their own light;

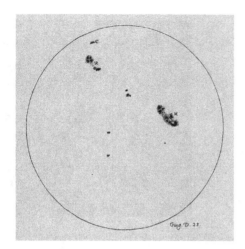 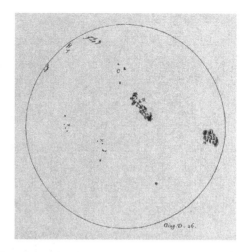

Figure 2.33: **Sunspot drawings.** GALILEO systematically observed sunspots in 1612. The drawings are from June 23 and 26, respectively. Note that the sunspots are identifiable and appear to move. This allows for a determination of the sun's *rotation period*.

they are not stars, as even KEPLER still believed. Most importantly, the observation falsified—by direct evidence—the Ptolemaic geocentric model of the universe. Due to the sun and Venus both being on circular orbits around the Earth, according to PTOLEMY, see Figure 2.12, the sun illuminates Venus "from behind", and thus from Earth we always see a large fraction of the dark hemisphere of Venus. Only new and crescent Venus phases can be produced in a geocentric model. In the Copernican model, all phases can emerge; see Figure 2.31. Therefore, GALILEO's observations of half, gibbous moon and fully illuminated Venus, Figure 2.32, are not compatible with PTOLEMY's predictions; the geocentric model is falsified. Note two things. First, the Copernican model passes this test, yet this doesn't mean that it is correct, it is just *not wrong*. We thus tentatively accept the theory. Theories are continuously subjected to further tests. If they don't pass, that is, do not agree with observations, they will be discarded or modified. COPERNICUS' theory itself was falsified by KEPLER. It could not reproduce the planets' positions to the accuracy of TYCHO's data. KEPLER therefore modified it, keeping the sun in the center, but replacing the circular orbits by ellipses. He formed a new theory, sometimes known as the Copernico-Keplerian model of planetary motion. Second, observing Venus's phases is only *indirect* evidence for the motion of the Earth. It is indirect because only in a heliocentric model do we predict the phases of Venus as observed by GALILEO. We don't observe the motion of the Earth itself, but its consequences (Venus's phases) *under the heliocentric assumption*. Over a century later, BRADLEY observed **stellar aberration**; see Section 3.2.1, which is a direct consequence of Earth's motion, with no other assumption necessary. BRADLEY's discovery therefore constitutes **direct** or **positive** evidence.

7. **The sunspots reveal the rotation of the sun.** GALILEO was not the first to discover that there are dark spots on the disk of the sun, nor did he determine the rotation rate of the sun very accurately. He was, however, a clever self-promoter, and his account of this discovery, published in the form of letters to MARK WELSER (1558–1614), a wealthy magistrate of Augsburg, in 1613 associated his name with the discovery. THOMAS

2.8. NEW METHODS: GALILEO

Figure 2.34: **Observing sunspots with a projection method.** Shown is SCHEINER's contraption to safely observe the sun by projecting a (Keplerian!) telescope's image onto a screen. HARRIOT had observed sunspots first in 1610 but did not publish, and JOHANN FABRICIUS (1587–1617) was the first to publish his sunspot observations. Yet, SCHEINER and GALILEO quarreled about who was first.

HARRIOT (1560–1621) in England had observed the sunspots as early as 1610, and determined the rotation period of the sun by carefully observing and timing the motion of the sunspots across the disk of the sun, finding a **synodic period** of 26.87 days[60] at the sun's equator. HARRIOT also discovered that the sun rotates slower by several days at higher solar latitudes. This so-called **differential rotation** was the first hint that the sun is a giant gas ball. At the very least, it showed that the hypothesis that the sun is rigid does not agree with observation. While HARRIOT did not publish his results, the Jesuit CHRISTOPH SCHEINER (1573–1650) had observed the sunspots with a clever projection method using a Keplerian telescope, see Figure 2.34. He published the results with the help of WELSER and claimed priority; a bitter argument with GALILEO ensued.

The reception of GALILEO's observations was initially rather positive. Of course, his discoveries and claims were surprising and extraordinary, so they needed to be independently confirmed. This was part of GALILEO's intentions when he wrote his short, accessible *Starry Messenger*. Indeed, Jesuit scholars soon confirmed some of his observations. JOHANNES KEPLER's confirmation of Jupiter's moons in his pamphlet *Narratio de Observatis a se quatuor Iovis sattelitibus erronibus*—with the first use of the word *satellite*—was a welcome endorsement by a respected scholar and did much to convince most of GALILEO's contemporaries that the objects seen through a telescope were real and not just products of the telescope itself, as some skeptics suspected.

2.8.3 Galileo as the Founder of Experimental Physics

We already mentioned GALILEO's research on falling objects in Section 1.2.7. His use of the inclined plane and other clever experimental methods earned GALILEO a reputation as the founder of experimental physics. The results of his research were a necessary ingredient

---

[60]Citing his result to four significant figures, his claim is that the uncertainty of his measurement is on the order of a few parts in $10,000$. The modern value is 26.24 days, so the *relative discrepancy* of his result was only $(26.87^d - 26.24^d)/26.24^d = 0.024 = 2.4\%$. However, this is a few parts in 100, so his use of four significant figures is misleading.

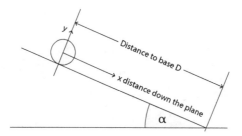

Figure 2.35: **The inclined plane.** A ball is rolling down an a plane inclined by an angle α with respect to the horizontal. It starts a distance D from the basis of the incline. Its acceleration and therefore its speed after some time depend on α. If α is increased the ball rolls faster. In the limit $\alpha \to 90°$ the ball is free falling vertically. GALILEO used a very small angle $\alpha = 2°$ to slow down the ball enough to be able to measure its position x as a function of time.

in NEWTON's universal explanation of planetary motion by gravity.[61] As KEPLER did with planetary motion, so did GALILEO with falling objects: Both asked the question *how* the objects move, not *why* the objects move. These two different questions define the fields of *kinematics* and *dynamics*, respectively. As an example, reconsider a ball rolling down a ramp, also known as an inclined plane, Figure 2.35. **Kinematics questions** about the motion of the ball are "How long does the ball take to reach the base of the ramp?" or "How fast is the ball rolling after half a second?" Obviously, these are "How?" questions. An example for a **dynamics question** is "What is the (net) force acting on the ball which makes it roll down the ramp?" This is a "Why?" question because it asks for the reason for the observed motion. It is much harder to answer. "How?" questions can empirically be answered by simply observing what the ball is doing: when is it where, and how fast is it moving at that moment. Only after the connection between kinematics and dynamics was understood by NEWTON could a consistent theory of motion and the forces causing it be developed. This theory is known as **classical mechanics**.

As trivial as measuring the motion of a ball on an incline sounds, as hard was the experiment in practice in GALILEO's era, when accurate clocks were not available. An important question at the time was how falling objects accelerate. Is the increase in the object's velocity proportional to the time elapsed, the distance traveled, or something else? The practical problem was that objects fall too fast. To overcome this obstacle, GALILEO's genius was indispensable. His circumvention of the roadblock did not only solve the problem, it defined a new way of designing and performing experiments in science. His new standard was in stark contrast to Aristotle's holistic method of casually observing nature at work and drawing qualitative conclusions, see Section 2.2.4. We saw in Section 1.2.7 that GALILEO used an inclined plane, Figure 2.35, to slow down the motion. Furthermore, GALILEO had to assume that other effects, like **friction** and **air drag** are small. As all scientists designing an experiment, GALILEO had a hunch how it would go and formulated a **hypothesis**: The ball will roll a distance d that is proportional to the time t elapsed **squared**, $d = Kt^2$, where K is a constant. To test this hypothesis without a stopwatch, he carved marks into the track at distances from the starting point that are in square ratios, that is, $1:4:9:16,\ldots$. For instance, if he carved the first mark at 1.5 inches, the next would be at $1.5 \times 2^2 = 6$ inches, the following at $1.5 \times 3^2 = 13.5$ inches, and so forth. With his preparation, the rolling ball

---

[61]Obviously, objects on Earth fall due to gravity. Newton extended this notion to *celestial* objects.

made "clicks" in equal time intervals, *if and only if* the hypothesis is correct. The human ear can identify even small deviations from equal time intervals very accurately. It was thus a small feat for GALILEO—the son of a musician for whom excellent timing was key—to confirm his hypothesis. He later reported a similar setup using water clocks in his last book *Discorsi e Dimostrazioni Matematiche Intorno a Due Nuove Scienze (Discourses and Mathematical Demonstrations Relating to Two New Sciences)*. Published 1638 and thus a year after DESCARTES's book on the scientific method, *Discours de la méthode*, it became his testament and established the new methods of experimental physics and the other empirical sciences. One was the careful design of experiments that narrowly focus on one pattern in nature. The blocking out of all other effects that could interfere is highly modern, and in its artificiality in stark contrast to Aristotle's approach. The open-minded, holistic observer of nature had become a determined examiner chopping up nature's patterns into tiny, well-defined bits to squeeze a maximum of data and insight out of a contrived situation. This may not sound very nice, but it is incredibly efficient, and as a method of science it has been used ever since.

2.8.4 Galileo and the Church

Contrary to common belief, GALILEO was not a martyr of science. His clash with the church authorities and the subsequent severe punishment are well known. But in the long quarrel neither side behaved very well, and the conflict could probably have been avoided altogether if GALILEO had acted more prudently and less stubbornly, and the church had operated less dogmatically.

The problem for the Catholic Church at the time was that it came under attack by the Protestant Reformation. It had to react to the growing number of people that left Catholicism for the new religion. As we saw earlier, having renounced Catholic doctrine, interpretation of the Holy Scripture was the only "weapon" of the Protestants. The Catholic Church counter-attacked with a Catholic reading of the bible. In this precarious time the doubts raised by Copernicans like GALILEO of interpretations of bible verses and Church doctrines based on Aristotelian views were increasingly viewed as hostile and dangerous. The initially benevolent position of church leaders like Cardinal BELLARMINE[62] changed as the drama unfolded over several decades. In 1616 GALILEO was questioned by BELLARMINE about his Copernican views and ordered not hold or teach them anymore, which he agreed to and followed for many years.

In 1623 MAFFEO BARBERINI (1568–1644) became POPE URBAN VIII. He was a former acquaintance of GALILEO and encouraged him to publish about his Copernican ideas, as long as he treated them as mere *hypotheses*, not as facts. It seems he was unaware around 1624, when he received GALILEO six times, of BELLARMINE's previous, anti-Copernican orders. GALILEO therefore dared to publish the *Dialogo sopra i due massimi sistemi del mondo (Dialogue Concerning the two Chief World Systems)* in 1632. In it he explicitly described the Copernican system as a reality, and was in gross violation of BELLARMINE's orders. Moreover, the dialogue was written in Italian rather than Latin, and hence accessible to every literate person. Furthermore, the advocate of the *Ptolemaic system*, called SIMPLICIUS (which can be interpreted as "simpleton"), was made to look like a fool, and, worse, could be identified with URBAN's position on the question. When the pope learned of the violated orders on top of

---

[62] He was one of the judges that condemned GIORDANO BRUNO (1548–1600) to be burned at the stakes for heresies, including but not limited to Copernicanism.

this, he had little choice but to unleash the **Holy Inquisition** onto GALILEO. He was put on trial in 1633, shown the instruments of torture, and forced to renounce the Copernican idea that the Earth moves. Legend has it that—after accepting the verdict—he mumbled *Eppur se muove (But it does move)*. He spent the rest of his days in house arrest in Arcetri. As harsh of a punishment as it seems for an old man, in view of what other so-called heretics endured, it appears rather light. Today we would maintain that nobody should be convicted because of his or her religious or scientific opinions whether they are right or wrong.

The whole drama was somewhat of an anachronism. Already before the time of the publication of the *Dialogo*, KEPLER's *Rudolphine Tables* had made it clear that the Copernico-Keplerian model was the superior theory. Yet, GALILEO refused to embrace KEPLER's elliptical orbits, and only mentions PTOLEMY and COPERNICUS in his *Dialogo*, leaving KEPLER and TYCHO's ideas out. GALILEO himself had falsified *Ptolemy's theory*, but TYCHO's system was still much alive, and embraced by many who couldn't accept a moving Earth, but agreed that Ptolemy's theory was dead. The battle of the world systems had already been decided, and yet it would take decades to convince many and centuries to persuade all.[63] This goes to show once more that, after all, science is made by humans. Its results are subject to verification by others, and therefore involve emotions and irrational convictions. The acceptance of a scientific theory is thus often significantly delayed, even if the case is clear, that is, even if its merits have been validated *logically*.

Concept Practice

1. Which is not a discovery of GALILEO?

    (a) Mountains of the moon
    (b) Venus's phases
    (c) The new star (supernova) of 1572
    (d) Sunspots
    (e) All are discoveries of GALILEO.

2. Why do Venus's phases show that it is not emitting light?

2.9 The Scientific Revolution

Some science historians view the publication of *De Revolutionibus* (1543) as the starting point of the *scientific revolution*, which, it is said, ended with the publication of NEWTON's *Principia* (1687). All of this has to be taken with a grain of salt. Nonetheless, these roughly 150 years catapulted us into modernity and have made our modern, rather sweet life possible. This was an extraordinary achievement, begging the question(s) as to why it happened when it happened in the form it happened.

2.9.1 How Revolutionary Was the Revolution?

In the section on Copernicus we saw why his revolutionary work was not possible in 1200, but likely in 1500. Now we ask the provocative question if there was a scientific revolution at all. It can be argued that the transition into modernity is only *apparently* dramatic, but

---

[63]GALILEO was exonerated by pope JOHN PAUL II in 1992.

2.9. THE SCIENTIFIC REVOLUTION

in reality went smoothly if intensely.[64] True, the ways that science was done and how its results were recorded changed dramatically from Copernicus to NEWTON. As such, the term revolutions seems adequate in its meaning of a complete and fundamental change. Given, the scientific revolution has not been as momentous and immediate as the American, French, or Russian Revolutions, but to insist on a sharp point in time would mean to be overly picky and *semantic*.

On the other hand, the historian DAVID C. LINDBERG [27] has pointed out that there was continuity in the methods used and questions considered, but a shift in the underlying ideas. The organic, purposeful, teleological universe of ARISTOTLE gave way to the modern mechanistic and lifeless universe of NEWTON. Hence, the scientific revolution is revolutionary mostly with respect to the change in metaphysical and cosmological outlook it created.

As you can see there are arguments for and against the revolutionary status of what happened between Copernicus and NEWTON. As often, a *pragmatic* approach grounded in factual knowledge of the controversy is best. We thus will use *scientific revolution* as a label we have agreed on.[65] For our quest to understand the expansion of the universe the different interpretations of what happened are largely unimportant. Our focus is on the achievements and inventions of science themselves.

2.9.2 The Scientific Method: Theory and Practice

In this spirit, we'll revisit our discussion of the *scientific method* begun in Section 1.2.1, because its name was coined in the seventeenth century. FRANCIS BACON (1561–1626) in his *Novum Organum* (1620) and *New Atlantis* (1626), and René DESCARTES (1596–1650) in his *Discourse* (1637) are usually quoted as having captured in words the spirit of the scientific endeavors of their contemporaries. The full title of DESCARTES's book is its program and executive summary. It reads: *Discours de la méthode pour bien conduire sa raison et chercher la vérité dans les sciences, plus la Dioptrique, les Météores et la Géométrie qui sont des essais de cette méthode (Discourse on the Method of Rightly Conducting One's Reason and of Seeking Truth in the Sciences, plus the Optics, the Meteors and the Geometry which are Essays [Applications] of this method.).* The book contains the famous *Cogito Ergo Sum* motto signifying the radical skepticism of empirical knowledge of DESCARTES as a *rationalist*. Also included are his influential rendering and justification of the *law of refraction* in optics and of the method of *analytic geometry*. While the latter were lasting contributions to science, it has to be said that the philosophical works of both DESCARTES and FRANCIS BACON did little to advance science [44], in the sense that no "real" scientist learned the scientific method from (these) books. Simply put, you learn how to do science by doing it.

More important for the progress of science was the establishment of scientific societies, the prime examples being the Royal Society of London and the French Academy of Sciences, founded 1660 and 1666, respectively. The former was conceived as a "College for the Promoting of Physico-Mathematical Experimental Learning" by CHRISTOPHER WREN,[66]

---

[64]I offer this vague analogy: When our first child was born, I was astonished how smooth the transition into parenthood was. I had imagined a *dichotomy*: Life without children is totally different from having children. Yet, anticipating the great event for nine month and the fact that newborns sleep twenty hours a day mellowed this bifurcation into a rather smooth transition.

[65]Incidentally, the first use of the term is recorded in 1799.

[66]Architect of St. Paul's Cathedral in London.

(1632–1723) and ROBERT BOYLE,[67] (1627–1692), among others. Its early members included ROBERT HOOKE[68] (1635–1703) and JOHN WALLIS[69] (1616–1703). ISAAC NEWTON became one of the first fellows of the Royal Society (FRS) in 1672 and its president in 1703. The society met weekly to discuss scientific issues and to run experiments. These activities were inspired by BACON's *New Atlantis*. The Royal Society's influence on the modernization of science was immense. In particular, publication, dissemination and defense of NEWTON's discoveries was crucial in this regard.

The curator of experiments at the Royal Society since 1662, ROBERT HOOKE made many important observations. In 1664 he discovered the Great Red Spot of Jupiter, a persistent atmospheric storm; see Section 1.6. He concluded from its motion around the planet that Jupiter rotates, thus making a rotating, anti-Aristotelian Earth more plausible. He augmented the evidence two years later with the observation of motion of some of Mars's surface features. GIOVANNI DOMENICO CASSINI (1625–1712), first director of the newly founded Paris observatory since 1671 and member of the Royal Society, subsequently determined the rotation periods of Jupiter and Mars. These examples show how much insight can be extracted from careful observation and interpretation of data. Remember that less than seventy years earlier, that is, before GALILEO's *Starry Messenger*, planets such as Mars were considered stars. Like stars, planets *were* featureless specks of light before 1610; it was not known if they shine by their own light—or what they are. Hook's observations made them "tangible" objects with properties similar to Earth's.

In 1669/70 HOOKE observed the star Etamin (γ *Draconis*) and claimed to have found **positive proof** of the Earth's orbital motion. HOOKE was able to measure the star's position very accurately, because Etamin passes close to directly overhead (zenith) in London. He cited a **parallax** of $30''$, that is, half a minute of arc, which was soon realized to be too large to be credible, and yet the value was still discussed over 150 years later when Bessel found the first true parallax (of 61 Cygni) to be merely $0.3''$, that is, a hundred times smaller; see Section 3.6.3. Incidentally, Bradley found a true movement of Etamin of $20''$ in 1728, which is due to **stellar aberration**, that is not a **parallactic** motion; see Section 3.2.1.

2.9.3 Planetary Science between Kepler and Newton

For the development of modern science the seventeenth century, basically the time between KEPLER and NEWTON, was crucial. Many modern concepts and subfields emerged during this time. On the other hand, the development was not nearly as straightforward as it is often presented in textbooks. The way from TYCHO via KEPLER and GALILEO to NEWTON is, in fact, a meandering path with many detours, dead ends and even steps back.

The textbook version of the "TYCHO-to-NEWTON-science-success-story" goes something like this: TYCHO observed the planetary positions with unprecedented accuracy. KEPLER took the data and distilled from them his three laws of planetary motion. NEWTON took these **kinematic formulae**, merged them with his **dynamic** laws of motion to invent the universal law of gravity. KEPLER's laws are then nothing but a special case of NEWTON's general theory of motion and gravity. This *executive summary* is correct, would garner many

---

[67] Discoverer of a relation between pressure and volume in gases (***Boyle's law***); see below.
[68] Famous for his law of elasticity, for example of springs. See also below.
[69] His *Arithmetica Infinitorum* (1656) on mathematical infinite series heavily influenced NEWTON's invention of calculus.

2.9. THE SCIENTIFIC REVOLUTION

points on exams if you regurgitate it—and yet it paints a distorted picture of how science is made and how it progresses. Let's go a little deeper and look behind the glamorous but foreshortened textbook version of science's alleged straight and inevitable march toward an understanding of the expanding universe.

We have already seen how hard KEPLER had to work to find the pattern of planetary motion in TYCHO's large stack of data. During his search KEPLER toyed with many hypotheses, found his solutions in a piecemeal fashion and documented them likewise. Only in hindsight were those of his achievements that stood the test of time put in a logical order. For example, historically KEPLER first found his *second* law, then the first, and his third law was hidden in a book with otherwise forgettable results published a decade after the first two. Even at NEWTON's times, the three KEPLER laws did not appear together in (text)books; NEWTON knew about KEPLER's third law years before he learned about the others.

There is a straight path from KEPLER's work and anticipatory insight to NEWTON, yet it was not how things played out in science history. Indeed, it is so surprising how KEPLER's successors deviated from the straight path and fell back to pre-Keplerian thinking that we must investigate. KEPLER himself had brought up the known decline of brightness of a light source with the square of the distance to the source as an analogy to the decline of the influence of the sun on the planets' motion. As we will appreciate from NEWTON's law of universal gravitation, Equation (2.8), KEPLER had hit the nail on the head. Furthermore, KEPLER had produced the superbly accurate *Rudolphine tables*, invaluable for the practicing astronomers (and, *horribile dictu*, astrologers) of the day. Still, people like ISMAEL BOULLIAU (1605–1694) in France, doubted if KEPLER's laws were entirely correct. In particular the second law prescribing the motion along the orbit was attacked. In *Astronomia philolaica* (1645) BOULLIAU replaced it with a Ptolemaic-looking construction in his so-called **conical hypothesis**. His main thrust was to find a *mathematical* rather than a *physical reason* for planetary motion and thus for KEPLER's laws. He therefore argued in precisely the opposite direction in which the real reason was to be found by NEWTON. Amazingly, BOULLIAU was otherwise a staunch supporter of COPERNICUS, KEPLER and GALILEO, and is credited for advocating an inverse square law of the gravitational force decades before NEWTON. Following BOULLIAU, the English JOHN STREETE (1621–1689) produced the influential *Astronomia Carolina, a new theorie of Coelestial Motions* in 1661 with tables rivaling the *Rudolphian* added in 1664. Indeed, both the first Astronomer Royal JOHN FLAMSTEED and ISSAC NEWTON studied the book in college. The latter gleaned KEPLER's third law from it.

Summarizing, the seventeenth century saw an unnecessary sequel to the epic "math-vs-physics" battle of PTOLEMY and ARISTOTLE. The two strands of scientific convictions that KEPLER managed to bring together were split apart by his successors before being unified for good by NEWTON. The reason why we do not hear much about detours and backward steps of science is simple. Once they have been recognized as worthless, they are irrelevant for the progress of science, and the scientist effectively and efficiently leaves them by the roadside—to be picked up by historians of science later if at all. The scientist has better things to do than to give account for what has let him or her to the present understanding of the universe. As the present understanding of the universe *becomes* the universe, history is rewritten just as marginal theories are forgotten due to their uselessness For the seventeenth century we undo this process so we can understand why the "making of science" is a nonlinear, meandering process. After all, the seventeenth century was unique in that science became modern—which happened, of course, only once.

The Cartesian Dead End

Worse than the deviation from the pure doctrine of KEPLER by BOULLIAU and others was the regress and denial initiated by DESCARTES. Above we saw the creditable work of René DESCARTES in optics and mathematics. Here, we get to know him as the author of what we might call the *Cartesian dead end* of planetary theory and cosmology. In the period around KEPLER's death (1630) DESCARTES invented a strange but intriguing cosmology called **vortex theory** that garnered a large following, in particular in his native France and on the European continent. This is largely due to the **psychological attractiveness** of his theory that avoided **spooky action at a distance**.[70] Unfortunately, DESCARTES's **vortex theory** worked only *qualitatively*, and, though many of the best tried their hand on it,[71] never panned out *quantitatively* DESCARTES's complacency with a qualitative understanding of the cosmos is reminiscent of ARISTOTLE, as this nonchalant quote attests: "I hope that posterity will judge me kindly, not only as to the things which I have explained, but also to those which I have intentionally omitted so as to leave to others the pleasure of discovery." DESCARTES pushed for a *physical* explanation of the universe rather than a *mathematical* one as manifest in KEPLER's three laws. DESCARTES hypothesized that the universe be filled[72] with three substances: luminous particles found in sun and stars, opaque particles for planets and other non-luminous objects, and particles that filled the spaces in between which were conveniently transparent. The basic idea was that motion—and anything else for that matter, even gravity—was explainable by continuous collisions of these particles. So the whole universe should be understood as an immense collection of particles that changed the direction of their motion by coming in physical contact with each other. Consequently, several objects circulating around a central body, for example the solar system, looked like a giant whirlpool or **vortex**, hence the name of the theory. Moreover, the vortex theory allowed objects to be at rest with local matter while in motion with distant bodies. In other words, it was a contrived compromise between PTOLEMY (stationary Earth) and Copernicus (moving Earth). It was Aristotelian in its rejection of the possibility of a vacuum, and in its *physical* explanation of the universe—planets move by constant collision of their particles. Its psychological attractiveness is therefore clear for anti-Copernicans or others who rejected, say, KEPLER's ellipses. Strangely enough, KEPLER was criticized for being a mathematician by some and for being a physicist by others, when he should have been praised for being both in one person. His ideas of the pseudo-magnetic force emanating from the sun, driving the planets around their orbits *was* physical, it was modern, it was a foreshadowing of NEWTON's ideas. Alas, a large part of the scientific community was intrigued by DESCARTES vision of a mechanistic universe as described in his book *Principia Philosophiae (Principles of Philosophy)* of 1644. The followers of DESCARTES largely ignored KEPLER's results, and ended up in scientific oblivion. On the continent, reception of NEWTON's correct ideas was much delayed as a consequence. We see once again that science is made by humans.[73] While perfectly rational and plausible, DESCARTES's theory led to nothing, and therefore was, except for temporary psychological

---

[70]This expression was actually coined by Einstein in a famous paper [16] in 1935 criticizing aspects of **quantum mechanics**. Since an **action at a distance**, that is, an influence of one object on another without physically touching it, ran counter the common sense of the seventeenth century, *spooky* here seems appropriate if anachronistic.

[71]CHRISTIAN HUYGENS (1629–1695) among others.

[72]Yes, literally filled, no *scary vacuum* anywhere.

[73]Latin: *Errare humanum est (To err is human)*.

2.9. THE SCIENTIFIC REVOLUTION

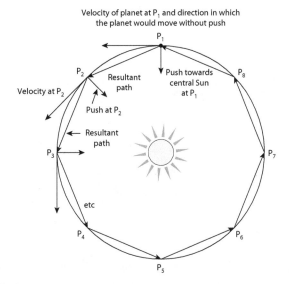

Figure 2.36: **Hooke's planetary hypothesis.** Robert HOOKE rationalized the "invisible" influence of the sun by demonstrating that a closed planetary path is possible. He assumed that the planet is subject to successive short blows or pushes toward a central sun. The planet starts at P_1 with a velocity perpendicular to the direction of the sun. When it receives a short push toward the sun, the resultant straight path leads it to P_2, where the next blow changes its inertial motion toward P_3, and so forth. A polygonic path inscribable into a circle results. When the number of blows is increased and the strengths of the blows is decreased, the number of sides of the polygon increases. Eventually, a smooth, circular path around the central sun ensues.

consolation, essentially useless and therefore forgettable.

Toward the Newtonian Synthesis

Even with the Cartesians vortex fervor raging, several important results and ideas emerged from the period after DESCARTES's death in 1650. NEWTON was able to synthesize many of them into his comprehensive theory of motion described in his *Principia* of 1687, which included, among many other explanations, the deduction of planetary motion due to the influence of universal gravity. Usually, much of the complex science history before NEWTON is glossed over, because NEWTON summarized it so well in his *Principia*. By looking behind this *façade*, we learn an important feature of science: that it is made in small, semi-continuous steps, with many scientists contributing often minuscule insights which nonetheless sum up to immense progress over the years.

In the decade after DESCARTES's death CHRISTIAN HUYGENS (1629–1695) made important discoveries. Around 1650 he formulated a concise theory of collisions, which is consistent with his Cartesian convictions. With his invention of the **pendulum clock** HUYGENS revolutionized experimental physics. Finally, good timekeeping was available. His clocks constructed around 1656 lost typically only a few seconds per day. This is a relative error of only

$$\frac{2s}{24 \times 60 \times 60s} = 0.000023 = 0.0023\%.$$

HUYGENS also developed the kinematics of circular motion late in the 1650s and finally published all of these results in his *Horologium Oscillatorum (The Pendulum Clock)* in 1673. The book contained the formula for the centripetal acceleration of an object in **uniform**

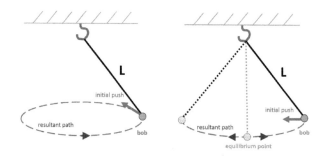

Figure 2.37: **Hooke's conical pendulum demonstration.** The two motions generated by HOOKE's conical pendulum. (a) Circular or elliptical motion when the pendulum receives a blow perpendicular to the direction of the equilibrium point (b) Straight (pendulum) motion if the blow is in the direction toward the equilibrium.

circular motion. As we will see, this formula was crucial for NEWTON when comparing the acceleration of the moon with the acceleration near Earth's surface to test his theory of gravity.

In 1666 HOOKE introduced an important *qualitative*, force-based model of planetary motion. It arguably led to NEWTON's *quantitative* explanation of the motion due to universal gravity. HOOKE explained the orbits of the planets around the sun as a compromise of two "tendencies." The tendency of objects to move in straight lines due to their inertia, and the tendency of the sun to attract the planets. His thoughts show how science progresses by carefully spelling out possible explanations of an observed pattern in nature. By noting that a planet should travel in a *straight* line due to its inertia (because no pushing or pulling by rods or ropes is observed), one has to admit by observing the *curved* motion of the planet in orbit that there must be some "invisible" force that pulls the planet toward the sun. HOOKE went on to construct the circular or elliptical orbit as a sequence of straight motion perpendicular to the direction toward the sun, and a short straight motion toward the sun. The resultant path is a polygon, Figure 2.36, that becomes more circular as distance between subsequent points is shortened, and in the limit turns into a *bona fide* orbit.

True to his role as curator of the Royal Society HOOKE illustrated his explanation with a convincing demonstration using a **terrestrial machine** to emulate celestial motion. The "machine" is a conical pendulum, that is, a mass (bob) suspended by a rope or rod from a point at a ceiling, Figure 2.37. The bob is always at a fixed distance L from the pivot point, but otherwise moving freely. The bob represents the planet in this physical model of planetary motion. Note that the net force on the bob is the sum of downward weight of the bob and the tension along the inclined rope or rod. The direction of the net force is thus toward the center, that is, the equilibrium point which the bob would eventually assume if left alone. The bob is initially held at rest in a position off the equilibrium point If the bob is set in motion by a quick horizontal push perpendicular to the central direction as in Figure 2.37(a), a circular motion results, even though there is a force which is always pulling the bob toward the center. If the bob is simply let go (or pushed toward the center), it will move toward the center, much like an unsupported stone will simply drop to the ground. This demonstration showed how different motions can result under the influence of the same central force, suggesting that planetary motion and free fall near the surface of the Earth might be explainable by the same physical mechanism. It also showed that planetary motion

2.9. THE SCIENTIFIC REVOLUTION

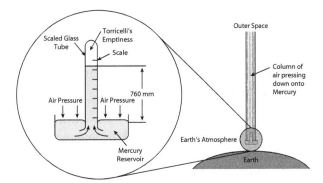

Figure 2.38: **Torricelli's mercury barometer.** A sealed glass tube full of mercury is put upside down into a reservoir full of mercury. Mercury flows out of the tube until the weight (a force!) of the mercury in the tube equals the weight of the column of air pressing down onto the mercury in the reservoir. The vacuum created above the mercury in the tube was dubbed the *Torricellian emptiness*. The contraption is a barometer, since the level of mercury changes with air pressure. In particular, higher air pressure will force more mercury into the tube. The mercury level can be read off at the scale. Standard air pressure is equivalent to a mercury column of height 760 mm.

can be seen as resulting from a **central force**, that is, a force that is always directed to a fixed point in space (here the equilibrium point). This was a very inspiring idea that was eventually realized by NEWTON when he postulated the central force of universal gravity, always pointing toward the sun, the gravitational center of the solar system.

2.9.4 The Vacuum Exists, Falsifying Aristotle and Starting Thermodynamics

Why was DESCARTES's (wrong) assertion that the heavens were filled with particles so comforting? What is so horrible about the existence of a vacuum? At fault is an ossified piece of ARISTOTLE's teachings. In reaction to the atomic ideas of DEMOCRITUS and others (Section 2.2.1) who postulated smallest particles drifting in empty space, ARISTOTLE postulated the existence of a substance, the **aether** or **quintessence**, filling all space to make motion possible. ARISTOTLE deemed motion without a medium impossible and diagnosed nature with a **horror vacui** or fear of a vacuum. With an overly philosophical argument, Platonists rejected the vacuum as something that is by definition *nothing*, and therefore does not and cannot exist.[74]

In reality, the existence of a vacuum is rather easily demonstrated. GALILEO's student EVANGELISTA TORRICELLI (1608–1647), for whom units of pressure are named,[75] did it by inventing the **mercury barometer**, Figure 2.38 in 1643. The barometer is simply a one meter long glass pipe containing mercury and sealed at the top. When put into a bowl full of mercury, the column drops to a height of 760 mm, creating a vacuum above it.[76] In fact,

---

[74] Another hilarious example of sophistic reasoning is the *Refutation of Bishop Berkeley* by SAMUEL JOHNSON (1709–1784). "After we came out of the church, we stood talking for some time together of BISHOP BERKELEY's ingenious sophistry to prove the nonexistence of matter, and that every thing in the universe is merely ideal. I observed, that though we are satisfied his doctrine is not true, it is impossible to refute it. I never shall forget the alacrity with which JOHNSON answered, striking his foot with mighty force against a large stone, till he rebounded from it—'I refute it thus.'[8]"

[75] 760 torr corresponds to standard atmospheric pressure.

[76] At the time, the empty space above the mercury was called *Torricelli's emptiness* A bitter fight between

the level of mercury in the pipe changes with the air pressure, because higher air pressure pushing on the mercury in the bowl forces more mercury into the glass pipe. This simple contraption is therefore an instrument to measure air pressure, that is a **barometer**. In essence, it compares the force of a column of mercury with the force of a column of air, both pushing onto the mercury in the bowl.

TORRICELLI's strikingly simple experiment begs the question why nobody did it earlier. As often in science, it was inspired by a practical problem. The Grand Duke of Tuscany, FERDINANDO II (1621–1670), was obsessed with new technology and wanted water pumped for an impressive fountain to entertain his guests. At the time, water was pumped with suction pumps (for example for mining purposes), but it was found that they cannot raise water higher than about 10 meters, and the Grand Duke needed water raised more than 15 meters. Unable to have it his royal way, the Grand Duke called on GALILEO, his court mathematician. Recall that GALILEO was under house arrest for Copernican heresy at the time, and died in 1642. His student TORRICELLI took over and clarified the situation by using a liquid thirteen times as dense as water, namely mercury. His choice was motivated by the fact that the pipe in the experiment could be thirteen times shorter, namely $10\,\text{m}/13 \approx 760\,\text{mm}$. This line of research led to the invention of the air or **vacuum pump** in 1648 by OTTO VON GUERICKE (1602–1686) who demonstrated the power of air pressure in 1654 with dramatic experiments in front of a large crowd including the Holy Roman Emperor: Four horses could not separate two evacuated metal hemispheres, held together only by outside air pressure. GUERICKE also showed that light travels in vacuum, while sound does not.

As the last example for groundbreaking scientific activities during the scientific revolution, we consider the work of ROBERT BOYLE (1627–1692), a founding member of the Royal Society of London. He started out as an **alchemist**, but became a champion of the new scientific method based on diligent experimentation and scrupulous documentation of results, including estimates of their uncertainties and a discussion of their interpretation. With ROBERT HOOKE he improved on GUERICKE's air pump. Subsequently, BOYLE entered into extensive experimentation with air pressure. Remember that pressure is a force per area; if you press down a piston of area $A = 1\,\text{m}^2$ with a force of $F = 1\,\text{N}$ (1 Newton), you exert a pressure of $p = F/A = 1\,\text{N}/1\,\text{m}^2 = 1\,\text{Pa}$ (one Pascal).[77] In a series of ingenious experiments he found that the volume and the pressure of a gas in a container are inversely related (at constant temperature). Quite intuitively, the volume of a gas drops as the pressure on it rises, and vice versa; see Figure 2.39. This became known as **Boyle's law** when he published it in 1662. It is a special case of the more general **ideal gas law**, Equation (3.6).

Thus, the **natural philosopher** BOYLE started research in what we would nowadays call **chemistry** and **thermodynamics**. The astronomical relevance of the latter comes from the fact that it explains phenomena from the boiling of water to the inner workings of the gigantic, hot gas balls we call stars.

---

two scientific sects, the *Plenists* and the *Vacuists*, raged for decades, on the question whether this space is full (*plenus*) or empty (*vacuus*).

[77] One Newton is not much of a force; it's the weight of a mass of about $0.1\,\text{kg}$. Pressing with this miniscule force onto a huge area of $1\,\text{m}^2$ produces a small pressure indeed. Normal atmospheric pressure is a staggering $100\,000\,\text{N/m}^2$.

2.9. THE SCIENTIFIC REVOLUTION

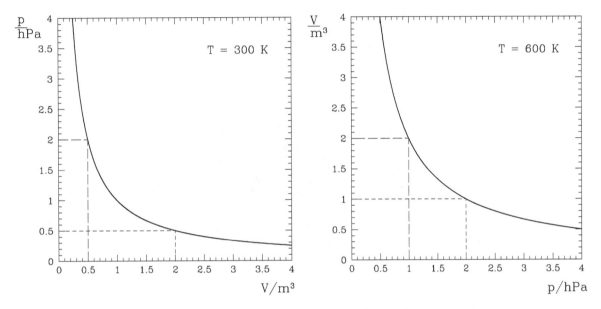

Figure 2.39: **Two aspects of *Boyle's Law*.** (a) The volume of a quantity (1 mole) of gaseous oxygen as a function of its pressure at temperature $T = 300\,\text{K}$. (b) The pressure of 1 mole of oxygen as a function of its volume at $T = 600\,\text{K}$. The two graphs *appear* to be the same, but they are very different. They depict different quantities in different units and two different situations. The temperature is kept constant throughout, but it is half as high in (a). The graph (a) suggests that the pressure on the oxygen was varied, and its resulting volume, that is, the volume at that pressure was measured. The volume appears here as a *function* of the pressure, the volume being the dependent, and the pressure the independent variable. In (b) the situation is reversed: Now the pressure is measured as the independent variable and the volume the gas occupies is varied. The descending curve is not straight—the two variables are not linearly related. Rather, they are mutually *inversely related*: As volume goes up, pressure goes down, and as pressure goes up, volume goes down. This is intuitive, since if you squeeze a compressible object (put more pressure on it) it becomes smaller (its volume decreases). BOYLE's insight was that the change happens in a special way, namely that the product of the two stays constant: $pV = const$. It was later found that the constant is different for different temperatures and also for different gases and different quantities of gas. Note that the two graphs would be redundant (but not the same!) if the temperature were the same in (a) and (b), since a graph can be read in two ways. The curve representing the relation between the two variables p and V can be used to find the p value for a chosen V value, and also to find the V value for a chosen p value. For example, to find the volume V_1 of 1 mole oxygen at temperature $T = 300\,\text{K}$ and at a pressure of $p_1 = 2\,\text{hPa}$ (2 hecto Pascal $= 200\,000\,\text{Pa} = 2 \times 10^5\,\text{Pa} = 2 \times 10^2\,\text{N/m}^2$), you'd use the (a) graph and draw a straight line parallel to the V axis starting at $p = 2\,\text{hPa}$. Where the (dashed) line hits the curve, you start a line parallel to the p axis, and read off the V value where it intercepts the V axis. To read off the pressure of 1 mole of oxygen at $T = 300\,\text{K}$ with a volume of $V_2 = 2\,\text{m}^3$, you go the opposite way, yielding $p_2 = 0.5\,\text{hPa}$. Note that the ratios of the volumes and pressures are inversely related $p_1/p_2 = 1/(V_1/V_2) = V_2/V_1$, which is clear from *Boyle's law*, since it is *equivalent* to $p_1V_1 = p_2V_2 = 1\,\text{hPa}\cdot\text{m}^3$. Here the constant in *Boyle's law* turned out to be $10^5\,\text{N}\cdot\text{m}$, due to the specific value of temperature, and the fact that we considered a quantity of 1 mole. The fact that we chose oxygen is of no consequence here, since it is considered an *ideal gas*. Hence, *Boyle's law* works for any other (ideal) gas which is useful, because the same formula can be used many times over. Effectively, thousands of experiments are described with one law; thousands of patterns in nature can be reduced to one: that the volume and pressure of a gas are inversely related.

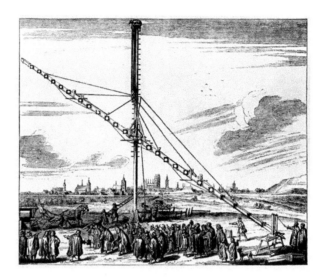

Figure 2.40: **Hevelius's 140-foot refracting telescope.** This woodcut is from HEVELIUS's 1673 book *Machina coelestis* (first part). The telescope wasn't much more than two lenses attached to a wooden pole, 140 ft (45 m) apart, with an elaborate hanging and raising mechanism. This was a very delicate and error-prone instrument which was difficult to use.

2.9.5 Observational Astronomy and a Theory of Light

As planetary astronomy got stuck with DESCARTES's **vortex theory**, another branch of astronomy which came into being in the seventeenth century ran into a roadblock. Telescopic astronomy, all the rage since GALILEO's *Siderius Nuncius* of 1610, developed feverishly. Larger and larger telescopes were constructed; see JOHANNES HEVELIUS's (1611–1687) 140-foot giant, Figure 2.40. A new type of scientific endeavor and literature was born. The sheer documentation of observations of celestial objects needed no mathematical expertise whatsoever, and became a favorite pastime of astronomers and amateurs alike. Even here the maturing of publications toward modernity is astounding. Compare the rather crude maps of the moon of GALILEO early in the century with HEVELIUS's, Figure 2.41 around 1650.

What's with the funny, overly long telescopes of the time? I have to confess that for a long time I held the plausible but wrong opinion that their extreme focal lengths were to maximize magnification, cf. Equation (2.4). Alas, it was just a band-aid to get around the debilitating image distortions of the single lenses used at the time, namely **spherical** and **chromatic aberration**. We will soon discuss the physics responsible for the latter, namely the color-dependence of **refraction** called **dispersion**. The image distortion was minimized by maximizing the objective lenses' focal length.[78] However, these overly long telescopes were prone to all kinds of problems, including instability due to wind. EDMOND HALLEY (1656–1742), upon seeing one of HEVELIUS's giants, described them simply as "useless."

This all changed in 1668 with the invention if the **reflecting** telescope or **reflector** by ISAAC NEWTON. Indeed, NEWTON achieved greatness in optics before his well-known work in physics and math, which we will discuss in the next section. The basis of the new type of telescope is the law of reflection, simply stating that a light ray is sent back by a reflective

---

[78] Later, **achromatic lenses** were used.

2.9. THE SCIENTIFIC REVOLUTION

Figure 2.41: **Map of the moon.** This map shows the state of the art of lunar observation in 1647. It is from JOHANNES HEVELIUS's *Selenographica*.

surface with the same angle as it was received: incident angle equals reflected angle or

$$\alpha = \beta, \qquad (2.5)$$

see Figure 2.42(a). The upshot is that curved mirrors, in particular those in the shape of a parabola, will reflect *any* light beam such that it goes through a focal point as in Figure 2.42(b). It is thus clear that a curved mirror can produce an image, very much like a lens produces an image. As a refractor, the reflector has a second lens, the eyepiece, which takes this image as its object and produces a much magnified image for the observer. There are different types of reflecting telescopes, but the basic design is the same for all of them. The **Newtonian reflector** depicted in Figure 2.43 uses a secondary, flat mirror to send the light toward the eyepiece perpendicular to the main mirror. All of today's large telescopes are reflectors, because lenses (which can be supported only along their rim) will bend under their own weight if too large and ruin the image. Thus, NEWTON had constructed a telescope with much reduced color distortions and much enhanced magnification.

His first reflector had a main mirror[79] of only one inch diameter, but his second, much improved reflector was impressive enough to garner the approval of the Royal Society which in turn elected NEWTON as a member in 1672. The reflecting telescope was thus his entrance ticket to the scientific establishment.

Even before the reflector, NEWTON had achieved much in optics. In his miracle year of 1665/66, when he avoided the plague and the Great Fire of London by relocating to the countryside, he started to develop his *theory of light and color.* An unsettled question since antiquity was why colors appear when (white) light travels through a curved piece of glass. It is plausible to assume that the glass somehow "adds" the color to the light, but this is a pre-scientific notion. NEWTON came up with the different hypothesis that white light somehow

---

[79]Telescope mirrors were not of glass until much into the nineteenth century. The **speculum metal** used initially had many drawbacks, in that it was not very reflective, and easily tarnished. Therefore, **refractors** stayed competitive for awhile.

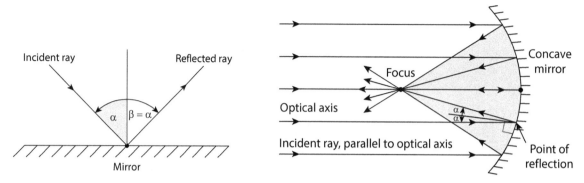

Figure 2.42: **Reflection.** (a) *The law of reflection*: A light beam is incident on a reflective surface like a mirror. If it's coming in at an angle α with respect to the normal to the surface, it will leave with the same angle $\beta = \alpha$. (b)*A curved mirror focuses light rays*: For an appropriately (parabolically) curved mirror, any light ray parallel to the optical axis will be reflected such that it goes through the focal point. The law of reflection is satisfied at any point of reflection. In particular, at the **optical axis** the ray is returning to where it came from, $\alpha = \beta = 0°$.

consists of all the colors of the rainbow, and that the role of the glass is to separate them. The glass possesses thus a property that distinguishes the colors, and reacts differently if red passes through it than if blue light passes through. This surprising claim ("White is all colors"), needed strong evidence to be acceptable for other scientists, and NEWTON delivered. With a set of two experiments he showed beyond doubt that it is the color-dependence of refraction or **dispersion** of the glass that produces the colors. First he used a prism to create a specific color, say yellow, from white sunlight, Figure 2.44. Then he sent the yellow light through a second prism, and no further dispersion occurred: the yellow light stayed yellow. The second experiment was more challenging, but even more convincing—even crucial.[80] NEWTON managed with an ingenious arrangement of multiple prisms (after the first which produced the colors) to reunite all the different colors—and they added up to white light! This result also explains the **chromatic aberration** of single lenses. A lens will send all rays to one focal point *only* if all the rays have the same color, that is, are **monochromatic**. If multicolored light—and white light is multicolored!—falls onto a lens, it will be focused at different points, and therefore blur the image.

After all these successes, it is maybe consoling to learn that even NEWTON didn't get everything right. When he went on to formulate a general *theory of light* in 1675 he posited that light is composed of many small particles (**corpuscles**), and is thus not different from matter. This is known as the **corpuscular theory** of light which turned out to be wrong. How come, then, that NEWTON's corpuscular hypothesis remained the leading theory of light until the Young interference experiment of 1803? HUYGENS had postulated light to be a (longitudinal) wave (like sound was known to be) in his 1678 *Traité de la Lumière (Treatise on Light)*. He had empirical evidence on his side, but there was a powerful argument against light being a wave. Namely, light was seen as always traveling in straight lines (light rays!) and waves should have shown their wavy nature by bending around obstacles like water waves do, see Figure 2.46(a). The other reason why NEWTON's wrong theory prevailed so long is thoroughly unscientific. After the publication of his *magnum opus*, the *Principia* in 1687, he

---

[80]An **experimentum crucis** (crucial experiment) shows decisively which of the contending hypotheses explaining a pattern in nature is superior.

2.9. THE SCIENTIFIC REVOLUTION

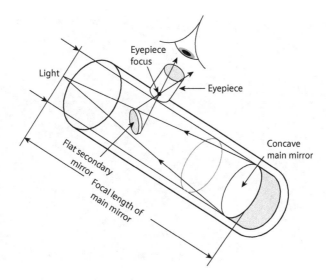

Figure 2.43: **Newtonian reflector.** Incoming light is reflected by the concave **main mirror** and sent to the flat **secondary mirror**. From there it travels perpendicular to the telescope's symmetry axis to the eyepiece. Like for the refractor, the reflector's magnification is the ratio of the focal lengths of the main mirror and the eyepiece, $M = \frac{f_{obj}}{f_{eye}}$.

became such an authority that other scientists thought twice before questioning NEWTON's theories. This role of NEWTON is reminiscent of the status of ARISTOTLE and PTOLEMY in previous centuries.

The true nature of light as a **transverse wave** of oscillating electric and magnetic fields did not emerge until almost two hundred years later. The *properties* of light, however, were observable and subject to experiments much earlier. The results of these experiments applied directly to and propelled observational astronomy. One of the disputed properties of light was its speed, assumed to be infinite since antiquity; light was thought to travel **instantaneously**. This hypothesis was falsified by the Dane OLE RØMER (1644-1710) using astronomical observations, see also Figure 3.20(a). Namely, he noted that *predictions* of the eclipses of Jupiter's Galilean moons were off by as much as 16 minutes when compared to observation. Rømer realized that this could be explained by a finite speed of light. The idea is that the signal "Jupiter's moon is eclipsed" has to travel to us from Jupiter's position which takes appreciable time if the speed of light is finite. The length of this time depends on the distance between Jupiter and Earth, which can vary by as much as as the diameter of Earth's orbit (2 AU); see Figure 2.45. As a crude estimate, assuming maximal time discrepancy, we get a light speed of 2 AU per 16 minutes, so

$$c \approx \frac{1\,\text{AU}}{8\,min} = \frac{1\,\text{AU}}{480\,\text{s}} = \frac{150}{480}\frac{\text{million km}}{s} = \frac{5}{16}\frac{\text{million km}}{\text{s}} = 315,000\text{km/s},$$

in fair agreement with the modern value of 300 000 km/s.

RØMER's similar value for the speed of light, and in particular its finiteness, was used by HUYGENS in his 1678 *Traité de la Lumière (Treatise on Light)* to *explain* the **law of refraction** by the following hypotheses.

- Light is a disturbance of a medium consisting of tiny particles.[81]

---

[81] HUYGENS's Cartesian convictions apparently served him well here.

Figure 2.44: **Newton's** *experimentum crucis*. Incoming light is refracted by the prism. The angle of refraction is dependent on the color of the light, a phenomenon known as the ***dispersion***, see Figure 1.10. White light is therefore decomposed into its "components," the ***spectral colors*** ranging from violet to red. Newton showed that this is the right explanation by sending monochromatic (yellow) light from the first prism and through a second prism, which did *not* decompose it any further. He also managed with a series of prisms to collect all colors and add them to reform white light.

- Each point of the medium reached by the disturbance becomes the source of a (secondary) spherical disturbance or wave (***Huygens's principle***, Figure 2.46(b)).
- The resulting disturbance is the sum of all secondary disturbances.

It then follows that at the boundary between two media of different density, there will be a change in direction of the incident wave in accordance with the ***law of refraction***, Equation (2.27) and Figure 2.46(b). The modernity of HUYGENS's idea was that it modeled an ephemeral phenomenon such as light in medium as a "terrestrial machine," here: tiny particles moving and colliding with each other. In other words, it enabled scientists to think rationally about seemingly abstract patterns in nature, and to make predictions as to how it should behave. Unfortunately, it also introduced the prejudice that light needs a medium to propagate in, the so-called ***aether hypothesis***. It took considerable effort in the late 1800s to falsify it. Electromagnetic waves, contrary to sound or water waves, can travel without a medium, that is, in vacuum.

Summarizing, we saw that scientists of the seventeenth century had similar if individual ideas about how to do science. Nonetheless, the scientific paradigm shifted toward precise observations and formalized logic—even estimates of uncertainties were beginning to appear. In other words, science was maturing and becoming modern.

Concept Practice

1. Who invented the pendulum clock? (a) Newton (b) Huygens (c) Rømer (d) Hooke (e) None of the above

2. What is a *terrestrial machine*?

    (a) The Earth itself

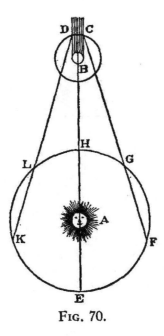

Figure 2.45: **Rømer's determination of the finite speed of light**. In Rømer's illustration, Jupiter B casts a shadow which its inner moon Io enters at C and exits at D. Io is thus eclipsed by Jupiter between C and D. The Earth moves around the sun counterclockwise, so it moves toward Jupiter and Io between F and G and away from Jupiter from L to K. If light travels at finite speed, the rotation period of Io (the time from one eclipse to the next) is systematically too small if moving toward Jupiter, and systematically to large if moving away from Jupiter. Between L and K, the extra light traveling time results in a delayed observation of the next eclipse. Hence, Io's orbital period *seems* longer.

(b) An apparatus that simulates the Earth's motion in space

(c) A contraption that emulates some feature of the cosmos

(d) None of the above

2.10 The Universal Explanation: Newton

ISAAC NEWTON (1642–1727) was born on Christmas day in the year GALILEO died[82] in the small hamlet of Woolsthorpe, England. He was arguably the greatest natural scientist in history, and without doubt the pivotal figure in finalizing the *scientific revolution* by lifting science squarely into modernity. As we have seen with other eminent scientists, he was, naturally, still of a mind-set he himself helped to change irrevocably, as his extensive and fruitless labors in *alchemy* attest. Yet, the fact that he was able to rise above the mystic fog of yesteryear's scientific convictions speaks rather for him than against him—as it does for his predecessors and successors in science and other fields who transcend their own persuasions in making progress. As a book is not enough to do justice to NEWTON's

---

[82] He was born before the Gregorian calendar reform was introduced in England (1752). While he was born when Christmas was celebrated in England in 1642, his Gregorian birth date is January 4, 1643. Since Italy adopted the reform right away (1582), technically NEWTON was born the year *after* GALILEO died January 8, 1642 (greg.), which is why authors often state that NEWTON was born *within* a year of GALILEO's death.

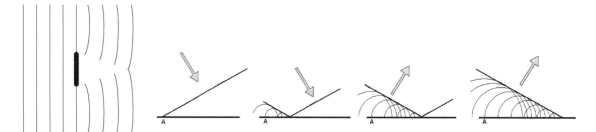

Figure 2.46: **Wave phenomena.** (a) Water waves bending around an obstacle. (b) *Huygens principle:* Every point reached by a wavefront becomes the source of a spherical **secondary wave**; the new position of the wavefront a moment later is the **superposition** of all these secondary waves. This bold **hypothesis** allowed HUYGENS to explain, among other things, the **law of refraction** Equation (2.3) and the **law of reflection**, Equation (2.5). The latter is illustrated in the four sketches. Since the radius of the secondary waves increases with time, the new wavefront is straight and inclined by the same angle α as the incident wavefront. Since a wavefront is perpendicular to the direction of the light ray, the arrows can be interpreted as the incident and reflected light ray, this is precisely what the **law of reflection** states in Figure 2.42. The modern explanation in terms of electromagnetic waves, that is, periodically changing electric and magnetic fields, is different, but not in contradiction with HUYGENS **phenomenological** description of the phenomenon.

discoveries, so a semester, much less a small part thereof, will not suffice to learn his science. We can therefore do little more than lip service to the man and his work. On the other hand, if you've studied some introductory physics, most of this section on mechanics will be all too familiar to you. In any case, rather than turning you into an astrophysics problem solver, the purpose of this chapter is to give you an appreciation of how much the universe has changed and expanded due to the work of ISAAC NEWTON.

We have discussed NEWTON's discoveries in **optics** in the last section, and focus here on NEWTON's theories of motion and gravity. They conclude our story of the problem of the planets, which it solved once and for all.[83] A bit of chronology is therefore warranted, because the solution of a two-thousand-year-old problem in basically twenty years (1666-1687) begs the question how NEWTON was able to succeed in such a short time. The short answer is that many of the ingredients of the solution were results scientists before NEWTON had already extracted from the patterns in nature. With many "correctish" ideas floating around, it was NEWTON's genius that was able to put them together, add to them as necessary, and synthesize them into an amazingly simple set of equations that describe the motion of the whole universe—from the falling apple on Earth to the rotation of the galaxies—truly a universal achievement.

Here is the extensive list of what had to happen to make this possible.

- HUYGENS came up with a theory of **uniform circular motion** describing the **kinematics** of an object rotating *around* a center, that is, a formula relating the acceleration of the object *toward* the center $a_{centripetal}$ to its velocity v around the circle and the radius r of the circle

$$a_{centripetal} = \frac{(\text{velocity of object going around a circle})^2}{\text{radius of the circle}} = \frac{v^2}{r}. \qquad (2.6)$$

---

[83] Of course, in science nothing is solved once and for all, and technically, NEWTON's solution was superseded by Einstein's gravitational theory called *general relativity*.

2.10. THE UNIVERSAL EXPLANATION: NEWTON

- HOOKE's intriguing idea that the inertia of an object together with a **central force** on the objects can create a circular or elliptic path if the object has a lateral velocity. HOOKE's demonstration in the form of a conical pendulum also makes it plausible that the same central force can explain circular motion around the center, for example of the moon around the Earth, *and* straight motion toward the center, for example of the falling apple toward the Earth.

- NEWTON was educated at Trinity College in Cambridge starting in 1661 and profited from the instructions of his talented teacher ISAAC BARROW (1630–1677) who recognized NEWTON's genius. JOHN WALLIS was also a professor at Trinity, and influenced NEWTON's progress in mathematics.

- In the years 1665 and 1666 the plague swept through England. In the summer of 1665 at least thirty thousand people died of the disease in London, and the university in Cambridge closed. NEWTON was forced to relocate to his hometown Woolsthorpe. Here, in the solitude of the countryside, within the year he made most of his scientific discoveries and inventions. The year is therefore known as NEWTON's miracle year or **annus mirabilis**. He invented differential and integral calculus, his theory of colors based on his experiments with prisms, and the inverse square law of universal gravitation. Virtually all ideas that he worked out and published in the next twenty years came to him during this amazingly short period. NEWTON returned to Cambridge Easter 1668, and rose through the ranks to become Lucasian Professor of Mathematics in 1669; his teacher Barrow had abdicated the Lucasian chair in favor of NEWTON, helping the cause of science tremendously in hindsight.

- As we already saw, the invention of the **reflecting telescope** made NEWTON a member of the Royal Society in 1672, and brought him into contact with the eminent British scientists of the day, like CHRISTOPHER WREN, JOHN WALLIS and EDMOND HALLEY. While the latter later became famous for predicting the 1759 return of the comet that now bears his name, all three had been unable to figure out the orbit of the great comet of 1664. Their vexing inability to solve the problem eventually became inspiration for NEWTON to try his hand on it as another comet appeared in 1680—with the known successful outcome. Incidentally, as late as 1680 NEWTON learned of KEPLER's second law of equal areas which helped him figure out the law of universal gravitation.

- In 1684 Halley visited NEWTON and asked of him the "fateful" question [44]: What is the shape of the orbit of an object moving under the influence of a force that falls off like an inverse square of the distance? NEWTON promised Halley a proof of a solution he had found already—and the rest is history. The problem of the planets was finally solved and published in NEWTON's *magnum opus*, namely the *Philosophiae Naturalis Principia Mathematica (Mathematical Principles of Natural Philosophy)*, aka the *Principia*, in 1687.

This list of fortuitous coincidences and accounts of NEWTON's genius beg another question: If he had indeed anticipated all his later results already in his *annus mirabilis*, why did it take him so long to publish the solution to the problem of the planets in his *Principia*? Well, the devil is always in the details, and indeed there were several issues with his ideas that had to be dealt with before publication. When publishing scientific results, one has to make sure that there are no loopholes in one's arguments. Otherwise scientific opponents will not accept the

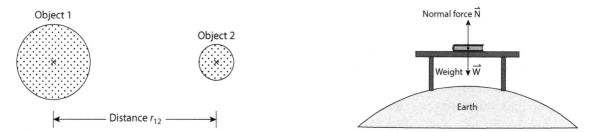

Figure 2.47: **(a) Extended objects can be idealized as mass points.** Extended objects consist of myriads of little particles which all are subject to the force of gravity. This Cartesian notion is correct even by modern standards. NEWTON worked hard to show that the gravitational force between two extended (spherical) objects is the same regardless of their radii; only their masses m_1 and m_2 and mutual distance r_{12} matters. Therefore, even large, close objects can be idealized as featureless point particles. This simplifies calculations and problem solving tremendously. **(b) Forces on a book on a table.** A book at rest on a table is subject to two forces opposite in direction and equal in strength. The *normal force* $\vec{N}$ is a *contact force* that arises because the book touches the table. The *weight* $\vec{W}$ is a force that arises due to the gravitational attraction proportional to the mass of the book and the mass of the Earth. It is a force at a distance—book and Earth do not touch. Both forces are shown acting on a point which is the center of mass of the book and the *idealization* of the book as a point particle. Note that $\vec{N}$ and $\vec{W}$ both act on the book, and therefore are *not* related by NEWTON's third law, even though they are equal and opposite.

new—and often radically different—ideas and concepts brought forth. While convinced that he was generally on the right path, NEWTON saw the following problems with his hypothesis of universal gravitation.

- To test the inverse square law of gravity, NEWTON had to compare the acceleration of moon in orbit and apple on Earth (see below). He had assumed circular orbits for simplicity. The generalization to elliptical orbits was not straightforward.

- Being under the spell of DESCARTES's corpuscular **vortex theory**, NEWTON had to address the fact that real, extended objects are not point-like, but made out of myriads of little particles which all individually feel the force of gravity. It was a tough problem to show—with calculus methods he had freshly invented—that the gravitational field of an extended object of a certain mass is equivalent to the field of a point-like object if all the mass is concentrated at the center of the extended object, that is, at the position of the idealized point particle, see Figure 2.47(a).

- If all objects attract each other gravitationally, then, for example the moon's orbit is not just due to the gravitational influence of the Earth, but also of the sun and the other planets. This leads to a very complex problem, and the question arises if and why one can approximate it by treating it as a two-body (Earth-moon) problem, that is, by omitting all other influences.

As you can see, it can be a frustratingly long way from the initial "epiphany" of having the right idea, to a bullet-proof mathematical formulation of the idea. Having thus motivated the historical genesis of NEWTON's theory, we will describe in detail what the theory entails, thereby hopefully elucidating its truly universal character.

2.10.1 Newton's Three Laws of Motion

NEWTON's three **laws** or **axioms** of motion are easy to memorize. On the other hand, it

2.10. THE UNIVERSAL EXPLANATION: NEWTON

is very hard to learn how to use them, that is, apply them to problems in mechanics like the inclined plane or circular motion. Newton's laws are thus a bit like the rules of *chess*: Knowing how the chess pieces move doesn't mean you are a good chess player—it takes years of practice to become a chess master. NEWTON's laws are more correctly referred to as axioms, meaning assumptions that have to be accepted and cannot be proved. The sole justification for these assumptions is their empirical success. *If* one uses the axioms to figure out (*predict*) how an object moves, *observes* the actual motion, and finds that the two *agree*, one has produced evidence for the correctness of the axiom, but not a proof. Someone could do another test and may find disagreement, in which case the axioms would have to be abandoned or modified. Because there have been millions of cases over the centuries in which NEWTON's axiom proved correct and therefore useful, it is extremely unlikely that this will change, but not impossible.[84]

The First Law (Law of Inertia)

An object at rest will remain at rest, and an object in motion will remain in uniform motion, **unless** *a net force acts on it.*

It is clear why this axiom is also called the **law of inertia**. According to this axiom, and long acknowledged by other scientists, objects have the tendency to resist change of their motion. This tendency is called *inertia*, a term often also used metaphorically in everyday language for a person slow to adapt to changing situations. The importance of the first axiom is its anti-Aristotelian assertion that not just rest, but also motion at constant speed (uniform motion) is "natural." An "unnatural" change of the motion, either in speed or direction, has to be explained.[85] The action of a net or **unbalanced force** is such an explanation. Typically, there are several forces acting on objects, like gravity, friction, and the normal force of a hard surface. Often, they act in different, even opposite directions and therefore their influences cancel each other—at least partially. Only if the sum of the **force vectors**[86] is nonzero, that is, only if there is a **net force**, will the velocity change. As a seemingly trivial but instructive example consider a book lying on a table, Figure 2.47(b). From trivial observation we glean that the book is at rest—and stays at rest. NEWTON's first law then tells us that there is no net force acting on the book. The novice is typically perplexed when realizing that gravity nonetheless does act on the book. How can there be a force acting on the book when the net force on it is zero? There must be another force that has the same strength as gravity but opposite direction. Indeed, the normal force, a contact force between surfaces, is pointing up while gravity points down, and the two add up to zero. No paradox here, but the situation is admittedly confusing for the uninitiated—and also frustrating, since the physical system "book on table" is so utterly simple. Hopefully you get the idea that part of NEWTON's genius was to recognize what pattern in nature needs explanation, and which doesn't—often quite contrary to common sense.

---

[84] To extremely fast or small objects, NEWTON's laws do not apply. These situations are outside of the realm of validity of NEWTON's theory and have led to the development of *special relativity* and *quantum mechanics*, both of which are non-Newtonian theories.

[85] If this seems rather Aristotelian, note that the categories of what is natural and what is not have changed significantly in NEWTON's version. Also, there is nothing wrong with categories per se, as long as they prove useful.

[86] Forces have a strength *and* a direction and have to be added such that all of these properties are consistently summed. These rules are called **vector algebra**, and have a straightforward geometrical meaning; see [28].

The Second Law

The rate of change of an object's motion is directly proportional to the net force $\vec{F}$ acting on it and is in the direction of that force.[87]

NEWTON's original version[88] is slightly more general, but in most cases the mass m of the object is constant, and therefore its acceleration $\vec{a}$ is indeed proportional and in the same direction as the force $\vec{F}$. This is the stereotypical NEWTON law which, in its form $\vec{F} = m\vec{a}$, everyone can recite in their sleep, but hardly anyone will truly understand. Full disclosure: it took me years to fully grasp its meaning, all the while being able to correctly apply it in calculations. What I failed to appreciate is the fact that the law equates two very different things. On its left side is a **dynamical quantity**, namely the force explaining why a certain motion happens. On the other side is a product of a property of the object itself, its mass, and the **kinematic quantity** acceleration. The latter is the (rate) change of the object's velocity, and thus tells us how fast (in time) and in which direction the object's velocity changes. In other words, the right side of $\vec{F} = m\vec{a}$ tells us *how* the object moves, and the left side tells us *why* it moves in this way. The amazing thing is that the two sides are equal. Therefore, we can compute the motion from the force if we know the force acting on the object, **or** what force must act on the object, if we observe how it moves—but not both! All kinds of motion problems can be solved with this equation: how the baseball flies, how the moon orbits, how the tides change. Of course, one has to know the forces these objects are subject to. The origin of the forces is the subject of the third axiom.

The Third Law

Given two interacting objects, object 1 exerts a force on object 2 that is equal in strength and opposite in direction as the force exerted by object 2 on object 1.

Sloppily speaking, everything that gives a push receives a push in the opposite direction. For example, if a white billiard ball collides with a black one, it pushes it and forces it to move forward. But it itself also receives a push in the opposite direction which slows it down. If you think about it, this is where all forces come from, they are mutual interactions between (at least) two objects. Most of them are contact forces transmitted by stick or ropes. Initially, noncontact forces like gravity or the electrostatic force were considered weird because they have a **spooky action at a distance**. It's no surprise that the noncontact *magnetic force* between permanent magnets (which can be studied and experienced in many play-like situation or tabletop experiments) was described and explored before the more ubiquitous but extremely weak gravitational interaction between masses.

This all sounds very straightforward, but again it is extremely hard to get used to the consequences of this simple law in practice. Introductory physics students spend considerable time grappling with the concept, and misunderstandings and difficulties are legion and well-documented [1]. Two examples might suffice to make the point.

---

[87] As is standard in *vector algebra* and intuitive, I'm putting little arrows over those quantities which, like the force $\vec{F}$, have a direction *and* a length.

[88] *Lex II: Mutationem motus proportionalem esse vi motrici impressae, et fieri secundum lineam rectam qua vis illa imprimitur.*(Law II: The alteration of motion is ever proportional to the motive force impress'd; and is made in the direction of the right line in which that force is impress'd.)

2.10. THE UNIVERSAL EXPLANATION: NEWTON

- If a truck and a mosquito collide on a highway, which exerts a greater force on the other? The answer is *neither*, because both exert the same force following NEWTON's third law. Yes, this sounds weird, and yes, it is true!

- If you push on a wall, the wall pushes back on you. If you push harder, the wall pushes harder back. Neither the wall nor you move. Strange? Yes. True? You bet! Just apply NEWTON's first law. Consider yourself the object that doesn't move. You are at rest and stay at rest. Hence, there is no net force on you. But the wall pushes on you! That means that there must be another force pushing on you in the opposite direction. Ah, you know how this goes: You push on the wall with a force of equal strength and opposite direction and the two forces add up to zero. Voilá, no net force! Sorry, but this is wrong. The pitfall is that you cannot add forces that are exerted on different objects. The force by you is exerted **on the wall**, but the force of the wall is exerted **on you**. In reality, the balancing force is the frictional force of the ground you stand on. It is a force **on you** and compensates the other force **on you** by the wall.

Isn't it frustrating how close attention you have to pay to get this right? Everything has to be done exactly how prescribed otherwise you end up with garbage. And in the last example, all you've explained is why two objects stay at rest! You might see now why it took a genius like NEWTON to figure this out. The math involved is not hard at all, but the distinction between what is essential in this trivial situation and what is not takes experience to figure out—and a NEWTON to invent.

NEWTON, speaking of his achievements, invoked an old saying: "If I have seen further than others, it is because I stood on the shoulders of giants." Thus, he payed reverence to the great scientists before him who invented or clarified concepts or simply tossed around interesting new ideas. NEWTON generalized and formalized some of them into his first two laws, but the third was entirely his own invention. *Invention* may be a strange word to use here, but it stresses the fact that concepts and connections between quantities in nature are not discovered, but made up by the human mind. They are nevertheless not arbitrary or relative, as some constructionists and social relativists might have it. Note that arguably for every concept that proves useful, there are thousands which are not, much like there are thousands of unsuccessful college-dropouts for every Bill Gates. As the double Nobel laureate LINUS PAULING (1901–1994) once quipped: "The way to get good ideas is to get lots of ideas and throw the bad ones away."

2.10.2 Newton's Law of Universal Gravitation

Even with all that axiomatic artillery, NEWTON would still be unable to conquer the problem of the planets. His brilliant solution is the ultimate *synthesis* and the first truly *unifying* scientific theory, as we shall see.

From the outset it seems that NEWTON with his laws of motion should be able to explain planetary motion because it is simply—motion. However, until the force is known that drives the planets around their orbits, we cannot solve this problem. You might say that we know since KEPLER *how* the planets move, and could use NEWTON's second law to figure out the force that explains *why* they move in this way. Unfortunately, to solve the equations in three dimensions without other insight is hard (though not impossible). Historically, the solution was found in another, meandering way typical for man-made science. We saw that KEPLER

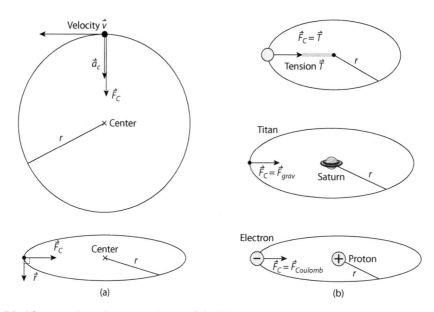

Figure 2.48: **Uniform circular motion.** (a) Schematic bird's-eye view and side view. The motion of a particle in uniform circular motion is completely described by the radius r of the circle and its speed v. The resulting centripetal acceleration a_c can be calculated from Huygens's Equation (2.6), and is directed toward the center of the motion. The motion is the result of a central force F_c and the inertia of the particle. (b) Several actual circular motions. *Top:* The motion of a bowling ball on a string connecting it to the center of the circle. The necessary force to supply the centripetal acceleration a_c is supplied here by the tension of the rope. *Middle:* Saturn orbited by its moon Titan discovered by Huygens in 1655. The necessary centripetal force is here supplied by Newton's universal gravity proportional to the masses of Saturn and Titan. *Bottom:* A negatively charged electron orbiting a positively charged proton in a classical model of the *hydrogen atom*. The necessary centripetal force is here supplied by the electrostatic or Coulomb force proportional to the electric charges of the two elementary particles. Note that the strength of the necessary force is the same if either the radius and velocities are the same in all cases, or if the ratio v^2/r is the same. The latter is nothing else than the centripetal acceleration a_c. For example, a bowling ball being 1 million times closer to the center of the circle than Titan to Saturn, but rotating $1,000 = 10^3 = \sqrt{10^6} = (10^6)^{1/2}$ as fast as Titan, would have exactly the same acceleration as Titan. Very roughly, Titan is 1 million kilometers from Saturn, and orbits in 16 days, so its velocity is $v_{Titan} = 2\pi \cdot (1 \text{ million km})/16 \text{ days} = \frac{\pi}{8}$ million km/day. The bowling ball on a 1 km rope and rotating with $v_{bowling} = v_{Titan}/1,000 = \frac{3141.5}{8}$ km/day would have the same acceleration $\frac{v_{Titan}^2}{r_{Titan}} = \frac{(1,000 v_{bowling})^2}{10^6 r_{bowling}} = \frac{(10^3)^2}{10^6} \frac{v_{bowling}^2}{r_{bowling}} = \frac{v_{bowling}^2}{r_{bowling}} = a_c$. Note that the bowling ball thus rotates once per $16/1,000$ days $= 0.016$ days $= 23$ minutes. If you didn't follow this calculation, you can still get the punch line: all we said is independent of the mass of the bowling ball and Titan. In particular, if instead of Titan our bowling ball would rotate around Saturn, it would have the same centripetal acceleration. But in this case gravity could *not* supply the necessary centripetal force, because its strength is a factor $\frac{m_{bowling}}{m_{Titan}} \approx 10^{-23}$ smaller!

2.10. THE UNIVERSAL EXPLANATION: NEWTON

had tried to explain the motion of the planets by considering a force analogous to magnetism, convinced that the sun had an influence on the planets that is smaller for a larger sun-planet distance. In the same vein, HOOKE with his **conical pendulum** had demonstrated the interplay of inertia and a **central force** to create a circular motion. As you can see, lots of plausible ideas were considered, but to write down anything resembling a solution took the genius of ISAAC NEWTON. Even for him this solution came in two steps. First he became convinced during his *annus mirabilis* that the force law describing the "influence" of the sun on the planets must decay with the distance squared, not just with the distance. The source for this realization was KEPLER's third law, Equation (2.2), which implies that the period that a planet with a *slightly* larger orbit takes *much* longer to revolve around the sun. Thus, a dramatic change of the sun's influence was necessary; a linear decline wouldn't cut it. If you are dissatisfied with this *qualitative* argument—and not afraid of some algebra—here is the *quantitative* argument. The ingredients of the argument are as follows.

1. The simplifying *assumption* that the orbits are circular.[89] The semi-major axis of a circular orbit is its radius $a = r$.

2. The fact that the distance d traveled by an object at constant speed v in a time t is $d = vt$. We'll use it in the equivalent form $t = d/v$.

3. HUYGENS's expression of the centripetal acceleration of an object moving with constant speed v around a circle or radius r: $a_c = v^2/r$.

4. The proportionality of force and acceleration, now known as NEWTON's second law $F = ma$, but traceable to GALILEO's free-fall experiments which established that the (force of) gravity near the Earth's surface results in a constant acceleration; see Section 2.8.3.

5. KEPLER's third law stating that the period squared of a planet's orbit is proportional to its semi-major axis a cubed: $P^2 = Ka^3$, for circular orbits $P^2 = Kr^3$. Also note that the time t traveling around the orbit is the period, so we can set $P = t$.

Therefore,
$$P = t = \frac{d}{v} = \frac{2\pi r}{v}$$

Using KEPLER's third law we get the relation
$$P^2 = \frac{4\pi^2 r^2}{v^2} = Kr^3.$$

But
$$4\pi^2 \frac{r^2}{v^2} = 4\pi^2 \frac{r}{a_c} = Kr^3,$$

and when we invert the equation it remains true, so the acceleration is with $K' = 4\pi^2/K$
$$a_c = \frac{K'}{r^2}.$$

Since the acceleration is proportional to the inverse square of the distance to the center, that is, radius, so is (according to $F = ma$) the force that explains why the object moves

---

[89]This is not a withdrawal back to Copernicus—it just makes the calculation simpler. NEWTON later worked out that the argument remains valid for elliptical orbits.

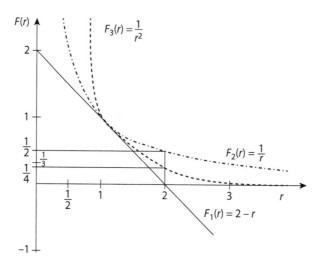

Figure 2.49: **Several monotonically decreasing functions.** It is intuitive that the force of gravity between two objects should decline with growing distance between them. But which is the right decline? Plotted here are three functions $F_1(r), F_2(r), F_3(r)$ of an independent variable r. The first represents a *linear decline*; it is a straight, descending line: $F_1(r) = 2 - r$. The second represents an *inverse* relationship: $F_2(r) = 1/r$, and the third an inverse squared relationship: $F_3(r) = 1/r^2$. Clearly, the first is not a good choice. The strength of a force is always positive (or zero), but the first curve dips below zero for large distances r. We want a large force for small r and a zero force for very large r, so both F_2 and F_3 are viable alternatives. One needs to be ruled out by observations. This is possible because they make different predictions. In particular, $F_2 > F_3$ for any distance $r > 1$.

in a circular fashion. In short, KEPLER's third law implies an inverse square law of the gravitational force.

Thus, the most important step was done: NEWTON had created a plausible hypothesis that could be tested. Of course, the testing is the nerve-wrecking part, because, as we said above, of one thousand good ideas, 999 do not pan out. NEWTON faced another problem: How exactly was he to test his hypothesis? In yet another stroke of genius, he imagined that the gravity near the surface of the Earth that makes, say, an apple fall to Earth, extends to the orbit of the moon. Gravity is a force that acts over a distance, so why should it stop? Naturally, its influence on an object would get weaker the further away from the Earth the object is, and NEWTON had this hypothesis that the strength of the force, that is, gravity, weakens like the distance squared. That is, $2, 3, 4, \ldots$ times further out the force would be $\frac{1}{4}, \frac{1}{8}, \frac{1}{16}, \ldots$ of its initial strength. This *epiphany* immediately suggested a test. The influence of gravity on objects near the Earth's surface produces a constant acceleration as GALILEO had shown, known as $g = 9.8\,\text{m/s}^2$. Remember that an acceleration is a change of velocity in a certain amount of time, so g's value tells you that an object near the surface of the Earth changes its velocity by $9.8\,\text{m/s}$ in each second it is unsupported, that is, only gravity acts on it, that is, it is in free fall. Note that, contrary to ARISTOTLE, also objects tossed *upward* are in free *fall*. This notion that the falling and the rising object are described by the same physics and hence by the same math equations is hard to fathom. It takes students in introductory physics courses quite a while to get used to. It doesn't help that the falling object gets faster, and the rising object gets slower. But, if you think about it, the *change in velocity* (as a directed quantity: negative when rising, positive when moving downward) is the same in both cases, namely $9.8\,\text{m/s}$ per second. For instance, your coffee

2.10. THE UNIVERSAL EXPLANATION: NEWTON

cup accidentally falling off a one meter table, will accelerate with g toward the ground, and acquire a large enough velocity to shatter upon impact. Note that it is going to fall for much less than one second. Incidentally, to calculate its falling time, or its final velocity just before it hitting the ground, are **kinematics** problems for beginning physics students.[90]. NEWTON's unifying idea was that there is, in principle, no difference between the process of the coffee cup falling to the ground, and the moon going around the Earth. At first glance, this seems preposterous: In which way are these two processes that look so different in essence the same? Again we've come to a **Gedankensprung** [91] that is basically impossible to follow for the beginner, takes years to master for an expert-in-training, and took a genius like NEWTON to discover. NEWTON recognized that also the moon is not supported, so it is free-falling toward the Earth, like the coffee cup. Why does it never hit the ground? Because, other than the coffee cup, the moon has a velocity perpendicular to the ground. Due to its inertia it tries to keep it moving in a straight line (NEWTON's first law), and not toward Earth. The Earth's gravitational attraction changes the direction of its velocity according to NEWTON's second law. You remember that change of velocity is (proportional to) acceleration and acceleration is proportional to force. Therefore, the Earth's gravitational force creates an acceleration (at the orbit) of the moon, the so-called **centripetal acceleration**; see Figure 2.48. Thus, the test is fairly simple: Compare the acceleration at Earth's surface, that is, $g = 9.8\,\text{m/s}^2$ to the acceleration of the moon which is one moon orbit radius away. To perform the comparison, we will make a simplifying assumption or approximation. To most beginners, this sounds suspiciously like cheating, but it isn't. What's more, assumptions like these are extremely important in science, and whoever does not let go of the idea of infinite precision, and refuses the liberating powers of the approximation, will never become a good scientist.[92] Also remember that we want to check *quickly* if the hypothesis of the inverse distance square behavior of gravity has *any* chance of working. We would be ill-advised doing a full blown calculation taking all effects into account before we are sure that this idea pans out in principle. Remember that with ideas we have a less than 1 in 1,000 chance of succeeding, so in 999 cases we would have wasted lot of time, effort and paper. The approximation in the present case is to treat the orbit of the moon as a circle. There is plenty of good arguments that this is a good approximation, for example the known small deviation of the moon's orbit from a circle ($e = 0.055$). This simplifies our reasoning, because we can use old results of HUYGENS, who figured out everything about circular motion there there is to be figured out. This recycling of old results is another essential feature of science. We cannot afford to reinvent the wheel, as the saying goes, so old solutions are a good source for time saving. This eagerness for result recycling also explains the scientists' obsession with **generalized solutions**. Most beginners loathe the seemingly excessive use of letters of the alphabet for **variables** instead of plugging in the given numbers. However, this avoids having to redo calculations for every specific situation—what works for the cup, works for the apple, works for the stone. Rest assured, this is *not* obvious, and in the case of falling objects, too, the genius of GALILEO to

---

[90] Just to show you that there is no black magic involved, let's check my claim that a fall from $1\,\text{m}$ takes less than a second. If you refuse to argue with the kinematics equations at constant acceleration, as a physics student would, let's do a **proof by contradiction** Assume the fall would take one second. Since the cup starts from rest, its final speed is $9.8\,\text{m/s}$, since its speed changes by g seconds per second. Since its speed increases linearly, the average speed is half the final speed. However, traveling $4.9\,\frac{m}{s}$ for a second would mean it traveled a distance of $4.9\,\text{m}$, contradicting the assumption. Hence, its falling time must be much less than a second.

[91] German: *Gedanken = thoughts, Sprung = jump, leap*.

[92] Pardon me for sounding like a pastor in the Church of the Holy Scientist.

Distance r relative to reference r_0	$\frac{1}{4}$	$\frac{1}{3}$	$\frac{1}{2}$	1	2	3	4
Strength I relative to reference I_0	16	9	4	1	$\frac{1}{4}$	$\frac{1}{9}$	$\frac{1}{16}$

Table 2.1: **The dependence of the strength of a quantity falling off or declining according to an inverse square law.**

recognize that all objects fall with the same acceleration $g = 9.8\,\text{m/s}^2$ close to the Earth was essential.[93] Without further ado, we use HUYGENS formula for the centripetal acceleration of *any* object in uniform circular motion due to *any* force[94] Equation (2.6), $a_c = \frac{v^2}{r}$. You see that the mass or type of the object doesn't matter, and neither does the force or mechanism which forces it to go around in circles. The only thing that matters is *how* it moves: how fast and how close to the center of the circle. The motion is completely specified by these two *kinematic* quantities v and r. The rest is number crunching. If the relevant distance is the distance between coffee cup or moon from the Earth's center,[95] then the acceleration due to Earth's gravity of the moon in its circular orbit should be a factor "*distance to moon in Earth radii*" **squared** smaller than that of the coffee mug near the surface of the Earth. The latter is precisely one Earth radius away from the center of the Earth, while NEWTON knew from ARISTARCHUS that the moon is about 60 Earth radii away. The moon goes around the Earth in a sidereal month or 27.3 days, and it travels a distance d that is, the circumference of the orbit with a radius $R_{orbit} = 60 R_{Earth}$

$$d = 2\pi R_{orbit}.$$

The moon's velocity is thus

$$v_{moon} = \frac{\text{distance traveled}}{\text{time elapsed}} = \frac{\text{circumference of moon's orbit}}{\text{one sidereal month}} = \frac{2\pi \times 60 R_E}{27.3\,\text{days}} = 13.8 \frac{R_E}{\text{day}}, \quad (2.7)$$

so we can compute its acceleration to be

$$a_{c,Moon} = \frac{v_{Moon}^2}{60 R_E} = 0.00271\,\frac{\text{m}}{\text{s}^2}.$$

Plugging in the numbers and comparing the two relevant accelerations we get

$$\frac{a_{c,Moon}}{g} = \frac{0.00271\,\text{m/s}^2}{9.8\,\text{m/s}^2} \approx \frac{1}{3600} = \frac{1}{60^2}.$$

As hypothesized, the force of gravity loses strength proportional to the square of the distance. This seems a peculiar way of weakening influence; see Figure 2.49, but it is the only one that makes sense in a three-dimensional world, and therefore we will encounter this type of decline over and over again as a pattern in nature. Here is why. Any kind of "influence," be it a force or a signal, must somehow get from the "influencer" to the "influenced." The "influencer" must therefore send out some sort of a signal, say a bunch of little messengers

---

[93] Assuming, idealizing, approximating that air resistance is not important is a manifestation of GALILEO's genius here.
[94] Could be gravity, electrostatic, even a rope or rod would do; see Figure 2.48(b).
[95] Seems clear to us, but NEWTON worked very hard to convince himself that it is true.

2.10. THE UNIVERSAL EXPLANATION: NEWTON

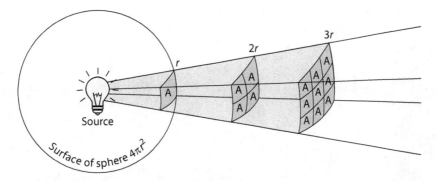

Figure 2.50: **A signal spreads out over the surface of a sphere.** The signal gets weaker proportional to the inverse area of the sphere. The area of a sphere grows like its radius *squared*, Section 1.2.6. If the source is characterized by P (power, luminosity, number of "messengers" per time), the ***intensity*** or strength I goes like $I(r) = \frac{P}{4\pi r^2}$. In other words, it follows an inverse square law, $I \propto 1/r^2$, see also Table 2.1.

that convey the message "you have been influenced by the influencer." The influencer will continuously send out a fixed number of these "messengers." As they move away from the source (that is the "influencer") they spread out over the surface of a sphere, sometimes called the sphere of influence. Therefore, the number of messengers per area goes down, and the influence gets weaker. Since the surface of any three-dimensional shape grows like the square of its size, see Section 1.2.6, the influence weakens like like the square of the distance to the "influencer," see Figure 2.50. This general scheme works for all "influences," and so the strength of the gravitational force between massive objects, of the electrostatic force between charges, and even the apparent brightness of a light source go down like the distance to the source squared.

NEWTON's idea, formulated as a hypothesis, and his subsequent successful test of it seems to be the textbook example for the *scientific method*—but it is not. Indeed, it is always presented in this straightforward way, as if it was always clear to NEWTON that this was the right thing to do, or at least that he had, due to his genius, stature or confidence, the "right stuff" to make this scientific progress happen. However, it was not so. At this point, early in his career and less than 30 years old, he came close to failure. He had considered the choice $1/r^2$ for the force law before, but used bad data for the distance to the moon, and therefore screwed up the test.[96] The acceleration of the moon was not $1/60^2$th of $g = 9.8\,\mathrm{m/s^2}$, and his hypothesis was falsified. Luckily for him and modern science, he reconsidered the choice when better data was available, and finally found the ratio of two accelerations to agree with the hypothesized ratio $\frac{1}{60^2}$.

To figure out the dependence of the gravitational force between two objects on the distance was the biggest roadblock on the way to uniform gravity. The rest is pretty straightforward. Clearly, the strength of the force should be proportional to the masses involved ("more mass, more force") and the mathematical formula should be symmetric in the masses since NEWTON's third law implies that the force of object 1 on 2 is equal in strength to the force of object 2 on 1. Curiously, an apple attracts the Earth gravitationally with the same strength

---

[96]This is unfortunately pretty common. KEPLER before NEWTON, and EINSTEIN after him came to similar false falsifications of their grand hypotheses, that is, of the theories for which they are rightly celebrated. If scientists prevail, it is usually due to their unshakable convictions, *despite* evidence to the contrary. Of course, these convictions themselves turn often out to be wrong—progress in science is made in mysterious ways!

Figure 2.51: **Jupiter with moons.** From photos like this, one can glean the angular distance of the moons from Jupiter. By carefully recording the movements of the moons like in Figure 2.30(b), we can determine the maximal angular separation of a moon from Jupiter.

as the Earth attracts an apple. In math symbols NEWTON's *law of universal gravitation* is thus
$$F_{grav} = G\frac{M_1 \times M_2}{r_{12}^2}, \qquad (2.8)$$
where r_{12} is the distance between objects 1 and 2, M_1 the mass of object 2, M_2 the mass of object 2, and the *gravitational constant* G is a proportionality constant that needs to be empirically determined, that is, measured. The trouble for NEWTON was that the gravitational constant was not known nor easy to measure, because it is incredibly small. Surprisingly, gravity is an incredibly weak force, only our vicinity to an astronomically large mass (Earth) gives us the illusion that it is strong. Therefore, NEWTON could not determine the absolute mass of the Earth.

NEWTON could, however, compare the mass of the Earth to that of other planets and the sun, that is, he could determine the *relative* mass of the Earth. That turned out to be interesting enough, since these masses are dramatically different. The calculation is a classic *application* of NEWTON's gravitational theory, so allow me to present it here even through some algebra is necessary. Before delving into it, we should have a feeling about the answer. When NEWTON wrote his *Principia*, six moons or satellites of planets were known: our own moon, the four Galilean moons of Jupiter, and Titan, the largest moon of Saturn, discovered 1655 by HUYGENS. Also keep in mind that the planets can be considered satellites of the sun. The intuition is clearly that a more massive object will pull harder on its satellites. Therefore, the period of the satellite is going to be the shorter the more mass the object has. For instance, Jupiter's moon Io has a distance to Jupiter comparable to the Earth-moon distance, but it zips around Jupiter in less than two days. Its period is thus more than an order of magnitude (10×) shorter than our moon's period, the month. Hence, Jupiter must be much more massive than the Earth. These kinds of considerations are routinely invoked by scientists to cross-check results, especially when they are unexpected. Remember that the Earth had only recently be dethroned as the supreme celestial object, so to find that it is much lighter than even the other planets (let alone the sun) was a bit of a shock.

NEWTON didn't have much to work with: He knew neither the masses of the satellites, nor the value of G, nor an accurate value for the Earth-sun distance or *astronomical unit*(AU). He used a crude estimate for the latter which was ten times larger than ARISTARCHUS's value, but still about a factor two too small. What about the distances within the solar system?

2.10. THE UNIVERSAL EXPLANATION: NEWTON

NEWTON used KEPLER's third law to figure out the *relative* solar distances (measured in AU) of Jupiter and Saturn from their easily measured sidereal periods (in Earth years)

$$a_{Jupiter} = P_{Jupiter}^{2/3} = (11.9)^{2/3} = 5.20, \tag{2.9}$$

$$a_{Saturn} = P_{Saturn}^{2/3} = (29.5)^{2/3} = 9.55, \tag{2.10}$$

that is, Jupiter is 5.20× and Saturn 9.55× as far from the sun than the Earth. The distances of the satellites from their planets can be gleaned from the *apparent* or angular separation from the planet and the known distance of the planet (from the Earth). As an example consider Figure 2.51 showing Jupiter at opposition with its moons. Io in the photo is 2.31' (minutes of arc) off the planet's center. When Jupiter is at opposition, its distance from Earth[97] is d_{EJ} = 5.20 AU − 1.00 AU = 4.20 AU. From a distance of 4.20 AU, an angle of 2.31' translates into a distance of 0.00282 AU; see Figure 1.8. Recall that this means that 0.00282 AU *appears* from a distance of 4.2 AU under an angle of 2.31', *regardless* of the actual length of the AU, because only the ratio of the two distances enters the calculation and the units cancel.

Finally, here is the calculation. Since satellite masses are unknown, we find that equating NEWTON's second law with NEWTON's gravitational force allows us to divide out the satellite mass m_{sat}

$$F_c = m_{sat} a_{c,sat} = G \frac{m_{sat} m_{planet}}{r_{sat,planet}^2},$$

to get the **centripetal acceleration** $a_{c,sat}$ of the satellite. Its value can be determined a la HUYGENS, Equation (2.6), to wit

$$a_{c,sat} = \frac{v_{sat}^2}{r_{sat,planet}} = G \frac{m_{planet}}{r_{sat,planet}^2}.$$

Thus, the left side of the equation is known, since the orbital radius and the observed period yield the velocity, see Equation (2.7). Unfortunately, the unknown G still appears in the equation, which is why NEWTON compared centripetal accelerations, and hence masses of planets A and B:

$$\frac{a_{c,sat_A}}{a_{c,sat_B}} = \frac{G \frac{m_{planet_A}}{r_{sat_A,planet_A}^2}}{G \frac{m_{planet_B}}{r_{sat_B,planet_B}^2}} = \frac{m_{planet_A}}{m_{planet_B}} \frac{r_{sat_A,planet_A}^2}{r_{sat_B,planet_B}^2}.$$

Solving for the mass of planet B in units of mass of planet A (Earth), NEWTON arrived at the following values [44]: the sun is 169,282 times more massive than the Earth (modern value: 332,950×), Jupiter 159 (318) times, and Saturn 56 (95) times. These values are off due to the underestimated Earth-sun distance, but NEWTON's conclusion that the sun is 1067 (1048) times more massive than the largest planet Jupiter was right on. These early results revealed a new pattern in nature: a hierarchy of masses. This generated two questions we will consider in Chapters 4 and 5, respectively: Why are some planets (Earth) 100 times lighter than other planets, and why are stars one thousand times more massive than even the most massive planets? There are good, physical reasons for both of these facts as we shall see in due time. For now, note how answers generate new questions. This is how science progresses.

---

[97] A planet in *opposition* is opposite of the sun in the sky, Section 1.6. At opposition, Earth stands between the planet and the sun such that all three form a straight line.

2.10.3 Successes and Limitations of Newton's Theory

Up to NEWTON, the universe was split into two parts, the lowly sublunar realm, and the heavenly superlunar realm, where totally different laws and explanations applied according to ARISTOTLE. NEWTON unified the universe by taking down the boundary between the two parts. Since NEWTON there is only one universe, and the laws of nature are the same everywhere. What a relief! Now the whole universe can be explained by relations we can discover and experiments we perform on our own planet under the assumption that the laws of nature are **universal**. While this is a bold **assumption**, to this day we have not seen anything in the universe that couldn't be explained by **terrestrial** physics or chemistry. In some sense our home planet[98] holds the key to and blueprint of the whole cosmos!

The **universality** of NEWTON's laws is truly astounding. Among others NEWTON was able to explain the following in his *Principia*:

1. *The motion of the planets* around the sun and the moon around the Earth.

2. *Projectile motion* from the falling apple to the pitched baseball, from the cannonball shot to the satellite launched, see the description of **Newton's cannon** below.

3. *The tides* on Earth as caused by the moon and the sun, including two high tides per day—something GALILEO wasn't able to explain.

4. *The oblateness of Earth:* The Earth's diameter is smaller from pole to pole; it is flattened due to its rotation; DESCARTES had predicted the opposite, and a costly, decade-long expedition proved NEWTON right.

5. *The precession of the equinoxes:* Remember that the Earth's axis is rotating itself—around the pole of the ecliptic as HIPPARCHUS discovered. NEWTON was able to show that this ultra slow rotation (one revolution in 26,000 years) is a consequence of the gravitational forces of the other planets acting on the nonspherical, oblate Earth.

6. *Kepler's laws*: They follow mathematically from NEWTON's, if the latter are applied to the (idealized) situation of exactly one planet orbiting the sun (two-body problem).

7. *Comets:* They have parabolic, hyperbolic, or highly eccentric elliptic orbits; the latter do return, they are periodic; the first telescopically detected comet (of 1680) was explicitly treated in the *Principia*, and its orbit was determined from observations, some of which NEWTON performed himself.

8. *Celestial mechanics:* In general, the forces on planets are not *central* because not just the sun, also the other planets exert forces on each other. Though the **many-body problem** is not **analytically** solvable, approximate solutions can be made systematically; eventually, this branch of astronomy, would lead to the **prediction** of the planet Neptune from unexplained deviations of the planet Uranus from its orbit.

NEWTON knew that some of the results of his *Principia* would be hard to accept for his contemporaries. For instance, his unification of sub- and superlunar science against the Aristotelian tradition with the postulate of an invisible[99] force that acts at a distance called gravity was a hard pill to swallow. NEWTON came up with a powerful illustration that

---

[98]And, for that matter, any other planet or part of the universe.

[99]In the sense that there is no visible rope pulling or stick pushing or other object touching the affected body.

2.10. THE UNIVERSAL EXPLANATION: NEWTON

demonstrated that there was a continuous transition from **projectile motion** close to the surface of the Earth to the **orbital motion** of a satellite in outer space. **Newton's cannon**, as it became known, is an artillery gun situated at a high mountain. The cannonball is fired with increasing amounts of gunpowder, resulting in an increased muzzle speed. The resulting path of the ball is due to Earth's gravity alone, always directed downward, that is, toward the center of the Earth. With little gunpowder, it is the familiar parabolic path of projectile motion. When cannonball leaves the cannon, it is in **free fall** toward the ground. As the amount of gunpowder is increased, its range becomes larger, until, at a critical value, it is flying so fast and far that it never reaches the ground (due to the curvature of the Earth). It returns to the starting point, and, in fact, continues to move (if the cannon is shifted out of the way): It is *in circular orbit* around the Earth. If even more powder is used, an elliptical orbit ensues. There is a "terminal" amount of gunpowder. When used, the initial speed of the cannonball is large enough to ensure that the Earth's gravity, getting weaker with the inverse square of the distance, cannot "turn it around" anymore. Launched at this **escape velocity**, the cannonball irrevocably leaves the gravitational field of the Earth to escape into *outer space*. Here then is, in essence, already the vocabulary of **space flight** as anticipated by NEWTON.

On the grounds of the success of these incredibly diverse applications, NEWTON's theory was accepted. There was ample empirical evidence that it was useful. Hence, it could not be rejected, even through it did not fit into the leading cosmological world view of the seventeenth century, namely DESCARTES's corpuscular universe, where everything should be *physically*[100] explained by the collisions of small particles. NEWTON himself was aware of the lack of **physical explanation** in his theory which only used *mathematical* relations between measurable quantities like forces, masses, velocities, and so forth. NEWTON saw it as a flaw of his theory, in particular he was dismayed that he could not *explain* gravity in DESCARTES's sense; he merely postulated a formula that described how it depends on measurable quantities like the distance between the two gravitating bodies. In hindsight, restraining his theory to measurable quantities was something very modern, and an advantage of his theory. Often science is the art of doing what is possible. DESCARTES and his followers had the worthy if lofty goal of explaining the cosmos physically. Alas, it couldn't be achieved. A humble **mathematical description** in lieu of a *physical explanation* was all that could be done, and, as we know now, all that *had* to be done. This restriction turned out to be very fruitful indeed,[101] but what a blow to the scientific paradigm of the day. There was much *trauerarbeit*[102] to be done during the extended mourning period in which the Cartesian universe became the Newtonian universe.

---

[100]Here *physical explanation* means to find a *mechanism* that brings forth the phenomenon to be explained. For instance, DESCARTES explained the rotation of the planets around the sun from their corpuscles colliding with the corpuscles filling the cosmos.

[101]The story was repeated when Heisenberg invented *quantum mechanics* in 1925, insisting that only measurable quantities should enter the theory [20].

[102]Literally "grief work," here: mental work to be done to overcome mourning and to accept a disturbing fact like the loss of a job, a relative, or a fond scientific theory.

Concept Practice

1. Which exerts more gravitational force: the Earth on the moon, the moon on the Earth or neither? Explain.
2. If the sun suddenly would double its mass, how would the its force exerted on the Earth change?
   (a) Double (b) Quadruple (c) Half (d) Quarter (e) No change
3. If the sun suddenly would double its mass, how would the force of the Earth on the sun change?
   (a) Double (b) Quadruple (c) Half (d) Quarter (e) No change
4. If the Earth suddenly would be twice as far from the sun, how would the force of the Earth on the sun change?
   (a) Double (b) Quadruple (c) Half (d) Quarter (e) No change (f) something else
5. Explain why NEWTON's law of gravity is *universal*.
6. In which way is NEWTON's theory a *unification*? What did he unify?
7. Why is NEWTON's work considered a *synthesis*?
8. Ponder the difficulties that NEWTON's contemporaries had with gravity as an *action at a distance*.
9. Write down an equation that relates the masses of Earth, sun, Jupiter and Saturn as computed by NEWTON, that is, express all other masses in units of the Earth's mass.

2.11 Summary and Application

2.11.1 Timetable of the Emerging Expanding Universe

The main steps toward an understanding of the expanding universe discussed in this chapter are described below. This schedule has to be taken with a grain of salt. Sometimes the moment when a new concept emerges is not easily pinned down. Even when published at a definite date, a new view of the universe will take time to be accepted—and linger long after its falsification. Often, different communities "live" in incompatible universes at the same time, for example Copernicans and anti-Copernicans around 1600 CE, plenists and vacuists around 1650 CE.

- Earlier than 600 BCE—**The holy, mysterious universe**: In prehistoric times, the universe is viewed as mysterious and therefore incomprehensible. It is often modeled after the local topography; see the Egyptian universe resembling an "upside-down" Nile River valley.

- Since c. 600 BCE—**The understandable universe**: The archaic Greeks (800–480 BCE) turn the cosmos into something that can be understood. Several interpretations of "understanding" are explored:

  – Substantive understanding—"The universe is elements." This view held that the universe can be modeled after the properties of substances that were thought to be elementary, like *fire, Earth, air*, and *water*. The latter, for instance, can be solid, liquid or gaseous and therefore might explain solid, liquid and gaseous things in the universe.

  – Mathematical understanding—"The universe is numbers." The Pythagoreans thought that abstract math was the fabric of the universe. They looked for simple mathematical truths or *harmonies* in nature.

  – Physical understanding—"The universe is atoms." DEMOCRITUS (c. 450 BCE) held that the universe contains smallest units, the un-cuttable atoms, that can be combined in various ways to explain all the things and patterns we experience.

2.11. SUMMARY AND APPLICATION

- c. 380 BCE—**The circular, ideal universe**: Uniform circular motion is the only possible motion in the universe, according to PLATO (428–348 BCE).

- c. 350 BCE—**The homocentric universe**: The motion of the sun and the planets is in the form of spheres, four for each planet. They are centered on the stationary Earth and therefore homocentric. The last and farthest sphere is that of the fixed stars, turning once every sidereal day—EUDOXUS (c. 406–350 BCE).

- c. 330 BCE—**The Aristotelian geocentric universe**: There are more homocentric spheres (about 55 total) to explain all known motions in the universe. These *crystalline spheres*—and everything *superlunar* for that matter—are made out of the fifth element, *quintessence*. In contrast, the *sublunar* realm and its phenomena are explained physically comprehensively by the properties of the four elements *water, Earth, air and fire*—ARISTOTLE (384–322 BCE)

- 300 BCE–150 CE—**The measurable universe**: The distance to the moon and the sun (Aristarchus), the radius of the Earth (ERATOSTHENES), the precise positions of the stars and the precession of the equinoxes (HIPPARCHUS) are measured with little attention to underlying cosmological assumptions. The Hellenistic period is a pragmatic era.

- 150 CE–1600—**The Ptolemaic geocentric universe**: The universe is explained *mathematically* by a complex geometric algorithm involving epicycles, deferents, equants and eccentric circles. It is finite and not very large.

- 1473–c. 1650—**The Copernican heliocentric universe**: Planets move around the sun in circles. The universe is finite and large. Stars are very far away because no *parallactic motion* is observed.

- 1588–c. 1650—**The Tychonic hybrid universe**: Planets move in circles around the sun, which moves in a circle around the Earth.

- 1609–1687—**The Keplerian universe**: Planets *orbit* the sun in ellipses. The universe largely empty, but not necessarily finite. The universe is *described mathematically*, but the search for a *physical explanation* of planetary motion is beginning.

- c. 1630–1740—**The Cartesian, corpuscular universe**: The universe is completely filled with particles (corpuscles). There is no action at a distance. Everything is *explained physically* by continuously colliding corpuscles. The universe is infinite and contains infinitely many worlds and solar systems; it has no center.

- 1687–1905—**The Newtonian universe**: Planets, stars, and everything else moves under the influence of a universal force (gravity) and general rules of motion (Three Newton Laws). While *gravity* explains KEPLER's laws, it is "just" a mathematical explanation. NEWTON is frustrated because he cannot figure out what gravity physically *is*. The Keplerian ellipses are a good approximation to the actual motion of a planet because the sun's gravitational pull is much stronger than that of the other planets. The universe is mostly empty. It is a stage on which motion happens—created by *absolute space* and *absolute time*. The effects of gravity and other forces are felt *instantaneously* everywhere in the universe.

Concept Application

1. In which sense is Ptolemy's theory better than ARISTOTLE's? In which sense is ARISTOTLE's theory better than Ptolemy's?

2. Why does the deferent of the inferior planets rotate once a year?

3. How big is the epicycle of Venus?

4. What reason did PTOLEMY give when he declared Saturn to be the farthest planet?

5. Why doesn't the sun have a major epicycle in Ptolemy's model?

6. Does the moon have a major epicycle?

7. Does the moon have a minor epicycle?

8. Why must the line from a superior planet to the epicycle center always be parallel to the line Earth-sun? This is not the only way to preserve the appearances. Construct an alternative to Ptolemy's choice.

9. Discuss why at COPERNICUS' times both sides could argue that their model was "more natural."

10. Use Figure 2.13 to determine the size of Venus's orbit (assumed to be circular) from the configuration at maximal elongation. The distance between the Earth and the sun was largely unknown at COPERNICUS' times (ARISTARCHUS had underestimated it to be about twenty times the distance to the moon), so we just give it a name: one astronomical unit or 1 AU. It follows that COPERNICUS could only determine the radius of Venus's orbit *relative* to the size of the Earth's orbit. Is the situation shown eastern or western elongation? Why?

11. Determine the size of Mars's orbit according to the method for exterior planets. Use the figure below (Figure 2.17), and the following data.

12. Given that Copernicus found the size of Mercury's orbit to be about 0.38 AU, that is, roughly three times smaller than Earth's, find the angular size of the sun in Mercury's sky. As you know, the sun (as the moon) appears under an angle of half a degree in the Earth's sky.

13. Discuss the psychological advantages of TYCHO's model.

14. In which ways was TYCHO progressive? In which ways was he conservative?

15. Discuss the modernity of TYCHO's method of observing and data taking.

16. A mystery plant has a period of 2 years. Use KEPLER's third law to estimate the semi-major axis a. exactly 2 years, about 3 years, about 4 years, about 8 years.

17. Discuss in which way KEPLER was standing with one foot in the past, and with one in the future.

18. Construct the following ellipses with rope, pencil and pushpins.

    (a) An ellipse with $e = 0.6$.
    (b) An ellipse with $a = 7cm$.
    (c) An ellipse with $e = 0.2$ and $a = 5cm$.

19. Could GALILEO have observed Mercury's phases? Why or why not?

2.11. SUMMARY AND APPLICATION

20. Does Mars show phases? Explain.

21. Redraw the diagrams, Figures 2.31, for the case of Mercury. What is different, what stays the same?

22. For each of the major discoveries of GALILEO, make a list of the problems it poses for an Aristotelian scholar.

23. Use the sunspot drawings in Figure 2.33 to determine the rotation rate of the sun.

24. Estimate the gravitational force of the Earth on the moon by an order of magnitude calculation. By using scientific notation and rounding all numbers to one significant figure, you should be able to do this in your head.($M_E \approx 6 \times 10^{24}$ kg, $M_M \approx 7 \times 10^{22}$ kg, $d_{E,M} \approx 4 \times 10^8$ m, $G \approx 7 \times 10^{-11}$ N·m^2/kg^2)

25. In a discussion, someone claims the gravitational force of the sun on the Earth is weaker than that of the moon, because the sun is about 400 times farther and the inverse square law means that distance is the most important factor in NEWTONs gravity law. Use the numbers from Prob. 24 and $M_{sun} \approx 2 \times 10^{30}$ kg to either support or falsify the claim quickly.

2.11.2 Activity: Introducing the Parallax

The parallax is an important concept in astronomy, because it allows for the determination of an object's distance by measuring its position. This is possible because of the parallactic effect: an object appears in front of a different distant background when viewed from two different viewpoints, for example A and B in Figure 2.2(a). The distance of the object follows from the geometry of the triangle with corners A, B and object. The distance between A & B is called the baseline. Relative to it, we can measure the angles α at A and β at B. Any triangle of which we know at least two angles and one side is completely determined. In particular, the other sides of the triangle (the distances to the object from A or B) can be calculated.

Stretch out your arm and look at your thumb with your left eye closed, then close your right eye. Repeat with your thumb close to your face.

1. In this exercise, what constitutes the object, the baseline, the background, etc.?

2. When the thumb is closer to your face, which statement is correct? (Select all that apply.)
   (a) The thumb is bigger.
   (b) The distance to the thumb is bigger.
   (c) The shift of the thumb in front of the background is bigger.

3. In this example you can tell that the thumb is closer because it appears bigger. Can you use the same argument for astronomical observations, for example "Jupiter is closer to use since it appears bigger"? Explain.

4. Fill in the blank: The bigger its shift (parallactic angle) in front of a distant background, the _____ the distance to the thumb (or any other object).

5. The lunar parallax shown in Figure 2.2(c) can be used to determine the distance to the moon. As shown, the moon appears shifted with respect to the stars (here: the Pleiades cluster) when viewed from different locations on Earth at the same time.

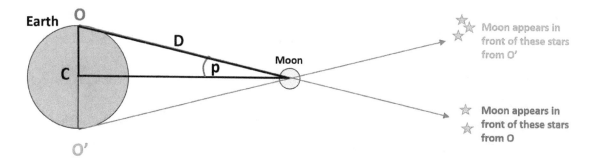

Figure 2.52: **Geometry of the lunar parallax.**

6. We know that the moon appears under an angle of 0.5°. Use this to determine the shift of the moon in front of the stellar background when viewed from the North versus the South Pole.

7. The parallactic angle is half the angular shift of the object. Here it is: $p = $ _____.

8. Fill in the blank: For this lunar parallax, the baseline is the _____ of Earth.

9. What is the significance of the fact that the shift from pole to pole is the same as from two points on the equator 180° apart from each other?

10. Using some trigonometry, we see from Figure 2.52 that p is the angle at the moon in a right triangle with corners at the moon, the observer and the center C of the Earth. The triangle side OC is half the baseline, that is, the radius R_E of the Earth. Therefore the hypotenuse is the distance to the moon D and computed by: $D = \frac{R_E}{\sin p}$. Use $R_E \approx 6400$ km to compute the distance to the moon.

2.11.3 Activity: Copernicus' versus Ptolemy's Explanation of Retrograde Motion

We saw that planetary motion as observed in the sky constitutes a strange pattern. It is much harder to explain than the motion of the sun and the moon, because the planets move mostly like the sun and the moon, but about once a year slow down and reverse the direction of the motion with respect to the stars (not with respect to the ground!), showing so called retrograde motion. See Figure 1.28 for a typical pattern (Mars in 2003). Note that Mars is largely following the *ecliptic*, which is inclined 23.5° with respect to the *celestial equator*.

Ptolemy's Explanation: PTOLEMY tried to explain this pattern of motion by assuming that the planets are moving on spheres around the Earth, which explains why Mars largely follows a circle (the *ecliptic*) and moves eastward with respect to the stars—towards larger *Right Ascension* (RA). Note that Mars also moves with respect to the horizon westward, but this is explained by the *daily* rotation (of the Earth for us, of the the celestial sphere as PTOLEMY believed). PTOLEMY explained the retrograde loop by an *ad hoc* introduction of a smaller *epicycle* that is attached to the Martian sphere or *deferent*, see Figure 2.9. Mars is assumed to move around the smaller circle, which can explain why it sometimes moves backwards, that is, towards smaller RA (roughly westward).

1. In Figure 1.28(b), when is Mars moving in a retrograde fashion?

2.11. SUMMARY AND APPLICATION 171

Date	Right Ascension (RA)
Aug 26, 2007	70°
Sep 25, 2007	88°
Oct 25, 2007	100°
Nov 23, 2007	103°
Dec 24, 2007	93°
Jan 23, 2008	84°
Feb 22, 2008	86°
Mar 23, 2008	98°
Apr 22, 2008	113°

Table 2.2: **Simplified position of Mars in 2007 and 2008.**

2. For how long is Mars moving retrograde?

3. How large is its retrograde loop? (In degrees? In million km? Can we tell?)

4. Qualitatively sketch the alleged epicyclic motion of Mars by drawing a dot on a small circle (epicycle) which rolls around a big circle (deferent). Trace the motion of the dot as this contraption moves.

5. What can we see in the sky of the back and forth motion of Mars? Use words like closer/farther, prograde/retrograde.

6. Does this theory work, that is, does it describe the appearances of Mars that we actually see in the sky? Explain.

Copernicus' Explanation: PTOLEMY's explanation works, but it seems awkward and *ad hoc*. To replace it with a simpler theory, COPERNICUS went back to ARISTARCHUS's heliocentric model of the solar system, where both the planets and the Earth orbit the sun in circles. In Table 2.2 you'll find some (simplified) data that we will use to see if COPERNICUS's explanation works.

7. When is Mars moving retrograde?

8. Remember that RA is the celestial longitude counted from the **vernal equinox**, that is, the direction in which sun is standing on March 21. This direction is marked in Figure 2.53. Keep in mind that the vernal equinox is infinitely far away, so for some other date you'll have to put your protractor parallel to that direction to count the RA angle again from the vernal equinox. This is done in Figure 2.53 for some fictitious planet at RA 45° on November 21.

9. On a new sheet of paper, draw your own version of the Earth's orbit as shown in Figure 2.53. Then draw in the direction in which Mars is seen from Earth on the nine dates in the Table 2.2.

10. Where is Mars along these lines of sight? For that we would have to know the actual shape, size and orientation of Mars' orbit. COPERNICUS did not know, but we know since KEPLER that Mars's orbit is a slightly eccentric circle (an ellipse) with a radius about 1.5 times bigger than the Earth's orbital radius. Draw in Mars's orbit.

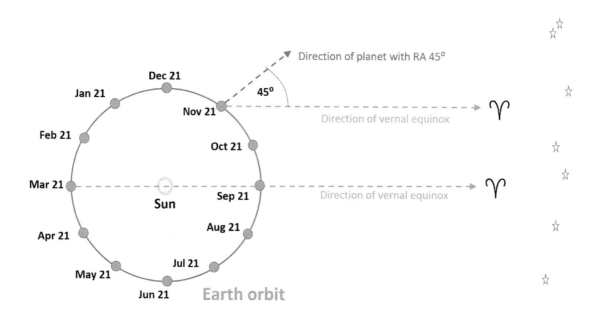

Figure 2.53: **Horizon View on December 1 facing South.**

11. How is Mars moving along its orbit? Always in the same direction, or moving back and forth?
12. Why is the Copernican theory a better explanation of the pattern we see in the sky?

2.11.4 Activity: Kepler's Laws

The *problem of the planets* was finally solved by JOHANNES KEPLER in 1609 in his book *Astronomia Nova*. He had wrested the laws of planetary motion from the accurate data of TYCHO BRAHE. By *triangulation* he was able to infer the true position of the planet Mars in space, and thus to construct its orbit. This orbit turned out to be *eccentric*: KEPLER had to use the geometrical shape of an ellipse to describe it. Thus his first law was established: **Planets move around the sun in *ellipses*, with the sun in one of the two foci.** The second law or *equal area law* then tells us how the planets *move* along this orbit: **The line connecting planet and sun moves so that it sweeps out equal areas in equal times.** An immediate consequence is that planets move faster when they are closer to the sun. Note that KEPLER did not explain **why** planets move faster (NEWTON achieved this over half a century later with his universal law of gravity)—he just stated **how** they move. But that was all astronomers needed to compute and predict the position of the planets in their orbits and in the sky for an point in time—be it past, present, or future.

1. Draw an ellipse on a piece of blank paper using a loop of string, two pushpins and a pencil like in Figure 2.23. Mark the position of the sun in one of the two focal points, that is, at the position of one of the pushpins.

2. Draw in the *major axis* which connects the *perihelion* (closest point to the sun [Greek: Helios]) and the *aphelion* (farthest point from the sun). This is called the **line of the**

2.11. SUMMARY AND APPLICATION

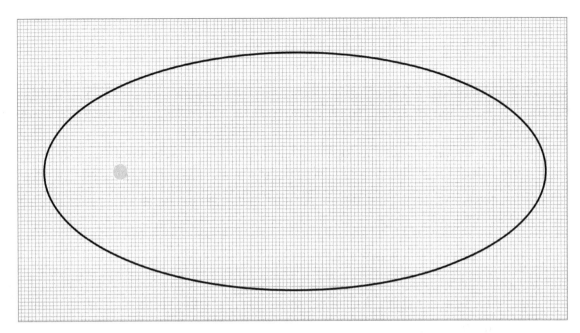

Figure 2.54: **A very eccentric ellipse.**

apsides. Its orientation in space is one of the six so-called **orbital elements** which specify the orbit completely.

3. The two numbers that specify an ellipse are its **semi-major axis** a and its **eccentricity** e. As the name suggests, the **semi-major axis** is half the long axis of the ellipse. Measure a.

4. The eccentricity tells us how eccentric an ellipse is. A circle is not eccentric at all, and therefore has eccentricity $e =$ _____. To determine the eccentricity of your ellipse, measure the distance between the focal points with a ruler. Geometry tells us that this distance is $d_{ff} = 2ea$. Calculate e.

5. What is the average distance of the planet from the sun?

6. Draw in another ellipse with the same string loop, but change the eccentricity by increasing the distance between the pushpins. Draw the ellipse so that the lines of apsides is oriented differently. Make sure that one of the focal points of the second ellipse coincides with the sun(focus) of the other ellipse. Why?

7. Describe the second ellipse: Is it bigger or smaller than the first one? Is it less or more eccentric? Do you think a planet going around the second ellipse will orbit the sun in more or less time?

II. Second Law

8. Draw in the line of apsides and mark peri- and aphelion of the ellipse in Figure 2.54.

9. Mark off about 16 intervals of equal distance around the perimeter of the ellipse by halving the distance between perihelion, aphelion and the point halfway between them. Then half them again.

174 CHAPTER 2. INFERRING & THEORIZING: PLANETARY ASTRONOMY

10. How does the planet move along its orbit? Will it take equal time to move the equal distances between the tick marks? Why or why not?

11. Shade an area in the ellipse that is bounded by the sun, the aphelion and one of the tick marks close to it.

12. Now try to find an area of equal size around the perihelion. How many tick marks do you have to include to make the two areas equal in size?

13. Kepler's second law tells us that the sun-planet line sweeps out equal areas in equal times. Does the planet therefore move equal distances in equal time? Why or why not?

14. How much faster does your planet move when it is closest to the sun compared to when it is farthest from the sun?

2.11.5 Activity: Telescopes

Telescopes are light collectors. In the construction of telescopes, we use the fact that light rays can be focused by either a lens or a mirror. There are two main types of telescopes.

I. Refracting Telescopes

1. Do refractors work with lenses or with mirrors?
2. Which fundamental law of optics do refracting telescopes use?
3. Draw the simplest form of a refracting telescope.
4. Compare the light collecting power of a refractor with a 50 mm diameter main lens (objective) with that of a telescope with a 25 mm diameter lens. How much more light does the larger telescope collect?
5. The magnification of a telescope is the ratio of its objective's lens compared to its eyepiece lens. Calculate the magnification of a $f = 2000$ mm refractor if used with a 40 mm eyepiece.
6. What is the magnification when a shorter focal length eyepiece of $f = 20$ mm is used?
7. Formulate a rule that shows how magnification changes with the focal length of the eyepiece.

II. Reflecting Telescopes

8. Do reflectors work with lenses or with mirrors?
9. Which fundamental law of optics do reflecting telescopes use?
10. Draw the simplest form of a reflecting telescope.
11. Compare the light collecting power of a reflector with a 100 mm diameter main lens (objective) with that of a telescope with a 25 mm diameter lens. How much more light does the larger telescope collect?
12. The magnification of a telescope is the ratio of its objective's lens compared to its eyepiece lens. Calculate the magnification of a $f = 2000$ mm reflector if used with a 50 mm eyepiece.

2.11. SUMMARY AND APPLICATION

13. What is the magnification of a $f = 1000\,\text{mm}$ telescope when used with the same 50 mm eyepiece?

14. Formulate a rule that shows how the magnification changes with the focal length of the telescope, that is, the focal length of its main optical surface.

2.11.6 Activity: Venus Phases

Galileo's observation that Venus shows phases like the moon was crucial for the development of modern astronomy, since it can be used to decide whether the **helio-** or the **geocentric model** describes our solar system correctly, see Figure 2.31. Here we explore how this is possible.

1. Draw Venus, Earth and the sun in a **heliocentric model**, where both planets orbit the sun.
2. Draw in Venus at 4 to 8 different positions along the orbit as open circles.
3. With the sun at the center of the orbit, draw in the illuminated side of Venus at these positions.
4. How much of these illuminated sides do we see from Earth at the different positions? Label these phases with the names familiar from the moon, that is, new Venus, full Venus, etc.
5. Repeat for a **geocentric model**. Note that the sun is always on the far side of Venus, since Venus's orbit is an epicycle that lies between the Earth and the sun.
6. How are the predictions of the two theories different?
7. By observing which phases of Venus can we rule out the geocentric model?
8. Could the heliocentric model have been falsified at all? Why not or which phase would have to be observed?

2.11.7 Activity: Newton's Three Axioms

Newton's three laws of motion are easily stated and memorized, but hard to apply. Here we confront several of their strange but consistent consequences.

1. A book is at rest on a table. The force of gravity acts on it. Newton's first law states that objects subject to a force are accelerated, but the book remains at rest. Explain how this is consistent.

2. A force F acts on an object and accelerates it. How does the acceleration change if ...

    (a) the strength of the force is doubled?
    (b) the strength of the force is halved?
    (c) the mass of the object is doubled?
    (d) the mass of the object is halved?

3. A truck and a mosquito collide on a highway.

    (a) Which if any experiences a greater force?
    (b) Which if any experiences a greater acceleration?

2.11.8 Activity: Newton's Law of Universal Gravitation

1. The law of universal gravitation states that the gravitational force of an object A on an object B depends on three quantities—name them.

2. How does the force between two objects *scale* with the distance between the objects?

3. Explain why this law is called *universal*.

4. If you'd place a spaceship at rest exactly between the Earth and the moon, what would happen?

5. The Earth is roughly 100 times more massive than the moon. Compare the gravitational force of the Earth on the moon with the force of the moon on the Earth. Is this surprising?

6. The moon is about 400 times closer than the sun, and roughly 30 million times lighter. How much stronger is the gravitational force of the sun on the Earth compared to the force exerted by the moon on the Earth?

7. The law of universal gravitation states that the mutual force between two objects is directly proportional to the masses. On the other hand, Newton's second law of motion tells us that it becomes harder to accelerate an object if it is more massive. The acceleration of an object (A) in the gravitational field of another (B) thus is independent of the the mass of A (but not of the mass of B). Therefore, if we'd replace the moon with a baseball, the baseball would also orbit the Earth in a month (27.3 days). On the other hand, close to the surface of the Earth, the baseball falls to the ground with the familiar acceleration $g = 9.8\,\frac{m}{s^2}$, due to the fact that its distance r to the Earth('s center) is $r = R_E$. Given that the moon's orbital radius is about 60 Earth radii, what is the acceleration of the moon in its orbit?

2.11. SUMMARY AND APPLICATION

Image Credits

Fig. 2.2a: Copyright © 2006 by Booyabazooka, (CC BY-SA 3.0) at https://commons.wikimedia.org/wiki/File:Parallax_Example.svg.

Fig. 2.2b: Source: https://commons.wikimedia.org/wiki/File:Stellarparallax2.svg.

Fig. 2.2c: Source: https://commons.wikimedia.org/wiki/File:Lunarparallax_22_3_1988.png.

Fig. 2.6a: Source: https://commons.wikimedia.org/wiki/File:Earth_precession.svg.

Fig. 2.6b: Copyright © 2006 by Tauolunga, (CC BY-SA 2.5) at https://commons.wikimedia.org/wiki/File:Precession_N.gif.

Fig. 2.15: Source: https://commons.wikipedia.org/wiki/File:Copernican_heliocentrism_diagram-2.jpg.

Fig. 2.18: Source: https://commons.wikimedia.org/wiki/File:Tycho-Brahe-Mural-Quadrant.jpg.

Fig. 2.19a: Copyright © 2006 by Francisco Javier Blanco Gonzlez, (CC BY-SA 3.0) at https://commons.wikimedia.org/wiki/File:Refracci%C3%B3n.png.

Fig. 2.19b: Source: http://astronomy.nju.edu.cn/ lixd/GA/AT4/AT405/IMAGES/AACHCMU0.JPG.

Fig. 2.22: Source: http://n1.xtek.gr/ime/lyceum/?p=lemma&id=755&lang=2.

Fig. 2.26b: Source: https://commons.wikimedia.org/wiki/File:2004_Venus_Transit.svg.

Fig. 2.27b: Copyright © 2011 by Theresa_knott, (CC BY-SA 3.0) at https://commons.wikimedia.org/wiki/File:Pencil_in_a_bowl_of_water.svg.

Fig. 2.30a: Source: https://www.loc.gov/teachers/classroommaterials/primarysourcesets/understanding-the-cosmos/pdf/sidereusNuncius.pdf.

Fig. 2.30b: Source: https://santacroceinflorence.files.wordpress.com/2014/02/galileonotebook.gif.

Fig. 2.31a: Source: https://commons.wikimedia.org/wiki/File:Phases-of-Venus-Geocentric.svg.

Fig. 2.31b: Source: https://en.wikipedia.org/wiki/File:Phases-of-Venus2.svg.

Fig. 2.32a: Source: https://writescience.files.wordpress.com/2015/07/galileo_12.jpg.

Fig. 2.32b: Copyright © 2004 by Statis Kalyvas - VT-2004 programme.

Fig. 2.33a: Source: http://galileo.rice.edu/images/things/sunspot_drawings/ss623-l.gif.

Fig. 2.33b: Source: http://galileo.rice.edu/images/things/sunspot_drawings/ss626-l.gif.

Fig. 2.34: Source: https://history.nasa.gov/SP-402/p9a.jpg.

Fig. 2.40: Source: https://commons.wikimedia.org/wiki/File:Telescope_140_foot_Johann_Hevelius.jpg.

Fig. 2.41: Source: https://commons.wikimedia.org/wiki/File:Hevelius_Map_of_the_Moon_1647.jpg.

Fig. 2.44: Copyright © 2015 by Sascha Grusche, (CC BY-SA 4.0) at https://commons.wikimedia.org/wiki/File:Newton%27s_Experimentum_Crucis_(Grusche_2015).jpg.

Fig. 2.45: Source: https://commons.wikimedia.org/wiki/File:Illustration_from_1676_article_on_Ole_R%C3%B8mer%27s_measurement_of_the_speed_of_light.jpg.

Fig. 2.46b: Source: https://commons.wikimedia.org/wiki/File:Reflexion_im_Wellenmodell.png.

Fig. 2.51: Copyright © by Jan Sandberg.

Chapter 3

Deducing & Checking: Between Newton and Einstein

The century after NEWTON's *Principia* (1687) saw scientists busy with applying his theories of motion and gravity to pretty much every natural phenomenon. In other words, they were *deducing* specific *predictions* from a general theory. After a while it became clear that there are phenomena for which different concepts and frameworks are necessary. Electricity and magnetism come to mind, and the mathematical descriptions of these phenomena culminated in a unified theory known as—no surprise here—*electromagnetism*. At the same time, the study of heat gave birth to another distinct theory of natural phenomena known as *thermodynamics*.[1] These non-Newtonian theories, like any new hypothesis, had to be constructed *inductively*. Every new general rule we "make up" is abstracted from many, but isolated data points. For example, we assume that a correlation between quantities that we observed Monday and Wednesday in our laboratory, will also hold on Tuesday and in someone else's lab. This assumption is then subject to further testing.

One thing to appreciate in this chapter is the dramatic increase in accuracy of measurements. We saw that ARISTOTLE and his contemporaries were generally satisfied with *qualitative* agreement between theory and observation. This had changed in the *scientific revolution*, where strict *quantitative* agreement started to become the norm. Recall how important the few arc minutes (a small fraction of the apparent lunar diameter) discrepancy between Tycho's observations and Kepler's initial, improved Copernican theory were—important enough to throw out the circles and bring in the ellipses! Now imagine how different things must have looked with the ever improving telescopic tools of the eighteenth and nineteenth centuries. Discrepancies between theory and observation started to stick out like a sore thumb. Suddenly there were many unexplained effects, manifested in deviations of the motion of celestial objects from predictions. At the beginning of the eighteenth century, effects as small as a sixtieth of a degree could easily be seen. By the nineteenth century, telescopes combined with micrometer screws allowed effects over a hundred times smaller to be picked up. If you look that closely at nature, you'll have a lot of minute details to consider. On that *scale* a lot of things can contribute to an observed behavior. Therefore,

---

[1] Later, LUDWIG BOLTZMANN and others realized that the laws of thermodynamics follow from NEWTON's laws of mechanics applied to an exceedingly large number of particles, that is, that thermodynamics "is" *statistical mechanics*.

180 CHAPTER 3. DEDUCING & CHECKING: BETWEEN NEWTON AND EINSTEIN

the descriptions, explanations and methods—not the least the mathematical tools—of the scientists became a great deal more sophisticated and specialized during this period.

Guiding Questions of this Chapter

3.1 What characterizes the century after NEWTON in terms of its scientific methods, goals, and its general cosmological outlook? By which measurements was a universal acceptance of *Newton's theory* finally achieved?

3.2 How does the Earth's motion in space as a physical object change the perspective of an observer on its surface?

3.3 How can we measure distances in and other properties of the universe while confined to the Earth's surface?

3.4 How did chemistry and the investigation of the microstructure of nature get its start?

3.5 How was the understanding of the universe expanded beyond Saturn, below the equator, and before the existence of Earth?

3.6 How was science in the nineteenth century different from its eighteenth-century progenitor? How can distances to the stars be measured if they are about a hundred thousand times farther than the sun?

3.7 What is light? Why are all of its properties important for astronomy?

3.8 What is the importance of energy conservation and entropy maximization for astronomy? Are stars subject to the laws of physics?

3.9 What does the starlight tell us about the star that emits it? How much can we understand about stars just by observing their light?

3.1 The Enlightened Century after Newton

3.1.1 Enlightenment and the Age of Reason

The century after NEWTON was arguably a time of **deduction**. Namely, NEWTON's *general* laws were used to deduce *special* events, that is, to predict many hitherto unobserved patterns in nature. For instance, EDMOND HALLEY in 1705 used NEWTON's laws to conjecture that the comets of 1682, 1607, and 1531 were one and the same object, and to predict that it would return fifty-three years in the future, namely in 1758—incidentally sixteen years after his own death.

With the establishment, or rather *self-organization*, of the "scientific way of doing research" driven by astronomy,[2] there was an explosion of activity in virtually all areas of science. Therefore, we cannot hope to do justice to most of the ways in which our (mental) universe has expanded since. We will focus on the developments relevant to our understanding of the expansion of the cosmos.

The roots for the development both of the sciences and the associated interpretation or philosophy were laid in the previous century. Namely, DESCARTES had put forth his strong

---

[2]It is paradoxically easier to find out about things very far away than about those close. This remains true today; see weather predictions vs. landing a space probe on Mars. The reason being the isolation of objects in the vacuum of outer space.

3.1. THE ENLIGHTENED CENTURY AFTER NEWTON

views of the supremacy of the human mind—epitomized in his credo "Cogito ergo sum" mentioned before—and aptly called **rationalism**. DESCARTES believed that man could figure out the cosmos purely by *a priori* reasoning. The laws of nature in his view could be obtained by using mathematics and abstract logic. Meanwhile, JOHN LOCKE (1632–1704) held the opposing view that knowledge comes primarily from **sensory experience**, that is, from sources *outside* the human mind. LOCKE was the founding father of the philosophical field of **empiricism**. In the eighteenth century a compromise was forged out of these opposite persuasions. Of course, the importance of **empiricism** for the natural or empirical sciences and the **scientific method** is rather obvious.[3] However, the eighteenth century became known as the **age of reason** because *reason* in its more general meaning (as implied in words such as reasonable, rational or logical) became its unifying theme. Reason was held in high esteem, and applied to all human endeavors, not just science. In particular, the laws of nature were thought to be manifestations of reason, and, as such, God's will would be exposed by the light of nature.[4] Incidentally, in French the eighteenth century is known as the *siècle de lumières* or *Century of Lights*, while the corresponding German term **Aufklärung**—often translated as **enlightenment** but arguably better as *clarification*—emphasizes the *ongoing* act of figuring things out by reason and experiment. Not surprisingly, church dogmas were questioned and ideals such as liberty, equality, fraternity,[5] progress, and tolerance were advanced. The inherent *optimism* of the movement resulted from the hope that reason could be used not just to understand nature and the universe, but also to improve man's lot in life and to bring about a betterment also in moral issues. IMMANUEL KANT (1724–1804) penned in 1784 "*Aufklärung ist der Ausgang des Menschen aus seiner selbstverschuldeten Unmündigkeit. Unmündigkeit ist das Unvermögen, sich seines Verstandes ohne Leitung eines anderen zu bedienen.*"[6] This speaks to the necessity to educate oneself and fellow humans, while implying that in the past all too often the human mind was imprisoned by external authorities. The dominance of church doctrines in past centuries comes to mind, exemplified in Galileo's trial before the Roman Inquisition in 1633.[7] **Sapere aude**[8] is therefore the motto of enlightenment, and people like JEAN LE ROND D'ALEMBERT (1717–1783) and DENIS DIDEROT (1713–1784) set forth to publish an all-encompassing encyclopedia[9] of human wisdom, the *Encyclopédie, ou dictionnaire raisonné des sciences, des arts et des métiers (Encyclopedia, or a Systematic ("reasoned") Dictionary of the Sciences, Arts, and Crafts)* (1751–1772), with the aim to "change the way people think."(Diderot [22, p. 610])

There were, however, two main problems with these lofty goals. First, the laws of nature

---

[3]The Greek root of the word is *empeirikos* meaning *experienced*, which traces back to the verb *peiran* meaning *to try* or *to experiment*.

[4]These were NEWTON's words. His contemporary ROBERT BOYLE wrote "*to know God's ends ... is not a presumption, but rather ... is a duty. For there are things in nature so curiously contrived ... that it seems little less than blindness in him, that acknowledges a most wise author of things not to conclude that they were designed for this use.*" Thus, the notion that the universe had been planned by God was widely held until called into question by CHARLES DARWIN (1809–1882) in the mid-nineteenth century.

[5]The motto of the French Revolution (*liberté, égalité, fraternité*) is symbolized by the blue, white, and red vertical bands of the French flag or *tricolore*.

[6]Enlightenment is man's emergence from his self-imposed nonage. Nonage is the inability to use one's own understanding without another's guidance.

[7]We saw in Section 2.8 that the story is not quite as one-sided as it is often portrayed, and certainly not solely a clash of religion and science.

[8]Latin: Dare to know.

[9]Its modern analogue is arguably the free online encyclopedia *Wikipedia*.

are purely descriptive and therefore cannot tell right from wrong. In other words, they cannot function as a moral compass. Second, the world view by the end of the eighteenth century was staunchly deterministic. The French physicist and mathematician LAPLACE had gone so far as to say that "[w]e may regard the present state of the universe as the effect of its past and the cause of its future." In other words, the past and future of the universe were seen to be *calculable* as they necessarily seemed to follow from the laws of classical mechanics if the positions and velocities of all parts of the cosmos are known.[10] This statement of **causal** or **scientific determinism** was in stark contrast to the proposed **free will** of man, creating a paradox that could not be resolved within enlightenment movement. Instead, moral value was ascribed to the *pursuit* of science itself, and scientists came to see themselves as serving mankind, and as being "men of letters" serving the "republic of letters" as exemplified by the **encyclopedia**. LAPLACE's statement foreshadows another outcrop of the eighteenth century. Namely, the construction and perception of the universe as a perfect mechanical clockwork led to thoughts as to the *expendability of its maker* in the next century.

3.1.2 Newton Is Finally Accepted

The struggle to find the right **worldview** was not over with the publication of NEWTON's *Principia* which detailed his laws of motion and universal gravity. It may make sense to us today that NEWTON's sweeping and correct treatment of so many of nature's phenomena was the last word and definitive explanation of these patterns. His contemporaries would not have agreed. Sure, on the British isles NEWTON's success and the acceptance of his theories was fast and complete. On the European continent, however, there was seemingly stubborn resistance which can be understood for the following reasons. For one, NEWTON's modernist mathematical worldview alienated some contemporaries who saw NEWTON's *hypotheses non fingo*,[11] approach as lacking because it was not psychologically rewarding. They felt that a mere mathematical description of the universe—the modern approach—had to be augmented by a physical mechanism, a "true reason" for this behavior of nature. And precisely this void was filled, as we saw in Section 2.9.2, by DESCARTES' **vortex theory**. While DESCARTES' theory was cumbersome to apply and suffered from some arbitrariness, it nonetheless made predictions, and some were in contradiction with NEWTON's results. New observations has thus to be made to decide which of the two theories should be accepted. The fight in the opening decades of the eighteenth century over the acceptance of NEWTON can be seen as part of the normal way in which scientists come to a useful and accurate description of the universe—by constructing explanations of the patterns in nature that can be verified by observations.

The Shape of the Earth Is Predicted and Checked

A good illustration of the meandering scientific opinion in the early eighteenth century is the story behind the measurement of Earth's shape. To first approximation, of course, the Earth is a sphere, but there must be a deviation from this ideal shape, as NEWTON himself

---

[10] An intellect with that knowledge was dubbed *Laplace's demon*.

[11] *Latin: I do not make hypothesis*—in the sense that only mathematical but no physical explanations are given. For instance, NEWTON offers no explanation as to why the sun—separated by hundreds of millions of kilometers of empty space from the planets—should have a gravitational influence on them. He just decreed that it *does* and how this force depends on the distance.

3.1. THE ENLIGHTENED CENTURY AFTER NEWTON

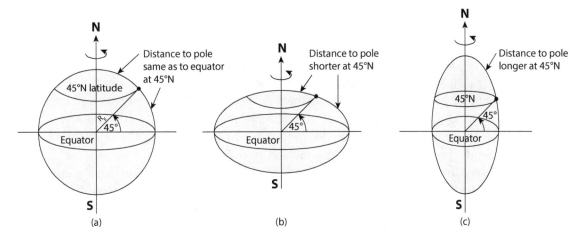

Figure 3.1: **The shape of the Earth.** (a) If the Earth were perfectly *spherical*, the distance between two latitude parallels would be the same everywhere. (b) For an *oblate* Earth (bulging out at the equator), the distance between two latitude parallels is bigger near the equator and smaller near the poles. (c) For a *prolate* Earth it is the opposite. In particular, the distance to the pole is greater than the distance to the equator for a point on the 45° latitude parallel.

had shown. Namely, he had explained the **precession of the equinoxes** by the forces (or rather torques) that the sun and planets exert on the Earth. These torques would change the inclination of the Earth's axis, were it not for the **conservation of angular momentum**. The latter in a sense redirects the torques, and the "compromise" is a change in the direction of the axis rather than its inclination. The point here is that there would not be any torques in the first place, if the Earth were perfectly spherical. NEWTON assumed a bulge around the equator to account for the torques, that is, he postulated the Earth to be **oblate**, Figure 3.1(b). On the other hand, DESCARTES' deliberations pointed to a **prolate** Earth.[12] Initial measurements carried our by JACQUES CASSINI[13] (1677–1756) in 1720 favored DESCARTES' theory. You might wonder how the shape of the Earth can be measured at all before the advent of space probes. The idea is that on a deformed Earth, the distance between two latitude parallels is different at different latitudes, see Figure 3.1. The Earth is very close to being a perfect sphere, so that it was hard at the time to make measurements which were accurate enough to distinguish between the two theories. The discrepancy between them is biggest if one compares the distance per latitude degree close to the equator with the number obtained close to the pole. Two expeditions were sent out in the 1730s. One to *Lapland* near the arctic circle, the other to *Peru* near the equator. One returned after several years, and the other after over a decade! The result was unequivocal: NEWTON was right, the Earth is oblate, its equatorial diameter being greater than its polar diameter, if by only 43 kilometers, or $\frac{43\,\text{km}}{12\,756\,\text{km}} = 3.3\%$.

---

[12] Recall the heads of the *Sesame Street* characters: ERNIE's is *oblate*, while BERT's is *prolate*.

[13] He was the son of Giovanni Domenico Cassini. His *Traité de la grandeur et de la figure de la terre* described his measurement of the arc of the meridian across France in 1713, which led him to conclude that the Earth has a prolate shape.

The Moon Is Highly Perturbed

It might seem that this test of NEWTON's theory couldn't have gone any other way, but this is in hindsight. At the time there were other deviations of NEWTON's theory from observations which cast serious doubts on its correctness. One such discrepancy was the motion of the moon. Again, to first approximation this pattern in nature was well understood; see Section 1.5. But as the quality of the observations improved with the advent of the telescope, so did the aspirations of the astronomers. In particular in the seafaring nations—such as England—predicting the moon's motion with high accuracy was seen as the key to solving the **longitude problem**.[14] Since the moon changes its position with respect to the stars rather quickly (about one apparent diameter per hour) navigators would measure this motion at sea, and compare it to predictions for a reference location such as Greenwich. The discrepancy could then be used to determine the longitude of the observer; see Figure 3.2. However, predicting the moon's precise position proved to be difficult. In 1693 EDMOND HALLEY had discovered a particularly distressing aberration of the moon from its average path known as the **secular acceleration**.[15] This was troublesome because it called into question the stability of the Earth-moon system and seemed to contradict NEWTON's laws of motion. Furthermore, aberrations like this made it impossible to navigate accurately by the moon. TOBIAS MAYER[16] (1723–1762) made significant progress in 1752 when he published his **lunar tables**, which allowed to determine the lunar position to five minutes of arc.[17] This in turn and made it possible to determine the longitude at sea to about half a degree.[18] LAPLACE found in 1776, that is, 99 years after NEWTON's *Principia*, that the secular acceleration of the moon can be understood with Newtonian mechanics, and is not troubling at all. LAPLACE considered the motion of the Earth, moon and sun a three-body problem, which has no closed solution and must be treated **perturbatively**, that is, approximated by the assumption that the influence of the sun on the moon is smaller than that of the Earth on the moon.[19] Eventually, it was realized that the motion of the moon is very complex, and that not just the sun but also the other planets contribute to the moon's deviation from its simple elliptic

---

[14] The *longitude problem* was to find a simple and practical method for the accurate determination of a ship's longitude at sea. Not knowing the position at sea can be fatal, exemplified at the time by the *Scilly naval disaster of 1707* in which 1550 sailors lost their lives. As a reaction, British parliament in 1714 passed the *Longitude Act*, promising £10,000 ($1.3 million in 2015) for a method to determine longitude within one degree, that is, an uncertainty of 60 nautical miles or 110 kilometers.

[15] It is not big (about $10''$ per century), and HALLEY discovered it by comparing ancient Babylonian eclipse observation with modern accounts. The point is its **secularity**, that is, its unrelenting increase.

[16] He was called the greatest astronomer of (not only) the eighteenth century by the French astronomer and science historian JEAN BAPTISTE JOSEPH DELAMBRE (1749–1822) and died during the occupation of Göttingen by French troops in the Seven Years' War.

[17] Incidentally, this was the same year when ALEXIS CLAUDE CLAIRAUT (1713–1765) published *his* lunar theory, decidedly based on NEWTON, as the title makes unmistakably clear: "*Théorie de La Lune: Deduite Du Seul Principe de L'Attraction Reciproquement Proportionnelle Aux Quarrés Des Distances (Lunar Theory: Deduced entirely from the principle of attraction inversely proportional to the squares of the distances)*."

[18] He was awarded £3,000 from the British government posthumously in 1765 for this achievement. It was sent to his widow who transmitted his manuscript *Theoria lunae juxta systema Newtonianum* on Newtonian lunar theory to the *Board of Longitude*.

[19] This story has several twists beyond the scope of this book which illustrate the meandering path of science. For starters, JOHN COUCH ADAMS (see Section 4.1.1) 77 years later found an error in LAPLACE's calculations rendering the explanation 50% too small. Then again, the moon is actually slowing down, yet seems to accelerate due to the gradual lengthening of the Earth's sidereal day due to **tidal friction**. As I said—too much information.

3.1. THE ENLIGHTENED CENTURY AFTER NEWTON

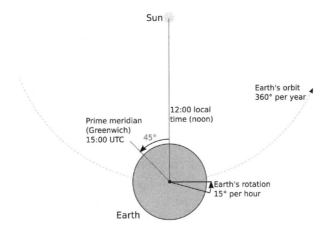

Figure 3.2: **Longitude determination.** The longitude of an observer, say aboard a ship on the ocean, relative to a reference position (usually Greenwich) can be obtained by comparing local time (determined for example by the fact that the sun is at its highest daily position) to a reference time (for example by carrying a *marine chronometer* showing Greenwich or *universal time*). In the figure, the observer's local time is three hours behind Greenwich time. Since the Earth rotates $\frac{360°}{day} = \frac{15°}{hour}$, the observer can conclude that she is situated 45° west of the *prime meridian*. Instead of a chronometer, the tabulated position of the moon can be used. The moon's position in the sky is determined and the *reference time* at which it assumes this position can be determine by lunar tables such as TOBIAS MAYER's; see text.

path. The moon is thus a highly perturbed celestial object—but its motion can be understood entirely on a Newtonian basis.

Halley Returns Triumphantly

As the last example for a successful prediction of NEWTON's theory consider the story of Halley's comet. EDMOND HALLEY's prediction in 1705 that the comet now bearing his name was to return late in 1758 according to NEWTON's planetary theory was much more than just a diligent execution of the *scientific method*. It was an exemplary *rational* explanation of a pattern in nature, very much in line with the spirit and belief of the eighteenth century; namely, that *everything* was explainable by the powers of the human mind and its reasoning abilities. As mentioned, this mind-set culminated in the construction of **Laplace's demon** predicting the fate of the universe, before it was realized that this view is untenable, even in principle.[20]

Recall that comets had for centuries be considered erratic and volatile phenomena, associated with superstition and considered bad *omens*. The ability to coolly analyze them, predict how they will move and even set the historic record straight,[21] was the capstone of a complete demystification of comets. We saw that Tycho had started it by concluding that the comet of 1577 was beyond the moon from its missing parallax, and confirming changes in the heavens contradicting ARISTOTLE. NEWTON had shown that comets must be solid, lest the sun evaporates them when they pass through the perihelion of their orbits.[22] It was HALLEY

---

[20] Undeterred, many erroneously believe to this day that scientists can compute and predict everything.

[21] The 1066 comet that allegedly heralded Harold's defeat at the hands of William the Conqueror was a predictable apparition of Halley's comet close to the sun.

[22] How do we know they do not evaporate? Indeed, "object constancy" is not obvious in the case of the

who applied NEWTON's theory to the comet's orbits and he was able to show that many comets have parabolic orbits, while Halley's comet is a **short-period comet** with a highly eccentric elliptic orbit. After consultation of the historic record, HALLEY was able to identify his with the comets of 1066, 1531, 1607, and others. Furthermore, he realized that the massive planets Jupiter and Saturn perturb and alter its orbit because of the gravitational force they exert on the comet. This made it difficult to accurately predict its return. HALLEY's 1705 suggestion of a late 1758 return of the comet was in fair agreement with observations, especially given the 53 year time lag. However, HALLEY had noticed that the period of the comet was variable and between 75 and 76 years. Due to an encounter of the comet with Jupiter in 1682, HALLEY thought a perihelion passage of early 1759 possible.

Astronomers started looking for the return of Halley's comet in 1757, and ALEXIS CLAUDE CLAIRAUT with collaborators embarked on the first large-scale numerical integration to more accurately predict the perihelion passage. In the end, it worked out beautifully. The predicted date was mid-April 1759 with an uncertainty of a month. The comet was first detected Christmas Day 1758, and observed until June the next year, so that an orbit could be determined from many such sightings. On this basis, it was determined that the perihelion passage had occurred March 13, 1759, thereby constituting a decisive check and confirmation of Halley's comet theory and Newtonian celestial mechanics.

Concept Practice

1. The Earth's shape is (a) irregular (b) oblate (c) prolate (d) perfectly spherical (e) None of the others
2. The moon's orbit is influenced by

    (a) ... the Earth only.
    (b) ... the sun only.
    (c) ... Earth and sun.
    (d) ... the planets.
    (e) Two of the above
    (f) None of the above

3.2 Physical Astronomy

Physical astronomy is concerned with the physical causes of celestial motions. Especially in the early eighteenth century, it meant to trace the observed motions of planets and moons back to their origin in universal gravitation. We already discussed NEWTON's explanation of the precession of the equinoxes by gravitational forces which the sun and planets exert on the **equatorial bulge** of the oblate Earth. Eventually, even minor movements such as the **nutation** of Earth's axis (described below) could be attributed to the Earth as a physical object subject to the gravitational forces of other members of the solar system. In the next section we will see how the physical motion of the Earth results in a very different kind of explanation of perceived stellar motion, namely the **aberration**[23] of starlight.

---

comets. NEWTON had correctly conjectured that the comets of 1680 and 1681 were the same comet, thus showing that comets stay intact during the time they are behind the sun and become unobservable for some time.

[23] Latin: *aberrare* = to stray, to wander off path.

3.2. PHYSICAL ASTRONOMY

3.2.1 Stellar Aberration: *Eppur Si Muove!*

By the late seventeenth century, accuracy of astrometric measurements, that is, the determination of stellar positions, had reached unprecedented quality, made possible by inventions such as the **micrometer eyepiece** (1640). However, one of the vexing counterarguments against Copernicanism was still not debunked. Namely, the physical motion of Earth around the sun had not *positively* been detected. If the Earth moves around the sun, then its position in space relative to the stars changes by a dramatic 300 million kilometers (2 AU) in the course of half a year, Figure 2.3.1(b). Hence, the sky should look (slightly) different from these two orbital locations. In particular, the **parallactic effect** should systematically and periodically change the position of nearby stars relative to background stars, see Section 2.3.1. The failure to observe such positional change had to be attributed to the immense distance of the stars relative to the Earth-sun distance. This was not implausible, since also the brightness of the stars is greatly reduced compared to the sun, which leads to the same conclusion, namely that they must be much farther out. Still, the non-observation was frustrating, and astronomers hoped for a double reward: Proof once and for all that *"Eppur Si Muove!"*[24] plus a trigonometric, direct measurement of the distances to the stars.

As early as 1674, ROBERT HOOKE had reported that the star γ in *Draco* changed its position periodically. In particular, it was 23 arcseconds more northerly in July than in October. HOOKE had picked the star for good reason: At London's latitude it passes directly overhead,[25] and hence his observations were free from corrections due to **atmospheric refraction**. His choice was fortuitous for another reason: The star is less than 15° off the **pole of the ecliptic** which is in *Draco*, and therefore shows what would come to be known as **stellar aberration** almost at maximum strength, as we will see. Independently, JEAN PICARD[26] (1620–1682) in 1680 and the first Astronomer Royal JOHN FLAMSTEED (1646–1719) in 1689 had reported that *Polaris* itself varied in position by as much as 40 arcseconds annually. Astronomers were nonetheless skeptical whether this was the long-sought parallax. One reason for this is obvious. The parallactic motion of the star is proportional to its distance from the observer. The reports of HOOKE, PICARD and FLAMSTEED on the other hand suggested that the observed effect is the *same* for *Polaris* as for γ *Draconis*. This is suspicious, because there is no reason why these stars should be at the same distance.[27]

In any case, the annual motion of stars was a pattern in nature that awaited explanation. So JAMES BRADLEY (1693–1762), third Astronomer Royal, and his collaborator SAMUEL MOLYNEUX (1689–1728) started a series of observations of γ *Draconis* in 1725 to settle the question. Indeed, they found that the star moved in a circular pattern, being 40″ further south in March than it appeared in September. To complicate things, they found that another star, *35 Camelopardalis*, also at declination 51.5° but 180° off in celestial longitude, was furthest south in *September* if only by 19″ compared to March. This pattern defied all previous explanations. So—in standard science fashion—the two kept up observations of the stars not

---

[24] "And yet it [the Earth] moves!" is what Galileo allegedly muttered after his condemnation by the inquisition.

[25] Celestial latitude (declination) of γ *Draconis* is 51.5°; London's geographical latitude is also 51.5°.

[26] He measured the size of the Earth with an accuracy comparable to modern values with the help of recently invented instrumentation such as the cross-wired eyepiece. He also had the good idea to attach a telescopic sight to his quadrant, thus improving its accuracy by a factor 24 over TYCHO BRAHE's.

[27] Recall that the celestial sphere is just a convenient tool to describe the position of stars, but not real in any sense.

188 CHAPTER 3. DEDUCING & CHECKING: BETWEEN NEWTON AND EINSTEIN

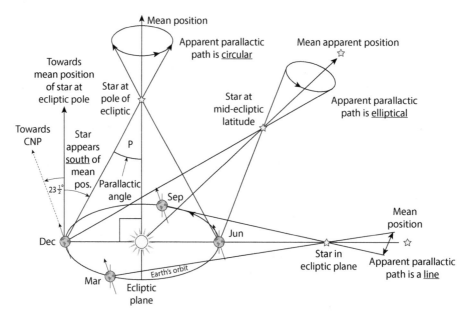

Figure 3.3: **Stellar parallax due to orbital motion of Earth.** An observer on Earth sees the sky from different vantage points throughout the year. In particular, when observing on nights half a year apart, a parallactic effect with a baseline of 2 AU = 300 million km is experienced. Close stars will appear shifted with respect to the background of distant stars. The annual path of the star depends on its angular distance to the **pole of the ecliptic**. A star at the pole will display circular parallactic motion, a star on the ecliptic a linear motion, and a star in between an elliptic path. The southernmost point of such a path on the celestial sphere is reached December 21, because at the solstice the Earth axis (defining North) is slanted away from the sun, but the line of sight of the star is slanted toward the sun. Note that the Earth's axis does not change its direction; it points to the **celestial north pole** year round.

to be fooled into publishing wrong results or conclusions. The patience, gruesomely tried by Molyneux's untimely death, paid out. Several explanations or hypotheses were tested and rejected. First off, this was clearly not a parallactic motion which predicts a northernmost position in June, see Figure 3.3. Another plausible explanation was a periodic change of the Earth's rotational axis, known as **nutation.** Since the two stars are situated on opposite sides of the celestial sphere, this would explain why one reaches the southernmost extreme position just as the other reaches the northernmost extreme position. However, the north-south motion of *35 Camelopardalis* was only about half of the motion of γ *Draconis*, falsifying this hypothesis.

While BRADLEY's diligent observations enabled him to *calculate* and predict the motion for any star, the *reason* for the pattern escaped him for a while. According to an anecdotal story, BRADLEY realized the true nature of the observed pattern while sailing on the River Thames. Observing that the flags on the boat turned around whenever the boat changed direction, although the wind hadn't turned, he realized that an analogous effect would happen when the Earth changes its direction owing to its rotation around the sun. BRADLEY's classical explanation of stellar aberration is based on the **velocity**[28] of the Earth's orbital motion, and not, as the parallax hypothesis would have it, on the **position** of the Earth as it moves around the sun. The gist of the argument is that the light from a distant star seems to come from a different direction, depending on the **reference frame** of the observer. In

---

[28]The velocity (vector) of an object points in the direction of its motion.

3.2. PHYSICAL ASTRONOMY

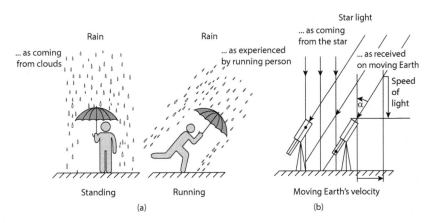

Figure 3.4: **Stellar aberration.** (a) Analogy for stellar aberration: A person in the rain has to adjust her umbrella if running. (b) BRADLEY's corpuscular explanation of stellar aberration. If the starlight is modeled as "drops of light" falling into the telescope, then the direction of the telescope has to be adjusted by the *aberration angle* α so that light falls *through* the telescope as it is moving with Earth relative to the star. The modern explanation is different, but yields basically the same result since the velocity of Earth is so small compared to the speed of light. In other words, BRADLEY's explanation is a good *approximation*.

other words, things look slightly different in the rest frame of the sun than they do in the moving frame of the Earth. The reason is that the speed of light is finite.[29] As an analogy, consider a person in the rain, Figure 3.4(a). Assuming the rain falls straight down, the direction of the umbrella has to be adjusted for maximum coverage if the person is moving. BRADLEY's explanation of **aberration** assumes that starlight is analogous to rain drops,[30] Figure 3.4(b). By the time the light reaches the eyepiece at the lower end of the telescope, Earth and telescope have moved by a distance proportional to the ratio of the orbital speed v of the Earth and the speed of light c. The effect is thus proportional to $\frac{v}{c}$, and allowed BRADLEY to measure the speed of light! The angle by which the position of a star differs in the rest and the moving frame is $\phi \approx \frac{v}{c}$, and BRADLEY had measured this **aberration constant** to be $20.5'' = \frac{20.5}{3600}\frac{\pi}{180} rad \approx \frac{1}{10,000} rad$. Therefore, he concluded that the speed of light is about 10,000 times faster than the orbital velocity of Earth. Nowadays we know that $v = 30\,\text{km/s}$, so the speed of light comes out very accurately to $c = 300\,000\,\text{km/s}$. How do we know the orbital speed of Earth? Via the fact that the Earth moves a distance equal to its orbital circumference in precisely a year. However, BRADLEY did not know the distance to the sun—at least not accurately. Therefore, he could have expressed the speed of light in astronomical units per second, but he chose equivalently, just as his predecessor Ole Rømer, Section 2.9.5, to calculate the time it takes light to traverse one astronomical unit. In other words, he determined how long light from the sun takes to reach us on Earth.

---

[29] If you think that this sounds almost like a tale from *relativity theory*, you are absolutely right. There is a rich and interesting connection between *stellar aberration* and EINSTEIN's construction of the theory of *special relativity* in 1905, which gave the first correct explanation of *stellar aberration*.

[30] In other words, he uses NEWTON's *corpuscular theory* of light. As the latter was challenged by Thomas Young's interference experiment showing the wave nature of light in 1801, so was BRADLEY's explanation. What followed was a series of hypothesis involving the *aether* (the medium in which lightwaves allegedly travel) until EINSTEIN explained aberration with his *relativity theory*.

Assuming the Earth's orbit to be circular, we get

$$v = \frac{\text{orbital circumference}}{1 \text{ year}} = \frac{2\pi \,\text{AU}}{365.25 \times 24 \times 3600\,\text{s}} \approx 2 \times 10^{-7}\,\frac{\text{AU}}{\text{s}},$$

hence quoting the speed of light to be $c \approx 2 \times 10^{-3}\,\frac{\text{AU}}{\text{s}}$, or 1 AU in 500 seconds (8 minutes and 20 seconds), in fair agreement with RØMER, who reported 11 minutes per AU. This underlines once more how important it was for the astronomical community to measure the distance to the sun reliably. This was attempted half a century later on HALLEY's suggestion with the transit of Venus; see Section 3.3.1.

Note that BRADLEY's description of stellar aberration also explains why the star *35 Camelopardalis* shows only half the effect. Namely, it is farther from the pole of the ecliptic, and therefore its **aberration ellipse**, Figure 3.5, has a minor axis that is only half as big as that of γ *Draconis*. The discovery of aberration is thus really a twofold story. Namely, the detection of a new pattern in nature and its explanation on the basis of existing knowledge go hand in hand. Note that in the end **stellar aberration** became the first **positive proof**[31] of the motion of the Earth.

3.2.2 The Earth as a Spinning Top

Ultimately, BRADLEY's seemingly destructive **falsification** of the hypothesis that the annual motion of stars like γ *Draconis* is due to the **parallax**, was very constructive. It led to an understanding of the importance of relative motion and reference frames. It consolidated the view that light travels at a finite, measurable speed. Of course, it also validated COPERNICUS, if such validation was still necessary at this time. What's more, the failed hypothesis that the orientation of the Earth's axis undergoes periodic changes led to the discovery of **nutation**, Figure 3.6(a), basically a "nodding" motion of the Earth's axis. The data that BRADLEY scrupulously recorded over a period of 20 years to bolster his interpretation of **stellar aberration**, led to the discovery that indeed the Earth's axis moves in a slightly elliptical path around its mean direction. BRADLEY was able to show from his observations that the period of this oscillation is 18.6 years. This in turn implied most directly that the moon must be responsible for the nutation, since its orbit—inclined by 5.2° with respect to the ecliptic—rotates in 18.6 years around the ecliptic, Figure 3.6(b). The magnitude of the effect is smaller than that of stellar aberration, and the **constant of nutation** is 9.2 arcseconds, that is, less than half of the **constant of aberration**.

While BRADLEY presented an intuitive understanding of nutation, the details were worked out by the mathematicians around 1760, most notably LEONHARD EULER (1707–1783) and JEAN LE ROND D'ALEMBERT (1717–1783). In their works, the Earth was treated as a spinning top, and therefore as a **terrestrial machine**. The minuscule aberrations in stellar astronomy were explained by a model of the body of Earth itself together with the gravitational forces acting on this body. In other words, it was realized that the positional observations of the night sky have a **physical reason**. In Chapter 2 we saw the clash between **mathematical** and **physical astronomy**. The rise of the latter since the mid-1600s is the foundation of the belief of the eighteenth century that nature—and everything else for that matter—can be understood by the human intellect. In this approach the laws of NEWTON and the

---

[31] Science jargon: direct evidence, direct consequence of the postulated mechanism.

3.2. PHYSICAL ASTRONOMY

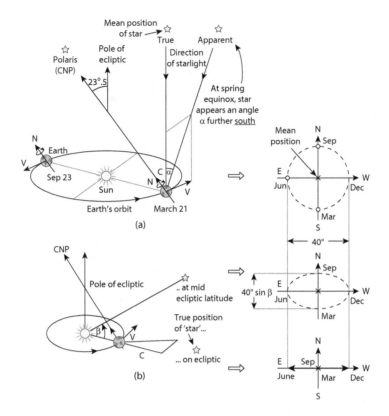

Figure 3.5: **Aberration ellipse.** Due to aberration, stars deviate from their mean position as the Earth moves around the sun. The path of the star around its mean position is due to the changing direction of Earth's orbital velocity. In March, velocity points away from the axis' tilt, and the star is south of its mean position. In September it's the opposite. (b) The shape of the apparent path depends on the **ecliptic latitude** of the star. Stars on the ecliptic show only east-west aberration and their path is a straight line. At the pole of the ecliptic, a circular path exhibits full east-west and north-south aberration as in (a). In between, an elliptical path ensues, like the one of *35 Camelopardalis* which triggered BRADLEY's comprehension of the phenomenon.

infallible logic of mathematics are combined and applied to explain nature. The universe is constructed as a **terrestrial machine**, based on models and analogies gleaned from our immediate vicinity on Earth.

While the details of the theory of the spinning top are firmly beyond the scope of the book, we can appreciate the scientific advances in the understanding of the Earth itself, paradoxically due to observations of the *sky*. From ERATOSTHENES's deduction of spherical shape of the Earth and its radius (Section 2.5), via COPERNICUS's postulate of the moving Earth, to the prediction of NEWTON that the Earth must be oblate to explain the **precession** of the equinoxes, people came to realize that the Earth does not constitute an independent, unshakable basis of reference for observations. When EULER went on the ascertain that moon, sun and other planets influence the motion and rotation of the Earth, the Earth-based observer lost much of her status as an independent spectator. Indeed, the astronomer has to carefully *reduce* observations to take into account that the reference frame itself, epitomized by the position and direction of the Earth's rotational axis, are in constant motion.

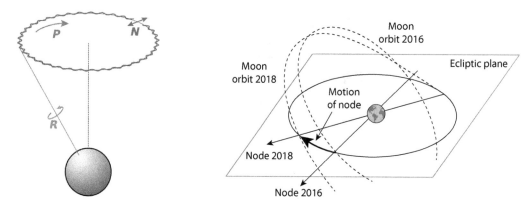

Figure 3.6: **Nutation due to lunar orbit rotation.** (a) BRADLEY discovered a "nodding" of the Earth's axis by noticing small, periodic changes in stellar positions. Recall that the Earth's rotates (R) in 23h 56m 4.1s around its axis which very slowly *precesses* (P) around the pole of the ecliptic in 26,000 years. Also shown is the nutation (N) of the axis. Periods and amplitudes are not to scale. (b) Since the nutation period is 18.6 years, and the orientation of the lunar orbit as represented by the position of the ***ascending node*** (the point where the orbit crosses the ecliptic plane) rotates in 18.6 years around the ecliptic, the cause of the effect was readily exposed.

Concept Practice

1. What is *stellar aberration*?

    (a) The same as *stellar parallax*
    (b) An optical flaw in lenses
    (c) The proper motion of star, that is, with respect to other stars
    (d) An apparent motion of stars due to Earth's changing orbital velocity
    (e) None of the above

3.3 The Scale of the Cosmos Is Determined

3.3.1 Measuring the Astronomical Unit

The construction of the universe starts with measurements on Earth. This statement is both provocative and bordering on triviality. It means that in order to determine distances between celestial objects while confined to the Earth, we must compare celestial with terrestrial distances. This is trivial in the sense that every length measurement is a comparison,[32] but practically, we cannot compare distances in space in this way, and must resort to other methods. Trigonometry offers a way, because the size of a triangle can be deduced from measuring one on its sides (the **baseline**) and two angles. We explained the method in Section 2.3.1, and here just emphasize the fact that the baseline has to be on Earth to be measurable at all. As a period example for the use of this ***trigonometric method*** consider the determination of the distance to the moon by NICOLAS LOUIS DE LACAILLE (1713–1762) and J.J. DE LALANDE (1732–1807). They determined the parallax of the moon, that is, the difference of moon's position relative to the fixed stars, see also Figures 2.2(c) and 2.52. LACAILLE was at the Cape in South Africa in 1752 to compile his great southern star catalogue, when his

---

[32] For instance, we compare the size of an object with the length of a **yardstick**, and if the object's size is half the length of a yardstick, we say that its size is 0.5 yards.

3.3. THE SCALE OF THE COSMOS IS DETERMINED

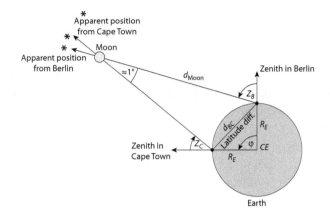

Figure 3.7: **The lunar parallax determines the distance to the moon.** Observing at the same time from two different locations, for example LACAILLE in Cape Town, LALANDE in Berlin, the moon appears shifted with respect to the background stars. Measuring the moon's zenith distances Z_B and Z_C from Berlin and Cape Town yields the **lunar parallax** ($51'' \approx 1°$). The radius of the Earth R_E and the latitude difference φ are known and allow for the computation of the distance between the observers d_{BC}. Then one side and all angles of the triangle BCM are known, and finding the distance d_{Moon} becomes a purely geometrical problem.

pupil LALANDE observed the moon concurrently from Berlin, Germany. They found that the moon observed from the Cape was shifted by 57' or almost one degree when compared to observations from Berlin. The Cape and Berlin are almost at the same longitude, but about 90° different latitude: Berlin is about 50 degrees north, the Cape about 40 degrees south of the equator. The geometry of the situation is depicted in Figure 3.7. Practically, the men were measuring the zenith distances[33] Z_B and Z_C, which they found to differ by about one degree. They could therefore determine the distance to the moon from either observer as the length of one side of the **isosceles triangle** formed by the moon, Berlin and the Cape. The **baseline** for this measurement is the straight line between the two towns and easily found because the right angle between the locations allows for a straightforward application of the **Pythagorean theorem**. Expressed in Earth radii it is $\sqrt{2}R_E$. An isosceles triangle with this baseline and opening angle of 1 degree has a side length of $60R_E$, so the distance to the moon is $60 \times 6400 \,\mathrm{km} = 384\,000 \,\mathrm{km}$, right on the modern value, and compatible with ARISTARCHUS's value obtained two thousand years earlier.

The big price, however, and the key to the **scale** of the universe, was the determination of the distance to the sun, the **astronomical unit**(AU). With it, the map of the solar system receives its scale,[34] and the **stellar parallax** an absolute baseline; see Figure 2.3.1. As pointed out, ARISTARCHUS value for the AU was twenty times too small, and a reliable method to measure it was overdue. The invention of the **micrometer eyepiece** in 1640 improved the accuracy of measurements considerably. In 1672 CASSINI—observing Mars from Paris—and a colleague—observing from French Guyana in South America—used the **trigonometric method** and determined the distance to the sun to be 138 million kilometers, about 11 million kilometers shy of the modern value, a 7.5% **relative discrepancy** according to Equation (1.4). BRADLEY in 1719 and LACAILLE in 1751 fared worse. Therefore, hopes were high to follow

---

[33] Recall that at locations of equal longitude the sun culminates at the same time, that is, they have the same orientation with respect to the celestial sphere.

[34] Remember that Kepler's third law just determines the *relative* distances, that is, distances in AU.

HALLEY's suggestion to use a once-in-a-lifetime chance to observe a transit of Venus in front of the sun in the 1760s from distant locations. The geometry of such a transit is depicted in Figure 2.26(b). Venus at **inferior conjunction** appears as a tiny dot on the sun's face, Figure 2.26(c), with its exact path depending on the observer's location. Venus is the closest planet to Earth, therefore this astronomical event offers the greatest accuracy of determining the distance to Venus and therefore to the sun from Earth. The trouble with **Venus transits** is that they are extremely rare. The first transit to be predicted (by KEPLER in 1627) was first to be scientifically observed by JEREMIAH HORROCKS (1618–1641) in 1639,[35] and the next one was not to come until 1761. The measurements made in 1761 were much less accurate than hoped for, owing to the unexpected optical **black drop effect** seen in Figure 2.26(a), making it hard to pinpoint the time when Venus reaches the limb of the sun. Therefore, much was riding on the next transit in 1769.[36] Britain sent famous CAPTAIN JAMES COOK (1728–1789) with a **marine chronometer** of JOHN HARRISON (see Section 3.6.2) to *Tahiti* in the South Pacific to observe.[37] Other countries followed suit, and more than 200 memoirs were published about the transit. Again, expectations were not met, although the new value for the AU was now firmly in the 150 million km range. Since the next transit was predicted to occur over a hundred years later in 1874, scientists discussed and reanalyzed the observations of the 1769 transit, and ENCKE came to the conclusion in 1824 that the sun is 153 million kilometers from the Earth, about 3.5 million kilometers in excess of today's accepted value, a 2.4% *relative discrepancy*.

3.3.2 Weighing the World: Quantifying Gravity

To figure out the **scales** of the cosmos, we need not only a fundamental length unit like the AU, but also a mass unit. Weighing things on Earth, that is, comparing the gravitational forces on an object with those on a reference object, does not get us very far in the celestial realm. We need to compare masses of astronomical objects. Again, the quest started at home: The Earth itself was to be weighed. This might have seemed an impossible task in the seventeenth century [40], but not so with the improved instrumentation and accuracy of the eighteenth century. The fundamental experiment, suggested by NEWTON and performed by NEVIL MASKELYNE[38] (1732–1811) a hundred years later, was to measure the angle by which a plumb line is deflected due to the gravitational force of an isolated nearby mountain, Figure 3.8. The deflection is relative to the zenith direction determined by an astronomical observation (position of a star directly overhead for a given location and time). Therefore, the plumb line which is attracted by the mountain suggests a larger latitude difference ϕ' than the true latitude difference ϕ gleaned from astronomical observations. MASKELYNE in 1774 measured north and south of Schiehallion, (a mountain in Scotland) and found $\phi' = 54.6''$, $\phi = 42.9''$. He attributed the discrepancy of 11.7 arcseconds to the gravitational attraction of the mountain. This was a solid test of the universality of NEWTON's laws, showing that not just celestial bodies, but also objects on Earth attract by gravity. But MASKELYNE did not stop there. Measuring the shape and dimensions of the mountain, and the density of its

---

[35] Horrocks used it to estimate the AU to be 96 million kilometers, about 36% too small, but significantly larger than the value previous observers obtained.

[36] The transits come in pairs eight years apart, but there are more than 110 years between two sets of pairs.

[37] In a secret mission following the transit observation, COOK discovered Australia in 1770.

[38] He was the fifth Astronomer Royal.

rocks, he concluded that the result could be explained under the assumption that the Earth, providing most of the gravitational force, is twice as dense as the mountain on average. This turns out to be correct as we will see in Section 4.2, and shows that the Earth's crust is half as dense as its deeper layers. This may not be too surprising to us, but in the day it falsified the widely-held belief[39] that the Earth is a hollow sphere.[40]

MASKELYNE's result can be used to estimate the mass of the Earth, which was only very vaguely known at the time.[41] The density of Schiehallion mountain can be estimated, since most rocks on Earth have a density of about 2.5 times that of water. Hence, the Earth is about five times denser than water according to MASKELYNE's measurements, and its volume is computable from its radius as determined initially by ERATOSTHENES, Section 2.5. Hence, we obtain

$$M_{Earth} = (\text{density}) \times (\text{volume}) = 5,000 \frac{kg}{m^3} \times \frac{4\pi}{3} R_{Earth}^3 \approx 20,000 \frac{kg}{m^3} (6.3 \times 10^6 m)^3 = 5 \times 10^{24} kg,$$

less than 20% off the modern value. Quite an achievement to weigh the Earth—while standing on it!

The **Schiehallion experiment** of 1774 had a big impact on the development of the sciences since it was realized that measurements of fundamental laws of nature could be made in a laboratory on Earth under well-defined and controlled circumstances, rather than having to passively observe an event in the heavens without the hope of carefully developing or refining the experimental method. In this spirit, the next fundamental measurement of the forces of the cosmos was undertaken by French physicist CHARLES-AUGUSTIN DE COULOMB (1836–1806) in 1784. Coulomb used a **torsion balance**, to measure the forces between **electric charges**. The repulsive forces between two like electric charges accelerates the rotating arm of the setup with one charge connected to the end of the arm, until the associated force is as great as the force due to the torsion of the wire holding the main arm of the experimental setup. The angle by which the balance has rotated is proportional to the strength of the electric force, which is thus directly measured. It turned out that **Coulomb's law**, detailing the dependence of the electric force on the charges q_1 and q_2 of two objects and their mutual distance r has exactly the same form as NEWTON's **law of universal gravitation**, cf. Equation (2.8), namely

$$F_{electric} = k_e \frac{q_1 q_2}{r^2}. \tag{3.1}$$

The proportionality constant k_e tells us how strong the repulsion or attraction between charges is. Empirically, it turns out to be $k_e \approx 9 \times 10^9 Nm^2/C^2$; the electric force is thus much stronger than the gravitational force.[42] We do not experience its strength because most objects are

---

[39]Even the great HALLEY subscribed to it.

[40]Probably not the inspiration for stories like Jules Verne's *Journey to the Center of the Earth* (1864), which was inspired by CHARLES LYELL's books on geology; see Section 3.8.2.

[41]PIERRE BOUGUER (1698–1758) had estimated the Earth mass about a factor two to three too large when he measured it as part of the French expedition to Peru to determine the shape of the Earth; see Section 3.1.2. In his measurement he recorded the period of a pendulum at different elevations. According to NEWTON's gravity law, the period depends on the gravitational constant G, which cancels if two measurements are compared. Incidentally, BOUGUER invented the **heliometer**, later crucially improved by JOSEPH FRAUNHOFER, Section 3.6.2.

[42]Naively, the electric force is stronger by the ratio of the numerical values of the two constants k_e and G, to wit $\frac{9 \times 10^9}{6.67 \times 10^{-11}} \approx 1.3 \times 10^{20}$.

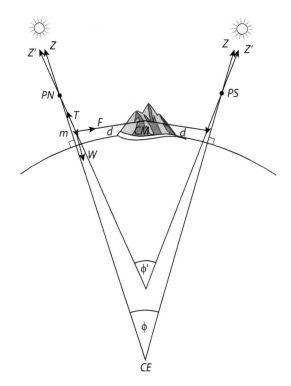

Figure 3.8: **Determining the mass of the Earth by measuring the deflection of a plumb line due to a mountain.** The direction to the Earth's center CE is determined by an *astronomical* measurement which yields the true zenith direction Z. Hanging a plumb line from pivot points north and south of the mountain, PN and PS, yields the directions Z'. The discrepancy between the two directions is indicative of the Earth's mass relative to the mountain's mass. The plumb bob of mass m is subject to the gravitational force F of the mountain, the tension T of the plumb line, and the weight $W = mg$, that is, the gravitational force of the Earth. Applying NEWTON's laws yields the relative masses of Earth and mountain.

electrically neutral, that is, they contain as many positive as negative charges; there is only one (positive) "gravitational charge," namely ***mass***.

Due to this weakness of gravity, the fundamental constant G in NEWTON's gravity law is very small and was not accurately measured until the very end of the eighteenth century. Not that we can deduce G from MASKELYNE's mass of the Earth above. The acceleration due to gravity g as measured, say, by the period of a pendulum, is due to NEWTON's gravity law at the surface of the Earth, that is a distance R_{Earth} away from the Earth's center. Thus

$$F = mg = G\frac{mM_{Earth}}{R^2_{Earth}},$$

or

$$G_{Maskelyne} = g\frac{R^2_{Earth}}{M_{Earth}} = 9.8\,\frac{\text{m}^2}{\text{s}^2} \times \frac{(6.38 \times 10^6\,\text{m})^2}{5 \times 10^{24}\,\text{kg}} = 7.9 \times 10^{-11}\,\text{Nm}^2/\text{kg}^2,$$

since the mass m cancels. This value is about 20% too large because MASKELYNE's value for the Earth's mass was 20% too small.

In 1798 HENRY CAVENDISH (1731–1810) measured G with a torsion balance by comparing the gravitational force between massive lead balls. His setup and measurement procedure were superb, and the accuracy of his result was not to be superseded for a hundred years!

Cavendish did quote the result as the mass of the Earth rather than as a value for G, and referred to it his achievement as having "weighed the world." His result was within 1% of the present day value—remarkable indeed!

Concept Practice

1. Why can't *Kepler's third law* be used to compute the length of the *astronomical unit*?

    (a) Because it is not based on Newton's gravity theory.
    (b) Because the gravitational constant G was not known at the time.
    (c) Because the planets are too far away.
    (d) Because it is a relation between *relative* sizes of orbits only.
    (e) None of the above

3.4 The Birth of Chemistry Dissolves Aristotle's Elements

3.4.1 Heat Is Not Temperature. Is Chemistry Physics?

For much of the eighteenth century, there was no clear dividing line between chemistry and physics. As scientists tried to quantitatively understand the universe, there was no clear sense of which quantities and properties to measure. One example important both for the development of the physical discipline of **thermodynamics** and for the nascent field of **chemistry** was the confusion about **heat** and **temperature**.

We have to understand that even after the sweeping anti-Aristotelian success in astronomy, where his theories were decisively falsified, ARISTOTLE's ideas on the rest of nature, especially what we now call biology, were generally sound and enduring. Also his idea that the properties of objects and substances can be understood by assuming that they are mixtures of the four elements **water, earth, fire**, and **air**—each endowed with distinctive properties—was not dismissed until the end of the eighteenth century. As it was hard for scientists to make sense of such varied phenomena such as heat, electricity, combustion and evaporation, they scaffolded their thoughts using the old Aristotelian framework. For instance, they thought of heat as connected to the Aristotelian element fire, and assumed that hot objects contained a large quantity of that element, which was released when the object cooled down. Furthermore, ARISTOTLE had considered heat to be a *quality*, not a *quantity*. Therefore, it could not be measured, that is, quantified. On the other hand, Galileo as early as 1592 had constructed a thermometer, and by 1641 the thermometer with an enclosed liquid (usually mercury or alcohol) was invented in which the expansion of the liquid allows for a quantitative measurement—but of what? Does a thermometer measure **heat** or **temperature** or both? A clear answer was not forthcoming. Only in the middle of the eighteenth century did it became clear that the two are different. It was JOSEPH BLACK,[43] (1728–1799), a physician and chemist, who pointed out that temperature and heat are related but not identical. Heat, like other phenomena as electricity and light, was modeled as an ***imponderable fluid*** in those days, implying that it flows from one location to another (like the heat of an iron rod warmed at one end by a fire) yet they do not have a detectable weight. BLACK explained that a thermometer measures the temperature of an object, which can be thought of as the **density** or intensity of the heat fluid, but not the *amount* of the fluid. Along those lines,

---
[43]He discovered *magnesium* and *carbon dioxide* as well as **latent** and **specific heat**.

Scale	Fahrenheit	Celsius	Kelvin
absolute zero	−459.67 °F	−273.15 °C	0 K
freezing point of water	32 °F	0 °C	273.15 K
boiling point of water	212 °F	100 °C	373.15 K

Table 3.1: **Temperature scales.**

DANIEL GABRIEL FAHRENHEIT (1686–1736) around 1730 found from experiments that heat is proportional to volume. This is plausible and intuitive, viz. two gallons of hot water would contain twice the heat as one gallon, while the water serves as a *container* for heat. However, it is incorrect, since the amount of heat contained in a substance depends on a specific property called its **heat capacity** C as BLACK theorized around 1760. Heat capacity can be measured by observing how much heat Q is necessary to change the temperature of a given chunk of substance by an amount $\Delta T = T_{\text{after heating}} - T_{\text{before heating}}$.

$$C = \frac{Q}{\Delta T}. \qquad (3.2)$$

As it turns out, water has an exceptionally high heat capacity,[44] and it takes an energy of 4200 Joules to increase the temperature of one kilogram water by just one degree.[45]

FAHRENHEIT had invented a temperature scale in 1728 based on human body temperature; see Table 3.1, which is in use virtually only in the United States today. Swedish astronomer ANDERS CELSIUS (1701–1744) developed his widely used **centigrade scale** in 1742[46] by subdividing the temperature range between the boiling point and freezing point of water into 100 parts. Illustrative for the confusion of the day is the fact that he decreed the freezing point of water to be 100° C. As odd as this may sound today, there is no reason why numbers should run up or down; this is just a *convention* we have become accustomed to. Inquiring further, you might wonder why one scale is preferable over another. While both humans and the properties of water are special in a sense, neither has absolute significance in the universe. As it turns out, there is such an *absolute*, namely the existence of a lowest temperature called **absolute zero**, suggested by the theory of heat engines, see Section 3.8. Kelvin in 1848 calculated this temperature by *extrapolating* the temperature dependent volume reduction (as in a gas thermometer) to zero volume, which happens at −273 °C[47]. The scientific temperature scale is therefore the **absolute Kelvin scale**, with an origin that is based on an absolute property of nature itself, rather than on the properties of certain substances existing in the universe.

As we will see in Section 3.8, the way to a true understanding of heat was via the new concept of **energy**, which eventually opened the door to an understanding of stars and their energy production mechanism. But since energy is such an abstract concept, its construction had to progress along a painfully long-winded path. We cannot do justice here to this

---

[44] EDDINGTON used the heat capacity of water as an *upper bound* to model the interior of stars in 1926, see Section 5.3.1.

[45] For comparison, less than 400 Joules are necessary to warm a kilogram of solid copper by one degree; that is, an **order of magnitude** less!

[46] He was a member of the *Lapland expedition* which established the Newtonian oblate shape of the Earth.

[47] Absolute zero is now defined via the **triple point of water** to be at −273.16 °C.

3.4. THE BIRTH OF CHEMISTRY DISSOLVES ARISTOTLE'S ELEMENTS

fascinating part of science history, and only describe one milestone along this way, namely the realization by COUNT RUMFORD (1753–1814) that heat is not a fluid. In 1787 RUMFORD was working at the military academy in Munich, Germany, overseeing the boring of canons. RUMFORD observed that drilling into the cannon body produced large amounts of heat. He set up an experiment in which a cannon was submerged in water with the drills driven by horses. He noticed that the water could be brought to a boil by the boring process, and that the water continued to boil as long as the horses were turning the boring tool. His conclusion was that there must be an inexhaustible supply of heat, as the horses could keep at it for an arbitrarily long time. Here was for the first time a decisive experiment showing that heat is not a substance, but is connected to the motion of the atoms or particles of which the object consists. In other words, heat could be explained by a mechanical theory. It was the death blow to the fluid theory of heat which assumes that heat is a conserved and therefore limited quantity. As often in science, the inferior theory continued to be used because it was useful to explain several patterns in nature adequately and with more ease than the technically correct theory. Its final demise came when the theory of heat was extrapolated to the radiant heat supplied by the sun. It seemed too contrived to assume that a heat fluid should be able to travel through millions of miles of empty space to the Earth.[48] As we will see in Section 3.7, by then the wave theory of light suggested a *wave theory* of heat. Eventually, these theories of heat were replaced by *thermodynamics* and the *kinetic theory of gases*; see Section 3.8. A detailed description would lead us too far into physics. However, it is important for our understanding of the construction of the expanding universe to note the fact that these new, modern theories are substantially more abstract and mathematical than the older ones. Therefore, somewhat paradoxical, they furnish less of a physical *explanation* of the cosmos, and more of a mathematical *description* of the universe.

3.4.2 The Delayed Scientific Revolution

This section on *chemistry* may seem displaced in a book on astronomy, but the universe consists of 90% hydrogen, and it was the pursuit of chemistry which discovered it. Furthermore, the construction of the modern concept of atoms and elements is deeply intertwined with the process that shaped chemistry as an independent field of *natural philosophy*. In the time between the rise of modern science in the early seventeenth century, and the middle of the eighteenth century, the chemists were often also seen as physicists, and the profession was largely the business of physicians and pharmacists.[49] Our goal here is to understand how science progressed from the Aristotelian four-element theory to a modern theory of atoms that come in different flavors known as the *chemical elements*, of which hydrogen is the first and most simple.

The field of chemistry emerged as scientists sought to explain patterns in nature connected to substances, like their *acidity*, *salinity* or *metallicity*. Important processes that needed to be understood were the *combustion* of materials, *fermentation* and *distillation*, all of them important in the decades leading up to the *industrial revolution*. Even around 1750 both chemistry and physics were still based on the Aristotelian idea that objects contain vary-

---

[48] Note that the mechanical theory had a similar problem: If there is no matter in outer space, how can heat be a mode of motion of particles?

[49] Important chemists of the time like GEORG ERNST STAHL (1659–1734) and the aforementioned JOSEPH BLACK were physicians.

ing amounts of the four elements, which in turn explains their physical (density, malleability, melting point) and chemical (solubility, alkalinity) properties. How confusing and hampering these Aristotelian convictions must have been for the contemporary scientists is illustrated in the changing role of air, which started out as an **element** and ended up as a **state of matter** when reinterpreted as the gaseous or vaporous state of a substance. Likewise, *fire* was at times identified with contradicting concepts such as heat, temperature, energy, and caloric fluid. Adding to the confusion was the fact that both *air* and *fire* could be *fixed* into solids, and be released by them. For instance, the heating of objects was interpreted as the absorption of *fire* by the solid, and, perplexingly, *air* could be released from solids when heated, as in the roasting of *magnesia alba* to form *magnesia usta* and *fixed air*.[50] In Aristotelian terms, the addition of *fire* releases *air*. You can probably tell how hard it must have been for scientists to wrest quantitative insights or procedures from qualitative statements like this one. Indeed, these difficulties contributed to the fact that chemistry experienced its *revolution*, that is, its birth as a modern science, about a hundred years later than astronomy and physics. Much has been written about this "postponed scientific revolution in chemistry" [12]. It can be argued that what Newton's *Principia* (1687) was for physics, the discovery of the role of oxygen in the process of combustion and the naming of chemical substances by Antoine Lavoisier (1743-1794) around 1780 was for chemistry.

The chemical revolution happened in three stages [19]. First, it was realized that "air" is better described as a state of matter, labeled the **gaseous state**, to distinguish it on the one hand from the **solid** and the **liquid state**, and to set it apart from the **substance** air that surrounds our planet. Indeed, chemists—or rather pharmacists—at the time had been almost exclusively experimenting by mixing together solids and liquids, but not gases. Incidentally, the label and concept "gas" is an invention of the eighteenth century. Second, it was discovered that air is a mixture of several substances which are all in their respective **gaseous state**. Joseph Black once again was driving science forward with his experiments on magnesium mentioned above. In particular, he reserved the term "fixed air" for a special type of substance (carbon dioxide), which had specific properties, such extinguishing an open flame or killing a mouse in a closed container. It was then clear that this ***fixed air*** was unbreathable air, hence different from "naturally occurring air." This result gave rise to **pneumatic chemistry**, that is, research on different kinds of "air." Joseph Priestley (1733–1804) was one of the leading scientists in this field, and discovered in 1772 "airs" such as nitric oxide, ammonia, and "dephlogisticated air," aka oxygen. His label for the latter was informed by the leading theory of combustion at the time, the so-called ***phlogiston theory***. Georg Ernst Stahl had formalized an earlier theory from the 1660s and explained combustion by a postulated substance called ***phlogiston*** that would leave an object when burned. As the substance releases *phlogiston*, it is absorbed by the air. In turn, growing plants like trees take *phlogiston* from the air, and incorporate it, which explains why they burn well. According to this theory, air could only absorb a certain amount of *phlogiston* before becoming saturated. This assumption was used to explain why eventually the burning stops. Breathing was interpreted as transporting phlogiston out of the body, producing *phlogisticated* or *fixed air*. Therefore, *phlogisticated air* could not support life or fire. Following this logic, Priestley's discovery of oxygen was therefore not the end of the incorrect phlogiston theory. Rather, oxygen was interpreted as a special ***dephlogisticated air*** that could absorb a lot of phlogiston, and

---

[50]These obsolete names of magnesium carbonate ($MgCO_3$), magnesium oxide (MgO) and carbon dioxide (CO_2) were used before Lavoisier revolutionized chemical nomenclature.

3.4. THE BIRTH OF CHEMISTRY DISSOLVES ARISTOTLE'S ELEMENTS

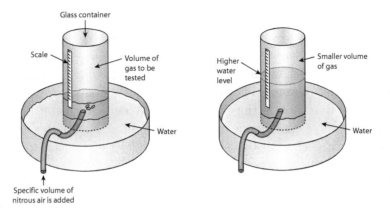

Figure 3.9: **Priestley's "good air" experiment.** The "air" to be tested is put into an upended glass over water. A volume of "nitrous air" (nitric oxide, NO) is added via a hose. The air volume decreases if it is common, breathable air or "good air." This can be seen by the rising water level in the glass.

therefore made a flame burn brighter, and a mouse in a air-tight box live longer. Already in 1766 HENRY CAVENDISH had isolated a gas produced by a metal-acid reaction and called it *inflammable air*, finding that it burned explosively and was much less dense than common air. He speculated he had discovered phlogiston, but it was actually *hydrogen* he had isolated.

Eventually, the time was ripe to throw out the *phlogiston theory*, and it was LAVOISIER who started to put the right ends together. He first showed in 1775 that the calx of mercury (HgO) released a gas that was not *fixed air*. How could he know? The answer came via a procedure that is typical for science, as we have seen many times over: Observing a pattern in nature and agreeing on a name or label for it. In this case the pattern had been observed by PRIESTLEY and was as follows. When a volume of air to be tested is held in an upended glass over water as in Figure 3.9, and one volume of "nitrous air" (nitric oxide, NO) is added to two volumes of air to be tested via a hose, its volume decreases by one-fifth if it is common, breathable air. If the air had been "spoiled" by combustion or respiration (that is, if carbon dioxide had replaced the oxygen) then the volume did not change at all. The volume would also be unchanged if the test gas is *fixed air*. The procedure then is to test the "air" and call it *common air* if the volume goes down by one-fifth.[51] When LAVOISIER performed the test and saw that the volume shrank by one-fifth, he claimed that the gas released by the calx of mercury was common air. However, PRIESTLEY in a serendipitous turn falsified that claim. He added more nitrous air, and found that the volume shrank by much more than one fifth. Also, a candle burned brighter in this "air." He thus showed that the substance released was pure oxygen. Armed with this knowledge, LAVOISIER in turn was able to explain the true chemical processes involved in combustion. When heated, substances combine with a part of common air (the oxygen) in an **exothermic reaction**[52] to form oxides.

After combustion, LAVOISIER sought to explain other chemical phenomena such as acidity. In 1778 he published results of experiments that aimed to show that the "dephlogistated air" was responsible for creating acidic substances, and thus named it *oxy-gen* from the Greek words meaning "acid former." Another step forward was the realization that the "in-

---

[51]In modern notation, common air is one-fifth oxygen, which combines with the nitric oxide to produce nitrogen dioxide (NO_2), which dissolves in water and thus escapes.

[52]Exothermic reactions *release* energy as new and stronger bonds between atoms are created. The difference in binding energy is released as heat or light.

flammable air" of CAVENDISH when explosively combusting in a container left a dewy residue, which turned out to be plain water. Hence, the "inflammable air" was renamed *hydro-gen*, that is, "water former" in Greek. Following these successes was a complete *rationalization* of chemistry including language and naming schemes used. During the enlightenment, language was viewed as a mode of reasoning, not unlike algebra which guarantees sound logic in mathematical calculations due to its stringent symbols. When LAVOISIER published his *Méthode de nomenclature chimique* exactly one hundred years after NEWTON's *Principia*, he laid the foundations of modern chemistry by establishing modern nomenclature[53] and replacing the phlogiston theory with the oxygen theory.

But much more had happened! The elements of ARISTOTLE had dissolved in the creation of a new science. *Air* was recognized as the *gaseous state of matter*, and the atmosphere as a mixture of different gases. *Water* had lost its elemental role, and was now recognized as a composite of hydrogen and oxygen (H_2O), which in turn were elevated to the rank of **chemical elements**. ARISTOTLE's *earth* was now seen as a mixture of different substances in the *solid state of matter*. The status of *fire* would eventually be clarified in the *physical* field of **thermodynamics**. In the end, it was not so much alchemy becoming chemistry, but the establishing of a new science by constructing a suggestive, rational, and therefore useful language to describe chemical processes and patterns in nature.

Concept Practice

1. What is the difference between *heat* and *temperature*?
    (a) There is none, but scientists used to think there is, so there are two names for the same concept.
    (b) Heat is a chemistry concept, temperature is a physics concept.
    (c) Heat is an energy transfer, temperature is a measure for the hotness or coldness of an object.
    (d) None of the above

2. The name of which element can be translated as "acid former"?
    (a) Hydrogen
    (b) Helium
    (c) Nitrogen
    (d) Oxygen
    (e) None of the above

3.5 The Universe beyond Saturn, before the Earth, and below the Equator

3.5.1 The Deep Sky

Back to astronomy. Up to this point, astronomers had almost exclusively focused on the solar system, considering stars mostly as convenient markers for orientation and the establishment of accurate coordinate frames. This changed in the late 1700s when it was realized that there are celestial objects which are neither stars nor part of the solar system. These are called **deep sky objects**. Avid comet hunter CHARLES MESSIER (1730–1817) made a list of these

---

[53] For instance, acids were named after their amount of oxygen, suggesting that *sulfuric acid* (H_2SO_4) contains more of it than *sulfurous acid* (H_2SO_3).

3.5. THE UNIVERSE BEYOND SATURN, EARTH AND EQUATOR

Figure 3.10: **Nebulae (I).** (a) The Great Orion nebula (Messier 42), an *emission nebula*. (b) The Horsehead nebula, a *dark nebula* discovered by WILLIAMINA FLEMING (1857–1911) in 1888. The dark horsehead is visible because it sits in front of a red emission nebula.

objects because they can be confused with a comet on first glance. His 1771 table of about 100 objects includes star clusters and what we now call galaxies. With the telescopes available at the time most of them just looked fuzzy, and many of them were referred to simply as **nebula** (plural: **nebulae**). Only later did it become clear that most are not nebulae in the strict sense. Given that several of the deep sky objects are visible with the naked eye, their late discovery is surprising. TYCHO BRAHE had discovered about half a dozen of these objects, but missed the two most prominent ones: the *Andromeda nebula*, also known as Messier 31 or M31, Figure 6.2(a), (re-)discovered[54] in 1612 and the *Great Orion nebula* (M42), an **emission nebula** in Orion's sword found by HUYGENS in 1656; see Figure 3.10(a). How dramatically the interest in the deep sky changed can be gleaned from the fact that in the first 170 years of telescopic observation their number grew from a few to one hundred, and the next twenty-five years added about 2500 more, mostly due to the systematic search program of WILLIAM HERSCHEL; see Section 3.5.3.

Keep in mind that at the time nothing but the *appearance* of the object could be recorded. Most of these objects were to faint to make out any internal structure. For instance, the Andromeda nebula was described as looking like "the flame of a candle shining at night through transparent horn" by its discoverer. None of its spiral structure was visible in the instruments of the day. Important information like that its center is yellowish, while the spiral arms are bluish was not available until the advent of astrophotography in the late nineteenth century. Incidentally, deep sky objects have such low surface brightness that they do not trigger the color receptors in our eyes. The reason is that the color receptors are less sensitive than the receptors helping us see intensity differences ("black-and-white"). The colorful photos of Figures 3.10 and 3.12 are a product of cameras with long exposure times. Only

---

[54] It appears already in tenth-century Arabic manuscripts and was later reclassified as the Andromeda *galaxy*; see Section 6.2.1.

Figure 3.11: **Two types of star clusters.** (a) The Pleiades *open star cluster* (M45) in *Taurus* with about 100 stars. The stars are embedded in a blue *reflection nebula*, part of the *interstellar matter* out of which the stars of the cluster formed. (b) Messier 13 (M13) in *Hercules*, a *globular star cluster* with hundreds of thousands of stars. The overall color of M13 is yellowish, characteristic of an older star cluster. Its spherical (that is, *globular*) shape is stereotypical.

the brightest stars and the solar system objects are bright enough for the eyes to record color information. The interest in deep sky objects was therefore understandably low until more information could be gleaned—when spectroscopic methods became available in the second half of the nineteenth century. In the late eighteenth century and early nineteenth century, much of what astronomers could do was collecting and cataloging the objects and describe their appearance with the hope that at some point in time this information would become valuable for those that wanted to try to understand them and explain these patterns in nature. Until then, the broad categorization of these objects was as follows. There appeared to be clusters of stars which were easily resolved by increasing the magnification of the telescope. These were labeled **open star clusters**. Often they contained white or bluish stars, like the Pleiades cluster (M45), Figure 3.11(a). Then there were much denser star clusters, were individual stars could be discerned on the fringes, but not at the center, regardless of the magnification. Due to their spherical appearance they were called **globular star clusters**. Overall, these clusters have a yellowish appearance.

Further, there were several types of fuzzy or nebulous objects that could not be dissolved into stars at all. Keep in mind that without knowing the distance to these objects, it was impossible to tell whether they were near and small or far away and large. The first subcategory of these fuzzy objects were nebulae that often appeared to have an elliptical shape, or even some internal, spiral structure. They would later be called **galaxies**, and we already encountered M31, Figure 6.2(a), as an example. With the largest telescopes of the mid-nineteenth century it was possible to study their structure further, see Section 3.5.4. A second category of nebulae had a disklike shape, which resembled the appearance of a planet at low magnification. Consequently, they became known as **planetary nebulae**. A good example is the *Ring nebula* (M57) in Lyra, Figure 3.12(a). Then there were bright, glowing, amorphous nebulae like the *Great Orion nebula* (M42), categorized as **emission nebulae**. **Dark nebulae** are quite the opposite, in the sense that a region of space seems devoid of stars or glowing matter, like the *Dustlanes nebulae* in Figure 3.10(b). Finally, there were

3.5. THE UNIVERSE BEYOND SATURN, EARTH AND EQUATOR

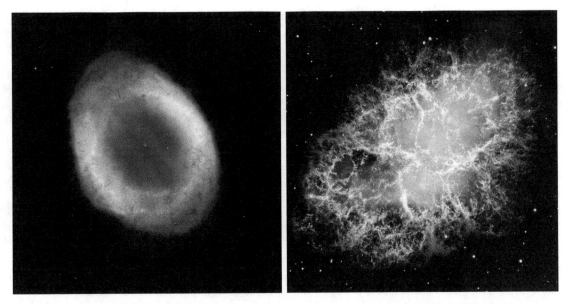

Figure 3.12: **Nebulae (II).** (a) The Ring nebula in Lyra (Messier 57), a *planetary nebula*. Its appearance is spherically symmetric and it features a dim, white, central object, the burned-out core of a star similar to our sun. (b) The Crab nebula (Messier 1) in *Taurus*, a *supernova remnant*, looks noticeably different with its jagged, irregular shape which is foretelling of its origin in a turbulent supernova explosion. To MESSIER, both objects looked nebulous and thus potentially confusable with comets, so they made his list.

wispy nebulae like the *Cirrus nebula* in Cygnus or the *Crab nebula* (M1) in Taurus discovered in 1731, Figure 3.12(b), which are now classified as **supernovae remnants**, although their origin remained unclear until the mid twentieth century.

Eighteenth century astronomers were fairly certain that all of these object were far out in space, and not part of our solar system, but it took a long time to figure out their roles in the expanding universe.

3.5.2 The Formation of the "Universe"

As early as 1750 ideas about the nature and formation of the Milky Way emerged. Recall that we perceive the Milky Way as diffuse "milky" clouds forming a band that extends all around the sky and is brightest in the constellation of Sagittarius, Section 1.1. It was shown by Galileo that these "clouds" can be dissolved with the help of telescopes into myriads of stars. Also remember that stars remained featureless specks of light, even after the advent of the telescope. Only after the discovery of **proper motion** of the "fixed" stars by HALLEY in the early part of the eighteenth century (Section 3.1.2) did stars "become" other suns but at vast distances. Thus freed from thinking about the stars as more or less attached to a sphere, astronomers were emboldened to start speculating about the spatial *distribution* of the stars.

The nature of the Milky Way's star-filled band was first explained by THOMAS WRIGHT (1711–1786). He hypothesized that the Milky way is a flattened, lenticular disk of stars. In his view, the sun is not at the center of the disk, which is the reason for the fact that the portion of the Milky Way in the southern hemisphere is brighter[55] and other irregularities in

---

[55] We thus must be north of the *galactic equator* which was first shown by KANT; see below.

Figure 3.13: **Galaxies.** (a) The Andromeda nebula (M31), a galaxy, showing spiral arms. (b) The elliptical nebula (M87 or Virgo A), a galaxy. It features a jet of material ejected from its core that makes it look a bit like a comet. To Messier, both looked nebulous and thus potentially confusable with comets.

its appearance. In his essay *An Original Theory of the Universe* (1750) he comes thus to the conclusion that the stars of the Milky Way are organized in a gigantic system of finite extent. The German philosopher IMMANUEL KANT knew of WRIGHT's work, misinterpreted it in a fruitful way, and went a step further. His reasoning was that if the Milky Way is an organized system of finite extent, then there must exist other such systems, which he called *island universes*. In his anonymously published book *Allgemeine Naturgeschichte und Theorie des Himmels oder Versuch von der Verfassung und dem mechanischen Ursprunge des ganzen Weltgebäudes nach Newtonischen Grundsätzen abgehandelt (Universal natural history and theory of the heavens or essay on the constitution and the mechanical origin of the whole universe according to Newtonian principles)* of 1755 he identifies the handful of known "nebulae" (see last section) with these island universes—or *galaxies* as we would say today.[56] Note that KANT's program gleaned from his book's title is both truly universal and Newtonian-mechanical. The *natural history* part of KANT's book is nothing short of a first stab at explaining the *formation of the universe*, although we would say today that KANT tried to explain the formation of the Milky Way, and extended his ideas to the formation of the solar system. The theory, often referred to the **Kant-Laplace theory** or **nebular hypothesis** postulates that the galaxy and the solar system formed from an immense gas and dust cloud. The flatness of the galaxy, and also the **coplanarity** of the planets' orbits are explained by a rotation of the gas cloud, which flattens due to its minute density to a much larger degree than the early Earth; see the section on the shape of the Earth. The sun and the planets then formed by collapses of parts of the gas cloud, Figure 3.14. LAPLACE later elaborated on the idea and resolved some of its more technical problems, but the modern theory of the star and solar system formation owes much to KANT's original ideas. In summary, these early ideas in **cosmology** and **cosmogony** were a great step forward: Rational thinking about the origin of our planet and, indeed, the whole cosmos became possible!

Swiss physicist *Johann Heinrich Lambert* (1728-1777) took the next logical step in his

---

[56] In that he was mostly correct, albeit that some of these nebulae are not galaxies but gas clouds or star clusters of our own galaxy.

3.5. THE UNIVERSE BEYOND SATURN, EARTH AND EQUATOR

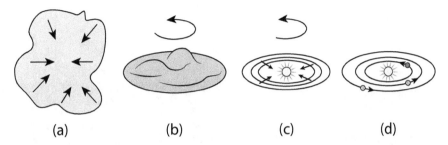

Figure 3.14: **The Kant-Laplace nebular hypothesis.** To explain features of the solar system, like the fact the all planets rotate in a plane and in the same counterclockwise sense, IMMANUEL KANT assumed that a giant gas nebula had contracted under the influence of gravity (a) and flattened into a disk due to rotation (b). Clearly, most mass will drift toward the disk's center, forming the (proto-)sun (c), while some material will remain in orbit around the sun. Out of this material the planets form some time later (d).

Kosmologische Briefe (Cosmological Letters) of 1761. He imagines a system of the star systems, thus forming clusters of galaxies, and then clusters of clusters of galaxies (what we nowadays call *superclusters*) all stabilized by rotation under the influence of Newtonian gravity. He thus constructed a hierarchical structure of the entire universe, an idea which basically stands true and tested today.

3.5.3 W. Herschel as the Father of Modern Astronomy

In the history of astronomy, FRIEDRICH WILHELM "WILLIAM" HERSCHEL (1738–1822) deserves a special place for several reasons. The son of a Hanoverian court musician came to England and earned fame by being the first since historic times to discover a new planet. On March 13, 1781, HERSCHEL spotted an object that he initially classified as a comet. On March 17 he wrote "I looked for the Comet or Nebulous Star and found that it is a Comet, for it has changed its place." Continuing to observe the object, astronomers were able to compute its orbit, which turned out to be nearly circular, with a semi-major axis of about double the distance of Saturn from the sun. This was no comet, but a planet! By discovering it, HERSCHEL had in effect doubled the size of the solar system, and opened up the universe beyond Saturn's realm. Also his personal life changed. KING GEORGE III (1738–1820) endowed him with £200 yearly for his discovery of Uranus, albeit on the condition that he'd move to Windsor so the king could peek through his telescope. This in turn allowed HERSCHEL to start an ambitious astronomical observation program. He was helped by his sister CAROLINE HERSCHEL (1749–1848) and his son JOHN HERSCHEL (1792–1871) who became eminent astronomers of their own right. Of particular importance to HERSCHEL were *double stars*, through which he hoped to find out about the distance and nature of stars. He recorded about 900 double stars, measuring their apparent separation and position angle with a *micrometer eyepiece*; see Section 3.9.1.

In 1784 HERSCHEL decided to figure out the shape of the Milky Way. Since HERSCHEL did not know about the true nature of the stars making up the Milky Way; see above, he had to make some bold *assumptions*. Namely, he assumed that the stars in the Milky Way are *uniformly distributed*, and that all stars are equally luminous, so that their apparent brightnesses differ only because they are at different distances from the observer, owing to the *brightness-luminosity-distance relationship*, Equation (3.4). In other words, he devised his own way of measuring the distance to stars, as none was available. In true scientific spirit,

Figure 3.15: **The telescopes of W. Herschel and Lord Rosse.** (a) HERSCHEL's 40-foot telescope. Note that the focal length, not the mirror diameter is 40 feet. (b) LORD ROSSE's 72-inch diameter, the so-called "Leviathan of Parsonstown." It was built into a castle, and could only be moved up and down, that is, its altitude but not its azimuth orientation could be changed. Therefore, one had to wait for the Earth's rotation to bring an object into the field of view, where it would remain only a short amount of time.

he'd rather make simplifying assumptions of questionable justification than not being able to carry out a crude but innovative foray into new territory. By measuring the number of stars of a certain ***brightness*** in a specific direction, he established that there are stars in that direction at a certain ***distance***. The number of stars he was able to see fell off sharply at a certain brightness, which he interpreted as having reached the outer edge of the Milky Way. In this way he was able to draw a first map of the Milky Way, Figure 3.16(b). Owing to his bold assumptions, and the unknown effect of yet to be discovered ***interstellar matter***, his map has many shortcomings. However, it is correct in that the Milky Way is shown to be extremely oblate—certainly not spherical—and that the sun is *not* at its center.

Relatedly, HERSCHEL was able to show, following earlier work of TOBIAS MAYER on ***proper motion*** of stars, that the solar system is moving within the Milky Way, relative to other stars. The effect of the motion of the solar system is analogous to driving in the snow on a highway at night. Snow flakes in the headlines seem to come from one point far ahead, and converge to a point far behind the car, Figure 3.16(a). By making a map of the proper motion of several dozen of stars, HERSCHEL was able to determine the ***apex*** or direction of the sun's galactic motion to be in the constellation of *Hercules*. He was unable to determine the velocity of this motion, because the distance to the stars was not yet known. He sought to find it by considering ***optical doubles***, that is, stars that appear binary, but really are two physically unrelated stars which happen to stand in the same direction when observed from Earth. HERSCHEL hoped that one of the stars would be a positional reference to measure the parallactic motion of the other. His original idea didn't pan out, but he accidentally discovered that some of the stars (so-called ***physical doubles***) rotate about each other. This was proof of the universal applicability of NEWTON's gravitational law which governs their motion: Not just planets rotate around stars, but also stars around stars. Later, double stars were used to determine the masses of stars, Section 3.9.1—an important input for understanding them. HERSCHEL enhanced his observations of double stars with

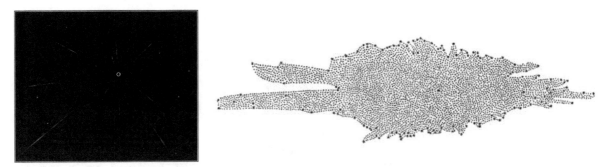

Figure 3.16: **Radiants and Herschel's first map of the galaxy.** (a) When stars' proper motions are plotted on the celestial sphere, they resemble the paths of meteors which all seem to emanate from a *radiant*, a point in space that is defined by the *relative velocity* of the observer and the observed objects. In the case of meteors, the Earth is moving toward a field of solar system debris. Herschel discovered that the sun is moving in the galaxy, that is, with respect to other stars. Both phenomena are reminiscent of driving in the snow on the highway. The snow flakes in the headlights all seem to come from a point ahead. Compare to *stellar aberration*, a similar directional change due to the observer's relative velocity. (b) Herschel's first, crude map of the Milky Way. The sun is the darker star somewhat off center. The shape is distinctly lenticular, not spherical.

photometric measurements. That is, he compared the brightnesses of the components of the doubles. He did so by pointing two telescopes of equal size at the two stars, and dimmed the image of one by reducing its aperture until the stars appeared equally bright. This very sensitive method yielded the far-reaching result that the components of double stars are often of different brightness. Since both are at the same (unknown) distance, Herschel had to conclude that their energy output or luminosities were different—thereby falsifying one of the assumptions of his own work on the galactic shape. It was now clear that stars are different, not just in color but also in energy production! This finding eventually led to the new field of *astrophysics*, to be fully established at the end of the nineteenth century.

In that vein, Herschel discovered that the spectrum of colors as produced by sunlight falling through a prism extends beyond its apparent violet and red boundaries. Thereby he opened up the field of *spectroscopy*. His experiment of 1800 was so simple that one wonders why nobody performed it earlier. He placed thermometers into the colored regions produced by sunlight and prism, Figure 3.17, and two control thermometers outside of the illuminated region. He found that the temperature of the colors increased from violet to red. As a further control he measured the temperature just beyond the red part of the spectrum, where no light was visible. To his surprise he found that the temperature was *highest* in this part. Herschel continued to experiment with this invisible light he called "calorific rays," and found that they were reflected, refracted, absorbed and transmitted just like the visible part of the spectrum. He had shown that there was a kind of rays or radiation that was invisible for the human eye, but otherwise not different from light at all! Today we call the light he discovered *infrared* radiation, from the Latin prefix *infra* meaning *below*. This light is, in a sense, redder than red. Within the year, Johann Wilhelm Ritter (1776–1810) discovered light that was more violet than violet. It is now called *ultraviolet* radiation, from the Latin prefix *ultra* meaning *beyond*. Ritter had heard of Herschel's discovery, and was intrigued by the fact that silver chloride blackened faster when exposed to violet light than when blue light is used. His logical conclusion was, that any "more violet" light should blacken it even faster, and he showed that ultraviolet light, invisible to the

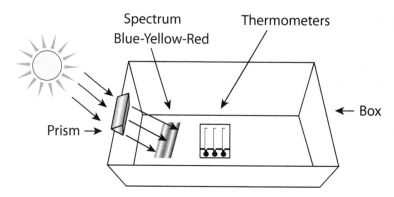

Figure 3.17: **Herschel's discovery of infrared radiation.** HERSCHEL used a prism to decompose sunlight into the spectral colors. He noticed that a thermometer placed in the region beyond red showed the highest temperature. He concluded that there must be radiation that is invisible to the human eye but is warming up the thermometer, which here functions as a *detector* of this type of light.

human eye, does the trick. In effect he had constructed a **detector** that reacted chemically to the otherwise undetectable radiation. For obvious reasons, he initially called it **chemical radiation**. RITTER's experiment foreshadowed the invention of **photography**, which became a major tool in astronomy by the end of the nineteenth century.

3.5.4 The Southern Sky and Spiral Nebulae

It is no accident that much of scientific and technology was developed in the northern hemisphere of Earth. After all, over two thirds of Earth's landmass is north of its equator, and population density was likewise historically much higher in Earth's northern temperate zone. Therefore, the knowledge of the **southern sky** was very limited, although a good part of the sky below the **Celestial Equator** can, of course, be seen from the northern hemisphere, see Section 1.4. Note that the southernmost object in MESSIER's catalogue of deep sky objects is M7, an **open star cluster** in the constellation of *Scorpio* 34.8° below the celestial equator. Much remained to be discovered below the equator in the eighteenth century, and young EDMOND HALLEY was one of the first to heed the call.[57] He recorded hundreds of southern stars in his years 1676–1678 on the South Atlantic island of St. Helena.[58] In Section 3.3.1 we saw that LACAILLE established a tradition of astronomy at the Cape of Good Hope in South Africa. He determined the distance to the moon from here in 1750, see Section 3.3.1. During his two-year stay LACAILLE catalogued 10,000 stars and introduced a dozen new (southern) constellations still in use today, like *Telescopium, Microscopium, Horologium*, certainly inspired by the scientific instruments of the day.[59]

It was again the Cape where JOHN HERSCHEL did much of his astronomical work on the southern sky from 1834 to 1838 with a private 21-foot telescope. Recording over 1,700 nebulae and 2,000 binary stars he contributed significantly to the systematic exploration of

---

[57] "Go South, young man!"—reminiscent of the phrase used during the American westward expansion.

[58] At 16° southern latitude it was deemed remote enough by the British to permanently exile and detain NAPOLEON BONAPARTE who died here in 1821.

[59] He also determined the Earth's radius in the southern hemisphere by measuring a 137 km arc of the meridian starting in Cape Town. Later, his findings that the Earth more flattened toward the south pole were questioned because his measurements could have been influenced by the gravitational presence of the iconic Table Mountain, an effect exploited by MASKELYNE as we saw in Section 3.3.2.

3.5. THE UNIVERSE BEYOND SATURN, EARTH AND EQUATOR

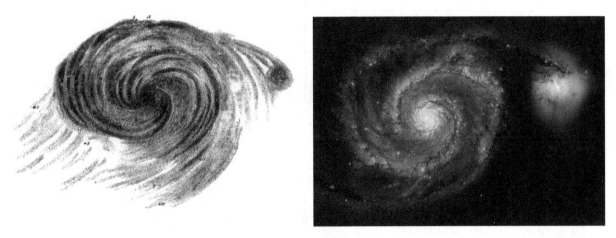

Figure 3.18: **The Whirlpool galaxy M51.** (a) Eyepiece sketch of LORD ROSSE from 1845. (b) Hubble Space Telescope image from 2001. The red spots in the spiral arms are *stellar nurseries*, that is, contracting gas clouds where stars are forming as suggested by the **Kant-Laplace theory**.

the sky below the equator. From there he continued the program of his father to shed light on the structure of the Milky Way. His general observation that the southern sky seems richer in stars was later corroborated. It is now believed that the solar system sits slightly north of the galactic disk; see Section 6.1. In a preview of the nineteenth century, we may add that the HERSCHELS' quest to understand spiral nebula was extended in the mid 1800s by LORD ROSSE (1800–1867) who owned the largest telescope of the time, Section 3.6.2. LORD ROSSE catalogued nebulae and discovered the spiral structure of the nebulae later classified as **spiral galaxies**. The most emblematic is the Whirlpool galaxy (M51) which he captured in an eyepiece sketch of exquisite detail, Figure 3.18. This work was continued by J.L.E. DREYER[60] (1852–1926). He compiled the *New General Catalogue of Nebulae and Clusters of Stars* (NGC) containing 7840 objects, a new version of JOHN HERSCHEL's *General Catalogue of Nebulae and Clusters of Stars*. NGC objects are labeled by their number in the catalogue after the prefix "NGC". As an example, consider the flocculent galaxy NGC 4414, depicted in Figure 6.10.

Concept Practice

1. Which of the following is *not* a *deep sky object*?

    (a) The Andromeda galaxy
    (b) Sirius
    (c) The Ring nebula
    (d) A globular star cluster
    (e) All of them are deep sky objects

---

[60] DREYER was the assistant of LORD ROSSE's son at Birr Castle.

2. Which of the following did Herschel *not* discover?

    (a) Uranus

    (b) Infrared radiation

    (c) Ultraviolet radiation

    (d) The apex of the sun's motion within the galaxy

    (e) He discovered all of them

    (f) He discovered none of them

3.6 The Industrialization of Science

3.6.1 From the Eighteenth to the Nineteenth Century

As the eighteenth century drew to a close, important and far-reaching scientific discoveries had been made. In many textbooks, though, the scientific history of this and the following nineteenth century are foreshortened, owing to the idolization of NEWTON and EINSTEIN. The time between these luminaries is often covered in textbooks by merely listing the discoveries made, thereby giving the false impression of an interrupted, episodic progress of science. Nothing could be farther from the truth. In fact, it is amazing just how continuous the steady refinement of scientific concepts and theories is. We have seen in Section 2.10 that this is true for NEWTON himself who was able to build his grand theory on the foundational ideas of others. While he and EINSTEIN arguably are the most eminent scientists of all time, it is always dangerous to put scientists on pedestals as untouchable authorities. After all, it's the science, not the scientists that really matters. We have seen how badly idolization played out in the case of ARISTOTLE. Also NEWTON's unquestioned authority stood in the way of progress, for instance his wrong belief that light can be described by a ***corpuscular theory***. We saw on the other hand, how the acceptance of NEWTON's correct theories of motion and gravity was delayed because of the somewhat nationalistic and psychological propensity for DESCARTES' ***vortex theory*** on the European continent. This "turf war," on the other hand, is a normal part of the scientific process: New ideas are carefully analyzed and vetted before being tentatively accepted. Ironically, the history of the military wars and political conflicts of the time, like the Seven Years' War (1756–1763) or the French Revolution (1789–1799) are taught in school, although their results were ***ephemeral***.[61] Meanwhile, the truly eternal story of scientific struggle and progress that makes our modern lives all but possible does often not get enough curricular consideration.[62]

The way that science was done changed considerably from the 18th to the nineteenth century. Initially, much new ground was broken by ***gentlemen scientists*** like BENJAMIN FRANKLIN, ALESSANDRO VOLTA (1745–1827) and LUIGI GALVANI (1737-1798) who pioneered the new field of electricity. By the middle of the nineteenth century, however, the majority of

---

[61] The territorial gains that earned the Prussian king FREDERICK II(1712–1786) his nickname "the Great" in the Seven-Years War were obliterated in the World Wars—by which time, incidentally, Prussia had ceased to exist. The French Revolution led to the rise of the tyrant NAPOLEON BONAPARTE, who led over half a million soldiers into death on the battlefields for vain reasons. And yet he quipped: "The true conquests, the only ones that leave no regret, are those that have been wrested from ignorance." It must also be said that the ideas of the French revolution were preemptively realized in the American War of Independence.

[62] See footnote 92.

3.6. THE INDUSTRIALIZATION OF SCIENCE

scientists were "institutionalized" at the universities,[63] although in astronomy amateurs kept making important contributions until at least the end of the nineteenth century.

The *industrialization of science* in the section heading can be understood in at least two different ways. First, the volume of observations and experiments picked up considerably,[64] and therefore scientist got efficient and improved their protocol, moving quite literally into a period of *mass production* of measurements. They grew, in a secondary sense of the word, truly *industrious*.[65] Secondly, this happened at a time when the *industrial revolution* was well underway. Since the invention of the steam engine by THOMAS NEWCOMEN (1664–1729) in 1712 and its crucial improvement due to JAMES WATT (1736–1819), this reliable and location-independent source of power[66] revolutionized the economy. Industry and technology formed a close *symbiosis*. This can be seen most clearly in the development of the new field of *thermodynamics* from the needs of industry. For a while in the early nineteenth century, much progress in physics came from studying steam engines, and from the need to make them more efficient. After all, this was the century of steamships and railways. New scientific results would thus often result in technological applications, and the revenues created would be fed back via funding for the sciences. Technology thus fed science, which fed technology. Feats unthinkable a century prior were achieved: The first transatlantic communication cable was lowered into the depth of the Atlantic Ocean in 1859, the same year that CHARLES DARWIN published his *Origin of Species*, establishing *evolutionary science* and calling into question long-held religious beliefs about the genesis of mankind. The Suez Canal opened in 1869 connecting the Mediterranean with the Red Sea and the Indian Ocean. Accelerating urbanization was the result of and propelled *industrial capitalism*. Concentrating scientists at the universities helped to streamline the education and ensure a direct communication of research results to the next generation of scientists.

In astronomy probably the biggest leap forward was the emergence of *astrophysics*, that is, the application of "lab physics" to the heavens. In particular, the invention of *spectroscopy*, the discovery of the *Doppler effect*, and their immediate application to the *line spectrum* of celestial objects furnished a *bona fide* revolution in the way astronomy was being done. The later part of the nineteenth century saw the beginning of yet another revolution, namely the eventual replacement of visual observations with *astrophotography*, which allowed an almost complete *objectification* of observations, freeing them from any *subjectivity* of the individual observer. Of course, scientific results always have to be confirmed by several observers or experimentalists to ensure their *intersubjective validity*. That is, scientific results are considered truly independent from an individual observer if confirmed by many (competent) researchers. Relatedly, a philosophical debate was raging throughout most of the nineteenth century, because physics and therefore astronomy or astrophysics were establishing many new patterns in nature that are not at all accessible to the human senses. Starting from HERSCHEL's infrared radiation (invisible to the human eye), things like the existence of atoms and molecules, the emergence of collective properties of states of matter such as pressure and temperature, the definition of such abstract entities as *energy* and the *elec-*

---

[63] In philosophy, the transition was even sharper, as virtually none of the great philosophers before KANT and all starting with him were affiliated with a university.

[64] As an example, compare MESSIER's eighteenth century catalog of about 100 deep sky objects to HERSCHEL's early nineteenth century catalog of 2500. That is more than an order of magnitude difference!

[65] From French *industrie* or Latin *industria* both meaning *diligence*.

[66] Water-powered machines such as grist mills were bound to rivers and streams.

tric and ***gravitational fields*** and the postulation of the ***luminiferous aether***, presented contemporary scientists and philosophers with ample ***epistemological problems***.

3.6.2 Telescopes and Instrumentation

It is easy to forget that virtually all discoveries in astronomy after 1600 were made possible by the dramatic improvements of telescopes and instrumentation. It is important to get a sense of how much our expanding knowledge of the universe was driven by the construction and evolution of observing and measuring devices. By getting to know some of the most important instruments, we will learn a great deal about how science is done—much more so than by solely memorizing the results emanating from diligent use of these instruments. Also we can appreciate how much insight was wrought from these imperfect and sometimes outrightly flawed devices. Finally, it gives us a sense that our knowledge of the cosmos is limited by the accuracy of the instruments we are using to explore it. For instance, the invention the telescope made it possible to discover patterns of nature like the ***phases of Venus*** or ***stellar aberration*** that would otherwise be inaccessible to the human senses, yet revolutionized our ***worldview***.

In Section 2.9.5 we saw that the first telescopes in the seventeenth century were used to *discover* new "things" rather than precisely *measure* their positions, brightnesses, and other properties. This was in part due to the awful distortions of the images of the early ***refracting telescopes***. A simple glass lens has a major flaw, Figure 3.19(a). Due to ***dispersion***, that is, the dependence of the refractive properties of the glass on the color of the light passing through it, the lens acts like a prism and color distortions like in Figure 3.19(b) occur. This is known as ***chromatic***[67] ***aberration*** and makes it hard to accurately observe celestial objects. To improve things, seventeenth century astronomers used very long focal length telescopes, Figure 2.40, which made precises determination of stellar positions almost impossible. A crucial improvement was the invention of the ***achromatic lens*** for which JOHN DOLLOND (1706–1761) obtained a patent as late as 1758.[68] The ***achromatic lens*** is a two-lens combination in which one lens cancels the dispersion of the other, Figure 3.19(b). With this improvement, good images could be obtained even at small focal length, which meant that ***refracting telescopes*** could now be pointed precisely and became good instruments to accurately measure the positions of stars on the celestial sphere. Indeed, in 1772 PETER DOLLOND refitted the ***great mural quadrant*** of the Greenwich Observatory with an achromatic objective (main lens) and an eyepiece with finer cross wires to increase its precision.

TYCHO BRAHE had made his most accurate positional measurements with a great mural quadrant, Figure 2.18—of course *without* a ***telescopic sight***—and it is high time to describe the associated measurement method. A mural quadrant is mounted on a wall (Latin: *murus*) so that it precisely faces the meridian, where celestial objects ***culminate***. Its arm with the ***sight*** can only be moved up and down (rotated around an East-West axis), thus allowing to determine precisely the altitude angle of an object as it goes through the meridian owing to the ***daily (rotating) motion*** of the Earth. From this measurement and the known ***latitude*** of the observatory, the ***celestial latitude*** of the celestial object can be determined, see Figure 1.18(a). At the precise time of the ***meridian transit***, the time of a ***sidereal***

---

[67] Greek: *chroma = color*
[68] His son Peter in 1763 invented the design of the modern ***apochromatic lens***.

3.6. THE INDUSTRIALIZATION OF SCIENCE

Figure 3.19: **Chromatic aberration and its cure.** (a) Chromatic aberration yields images in which objects have colored edges. (b) As a result, the image looks blurred and edges look color distorted (bottom) when compared to the image of an achromatic lens (top). (c) An *achromatic lens* is a combination of two lenses made of different types of glasses such as *crown glass* and *flint glass*. It corrects chromatic aberration because the opposing *dispersive properties* of the two glass types cancel—at least for two wavelengths. More complicated combinations will do an even better job.

clock is recorded. A *sidereal clock* shows "star time,"[69] that is, the **Right Ascension** or *celestial longitude* of the *vernal equinox*. Since the Earth turns 360° in a *sidereal day*, we can identify the *time* with an angle. Namely, 360° equals 24 hours, since the Earth turns 15° in one hour. Since **Right Ascension** is measured in hours from the equinox, the *celestial longitude* of the transiting star is thus most directly determined by the *sidereal clock*.

Two things come to mind. First, the Greenwich instrument must have been far superior to Tycho's quadrant due to the telescopic sight, which at magnification 62× allowed for a much finer measurement of the precise transition altitude and time. In fact, so fine a directional determination called for the best techniques in making the scale of the instrument. A whole trade developed around the fact that it is not easy to engrave $\frac{1}{4} \times 360 \times 60 \times 60 = 86,400$ ticks[70] into a perfectly shaped brass plate to count the individual seconds of arc. One of the finest instrument makers was John Bird (1709–1776) who in 1750 produced the great quadrant of Greenwich, Figure 3.20(c). Second, clock-making much improved in the eighteenth century.[71] Today, keeping accurate time is an almost trivial problem. In centuries past it has been an almost intractable one. It is said that Galileo in the early 1600s used his own pulse for timekeeping. Finally, John Harrison (1693–1776) invented the *marine chronometer*, a mechanical clock which kept time so precisely that it allowed to determine longitude at sea, see Section 3.1.2. He was awarded £8750 for his invention in 1773.[72]

These improvements and inventions made the *refracting telescope* a versatile instrument that was widely used during the 18th and 19th centuries. Dollond's achromatic lenses initially were limited to 3- to 5-inch diameter, but around the year 1800, larger lenses became available. Recall that telescopes are judged by their light-collecting power, which is proportional to the

---

[69] Recall that in Latin *sidus=star*.

[70] The $\frac{1}{4}$ factor comes from the fact that the instrument is a *quadrant* using only a quarter of a full circle; the *sextant* uses a sixth.

[71] It is known that Tycho often did not trust mechanical timekeeping devices and measured time by determining the *transit time* of stars of known celestial longitude.

[72] He thus came closest to winning the *prize* of the **Board of Longitude** which Tobias Mayer aimed for with his *lunar theory*—and which was never awarded in full.

surface *area*[73] of its main **optical surface** (objective lens for **refractor**, main mirror for **reflectors**). This ushered in the era of the **great refractors**. They were often the main instrument of the new observatories which were founded in large numbers in the nineteenth century. Often they were complemented by other instruments such as a **transit circle**, the modern version of the mural quadrant, or a **heliometer**, described below, Figures 3.20(b) and (d). The first **great refractor** was the 9.6-inch achromatic refractor built by FRAUNHOFER in 1826 for the *Dorpat Observatory* in Estonia. It was used by FRIEDRICH GEORG WILHELM VON STRUVE (1793–1864) to measure the parallax of Vega; see next section. The similar Fraunhofer refractor of the Berlin Observatory was used by Galle in 1846 to discover Neptune, Section 4.1.1. STRUVE had meanwhile moved to *Pulkovo Observatory* in St. Petersburg, Russia where the next **great refractors** were installed in 1839 and 1885 with 15-inch and 30-inch diameters, respectively. The era came to an end with the largest refractor at Yerkes Observatory installed in 1897 with 40-inch **aperture**. It was built by *Alvan Clark & Sons, Mass.*[74] **Lens sagging**[75] due to gravity acting on the heavy glass lenses made it impractical to construct even larger refractors. Also precision maneuvering became an issue. As a reference point, the 75 ton tube and support structure of the Yerkes **great refractor** has to be guided by the same, sub-arcsecond precision as the smaller telescopes.

Even in the late 1700s apertures of **reflectors** were larger than of refractors, but the reflectors had a crucial disadvantage, largely forgotten today. Namely, their mirrors were not made from silverized glass, but from **speculum**, a reflective metal alloy. However, speculum only reflects 66% of light (modern mirrors upward of 95%), and tarnishes in as little as several months, which meant frequent mirror "restorations" over the life of an old-style reflecting telescope. This changed with the invention of the method of coating a polished glass surface with a reflective layer. LÉON FOUCAULT in 1857 used a chemically deposited thin layer of silver, which is highly reflective but also tarnishes fairly fast. Modern reflector mirrors are **aluminized** since the associated vacuum-coating method was developed in 1932.

The history of reflecting telescopes was thus a rocky one. After the original ideas and designs by JAMES GREGORY (1638-1675), NEWTON and LAURENT CASSEGRAIN (1629–1693), Figure 3.21, with a large convex mirror and a small secondary mirror (concave, flat, convex, respectively) in 1672/3, no major advance was made until JOHN HADLEY (1682–1744) was able in 1721 to produce larger parabolic mirrors.[76] Strangely, the largest reflectors from the mid 1700s until 1917 were made by amateurs. Around 1774 W. HERSCHEL started to experiment with grinding and polishing mirrors, and eventually, though an amateur in a profession that was becoming increasingly institutionalized at universities, HERSCHEL had the best telescopes available at the time; see Section 3.5.3. In fact, he did not just grind his own mirrors but he had a profitable enterprise selling telescope components until he discovered Uranus and started collecting a royal pension. HERSCHEL constructed the largest telescopes of the time. His largest was a 40-foot focal length reflector situated in a house which could be rotated, Figure 3.15(a). It had a 49-inch aperture, but was relatively unwieldy, and HERSCHEL made most of his discoveries with an earlier, smaller telescope of 20 ft focal length and 18-inch

---

[73] The area itself is proportional to the *square* of the diameter of the telescope, which is most often quoted.

[74] Alvan G. Clarke discovered Sirius B, the first **white dwarf**, while testing an 18-inch **great refractor**.

[75] "Sagging" here sounds dramatic, but even a distortion by the width of a human hair can be fatal; the relevant distance here to judge optical flaws is the wavelength of (visible) light, which is exceedingly small.

[76] He showed an instrument with a 6-inch mirror to the Royal Society that year. In 1730 he invented the **octant** to measure altitude of celestial objects above horizon for navigation at sea.

3.6. THE INDUSTRIALIZATION OF SCIENCE

Figure 3.20: **Instruments.** (a) Ole Rømer invented the transit instrument around 1690. Through a telescope mounted on an East-West axis and with the help of precise clocks (A and C), the exact timing and elevation angle of a celestial object's crossing of the *meridian* was determined. From that measurement, the position of the object on the celestial sphere was computed with high accuracy. This was a time-consuming task. (b) The Groomsbridge *transit circle* of 1806. The full, finely graded circle (looking like a wheel in the etching) made it possible to read elevation angles with higher accuracy. To this end, the scale was read with a microscope fitted with crosshairs, as visibly sticking out in the left pillar of its mount. (c) Bird's Mural Quadrant as used at Greenwich Observatory. (d) A heliometer has its large (objective) lens cut in half. The halves can be moved by screws (visible as the rods extending from front to back on top of the telescope tube) to bring the two images created to an overlap. If two distinct stars are viewed this way, their precise angular separation can be inferred from the number of turns of the screws. The instrument was originally used to determine the solar diameter, hence the name.

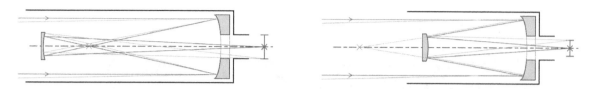

Figure 3.21: **Reflecting telescope designs.** (a) Gregorian Reflector (b) Cassegrain reflector. For the Newtonian reflector see Figure 2.43. Note that all designs have a central obstruction. The resulting images are dimmer, but they certainly do not have a "hole in the middle."

aperture. HERSCHEL's 40-foot telescope was the largest telescope until LORD ROSSE built a large reflector in Parsonstown, Ireland; see Section 3.5.4. His "Leviathan of Parsonstown," Figure 3.15(b), was a 72″ diameter reflector which remained the largest telescope until the 100″ Hooker telescope came online 1917 on Mt. Wilson, California. The latter was designed and founded by GEORGE ELLERY HALE (1868-1938) who was the mastermind behind no fewer than four of worlds largest telescopes. Namely, the Yerkes 40″ largest refractor (1897), the 60″ Mt Wilson largest fully directable reflector (1908), the 100″ Hooker reflector, and the Mt Palomar 200″(5 m) reflector (1948).

Modern telescopes often have segmented mirrors, and are of 10 m (400″) or larger aperture. The famous **Hubble Space Telescope** has a rather modest 2.4-meter mirror, but it sits above the distorting atmosphere, where *seeing* is superb. Incidentally, this is why all large modern telescopes are situated on the top of high mountains. Ideally, telescopes should also be as close to the equator as possible (Hawaii 20° N, Northern Chile, 30° S latitude), to see as much of the universe as possible (Hawaii: all but 20° around the SCP, Northern Chile: all but 30° around the NCP).

3.6.3 The Stellar Parallax and Precision Astronomy

Almost four decades into the nineteenth century, the distance to the stars was finally directly measured by BESSEL. Why so late? We saw that BRADLEY discovered **stellar aberration** over a hundred years earlier, and even detected the minute **nutation** of Earth's axis. The reason for the delay is—unsurprisingly—the exceeding smallness of the parallactic angle due to the immense distance to the stars. Incidentally, astronomers were well aware of that, which is why several previous claims of parallax discovery were readily falsified. For instance, GIUSEPPE PIAZZI (1746–1826), discoverer of the first **asteroid** and equipped at Palermo with a famous zenith instrument made by the premier English instrument maker JESSE RAMSDEN (1735-1800), in 1805 quoted parallaxes between 2″ and 10″ for Sirius (α *Canis Majoris*), Aldebaran (α *Tauri*), Procyon (α *Canis Minoris*) and Vega (α *Lyrae*). Note that these are all α stars, that is, the brightest stars in their constellation, or, for that matter, in the sky. This is, in fact, the reason that PIAZZI chose them. His correct reckoning was that brightness is correlated with closeness via the **brightness-luminosity-distance relation**, Equation (3.4). Still, the discovery just around the same time by W. HERSCHEL that stars have *different* luminosities, was planting doubts that brightness was the correct criterion to judge closeness. Both HERSCHEL and PIAZZI came to realize that there are many *dim* stars which large **proper motion**, and this convinced BESSEL to take large proper motion as a better criterion for closeness. Indeed, PIAZZI had already singled out the star 61 *Cygni*, because it had the largest proper motion of any star then known, moving 5.2″ per year. BESSEL used this

3.6. THE INDUSTRIALIZATION OF SCIENCE

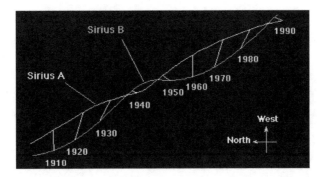

Figure 3.22: **Proper motion of Sirius.** Sirius is a close binary star. Its two components Sirius A and B move around the common center of mass which traces out a straight line with respect to distant background stars. The observed "wobble" of the much brighter A component gave away the existence of a dim companion.

star to perform his successful parallax measurements. Let us review how the position of a star changes during the year due to Earth's orbital motion,[77] and repeat some of the discussion of Section 2.3.1. Consider Figure 3.3. It shows that the apparent, so-called **annual parallax** or **parallactic motion** of the star is a projection of the Earth's orbit onto the celestial sphere. Note that the motion can appear circular (if the star is exactly at the pole of the ecliptic), as a straight line (if the star is in the plane of the ecliptic) or elliptical (if the star is somewhere between ecliptic and its pole). In other words, the parallactic angle p in Figure 3.3 is the angle under which the radius of Earth's orbit appears as seen from the star. The distance to the star d can be determined with trigonometry, since the base of the right triangle formed by Earth-sun-star is known, and the opening angle p is measured. Therefore

$$d = \frac{1\,\mathrm{AU}}{\sin p}.$$

It is convenient at this point to define a new **distance unit**, since the distance to the stars is so large (on the order of hundred thousand AU for close stars). One **parsec** (1 pc) is defined as the distance from which the Earth's orbital radius appears under an angle of one arcsecond (1″). By comparing the brightnesses of Sirius and the sun, already NEWTON had estimated the distance to stars and concluded that their parallactic angles are less than an arc second. If we measure p in arc seconds and express the distance to the stars in parsecs ("parallactic seconds") the above formula simplifies and becomes Equation (2.1)

$$d = \frac{1}{p},$$

which is a simple **inverse relation**. For instance, a star with half the parallax is two times farther out.

We went through this discussion of the size of the parallax for a reason. As a general rule, in science **estimates** need to be made before an experiment or observation is carried out. To be worth the effort, it has to be plausible that the size of the effect is large enough to be observable and not obscured by the uncertainty inherent in the method used. BESSEL was

---

[77]Compare this to BRADLEY's argument in Section 3.2.1 that **stellar aberration** is also due to the orbital motion of Earth, however via its changing direction of **velocity** and not its changing **position** in space.

confident that the accuracy of astronomical instruments had improved by the 1830s to the point that a discovery of the parallax was finally feasible—even if p is less than an arcsecond. He was not alone in this belief.[78] Indeed, there are several instances in the history of science when several researchers independently and simultaneously sense that the time is ripe to perform a crucial experiment. In fact, BESSEL had carefully studied old data, and part of his eminence as a scholar was his decades-long program to systematically reduce and improve on James BRADLEY's data. BESSEL could therefore calmly report in his parallax article [4] that BRADLEY's observations ruled out any parallax angle of the bright stars that was in excess of one arcsecond.[79] BESSEL had the advantage of working at *Königsberg Observatory* which had the only achromatic heliometer of FRAUNHOFER's design (see Section 3.6.2) available at the time. His observation program was as follows. He measured the angular distances of the two components of *61 Cygni* (a binary star) from two fainter—and therefore likely more distant—stars by bringing their images into coincidence with the micrometer screw of his heliometer. He therefore produced two independent measurements (of the two components) while observing from August 1837 to October 1838 on every clear night. You might estimate from this description how many nights he stood at the telescope—all to report a single number. But what a number it was: $0.3136'' \pm 0.0202''$—less than a third of an arc second! BESSEL converted it into a distance of 657,700 AU, so *61 Cygni* is 657,700 times farther out than the sun! Light takes 10.28 years to get to us from *61 Cygni*, and in our new unit **parsecs** we have[80]

$$d_{61} = \frac{1}{0.3136''} = 3.189 \,\text{pc}.$$

It was a breakthrough. The distance to the stars was now positively determined, and the method was of remarkable accuracy—BESSEL quotes relative uncertainty of only $\frac{0.0202}{0.3136} \times 100\% = 6.44\%$. But the method was not widely applicable! Only the nearest stars were eligible, and who was to tell which *are* nearest without applying the parallactic method first? Consequently, the number of known stellar distances only rose slowly: Ten years later 11 distances where known, the number was 34 in 1882, and in 1895 still not larger than 90. Nonetheless, a significant step had been made. From the solar system BESSEL had gone into interstellar space and produced the second rung of the **cosmic distance ladder**, the first rung being the distance to the sun. The distances to stars were now constructable, and therefore many quantities which used to be *apparent* became *absolute*. Well, with DR. SEUSS (1904–1991) we could say—that the universe grew 657,700 sizes that day! The precision of BESSEL's observations led to another important realization: The proper motions of some stars were not uniform. In the particular case of Sirius, Figure 3.22, the detection of this wobbling motion allowed for the prediction of an unobserved companion. When it was discovered in 1862 by telescope maker ALVAN G. CLARK (1832–1897), it opened a whole new can of worms. The companion was so dim and yet as massive as the sun. The first "unusual" star had been discovered, and it would take 70 years to explain this new pattern in nature as a **white dwarf**, the burned-out core of a star.

---

[78] Within the year of BESSEL's 1838 parallax of 61 *Cygni*, F.G.W. STRUVE had published the parallax of Vega, and THOMAS HENDERSON (1798–1844) that of α *Centauri*.

[79] He even discussed the anticipated elliptic parallactic motion of stars which are far off the ecliptic pole and diligently chose pairs of stars on opposite sides of the sky to rule out any undiscovered systematic effects.

[80] The modern values are $p_{61} = 0.28588 \pm 0.00054''$, $d_{61} = 11.41 \pm 0.02 ly = 3.498 \pm 0.007 pc = 721500 \pm 1400 AU$.

3.7. LIGHT IS ELECTROMAGNETIC RADIATION

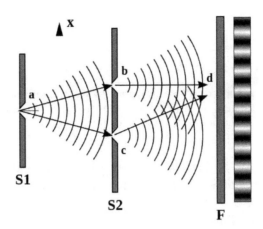

Figure 3.23: **Young's double slit experiment of 1803.** A wave (light or water) falls through a slit in a screen ($S1$). The resulting spherical wave falls onto a second screen ($S2$) with two slits, generating two waves which *interfere*. A distinct pattern of bright and dark spots is observed on the screen F. This was a crucial experiment, because only waves interfere; particles do not. The experiment falsified NEWTON's *corpuscular theory* of light.

Concept Practice

1. A star has a parallax angle of $p = 0.25$. What is its distance? (a) 0.25 pc (b) 0.5 pc (c) 1 pc (d) 2 pc (e) 4 pc (f) None of the above

3.7 Light Is Electromagnetic Radiation

The measurement of the distance to the stars via the *stellar parallax* (1838) and the discovery of Neptune (1846; see Chapter 4), as predicted by Newtonian *celestial mechanics* were the high-water marks of "pure" astronomy, as it was understood at the time. We now enter a stage in the unfolding cosmos where breakthroughs in other science fields had to be woven into the fabric of astronomy to make progress. Major thrust for the endeavor to understand the stars came from *spectroscopy*, that is analyzing light or *electromagnetic radiation*, and *thermodynamics*, loosely speaking the theory of *heat* and *energy*. These topics are the content of the next two sections.

3.7.1 Light Is a Wave (Again)

We have seen that BRADLEY, following NEWTON, in the eighteenth century considered *light* as a corpuscular phenomenon, that is, as consisting of little particles that carry information from the source to the observer. One of the crucial experiments at the very beginning of the nineteenth century was Young's *interference experiment*, Figure 3.23, that showed decisively that light is a wave phenomenon, as already Huygens had held in the sixteenth century.

The main idea here is that waves and particles behave differently, that is, they produce different, contradicting patterns in nature. For instance, particles *collide*, waves "move through" each other, Figure 3.24, since their disturbances simply add. Waves interfere, particles do not. What makes a wave a wave is that it obeys the principle of superposition (the sum of

Figure 3.24: **Waves interfere, they don't collide.** (a) A long string is shaken such that two disturbances of different amplitude move toward each other (b). When they meet, they simply add up (c) and thus effectively "go through" each other. This behavior is characteristic of waves.

two waves is another wave), which follows from the mathematical properties of the (wave) equation it fulfills. What makes a particle a (classical) particle is that it is localized, has a mass and moves according to NEWTON's laws.

You might wonder why the wave-particle debate about light isn't a purely academic pursuit. What are the practical consequences for astronomy, when the information from celestial objects gets to us regardless of whether we deem the messenger a particle or a wave? The basic idea is that by understanding the true nature of light, we can glean more information and devise a much more sensitive observational program than without this insight. In time, scientists were able to find out in detail how light as an electromagnetic radiation is sent out from an object. Thereby they were enabled to say much about the condition of and near the object in question at the point in time when the light was sent out. When the interactions of light with matter were better understood, corrections could be made for and conclusions could be drawn as to the combined effects of the regions of space the light had traveled to get to us. For instance, it was recognized that interstellar gas and dust clouds between us and an observed star can significantly redden and dim the received starlight, Figure 3.25, tricking us into attributing false properties to that star. If this invitation to understand light as an electromagnetic wave sounds too abstract, consider the following analogy. By learning to read music and play a musical instrument you can train your ear (and mind) to discern minute details of how music is performed. In this, you are like the scientist who learns how to "read" information encoded in the universe in the form of electromagnetic radiation. Like the music enthusiast able to tell the notes of a violin from those of a cello, and even able to identify famous violin players by their play, the astronomer will be able to identify the object that sent out the light. And indeed not just its material and temperature, but also its gas pressure and velocity, and many other things.

There are many different wave phenomena we observe in nature. Water waves, sound, and vibrations of a string all exhibit the following crucial properties unique to waves.

- Waves are types of motion in which information, momentum and energy are moving through space. This transfer is not achieved by physical movement of material from one place to another, but as a ***disturbance*** occurring in a periodic, repeating pattern. For instance, the water waves cause by a stone thrown in a pond will travel outward from the point of the initial disturbance. This in turn affects other parts of the pond—even through the water itself does not travel to these parts. The wave itself is not a physical object, but it can influence physical objects, such as a piece of cork bobbing up and down on the water surface, as the wave passes by.

- Since waves are periodic phenomena, they can be represented by sinusoidal functions, Figure 3.26. Note that a wave is a pattern in space *and* time: The disturbance will be different at a different points in space, and also at different times. Therefore, a wave

3.7. LIGHT IS ELECTROMAGNETIC RADIATION

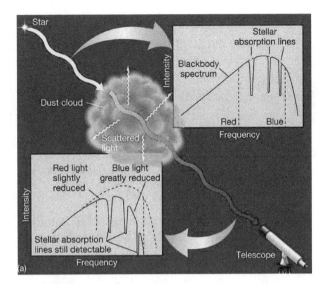

Figure 3.25: **Extinction of starlight.** As the starlight goes through interstellar medium it is absorbed and re-emitted. This scattering affects blue light stronger, and results in a reddening of the star's light. Since the dark lines of the star's spectrum are still identifiable, the effect can be corrected for, and in turn used to learn something about the interstellar medium.

can be characterized by its **wavelength** and its **frequency**. The former specifies the distance *in space* between two repeating points along the wave, like two crests or two troughs, and the latter tells us the time between two crests The only other information a perfectly periodic wave carries is the intensity of the disturbance, proportional to its **amplitude**; see Figure 3.26. The wavelength is a distance and therefore measured in meters, the frequency is inversely related to the **period** of the wave, and measured in inverse seconds or Hertz ($\frac{1}{s} = 1\,\text{Hz}$).

- You might wonder why I didn't mention the most apparent property of a wave, namely the speed at which it propagates. As it turns out, this is not independent information. For waves wavelength λ and frequency f are related due to the wave equation, to wit

$$v = \lambda \times f, \tag{3.3}$$

where v is the velocity of the wave. Electromagnetic waves travel at a very special, **absolute speed**, namely the speed of light, usually abbreviated as c from the Latin *celeritas* meaning *speed*. Lightspeed is the largest velocity in the universe, which is a pretty amazing statement. After all, couldn't you accelerate an object to an even higher speed? The fact that you cannot is one of the fundamental theorems of Einstein's **special relativity** theory. Without this rule, the whole fabric of space time would unravel, making things like time-travel and associated paradoxes possible. Note that therefore information travels at finite speed, and no causation can happen **instantaneously**.

- Waves are disturbances of a medium, but this statement has to be taken with a **grain of salt**. Namely, scientists drove themselves nearly mad in the nineteenth century to find the **aether**, that is, the alleged medium in which light waves travel. Used to the physics of other waves with an obvious medium (water waves, sound in air, vibrating string), they missed the forest for the trees. In fact, the electromagnetic fields making up a light wave are their own medium. If this sounds too much like a Münchhausen tale,[81] consider that the medium determines the speed of the wave. For instance, the vibrating string is characterized by its **tension** T (you adjust the tension on violin strings to tune the instrument) and its mass (density) μ. The more mass there is, the harder it is to move or accelerate the string, and the higher the tension is, the faster one piece of the string can affect its neighbor, so its plausible that the wave speed on the string should go up with the tension and down with the mass. Incidentally, it is

$$v_{string} = \sqrt{\frac{T}{\mu}}.$$

The important thing to note here is the wave speed being entirely fixed by the properties of the medium, that is, the string. Now, in the case of electromagnetic waves there is a completely analogous formula

$$v_{light} = \sqrt{\frac{k_e}{\mu_0}} \equiv c,$$

first derived by JAMES CLERK MAXWELL (1831–1879) in 1864.[82] We encountered the constant k_e as the electric analogue of the gravitational constant G in Coulomb's law, Equation (3.1). It tells us how strong the forces are between electric charges. Likewise, $\mu_0 = 4\pi \times 10^{-7} Tm/A$ tells us how strongly magnets interact with each other. In other words, k and μ_0 specify the electric and magnetic medium. Since electromagnetic waves are in this sense "self-sufficient,"[83] they can and do travel through the vacuum of space. Incidentally, the speed of light in any other transparent substance (air, glass, plastic) is slower, so technically, the speed of light is really the speed of light *in vacuum*.

- Electromagnetic waves are composed of electric and magnetic fields. The direction in which the electric field changes is called the **polarization** of the wave. This direction is perpendicular to the direction in which the wave propagates. This makes light a **transverse wave**, as opposed to sound, which has air density disturbances in the direction of movement making it a **longitudinal wave**, Figure 3.27(a). The polarization of light can encode extra information about the light source.

- Wavelength λ and frequency f of electromagnetic waves are inversely related—but not limited. Both vary over many orders of magnitude, while their product is always the same, $c = \lambda \times f$. Optical or **visible light** has wavelengths from 400 nm (blue light) to 700 nm (red light). Note that wavelength and color are **synonymous**, that is, saying the same thing. We can compute the associated frequencies, for instance[84]

$$f_{blue} = \frac{c}{\lambda_{blue}} = \frac{3 \times 10^8 m/s}{4 \times 10^{-7} m} = \frac{3}{4} \times 10^{8+7} \frac{1}{s} = 0.75 \times 10^{15} Hz.$$

---

[81] The Baron Münchhausen of eighteenth century prose pulled himself out of the swamps by yanking on his own hair.

[82] The story behind it is easily the most exciting tale of nineteenth century physics, but beyond the scope of this book. If you are interested in the connection between severed frog legs, compass needles and light, consult [32].

[83] According the MAXWELL's theory, a changing electric field produces a (changing) magnetic field and vice versa.

[84] Note that using **scientific notation** simplifies the calculation considerably.

3.7. LIGHT IS ELECTROMAGNETIC RADIATION

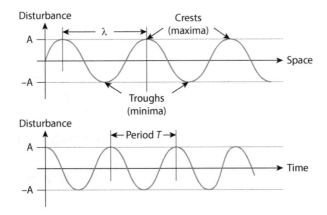

Figure 3.26: **A sinusoidal wave.** It is characterized by its wavelength λ, and its amplitude A. The time between two adjacent crests is called the period T, its inverse is the frequency $f = \frac{1}{T}$ of the wave. Wavelength and frequency are *inversely related*. Their product is the wave velocity $v = \lambda \times f$.

We described in Section 3.5.3 how HERSCHEL and RITTER discovered infrared and ultraviolet light beyond the visible range, but the full electromagnetic spectrum is much wider, ranging all the way from ultrashort wavelength, ultrahigh frequency **X rays** to long wavelength, low frequency radiowaves, Figure 3.27(b). Again, note that all of these are *electromagnetic waves*, and therefore all travel at the speed of light.

Initially, scientists in the nineteenth century were skeptical that light is just undulating electromagnetic fields. After all, the concept of a *field* that permeates empty space was abstract and had only recently been introduced by MICHAEL FARADAY (1791–1867). Furthermore, there was no source of the perfectly monochromatic[85] waves described above. All practical sources (candles, lightbulbs, the sun, hot pieces of iron) send out light rays with many different wavelengths; see the description of **thermal radiation** in the next section. After all, white light is, as already NEWTON had shown, Section 2.10, comprised of *all* colors of the rainbow. Not until HEINRICH HERTZ (1857–1894) had produced and received electromagnetic waves in a crucial experiment in 1888 was the hypothesis "light is an electromagnetic wave" universally accepted. Since this day, *light* **is** *an electromagnetic wave*.

3.7.2 Observing Light: Brightness and Doppler Effect

We have seen repeatedly how important it is to take into account the role of the observer in the observation. Different observers observe the same world but have a different view of it. Without a careful **reduction** of the recorded data taking into account the location and motion of the observer, it is not clear which patterns in nature are due to nature, and which are patterns created by the observer's way of observing. Consider a couple of examples. Our impression that the universe rotates around us (**diurnal motion**) could be due to the motion of the universe *or* the motion of the observer on a rotating planet. That it is the latter had to be established by careful analysis of the situation, and is not obvious. The retrograde motion of the planets could be due to the planets moving in the complicated way as PTOLEMY asserted *or* due to the observer on Earth moving with respect to the observed object, as COPERNICUS would have it. Earlier in this chapter we have seen that both the

---

[85] Of a single color, that is, of a single wavelength.

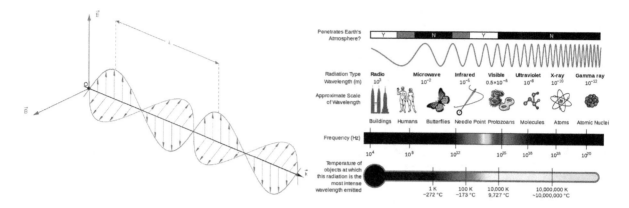

Figure 3.27: **Electromagnetic waves.** (a) Light, like other forms of electromagnetic radiation, consists of changing electric ($\vec{E}$) and magnetic ($\vec{B}$) fields. In space, the pattern repeats after the **wavelength** λ. The fields are mutually perpendicular and also perpendicular to the direction of propagation, characterized by the **wave vector** $\vec{k}$. This makes them **transverse waves**. (b) *The electromagnetic spectrum.* Electromagnetic waves travel at the speed of light. Their wavelength range from kilometers to the size of atomic nuclei. Likewise, their frequencies range from a few Hertz for the least energetic **radiowaves** to 10^{22} Hz for the most energetic **gamma rays**. Visible light occupies only a small range of wavelengths around 500 nm.

motion (velocity) of the Earth around the sun, and its change in location caused by this motion influences the way we observe stars' locations on the celestial sphere. The first effect leads to Bradley's **stellar aberration**, the latter to Bessel's **stellar parallax**.

Likewise, we expect that the status of the observer (location and velocity) will influence the way we observe light and radiation coming to us from celestial objects. Of course, the location of the observer defines the **distance** to the observed light source, and therefore will most directly affect the apparent brightness of the object. Since light spreads out from its source symmetrically, see Figure 2.50, its wavefronts will be concentric spheres of increasing radii. The signal of the source therefore spreads and thins out over the surface area of this sphere, which grows, as any area, like the **radius squared** (which is the distance to the source squared); see Section 1.2.6. Therefore, the observed brightness B of a light source falls off like the inverse square of the observer's distance r to it

$$B \propto \frac{1}{r^2}.$$

On the other hand, the light source will appear brighter if it is intrinsically brighter, that is, has a larger **luminosity** L. For instance, a 100 W lightbulb appears four times brighter from a distance of 10 meters than a 25 W lightbulb from the same distance. On the other hand, both appear equally bright when the 100 W lightbulb is placed at a distance of 20 m while the 25 W lightbulb remains at 10 m, because the brightness fall off by a factor $\left(\frac{10\,\text{m}}{20\,\text{m}}\right)^2 = \frac{1}{2^2} = \frac{1}{4}$. We conclude that there is a **brightness-luminosity-distance relation**, to wit

$$B = \frac{L}{4\pi r^2}, \tag{3.4}$$

where we have used the fact that the surface area of a sphere of radius r is $4\pi r^2$.

While the location of the observer relative to the observed lightsource has the obvious effect of reducing the brightness of the source, the velocity of the observer relative to the

3.7. LIGHT IS ELECTROMAGNETIC RADIATION

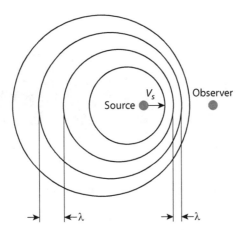

Figure 3.28: **The Doppler effect.** A moving source sends out waves of wavelength λ_{source}. If the source is moving toward an observer, he or she measures a smaller wavelength λ_{obs}. The effect is the same for light and sound, as both are **waves**. The difference $\lambda_{source} - \lambda_{obs}$ is proportional to the ratio of the speed of the source and the speed of the wave. Measuring the difference will give away the speed of the source.

source (or vice versa, as all velocities are *relative* velocities), has subtler effects, which have proven invaluable as a tool in the hands of the astronomer. The effect due to relative motion of source and observer was first postulated by Austrian physicist CHRISTIAN DOPPLER (1803–1853) in 1842 in an ill-fated attempt to explain the different colors of double stars. The Doppler effect applies to all wave phenomena, including light and sound. The basic tenet is that if a wave source moves toward an observer, the crests of the waves sent out appear closer together for the observer than if the source were at rest, Figure 3.28. Therefore, the wavelength observed is shorter than the wavelength sent out by the source. The effect depends on the ratio of the source speed v_{source} and the wave speed v, which is in effect a comparison of the relevant speeds involved in the measurement process. We can therefore calculate the observed wavelength λ_{obs} as a fraction of the wavelength sent by the source λ_{source}

$$\frac{\lambda_{obs}}{\lambda_{source}} = \frac{1}{1 \pm \frac{v_{source}}{v}}. \qquad (3.5)$$

If the source approaches the observer, we use the plus sign ($\lambda_{obs} < \lambda_{source}$ or **blueshift**), if it recedes from the observer we use the minus sign ($\lambda_{obs} > \lambda_{source}$ or **redshift**).

Typically, the Doppler formula is used to determine the source velocity v_{source} from a wavelength λ_{obs} observed "in the sky," and a reference wavelength λ_{source} measured in the lab. The wave speed is known *a priori*, for example $v = c$ for light in vacuum, and $v = v_{sound} = 443\,\text{m/s}$ for sound in air. The Doppler effect has been used to measure the rotational speed of the sun and planets, to determine the speed of molecules in gases, the radial velocity of stars, and has many other applications. We'll present an example for the Doppler shift in the next subsection, after we discuss a convenient method to measure the wavelength associated with a light source. While Doppler's original idea to explain the colors of binary stars by a wavelength shift did not pan out, his concluding remark that his hypothesis will ultimately determine the speed of remote stars was truly visionary. Moreover, the Doppler shift serves as an example how scientific work can be useful, even though it is initially undertaken for other reasons or misinterpreted. As long as the underlying science and math is sound, reinterpretation by a future generation of scientists or (re)application to a different problem can be fruitful. The

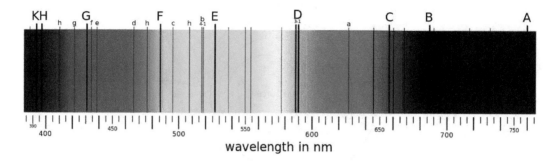

Figure 3.29: **Fraunhofer spectrum.** Shown is the spectrum obtained when sunlight falls onto a prism. The dark or *Fraunhofer lines* are seen on the *continuous spectrum* background. Note the prominent yellow double line D of sodium. Keep in mind that the sunlight passed through the Earth's atmosphere, so not all dark lines stem from the sun. For instance, the A and B lines are produced by terrestrial oxygen.

Doppler case also shows how great of a motivation astronomy was in the nineteenth century for basic research in *physics*.[86]

Incidentally, you can experience the *acoustic Doppler effect* when an ambulance approaches you (higher pitch sound) or has passed you (lower pitch sound). The ultimate application of Doppler's theory is the calculation of the Hubble constant governing the expansion of the universe via the recessional velocities of the galaxies.

3.7.3 Of Rainbows Dark and Bright Lines: Radiation

Even before NEWTON, scientists had studied light by sending it through prisms—or by observing rainbows[87]—thus showing that light of different wavelength or color is refracted differently as it enters another medium[88] by decomposing white light into all colors of the rainbow, Figure 2.44. It came as a surprise, therefore, when a new feature of the light produced by our primary light source, aka the sun, was found early in the nineteenth century.[89] Sunlight produces a continuous spectrum of visible lightwaves ranging from violet to red when sent through a prism. But superimposed on this colorful background, JOSEPH FRAUNHOFER (1787–1826) found hundreds of **dark lines**, Figure 3.29, which he systematically studied and catalogued beginning in 1814. Naturally, two questions arise: Why *does* the sun send out electromagnetic waves of so many colors, and why *doesn't* it send out waves of specific colors (the wavelength of the dark lines). While these question were not fundamentally answerable until the advent of **quantum mechanics** in the early twentieth century, practical laws *describing* and predicting the radiation of objects emerged much earlier. In fact, it is almost common sense that a hot object, like a rod of iron heated in a fire, will glow, and that its color as well as its brightness change depending on the its **temperature**. Iron starts glowing dimly red at about 800 K, and glows bright white at temperatures of about 1500 K, just below its melting point. This is the phenomenon of **incandescence**, namesake of the incandescent lightbulb which

---

[86] This was reiterated by MAX WOLF in 1912 [45] when he praised the applications of the Doppler effect and added that *spectroscopy* was equally beneficial for astronomy and terrestrial physics.

[87] The raindrops act like little prisms, decomposing white sunlight into colors.

[88] Much to the chagrin of lens crafters. Since different colors have different focal lengths, **chromatic aberration** plagued early refracting telescopes; see 2.9.5.

[89] The initial discovery was made by WILLIAM HYDE WOLLASTON (1766-1828) in 1804; FRAUNHOFER independently discovered the dark lines in 1814 and systematically studied them.

3.7. LIGHT IS ELECTROMAGNETIC RADIATION

has a glowing metal filament. The remarkable property of the radiation given off by a hot object is that it only depends on the temperature, but *not* on the material of the object. This radiation is called **thermal radiation**[90] because it can be explained by the **thermal motion** of electrically charged particles inside the object. Incidentally, this is a great example how different fields in science dovetail to decipher patterns in nature. Here, the **electromagnetic theory** contributes the fact that **accelerated charges** send out electromagnetic waves, while the **thermodynamic kinetic theory** affirms that the temperature of an object is a measure for the motion its constituents.[91] However, there are other sources of light. For instance, sodium heated over a **Bunsen burner**, Figure 3.30, produces a characteristic yellow light of a specific wavelength,[92] not a mixture of many wavelengths. In a crucial experiment, ROBERT BUNSEN (1811–1899) and GUSTAV KIRCHHOFF (1824–1879) studied this **line spectrum** of elements, Figure 3.30, and found to their surprise that adding sunlight (missing the yellow sodium frequency) to the yellow light of a sodium flame made the dark line in the solar spectrum *darker*. They of course realized that there is sodium on the sun but more importantly concluded that the sodium vapor **absorbed** light at exactly the same wavelength as it **emitted** light when heated. This insight led to the **Kirchhoff radiation laws**;[93] see Figure 3.40:

1. A solid or liquid object[94] of a certain temperature T produces a **continuous spectrum**, that is, gives off radiation at all wavelengths ("all colors of the rainbow").

2. Hot gas at low pressure will **emit** light only of specific wavelengths characteristic of the gaseous substance. These are the bright lines furnishing an **emission spectrum**. By analogy, the emission lines are the "fingerprints of the chemicals."

3. Cool gas at low pressure will **absorb** light of specific wavelengths characteristic of the gaseous substance. These dark lines of an **absorption spectrum** can be seen if the source of a continuous spectrum is viewed through a cloud of cool gas; see also Figure 3.25.

There are thus two very different spectra: **line spectra** where only specific colors or wavelengths appear, and **continuous spectra** where all colors appear, albeit some at greater intensity than others.

Kirchhoff went on to speculate about the precise dependence of the intensity I of the **continuous spectra** on the wavelength, that is, on the shape of the function $I(\lambda)$ now known as **Planck's curve**, Figure 3.35(a). In his article from January 1860 [25], he points out that it must be a smooth function which falls off to zero both at large and at small enough wavelengths. From our observation of the color of a hot iron rod, we can tell that an object must preferentially emit waves of a certain wavelength or color. This "preferred color" constitutes the peak of the curve, and the shift of the peak toward longer or "redder" wavelengths with decreasing temperature is clear. In Section 3.8.3 we'll see that this observation is the basis of **Wien's law**. An immediate astronomical application is the realization that white stars are

---

[90]If the object is *ideal* in some sense, for example a perfect absorber and emitter of radiation, its radiation is called **blackbody radiation**.

[91]In particular, all particles in an object above **absolute zero** move.

[92]Actually, it is a double line, Figure 3.31, so technically waves of two different wavelengths are emitted.

[93]Incidentally, KIRCHHOFF and BUNSEN were the first to discover new chemical elements with the method, namely the Group 1 **alkali metals** rubidium and caesium. The bright emission lines became their namesakes: *rubidus = deep red* and *caesius = sky-blue*.

[94]Only later it was realized that also highly compressed **gasses** produce continuous spectra.

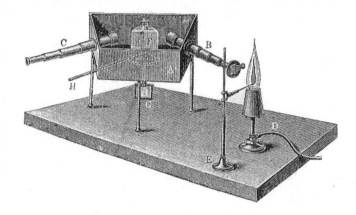

Figure 3.30: **Bunsen-Kirchhoff experiment.** A substance such as sodium is heated over a Bunsen burner. The light of the flame is collimated (B), sent through a prism (F), and observed at C. Different colors are refracted at different angles, so the measured angular position of a bright line yields its wavelength, and therefore a "fingerprint" of the substance.

hotter that red stars and bluish stars are hotter still. The next time you observe the night sky, try to tell which of the stars you see is the hottest!

KIRCHHOFF and BUNSEN had another *epiphany*, later retold by Heidelberg astronomer MAX WOLF (1863–1932) [45]. Namely, during the illumination of the famous castle[95] in June 1860, BUNSEN observed the fireworks of Bengal fire through a prism from the roof of his laboratory down the hill in the old city. He clearly discerned the green lines of barium and the red lines of strontium in the flames. BUNSEN then told KIRCHHOFF: "If we can tell from this distance which elements are glowing in these flames—why shouldn't we be able to determine out of which substances the celestial bodies are made?" Wolf adds: "Thus, the spectral analysis of the sun and the stars was born."[96]

Concept Practice

1. Compare red and infrared light. Which is a true statement?

    (a) Red light has a higher frequency.

    (b) Red light has a larger wavelength.

    (c) Red light has a larger velocity.

    (d) All are correct.

    (e) None of the above.

2. Consider the relation between brightness, luminosity and distance. Which is a true statement?

    (a) Brightness is independent of distance.

    (b) The more luminous an object is the brighter it appears.

    (c) The luminosity of an object falls off like the inverse distance squared.

    (d) None of the above.

---

[95] The so-called *Heidelberger Schloßbeleuchtung*.

[96] Contrast this with the quote by which astrophysics pioneer J. ZÖLLNER was confronted just one year earlier by physicist HEINRICH WILHELM DOVE in Leipzig: "What the stars are we do not know and will never know." [48]

3.8. THE RISE OF THERMODYNAMICS

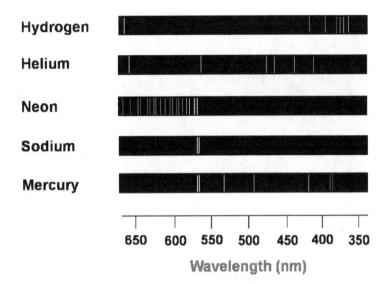

Figure 3.31: **Fingerprints of the elements.** The bright line or emission spectrum of several prominent chemical elements. The precise position or wavelength of these lines are measured with *spectrometers*, invented by BUNSEN and KIRCHHOFF, Figure 3.30. Note that the bright yellow double line of sodium sits exactly where its double dark line appears in the Fraunhofer spectrum, Figure 3.29.

3. A hot, glowing gas cloud produces a(n)
   (a) absorption (b) emission (c) continuous (d) Fraunhofer spectrum. (e) None of the above.

3.8 The Rise of Thermodynamics

We encountered in Secs. 2.9.4 and 3.4 the beginnings of **thermodynamics**. Out of a pragmatic study of the relations between quantities such as **volume**, **pressure**, and **temperature** of gases arose an understanding of the behavior of substances under given circumstances or boundary conditions. The simplest of such **equations of state** is the **ideal gas law** (CLAPEYRON, 1834), a generalization of Boyle's law we encountered earlier. It can be succinctly stated as

$$pV = Nk_B T, \qquad (3.6)$$

where p is the pressure, V the volume, T the absolute temperature, N the number of molecules in the volume, and $k_B = 1.38 \times 10^{-23} J/K$, the Boltzmann constant that makes all the units work out.

If you don't like the math involved, try to get at least an intuition of what is going on. This is important if you want to understand stars, and not hard, since the content of the ideal gas law is pretty intuitive, and hardly surprising. We saw earlier, Section 2.9.4, that the volume of a given amount of gas shrinks if its squeezed together, that is, the outside pressure[97] increases. What's new here is that the pressure (at constant volume) or the volume (at constant pressure) of the gas increases if the temperature increases. This is probably precisely what you expected, right? So no worries, just keep in mind that this relation breaks down for non-ideal gases, liquids, and so forth. By the way, *ideal* here means basically low density, that is, a "bit of gas" in a large container is ideal.

---

[97] And therefore the inside pressure, courtesy of NEWTON's third law.

232   CHAPTER 3. DEDUCING & CHECKING: BETWEEN NEWTON AND EINSTEIN

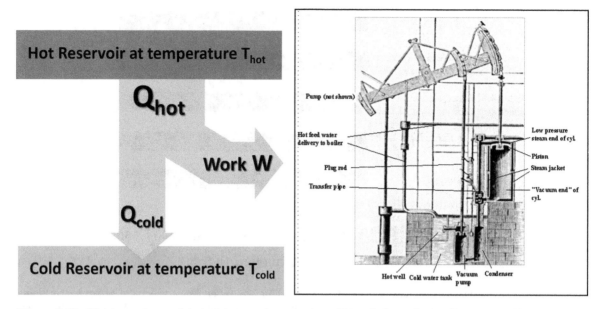

Figure 3.32: **Heat engines.** (a) An abstract heat engine: Heat Q flows from a hot to a cold reservoir of temperature T_{hot} and T_{cold}, respectively. Some of the internal energy of the hot reservoir is available as (mechanical) work W, but some heat has to be dumped into the cold reservoir since *entropy* cannot decrease. (b) Schematic steam engine as improved by JAMES WATT.

The dependence of steam (gaseous water) pressure of a fixed volume on temperature lends itself to be used in steam engines like in Figure 3.32(b), where a piston driven by steam does mechanical work: The expansion of the gas at high temperature drives the piston outward, its contraction when cooling down drives the piston back in. The improvement of the steam engine by JAMES WATT (1769 patent) made the ***industrial revolution*** possible, because contrary to wind and watermills, a steam engine can be installed anywhere. The amazing thing is that it does useful (mechanical) work by transferring heat from a hot to a cold ***reservoir***. In technical language, the steam engine is a ***heat engine*** that does mechanical work by using steam as its ***working fluid***. As an example consider the so-called ***Carnot cycle*** of Figure 3.33. A working fluid such as an ***ideal gas*** changes its volume and pressure, and therefore its energy content. The processes are described by thermodynamic laws, such as ***Boyle's law***, which we encountered in Section 2.9.4. Recall that ***Boyle's law*** describes how a gas changes its ***state*** when its temperature is kept constant. It is known as an ***isotherm***[98] process. Often, ***state changes*** happen so fast that no heat Q can be exchanged ($Q = 0$, though the temperature may change). These are known as ***adiabatic*** processes, and we will encounter them again, since they can be used to describe how the Earth's atmosphere and the interior of stars work.

After a long trial-and-error process it was eventually realized that there are certain restrictions or laws that are inherent in any heat engine, regardless of its ***working fluid*** or other physical realizations. In other words, these ***universal laws of thermodynamics*** apply to steam engines, car motors and stars alike. Around 1850 they were spelled out in detail and changed the way astronomers—and everyone else for that matter—thought about the universe. As often in science, once the laws were discovered they seemed almost obvious. The

---

[98]Greek: *iso = same* and *thermos = warm*.

3.8. THE RISE OF THERMODYNAMICS

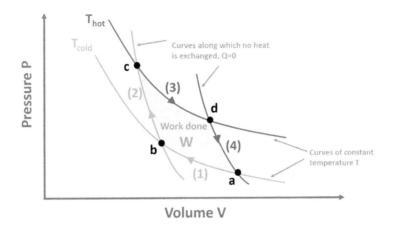

Figure 3.33: **Carnot cycle in a pV diagram.** The *thermodynamic state* of a *working fluid* is characterized by *state variables* such as pressure p, volume V, temperature T, and thermal energy E. Two of them are visualized in a *pV diagram*; see Figure 2.39, where the pressure of the fluid is plotted versus its volume. In the *Carnot cycle* the state variables change in a cyclical way, that is, after several changes they return to their initial values; the same state is recovered. Still, an amount of work ΔW equal to the shaded area has been done by transferring heat Q into and out of the system. The four processes making up the Carnot cycle are (1) (state a $\to$ state b) *isothermal compression* (volume decreases and pressure increases slightly as temperature stays the same); (2) (state b $\to$ state c) *adiabatic compression* (volume decreases slightly and pressure increases in a process so fast that the *working fluid* exchanges no heat with the surrounding materials); (3) (state c $\to$ state d) *isothermal expansion* (volume increases and pressure decreases as temperature stays constant); (4) (state d $\to$ state a) *adiabatic expansion* (back to the initial state: volume increases and pressure decreases while no heat is exchanged). Energy is conserved in each sub-process, because heat, work, and thermal energy are related by Equation (3.7).

two laws of immediate relevance for us are the *first and second law of thermodynamics*. The first law goes back to JULIUS ROBERT MAYER's (1814–1878) hypothesis of 1841 that the energy of a *closed (isolated) system* is simply constant. This is known as *energy conservation*. The abstract quantity *energy* can manifest itself in different forms. You know of the energy of motion (*kinetic energy*, from Greek *kinesis=motion*) proportional to the square of the speed v of an object of mass m, namely $E_{kin} = \frac{1}{2}mv^2$, or of the positional energy (*potential energy*) inherent in a *configuration* of objects, for example the potential energy of a spring of stiffness k stretched by an amount x, namely $E_{pot} = \frac{1}{2}kx^2$. There are also electric, chemical, and other forms of energy—which is quite confusing. To realize that all these different "things" or quantities are manifestations of the same concept *energy* was a milestone in science, and not easily won. According to Mayer, a *closed system* (picture a jug of hot water and steam packed in Styrofoam) can have all kinds of energy forms associated with it, but governing all the confusing transformations of one kind of energy into another is a very simply rule: Energy cannot be destroyed or created, only transformed; energy is conserved. As an analogy, think of energy as money. It can manifest itself as cash or a number in your bank account. Of course, in real life, your money is not part of a *closed system*. You spend money, and you earn money, so the amount of money you have (your net worth) will change over time. But since money itself in this analogy is not created nor destroyed, only transferred, there is the possibility of accounting for the money and its flow—in and out of you pocket and bank account. The analogous *balancing equation* in thermodynamics was

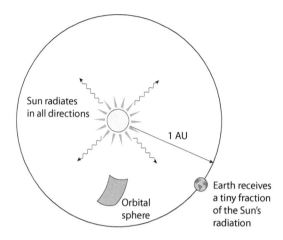

Figure 3.34: **Determining the solar constant.** The sun radiates in all directions. The Earth sits on a sphere of radius 1 AU through which all this radiation passes. It only a tiny fraction of the total radiation, namely the ratio between its cross-section and the surface of its "orbital sphere."

clearly formulated for the first time by RUDOLF JULIUS EMANUEL CLAUSIUS (1822–1888) and WILLIAM JOHN MACQUORN RANKINE (1820–1872) in 1850 and is known as the

First Law of Thermodynamics:
 *The energy of a closed system U known as **internal energy** or **thermal energy** is constant; it cannot change; it is neither destroyed nor created. If the system is **open**, the thermal energy can change in two ways: Either **work** is done by or on the system (energy leaves or enters the system), or **heat** is transferred from or to the system.*
We can write this down in mathematical form

$$\text{(change in thermal energy of system)} = \text{(heat transferred to the system)}$$
$$\text{minus (work done by the system)}$$
$$\Delta U = Q - W. \tag{3.7}$$

That's all there is. In our money analogy, your net worth changes either because you spend money or your boss transfers money into your bank account. Of course, these transfers can happen in to opposite direction: Someone gives you cash and increases your net worth, and you can transfer money out of your bank account to pay a bill; this will decrease you net worth. Technically, both **heat** Q and **work** W are *not* forms of energy, but **energy transfers**. In particular, **heat** is energy transferred from one body to another due to a difference in **temperature**.

The important message here is not so much the precise meaning of the physical quantities. Rather, the fact that such a balancing equation exists at all drove progress in science, technology and astronomy.

3.8.1 A First Stab at the Energy Source and Age of the Sun

The energy balancing equation of the universe, aka the ***first law of thermodynamics***, made it clear that energy is necessary to drive processes in the cosmos and here on Earth. This energy, as was soon realized, comes almost entirely from the sun. Indeed, the solar radiation received on Earth drive weather (wind and water energy), plant growth (fossil fuels), and,

3.8. THE RISE OF THERMODYNAMICS

quite directly, "solar energy," that is, the electric energy harvested in photovoltaic cells. Thus, with the exception of geothermal energy and nuclear energy, all our energy on Earth comes from the sun. The question arose: How does the sun produce the energy it radiates?

As always when something needs to be explained, it is indispensable that one measures precisely *what* needs to be explained. In 1838, CLAUDE POUILLET (1790–1868) devised a **pyrheliometer** to estimate the power (energy per unit time) radiated by the sun hitting a unit area on Earth. This is known as the solar constant. Its value is approximately 1,400 W/m^2, that is, each square meter on Earth receives the power equivalent of a large space heater. This means that a football field receives a power equivalent of 10,000 space heaters, and it becomes clear that the power radiated by the sun must be immense, since a football field is small compared to the surface area of the Earth, which is minuscule compared to the (imaginary) sphere of radius 1 AU, through the surface of which virtually *all* solar radiation will pass; see Figure 3.34(a). The sun continuously radiates a power

$$P_{sun} = 4\pi R_{Earth,Sun}^2 \times 1,400 \frac{W}{m^2} \approx 12(1.5 \times 10^{11}\, m)^2 \times 1.4 \times 10^3 \frac{W}{m^2} = 3.8 \times 10^{26} W. \quad (3.8)$$

In other words, in each second the sun dispenses 3.8×10^{26} Joules of energy. The first law of thermodynamics therefore posed the following dilemma for the contemporary scientists: If energy cannot be created, where does this enormous amount of radiated energy come from? How is it "stored" in and then "released" by the sun?

A naive hypothesis would be to assume the sun is a huge, hot stone, and that due to its enormous volume, it contains a lot of thermal energy, which it radiates thereby lowering its temperature. This is plausible for us, but does it agree with what what known around 1850? Well, it is not even clear how scientists would have computed the power P radiated by an object of given temperature T, because the formula describing this dependence, $P(T)$, namely the **Stefan-Boltzmann law**, was not discovered until 1879. We will look at this very important law below, and here state that the cooling time of the sun is so small, namely a few thousand years, that this "hot solar rock" hypothesis can be falsified immediately, because even the most conservative estimate of the Earth's age is much longer. MAYER himself interpreted his energy conservation law quite naturally as a balance equation: If the sun dispenses the energy, it has to somehow be replenished from the outside; the sun is not a closed system. His idea was to have an influx of matter, maybe asteroids or comets, whose kinetic energy would replenish the sun's energy. The amount of mass necessary is surprisingly small—only about one seventeen millionth of the sun's mass per year. But even this minuscule increase in mass would shorten the Earth's orbital period by a few seconds per year which could be ruled out easily with precision astronomy of the day.

The next idea was more prolific. The **Helmholtz-Kelvin contraction hypothesis** enjoyed about forty years of acceptance, before quietly dying and ending as a "unburied corpse" [15]. The origin is HERMANN VON HELMHOLTZ's (1821–1894) idea that a shrinking sun will heat up, according to the laws of thermodynamics, and therefore, by continuous contraction, will be able to radiate a large amount of energy for a very long time. His argument was crude because of his simplifying assumptions. Modeling the sun as a *homogeneous* sphere of radius R_S and mass M_S, one can compute its potential energy. In principle, one subdivides the sun into little parcels as in Figure 3.34(b), and adds up the energies which every parcel contributes because it sits a distance r_p away from the center of a sphere of mass $M_S(r_p/R_S)^3$.

The sum is the gravitational energy of the sun is

$$E_S = \frac{3}{5}\frac{GM_S^2}{R_S}. \tag{3.9}$$

Then he used the result Equation (3.2) that a body of mass M_S and specific heat capacity c_S will warm up by an amount ΔT (a temperature difference) if it receives a heat Q. HELMHOLTZ set this heat equal to the available potential energy $Q = E_s$, thus

$$E_S = M_S c_S \Delta T. \tag{3.10}$$

But how to estimate the heat capacity of the sun when nobody even knew what the sun was made of? Helmholtz made another bold assumption, namely that the sun's heat capacity has a "typical" value, namely that of water.[99] Note that HELMHOLTZ chose to make this assumption (which he knew was not correct) rather than doing nothing. Crude approximations often have to be made when too little is known. Clearly, nothing quantitative can be expected from such a theory, but a qualitative understanding is still better than no understanding at all. If the sun shrinks, that is, R_S becomes smaller, then according to Equation (3.9) the energy of the sun and therefore, according to Equation (3.10), its temperature goes up. If we plug in numbers as Helmholtz did, we'd find that if the solar radius shrank by only 0.0001 of its present value, the sun's temperature would increase by more than 2800 K. This in turn means that a contracting sun could radiate its present power for about 20 million years! Hence the **age of the sun** (and Earth) could according to the **contraction hypothesis** as large as

$$T_{contraction} \approx 20 \text{ million years}. \tag{3.11}$$

One objection comes to mind immediately: Wouldn't the decrease of the sun's radius be perceptible over the last two thousand years of astronomical observations? As it turns out, the solar diameter would decrease only by about 75 meters per year, which is not discernible. The theory was later refined by LORD KELVIN[100] (1824–1907), hence the name.

3.8.2 The Second Law, Heat Death and Other Scary Victorian Stories

Empirically, it was found that heat flows from a hot to a cold object and never the other way. So there must be an additional rule in thermodynamics, because the first law is completely symmetric with regards to the direction of the energy transfers. The additional rule is warranted, because otherwise it would be possible to extract a massive amount of work from the environment by reversing the flow of heat in the diagram 3.32(a). As heat flows from the colder environment with large internal energy to the hot system, work of arbitrary quantity could be done. This is a **perpetual motion machine of the second kind** and not possible in nature according to the additional rule known as the

Second Law of Thermodynamics:
*Heat does not flow **spontaneously** from cold to hot bodies.*

This law was implied in SADI CARNOT's (1796–1832) 1824 treatment of heat engines, formalized by CLAUSIUS in 1850, and written down in mathematical language in 1854. Note

---

[99] Which has one of the largest heat capacities of all substances, see Section 3.4.

[100] We will refer to WILLIAM THOMSON as LORD KELVIN, under which name he is more commonly known, although he was knighted rather late in life (1892).

3.8. THE RISE OF THERMODYNAMICS

that heat *can* flow from cold to hot bodies, as in a refrigerator, but mechanical work needs to be done to achieve this reversal of the natural flow. This is why you have to pay your electric bill for the energy delivered to your apartment to cool your food and drinks. The technical way to state the second law is to introduce another quantity (akin to internal energy) specifying the thermodynamic state of a system, namely **entropy**. It is then said that entropy always increases.

To consistently use the term **entropy** we should carefully and concisely define it—which is beyond the scope of this book. Here, it must suffice to note the dramatic consequence of the second law of thermodynamics: If heat only flows from hot to cold bodies, then eventually all hot bodies in the universe will cool down, and all cold bodies will warm up until all bodies have exactly the same temperature, and no mechanical work can be extracted anymore from any heat reservoir. In other words, the universe will come to a complete and inevitable standstill, dubbed the **heat death**. Although it was realized that this heat death of the universe will occur in the very distant future, the discovery of a terminal fate of the universe was a big shock not just for the scientists of the time. If true, it would put an end to notions such as the eternal universe. Moreover, a divine being influencing the universe would be quite powerless unless violating the second law—or at least standing *outside* the universe—but this would means turning the cosmos into an *open system*.

These considerations especially shocking because they appeared at a time in the **Victorian era**[101] when religious and other beliefs were confronted with several upsetting scientific insights, such as CHARLES DARWIN's (1809–1882) theory of evolution and natural selection, and CHARLES LYELL's (1797–1875) principles of geology, including the idea around 1850 that the world is older than 300 million years.[102]

3.8.3 Blackbody Radiation

Above we mentioned the importance of knowing how much an object of a certain temperature radiates. We also saw in Section 3.7.3 how Gustav Kirchhoff was able in 1860 to deduce that the amount of radiation an object gives off only depends on its temperature, not on its composition. He came to this conclusion by considering an **idealization**, namely an object with perfect radiative properties. He called an object which perfectly absorbs all incoming radiation a **blackbody**.[103] Fortunately, it was found that many objects—in particular stars and planets—radiate approximately like a blackbody; see Figure 3.35(b). The amazing property of blackbody radiation is that the total amount of energy radiated by the object per second per surface area depends on only one thing: its temperature! You could easily come up with a plausible list of other properties, like its mass, density, internal structure and so

---

[101] Queen Victoria reigned the United Kingdom from 1837 until 1901.

[102] The age of the world had (in)famously been computed in the seventeenth century by JAMES USSHER (1581–1656), archbishop of Ireland. He added up all genealogies in the Bible—assuming none had been left out. He came to the (wrong) conclusion that the first day of creation was Sunday, October 23, 4004 BC, thus claiming the universe to be about 6,000 years old. In 1750 the *encyclopédiste* BUFFON had reckoned the Earth's age to be about 75,000 years from estimating the cooling of its interior (reheating due to radioactive materials was, of course, unknown at the time), and from the formation of sedimentary rocks an even higher number of 3 million years was obtained not much later.

[103] To have a mental picture, imagine this blackbody to be either a perfectly black body, that is, a perfect **absorber** of radiation, or a perfectly isolated cavity in thermodynamic equilibrium with walls of uniform temperature.

forth, but none of them are important. What a relief! Of course, a larger blackbody has a larger surface area, so it radiates more. Hence, size matters.

Based on measurements of JOHN TYNDALL (1820–1893), JOSEF STEFAN (1835–1893) inferred that the power radiated is proportional to the temperature to the fourth(!) power[104]

$$P(T, A) = \sigma A T^4,$$

where T is the absolute temperature (in Kelvin), A the surface area of the blackbody, and $\sigma = 5.67 \times 10^{-8} \frac{J}{K^4 m^2}$ the Stefan-Boltzmann constant, which fixes the units. STEFAN's student BOLTZMANN got his name on the law and constant because he was able to *derive* it from the laws of thermodynamics in 1884. Stars and planets are (almost) spherical, so we can use a variant of the formula above

$$P(T, R) = 4\pi \sigma R^2 T^4, \tag{3.12}$$

and find that the total power radiated by a spherical object **scales** with the square of its radius R. For example, a star of twice the diameter radiates four times more. Note that a star of twice the temperature radiates 16× more! This remarkable law enabled STEFAN to determine for the first time reliably[105] the **effective temperature of the sun**. From the power output of the sun, Equation (3.8), and its radius $R_S = 698\,000$ km $\approx 7 \times 10^8$ m we can deduce using the Stefan-Boltzmann Equation (3.12) that

$$T_S = \left(\frac{P_S}{4\pi R_S^2 \sigma}\right)^{1/4} = \left(\frac{3.8 \times 10^{26} W}{4\pi \times (49 \times 10^{16} m^2) \times (5.67 \times 10^{-8} \frac{W}{K^4 \times m^2})}\right)^{1/4} \approx 5800\,\text{K}.$$

Note that this is the temperature of the **surface** of the sun, that is, the part of the sun called **photosphere** from which most electromagnetic radiation of the sun comes to us. At this point we have no idea what the temperature below the **photosphere** is. It will probably much hotter inside, and likely hottest at its core, but to say anything reliable, astronomers had to come up with a **model of the sun**, because one cannot look *into* the sun.

Most of the time we shall be content with this bulk property of blackbody radiation, namely that the *total* power radiated by a blackbody depends only on its size and its temperature. But we can extract much more from stellar radiation—which is a good thing, because radiation is pretty much the only information we have concerning stars. It turns out that blackbodies radiate their energies differently at different wavelengths; they are brighter at some wavelengths than others; see Figure 3.35. Since **wavelength** is another word for *color*, as we have seen in Section 3.7, we can appreciate that an object that radiates more at the red than at other wavelengths will appear red, and so forth.

The curve describing the dependence of the amount of radiation emitted at a given wavelength is known as **Planck's law**, Figure 3.35(a). The derivation of the mathematical formula describing the curve took much effort and time—and a surprising **assumption**. When Planck wrote down the function (depending on wavelength, but also temperature and size of the object) in 1900, he had to assume that the radiation was not given off continuously, but in small but finite packets, now known as quanta. His formula was the first to sport the incredibly

---

[104] Such a high power is highly unusual in physics, but can be traced back to the fact that the radiation emitted is in the form of electromagnetic waves traveling in three dimensions. It can be shown that the dependence in an arbitrary number of dimensions d is T^{d+1}.

[105] Estimates—or rather guesses—up to this time ranged from 1800 K to 100 000 K.

3.8. THE RISE OF THERMODYNAMICS

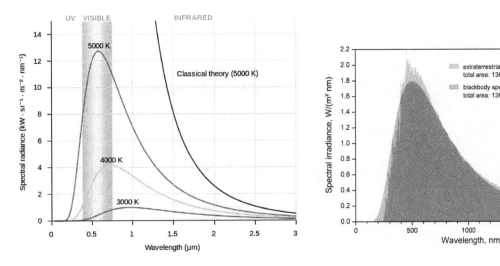

Figure 3.35: **Blackbody radiation.** (a) *Planck's law* is shown as a plot of the intensity as a function of wavelength for different temperatures. Note that the wavelength of the peak of the distribution lies on a hyperbola, since it decreases inversely with T according to *Wien's law* ($\lambda = \frac{0.0029 m \cdot K}{T}$.). The peak energy on the other hand increases with T. The area under the curve increases with the fourth power of T according to the Stefan-Boltzmann law. (b) The sun's ideal and real *blackbody radiation*.

small **Planck constant** $h = 6.6 \cdot 10^{-34} Js$. This was the birthday of **quantum mechanics**, and the end of **classical physics**. We'll come back to this fascinating *revolution* in Section 5.1.

Before we get lost in the math and advanced physics here, it's relieving to note that we will need only the following properties of **Planck's law** to understand the expansion of the universe; see Figure 3.35(a)

- The Planck curve goes to zero for very long and very short wavelengths; an object does not radiate substantially at very long and very short wavelengths.

- The Planck curve has a maximum or peak at some wavelength λ_{peak}.

- The Planck curve is *asymmetric* around the peak; the object's radiation drops off slower from the peak than it rises toward the peak; equivalently, the curve has a sharp rise and a long tail.

- The higher the temperature of the object, the more it radiates at *all* wavelengths; a hotter object is brighter than a cooler object (of same size) at all wavelengths.

- The hotter the object, the smaller the peak wavelength; a cooler object peaks "later," that is, at a longer or redder wavelength. Hence, a hotter object looks "bluer," and a cooler object looks "redder."

The mathematical expression of the last fact is known as **Wien's law** (1893)

$$\lambda_{peak} \cdot T = 0.0029 m \cdot K. \qquad (3.13)$$

Try to "get" the law: All it says is that peak wavelength and temperature are *inversely* related: the bigger one, the smaller the other. We contemplated this relationship in Section 2.10; the function $F_2(r)$ in Figure 2.49 represents an *inverse relationship* between the

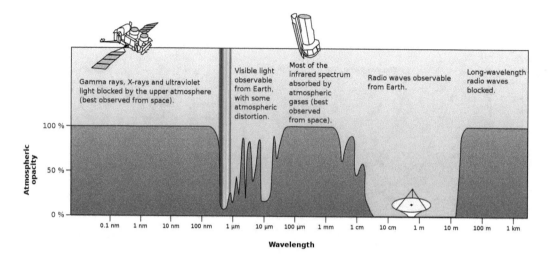

Figure 3.36: **Atmospheric absorption.** The atmosphere absorbs most forms of electromagnetic radiation (opacity ≈ 100%). Only visible light, some infrared, and most radiowaves can pass through the Earth's atmosphere and reach the surface.

independent variable r (here: T) and the dependent variable (here: λ). The value of Wien constant ($0.0029 m \cdot K$) tells us that the peak wavelength of a blackbody of temperature $1\,\mathrm{K}$ (three times colder than the universe[106]) is $0.0029\,\mathrm{m} = 2.9\,\mathrm{mm}$. Hence, the peak wavelength of the universe is three times shorter than that,

$$\lambda_{peak,Univ} = \frac{1}{3} \times 2.9\,\mathrm{mm} \approx 1\,\mathrm{mm}.$$

Electromagnetic waves of this wavelength are known as **microwaves**. They are part of the famous **cosmic microwave background** (CMB) we'll encounter in the cosmology section 6.4. Incidentally, microwaves can be used to heat your food because **water molecules** in the food preferentially **absorb** electromagnetic waves of this wavelength. Absorbing electromagnetic waves means gobbling up their energy, and having more energy means having a higher temperature. The fact that water readily absorbs microwaves means that the peak of the Planck curve of the cosmic microwave background cannot be observed from Earth, because water is also abundant in the Earth's atmosphere; the atmospheric H_2O molecules gobble up the microwaves before they reach the ground-based telescopes. The CMB was discovered "off-peak" at a wavelength of $\lambda = 7.35\,\mathrm{cm}$, at which the atmosphere is transparent, see Figure 3.36.

On the other hand, every object in the universe radiates! Even very cool objects can be identified due to their blackbody radiation. For an example, consider Figure 3.37. Humans are about $20\times$ cooler than the sun, and therefore radiate preferable at a peak wavelength that is $20\times$ longer than that of the sun, according to **Wien's law**, Equation (3.13). Also, using the **Stefan-Boltzmann law**, Equation (3.12), we conclude that humans radiate $20^4 = 160{,}000\times$ less per unit surface area than the sun. Of course, the surface of the sun is much larger than

---

[106] Outer space is not at **absolute zero**. Radiation left over from the **big bang** warms up the universe to $2.7\,\mathrm{K}$.

3.8. THE RISE OF THERMODYNAMICS

Figure 3.37: **Blackbody radiation.** Any object (here: a human) radiates at various wavelengths. (a) Our eyes are only sensitive to *visible light* (wavelengths between 400 nm [violet] and 700 nm [red]). (b) At a temperature of about 310 K (98 °F or 37 °C), humans radiate mostly in the infrared, that is, humans are *brightest* in IR. Note that some materials are opaque to some wavelengths, and transparent to others, as evident in the glasses and the plastic bag, the latter transparent, the former opaque to IR light.

that of a human, so the sun is brighter by another factor

$$\frac{\text{sun surface}}{\text{human surface}} = \left(\frac{4\pi R_{Sun}^2}{2(2 \times 0.5 + 2 \times 0.2 + 0.2 \times 0.5)m^2}\right)^2 = \frac{4\pi \times (7 \times 10^8 m)^2}{5m^2} = 1.2 \times 10^{18},$$

where we have **modeled**—or rather **approximated**—the human body as a box of $2m \times 0.5m \times 0.2m$. The comparison of the luminosities of a human with the sun is a little "out there," but perfectly possible and rational. Also, it illustrates how dim cold, small objects are, that is, how little they radiate.

Concept Practice

1. Blackbody A has a temperature three times higher than B of same size. They radiate power according to (a) $P_A = 3P_B$ (b) $P_B = 3P_A$ (c) $P_B = 9P_A$ (d) $P_A = 27P_A$ (e) None of the above.

2. Blackbody A has a radius three times higher than B of same temperature. They radiate power according to (a) $P_A = 3P_B$ (b) $P_B = 3P_A$ (c) $P_B = 9P_A$ (d) $P_A = 27P_A$

3. Blackbody A has a temperature three times higher than B. They are not of equal size. What can we conclude?

    (a) A radiates more than B.
    (b) B must be nine times larger than A.
    (c) The peak wavelength of B's Planck curve is three times as long as A's.
    (d) We cannot conclude anything because we don't know the relative sizes.

3.9 The Emergence of Astrophysics

3.9.1 Photometry Variable and Binary Stars

We saw that the advent of spectroscopy in 1860 afforded a new look at the specks of light we call stars. Up to that time, precision position measurements, and the careful calibration of stellar brightnesses (started by FRIEDRICH WILHELM ARGELANDER (1799–1875)) were the only two sources of knowledge about stars. This is not to be dismissed, since it yielded the distances to the stars via the **stellar parallax**, the determination of stellar masses via the Keplerian motion of the components of **double stars**, and the discovery of a whole new class of **variable stars**. The latter showed systematic, often periodic pulsation of their brightness, and therefore constituted a new pattern of nature that needed to be explained. Stellar pulsations happen on fairly short timescales lasting from a few days to several months. Most variable stars will change their brightness by a small amount, which makes it hard to discover these subtle changes in the night sky. Not surprisingly, there were only four variable stars known prior to ARGELANDER's systematic scan of the sky, in which he mapped an unbelievable 324,000 stars (*Bonner Durchmusterung*, 1863). For instance, *Mira* (Latin: *the wonderful*), a quasi-periodic variable which sometimes dims below the human perception threshold was discovered in 1596 and identified in 1638 as variable, while the variability of *Algol* (β Persei) was discovered in 1669. JOHN GOODRICKE (1764–1786) in 1784 realized that the **light curve**[107] of *Algol*, Figure 3.38, can be explained by the mutual eclipses of the components of this binary star.[108] *Mira*, on the other hand, physically changes, that is, its energy production and power output does change over time in a repeated pattern. In broad terms, there are two types of variable stars. **Intrinsic variables** are the true variables, since the stars themselves change physically. **Extrinsic variables** on the other hand, only appear to change, and do so because of external circumstances, such as being eclipsed by a companion. Both patterns can be used to find out more about stars. From the eclipsing binary we can get masses and sizes of stars, from an **intrinsic variable**, we might learn something about how stars behave, and thus about the stellar life cycle.

Measuring the brightnesses of stars, aka **photometry**, thus led to new insight. Surprisingly, even the apparent brightness of the sun was not measured until JOHANN KARL FRIEDRICH ZÖLLNER (1834–1882) invented the **astrophotometer**[109] around 1860.[110] Incidentally, the accurate measurement of celestial brightnesses or **astrophotometry** had only recently been established by the work of BOUGUER, JOHN HERSCHEL and ARGELANDER. Around the same time, GUSTAV FECHNER (1801–1887) found out that the human eye reacts **logarithmically** to stimuli. This makes sense, since the eye has an immense **dynamical range**, being able to function in near dark and at the brightest noon. FECHNER's discovery made the comparison of old brightness records to modern measurements possible. ARGELANDER meanwhile had devised a clever method to determine brightnesses of stars reliably even in the days before the invention of the **photometer**. ARGELANDER's method of **estimation by steps** determines a star's brightness by comparing it to neighboring stars of

---

[107] A light curve of a star is a plot of its brightness as a function of time.

[108] Goodricke also discovered two other variable stars: β *Lyrae* and the famous δ *Cephei*, of which later more.

[109] A *photometer* measures the amount of light coming from a source.

[110] If this late determination of the solar brightness seems outrageous, consider that the modern **germ theory**, that is, the notion that many diseases are caused by microorganisms such as bacteria, replaced the **miasma theory** ("bad air causes disease") only in the last decades of the nineteenth century.

3.9. THE EMERGENCE OF ASTROPHYSICS

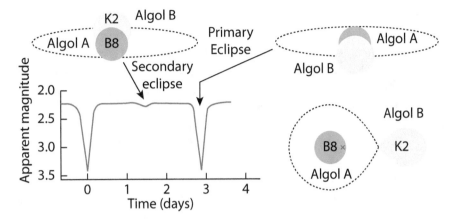

Figure 3.38: **Algol (β Persei) as an example for an eclipsing binary.** The *light curve* is shown in the lower left. The motion of the two components is also shown. When the brighter, blue B star eclipses the dimmer, orange K star, a small dip in apparent brightness occurs. When the brighter star is eclipsed, a larger dip in the light curve results.

slightly different brightness, and ultimately comparing to a set of stars with known brightness. Systematically monitoring stellar brightnesses led to the discovery of hundreds of **variable stars**.

3.9.2 Astrophysics, Spectroscopy, and Astrophotography

Astrophysics is the application of physics (and chemistry) to the study of celestial objects. Loosely speaking, astrophysics asks "What are stars?" rather than "Where are stars?" as the classical branch of astronomy does. The latter hit its high-water mark with BESSEL's discovery of the ***stellar parallax***, and is now known as ***astrometry***. ***Astrophysics*** is based on the assumption that the results and discoveries made in a laboratory on Earth are applicable to the stars. In other words, that physics, chemistry, and science as a whole are truly ***universal***—in the sense that their laws hold everywhere in the universe. We have seen that this anti-Aristotelian[111] view started with NEWTON's breakthrough synthesis and unification of gravitational patterns on Earth and outer space: As the baseball moves, so does the moon.

Like all changes, the emergence of astrophysics was met with resistance and took time to be accepted. Even the great BESSEL as late as 1846 stated that the role of astronomy is to measure the positions celestial objects as accurately as possible—and nothing else. Many false claims as to the impossibility of ever finding out the true nature of stars were made in the early part of the nineteenth century, and strange notions about the sun held sway. For instance, French philosopher AUGUSTE COMPTE (1798–1857), the founder of ***positivism***, stated in 1825 that its chemical composition would never be known,[112] while his compatriot physicist FRANÇOIS ARAGO (1786–1853) held that the sun was habitable. To contextualize these statements, recall that the temperature of the sun was not reliably determined until

---

[111]ARISTOTLE taught the strict separation of Earth and heavens as pertaining to the eternally divided sublunar and superlunar spheres, respectively.

[112]Recall the quote above that astrophysics pioneer ZÖLLNER met "What the stars are we do not know and will never know."

the **Stefan-Boltzmann law** was found in 1879. Even after the advent of **spectroscopy** it was a long way until the composition of the stars would be correctly deduced from the new data. For instance, Kelvin in 1862 stated that it is likely that the sun is made of the same elements as the Earth. After all, many of the Earth's elements such as sodium and oxygen had been found on the sun. To find the **relative abundances** of elements, however, it was necessary to understand the physics of the atoms in much more detail.[113] The great advances of physics and chemistry, in particular the electromagnetic theory of light and the understanding of spectra of the chemical elements and the dark lines of the sunlight, begged for an application on a universal scale—and so astronomy was complemented and expanded by the influx of "terrestrial discoveries."

On the observational side of this expansion we find the aforementioned **spectroscopy**. Anticipated by JOHN HERSCHEL as early as 1823, and started by KIRCHHOFF and BUNSEN in 1860, it quickly became possible to identify chemical elements in the sun and the stars by analyzing starlight falling through a prism and comparing the dark absorption lines with the "fingerprints of the elements," Figure 3.31. Early successes were the discovery of a new element in the sun, aptly named *helium* after the Greek sun god *helios*. It was identified by its characteristic yellow line at wavelength 587.49 nanometers in the spectrum of the **chromosphere** of the sun during a total eclipse. Note that Pierre Jules Janssen (1824–1907) found it as a *bright* line. Indeed, in the chromosphere of the sun elements are excited and **emit** light rather than **absorbing** it, see Section 4.6.[114] While being the second-most abundant element in the universe, helium as a **noble gas** is inert, that is, does not react easily with other substances, and thus escapes detection easily. Helium was finally **isolated** on Earth in 1895 by WILLIAM RAMSAY (1852–1916). Incidentally, Ramsay discovered all other non-radioactive noble gases (argon, krypton, xenon) and therefore expanded the periodic table by a whole new column or **group**.

The spectra of sun and stars with their thousands of dark lines are extremely complex, and it is no wonder that many new patterns were discovered that could not immediately be explained. In a *Herculean* effort most lines were eventually identified and attributed to known chemical elements, while some had to await the advent of *quantum mechanics* to be understood as very improbable but possible transitions between atomic states. It is no surprise, then, that new chemical elements were postulated, some of them prematurely. A case in point is the infamous **nebulium**, an element "identified" through its curious green emission lines in a planetary nebulae (NGC 6543 the "Cat's Eye nebulae") by WILLIAM HUGGINS (1824–1910). Decades later in 1927 these lines were explained as emissions of doubly-ionized oxygen, and nebulium was exorcised and taken off the periodic table. HUGGINS was on to something nonetheless. He and his wife were the first to take spectra of planetary and other nebulae. Seeing emission lines in some, and absorption lines in others, he forged a criterion to distinguish galaxies from other nebulae. Indeed, according to Kirchhoff's laws, Section 3.7.3, emission lines are characteristic of gas clouds, while absorption spectra (on a continuous "rainbow" background) are characteristic of solid objects shrouded by a gaseous layer, such as stars. It thus became clear that the Andromeda nebula with its absorption spectrum was likely made of many but unresolved stars, while the Orion nebula was a hot, glowing cloud

---

[113] CECILIA PAYNE used *quantum mechanical* results of MEGHNAD SAHA to show that the sun is mostly hydrogen and some helium in 1925; see Section 5.3.2.

[114] The existence of the bright line also showed that the chromosphere is *gaseous*, which had been previously been disputed.

3.9. THE EMERGENCE OF ASTROPHYSICS

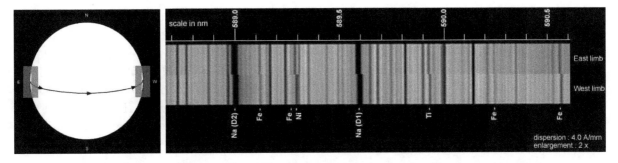

Figure 3.39: **Determining the sun's rotation period using the Doppler effect.** (a) the sun's eastern limb is rotating toward us, so its spectral lines taken through a slit appear at slightly shorter wavelength ("blue-shifted") when compared to the lines' wavelength as measured in the lab or measured through a slit at the western limb. (b) The amount of shifting $\Delta\lambda$ is larger for a faster rotating object. Hence, the rotation period can be computed from $\Delta\lambda$. Note that some of the lines are not shifted. They stem from the Earth's atmosphere.

of hydrogen and helium gas. HUGGINS also realized Doppler's dream by pointing out that the observed **redshift** of Sirius's spectral lines must make it possible to determine its **radial velocity**. Shortly afterwards in 1871 HERMANN CARL VOGEL (1841–1907) used the Doppler effect to measure the rotation of the sun, Figure 3.39. VOGEL went on to use spectroscopy to show that the variable star Algol (β Persei) is a binary star. In 1889 it thus became one of the first **spectroscopic binaries**[115] and since its variability is due to the two components' mutual eclipses, it is also known as an **eclipsing binary**, Figure 3.38. In sum, **spectroscopy** was a completely new way of observing the sky; it opened up another window to the heavens. On the theoretical side, **astrophysics** allowed to explain celestial phenomena with physics known from Earth and the laboratory. As we will see in the next section, astronomers could then try to treat stars like **gas balls**—expanded by their thermal pressure and held together by gravity. As such, they had to obey the laws of Newtonian mechanics and thermodynamics, like the **ideal gas law**, which stipulates relations between the temperature, the pressure, and the volume of a star. Thereby stars became understandable and ceased to be distant, featureless, enigmatic or mystic objects.

Today it seems almost unfathomable that until well into the second half of the eighteenth century, all astronomical observations were made and recorded visually or manually. Measurements of stellar positions, which are just lists of numbers or coordinates, are one thing, but to record the shape of a nebula, star cluster, galaxy, or even a line spectrum as an **eyepiece** sketch seems to us at least imprecise, if not unscientific because *sub-* and not *objective*. No other recording method was available, however, before the development of **photography**. Started by JOSEPH NICÉPHORE NIÉPCE (1765–1833) in 1826 and made popular by LOUIS JACQUES MANDÉ DAGUERRE (1787–1851) starting in 1839, photography found its way into astronomy rather slowly. The first photo of the sun, the moon, and the stars were produced in 1845, 1840, and 1850, respectively.[116] After the invention of the **wet collodion** in 1851, which increased the sensitivity of film between ten to a hundred times, **astrophotography** really

---

[115] Mizar A in *Ursa major* is officially the first one, discovered by PICKERING also in 1889.

[116] JOHN W. DRAPER (1811–1882) succeeded in imaging the moon before the sun because the latter is too bright. For LOUIS FIZEAU (1819–1896) and LEON FOUCAULT's first solar photo, the exposure time had to be limited to one-sixtieth of a second. W.C. BOND (1789–1859) photographed *Sirius* and *Castor* another five years later from the new Harvard Observatory he had established.

took off. Long exposure times required immaculate tracking to counter Earth's rotation, but once again expanded the senses of man. Previously undetectable features like the spiral arms of galaxies and the sevenfold larger extend of the Orion nebula became accessible, recordable, and objectively analyzable. One of the first bonanzas of astrophotography was the field of spectral classification of stars, made possible by taking thousands of photographs of stellar spectra, as we will see after the next section. It all started in 1872, when HENRY DRAPER[117] (1837–1882) took the first spectrum of a star (Vega) that showed absorption lines.

3.9.3 An Earthly Explanation of Stars: The First Stellar Models

We already saw that the development of thermodynamics gave rise to hypotheses concerning the energy production of the sun, like the **Helmholtz-Kelvin contraction hypothesis** of the 1850s. Due to the lack of alternatives, this theory was seen by the physicists for decades as the final word on stellar energy production. Even more so since two of the leading authorities of the physical sciences at the time had authored this hypothesis. Authority is a weak argument in science, but a strong one sociologically. Consequently, research by physicists concerning stellar energy production and the nature of stars was greatly hindered. Progress came from people outside the field [35], and therefore the direction in which our understanding of stars was furthered changed considerably—toward *meteorology*. The study of weather phenomena and the Earth's atmosphere may seem an unlikely ally of stellar astrophysics, but the idea is not that far fetched. First recall the **Kirchhoff laws**, Figure 3.40, which describe the production of spectra: A hot, dense, solid, or liquid material produces a **continuous spectrum**, a cold, dilute gas an **absorption spectrum**, and a hot, dilute gas an **emission** spectrum. Comparing these patterns with the solar spectrum as observed by FRAUNHOFER (dark lines in front of all colors of the rainbow, Figure 3.29) naturally (though not correctly) leads to the hypothesis that the sun is a hot, solid or liquid object producing a continuous spectrum, surrounded by a cooler, gaseous atmosphere responsible for producing the dark lines by absorption of light of specific wavelengths.

The development of the first models of stars as a hot gas ball which is influenced by its own gravity, as well as thermodynamic forces such as pressure, was aided by Kelvin's study of the vertical temperature distribution of Earth's atmosphere. In his work of 1862, Kelvin [38] had treated the atmosphere as a mixture of gases under the influence of Earth's gravity and mixed by **convection**, one of the three[118] major mechanisms of *energy transfer*. You might know **convection** from the **convection oven** in which you can bake a cake or turkey by circulating hot air around the food. The heat is thus transported by the air itself; the molecules of the gas contain (kinetic) energy which is moved toward and the transferred to the food. Another example for convection is the process of boiling water in a pot on a hot stove plate, Figure 3.41. The heat of the plate cannot directly influence the upper layer of water in the pot; the energy gets to the top by the transport of a hot water cell from the bottom where it is thermal contact with the hot plate.[119] This results in a circular rising and sinking of water masses. The same happens in the sun, as evidenced by the **granulation** of the sun's surface first observed by NASMYTH in 1860, Figure 4.24, and the zones and bands

---

[117] He was the son of moon photographer JOHN W. DRAPER.

[118] The others are **thermal conduction** and energy transfer via **electromagnetic radiation**.

[119] Actually, the water is in contact with the metal floor of the pot. The outside of the metal floor is in contact with the hot plate. Within the metal the energy is transferred by **thermal conduction**.

3.9. THE EMERGENCE OF ASTROPHYSICS

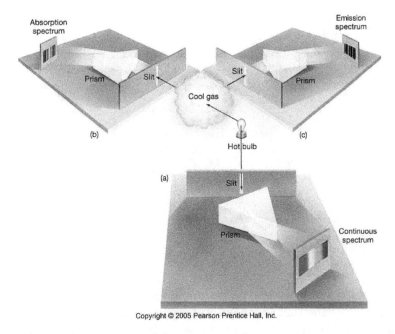

Figure 3.40: **Kirchhoff's laws.** The hot filament of a lightbulb gives of a continuous spectrum of *thermal radiation*. If the light passes through cool gas, part of it gets absorbed. This produces the dark Fraunhofer lines of a *absorption spectrum*. The absorbed radiation is eventually re-emitted and produces an *emission spectrum*.

of Jupiter's atmosphere, Figure 4.18. In the Earth's atmosphere, this mixing by convection happens so fast that the air masses do not have enough time to lose energy to the environment; no heat is exchanged in the process, which nonetheless changes the **thermodynamic state** of the gas. This is known in thermodynamics as an *adiabatic* change of state; see Section 3.8. As an aside, the thermodynamic *state* of a substance such as a gas is described by a set of quantities, for example its pressure, volume, temperature or internal energy. In an adiabatic change no heat is exchanged, but the temperature and other properties of the gas may change. As the hot air rises it cools, becomes denser, and eventually sinks back toward the ground. You can see why this cyclic process is considered a **convective equilibrium**. Naturally, we will encounter the term *equilibrium* often in astrophysics, because we'll have to explain patterns which are stable over exceedingly long times. The only way to sustain such processes for so long is via the *balancing* of forces or mechanisms involved. In other words, while one mechanism might move "things" in one direction, another moves them in the opposite direction, and a stable equilibrium ensues. Paradoxically, an effective "standstill" can ensue while things are moving around and change incessantly—if they do so in a balanced fashion.

Astrophysicists such as ZÖLLNER[120] took **solar granulation** and the observed formation of **sunspots** as evidence for the existence of strong convection currents in the **photosphere**. But they interpreted the **photosphere** as the gaseous atmosphere around a solid or liquid body of the sun. This was in accordance with the contemporary findings. Namely, the density of the sun was known to be higher than that of water, providing (false) evidence for a liquid or solid sun. In line with the astrophysical dogma ("the universe is like the Earth"), it was also

---

[120] Arguably, he was the first astrophysicist.

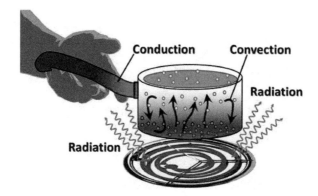

Figure 3.41: **Heat transfer.** Heat can be transferred or transported by **conduction**, **convection**, and **radiation**. Circumstances like the **temperature gradient** determine which will be the dominant transfer process in a specific situation.

(wrongly) assumed that the sun is made of the same substances as the Earth.[121] Moreover, it was unknown at the time that gases under high pressure can—even at high temperature—produce a continuous spectrum. Kirchhoff's laws, Figure 3.40 were taken as meaning that only solid or liquid objects do so.

Rarely is ignorance bliss, but here it may have emboldened J. HOMER LANE (1819–1880) to apply the lessons learned from Kelvin's treatment of the Earth's atmosphere to the *entire* sun, not just its **photosphere**. His was the first model of a star. In it the star appears as an adiabatic sphere of gas subject to its own gravity. Since this model serves as the prototype for basically all future star models, we describe the ideas here in more detail. Our treatment is more general than the original which allows us to refer to it when discussing more advanced models. Unfortunately, it involves some math which hopefully the reader is able follow and develop an appreciation of.

We note that LANE made no attempt to explain where the energy of the star comes from. This is a feature the model shares with many others developed as late as EDDINGTON's famous **stellar standard model** of 1926, described in Section 5.3.1. This pragmatic aspect of **scientific creativity** is not often appreciated. The motto is: If you can't explain it, assume it—and hope that (a) the exact explanation is not essential (that is, your model works) and (b) some future scientist will explain it. Success of an explanation is a strong argument. If it works, there must be something correct about it, right? Alas, not always—but often.

Before we start, we better agree on the "job description" for the star model. What should a stellar model tell us? Surely, we want to know how the mass is distributed within the star. We suspect that it is densest at the center and fluffy close to its surface, but we want a definite mathematical description of this fact. We are looking for the density ρ of the star as a function of the distance r from its center, that is, we want to find the function $\rho(r)$. To have any chance of understanding the energy production of the star (the legacy of the model for future generations of astrophysicists), we have to find the temperature T as it varies within the star. We expect the star to be hottest in its core and already know its temperature at the surface from the **Stefan-Boltzmann law**, but the precise function $T(r)$ is what we are

---

[121]Sodium, iron, magnesium, and other heavy elements are found in the sun's spectrum. Remember, though, that the spectrum is produced by the photosphere; at this point, there was no information available from the deeper layers of the sun.

3.9. THE EMERGENCE OF ASTROPHYSICS

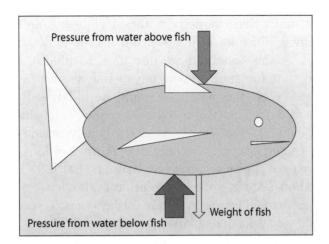

Figure 3.42: **An example for hydrostatic equilibrium.** The fish *floats* in the water, that is, it neither rises higher nor sinks lower. It is in *hydrostatic equilibrium*, because the forces on the fish add up to zero. The upward force of the water pressure from below is exactly as strong as the sum of the downward forces, namely the pressure of the water above the fish plus the weight of the fish.

after. Finally, the pressure p within the star is great insight because it tells us how matter behaves inside a star. In sum, we are looking for three unknown functions $\rho(r), T(r)$ and $p(r)$. Therefore, we have to come up with three equations to find them.

The first equation follows from the essential assumption that the star is stable. It must therefore be in **hydrostatic equilibrium**, see Figure 3.42. If the star is modeled as a "fluffy" gas cloud, it would contract under the influence of its own gravity. But it doesn't. Hence, NEWTON's first law tells us that there must be a balancing force such that the net force on any part of the star is zero. This force is the ***thermodynamic pressure***[122] p. The second equation comes from the properties of the stellar matter itself. Depending on the substance the star is made of, it will behave differently if temperature or pressure changes. For instance, a gaseous star will be compressible, but a solid star will not. We already said that the model is a gas ball, and we might as well reuse our previous knowledge—as scientist do so often—and *assume* it's an ***ideal gas***. So the second equation is the ***ideal gas law***, Equation (3.6).

We saw earlier (Figure 3.33) that substances behave differently if if their ***thermodynamic state*** changes under different conditions, for example constant pressure instead of constant volume. So we have to choose or assume a specific way in which gases are treated in our star model. If it doesn't work out, we might later have to change this or other assumptions to make it work, that is, to produce a model which agrees with observed properties of stars. Again, we'll make our life easier by choosing what others have used successfully, namely the ***adiabatic assumption*** that things happen fast enough such that no heat is exchanged in the process. This is the assumption that Kelvin made in his treatment of the Earth's atmosphere. You can see why it is tempting to follow authorities; we can later blame Kelvin if it doesn't work out. In Figure 3.33 the adiabatic curves 1 and 3 in the ***pressure-volume diagram*** are described by an equation that relates the pressure p of the ideal gas to its density ρ, namely[123]

$$p = C\rho^k. \tag{3.14}$$

---

[122]Recall that pressure is a force per unit area.

[123]This equation can be written as $pV^k = const.$, and this is very similar to the isothermal curves of ***Boyle's law***, Equation (2.39).

Here, C and k are constants that characterize the ideal gas. Note the **universality** of this equation. There are many different ideal gases, but if they undergo an **adiabatic process**, they are *all* described by this equation; we just have to plug in different numbers for C and k. If we believed that the star is made out of hot air, for instance, we would choose $k = 7/5 = 1.4$. However, at this point we would be ill-advised to commit to a specific gas, and we'll leave k as a *free parameter* in the model. The three equations can be put together to obtain an equation which determines the density inside the star as a function of the distance from its center, $\rho(r)$. It is called the **Lane-Emden equation**[124] after LANE and the Swiss meteorologist and astrophysicist ROBERT EMDEN (1862–1940), who in 1907 substantially expanded on LANE's work. Its solution is the density of the star as a function of the distance to the center. Of course, a density always must be positive, and this results in a restriction of possible values for the adiabatic coefficient k. Put bluntly, stars in this model are not stable if the wrong gas is chosen. One interesting prediction of Lane's model is that the equations describing a stable star made out of an ideal gas in **hydrostatic equilibrium** is that its temperature T is inversely proportional to its radius, that is, the smaller the star gets if it contracts, the hotter it becomes. This is interesting, because it affords an intuition for *variable stars*, see Sec. 5.5.4.

To show the predictive power of this crude gas ball model, let's compute the temperature at the center of the sun from the model's assumptions. The rough line of arguments is as follows. If the sun is a gas ball in **hydrostatic equilibrium**, then its gas pressure must be balanced by the force of gravity. In other words, the inward force of gravity provides the pressure at the center. NEWTON's law of universal gravitation can be used to compute the force that a column of star material of cross section A exerts on the center. The force per area is the pressure that exists at the center because the gas column presses down on it. If there is an equilibrium as assumed, then this pressure has to be balanced by an outward "counter-pressure." This pressure is the gas pressure, which is in essence due to the temperature existing at the center, since pressure and temperature are related by the **ideal gas law**. Therefore, the calculation is straightforward:

1. Compute the weight (force) of a column (cylinder) of star gas. This is going to be an expression involving the cross-sectional area A of the column.

2. Translate this into the pressure (force per area) at the center simply by dividing by the (unknown and arbitrary) area A.

3. Figure out how hot the gas must be to produce such a pressure by using a pressure-temperature relation such as the **ideal gas law**.

If we assume a constant gas density, then the center temperature necessary to withstand the immense gravitational pressure is about $T_{center} = 20$ million Kelvin. Compare this to the 5,800 Kelvin of the surface!

Lane's results were independently verified and extended by AUGUST RITTER (1826–1907), who also considered for the first time the consequences of star model *predictions* on the *evolution* of a star's properties. ZÖLLNER [47] in 1865 came up with the idea of the existence of a **stellar life cycle** from **photometric observations** in his lab(!) and of the changing

---

[124] If you need to know, it reads [35]
$$u'' + 2u'/r + \alpha^2 u^n, \tag{3.15}$$
with $u(r) = \rho^{k-1}$, $\alpha = 4\pi \frac{G}{C(n+1)}$, and $k = 1 + 1/n$ defines the **polytropic index** n.

3.9. THE EMERGENCE OF ASTROPHYSICS

brightness of *variable stars*. ZÖLLNER speculated that stars are born hot (blue) and then cool down to become red. This is plausible if one models the stars as a hot piece of iron, and "explains" the different colors of the stars as determined by their age, but in hindsight it is too simplistic a theory. Nonetheless, the **stellar evolution** hypothesis of ZÖLLNER was important and influential, because up to this point stars—and the universe—were assumed to be eternal and unchangeable, a relic of Aristotelian thinking that needed to be purged. In the spirit of ZÖLLNER's speculations, AUGUST RITTER's account of stellar evolution was equally plausible, crude, and (largely) wrong. According to him, stars start their life emitting little energy, but they contract and heat up. This makes them appear bluer and brighter. The contraction must stop when a maximal possible density (or minimal radius) is reached. From this time on they cool off, and become redder. AUGUST RITTER's ideas take into account the observed star patterns—there are bright and dim, red and blue stars—and influenced the next generation of astrophysicists, in particular HENRY NORRIS RUSSELL (1877–1957), the co-inventor of the great tool of **stellar classification**, the **Hertzsprung-Russell diagram**; see Section 5.2.

To summarize, a star model is a set of assumptions that predict the internal structure of a star. Its validity is checked by comparing the model's predictions to the observed properties of stars in outer space. The **Lane model**'s only "prediction" that could be checked at the time was the rather trivial fact that stars are stable, that is, do not collapse. Nonetheless, through LANE and RITTER's work, stars ceased to be abstract, eternal, unchanging, featureless specks of light, and because objects with properties which can be comprehended with the human mind. It was discovered that it is possible to construct models of stars *in the sky* that are based on processes observed *on Earth* and *in the lab*. The work of LANE and EMDEN showed that it might be possible to *understand* the processes inside of stars. The hope was that by measuring the parameters of a star such as its surface temperature, luminosity, spectrum, and mass, the **energy production mechanism** and an understanding of the **stellar life cycle** might be achievable. Indeed, these first and crude ideas about the thermodynamics of stars started a new line of astronomical research, because they implied that heaps of data had to be collected and carefully interpreted. This became the major thrust and prerogative of astrophysics in the last two decades of the nineteenth century.

3.9.4 The Classification of Stars and Their Properties

To find out more about stars, astronomers used the new tool—*spectroscopy*. Unlike names which are just arbitrary labels for stars, Section 1.1.3, and not based on actual properties of a star,[125] the spectrum of a star contains a lot of information. Initially, however, astronomers had no idea what the spectra were trying to tell them. As a first step, they tried to sort them according to a useful scheme. Of course, astronomers first had to agree on what was useful, and so several independent classification schemes were developed originally. For instance, the Italian priest and astronomer ANGELO SECCHI (1818–1878) decided that the broad features of the spectra were important, and proposed as early as 1862 three different groups of stars: yellowish stars like the sun, white stars with strong hydrogen lines like Sirius, and stars with no spectral lines. WILLIAM HUGGINS on the other hand, focused on the line by line differences of the spectra and came up with a different set of categories.

---

[125] An exception is the naming of *variable stars*. Star names starting with two capital letters of the end of the alphabet, like *RR Lyrae* are reserved for variables and therefore signify a physical property.

Spectral Class	Temperature	Color	Example
O	30 000 K	blue	
B	20 000 K	bluish	Rigel
A	10 000 K	bluish-white	Vega, Sirius
F	7000 K	white	Procyon
G	6000 K	yellow	sun
K	4000 K	orange	Aldebaran
M	3000 K	red	Betelgeuse

Table 3.2: **Spectral classes of stars, their surface temperatures and apparent colors.**

The modern classification was eventually developed at *Harvard University* owing to the invention of the **objective prism**—basically a lens shaped prism that is put in front of the telescope and decomposes every star's light into a rainbow of colors, that is, its spectrum. This allowed astronomers to produce hundreds of spectra by taking one photo. In other words, observatories could produce spectral data on an industrial scale. The epicenter of such activities was *Harvard Observatory* under director EDWARD C. PICKERING (1846–1919). To move things along, HENRY DRAPER's widow had bestowed her late husband's estate to fund photography and classification of star spectra in 1886. This enabled PICKERING to acquire an objective prism and "mass produce" stellar spectra. The spectra were of low dispersion but had enough detail to distinguish differences between stars. They were thus ideally suited to generate a classification scheme for stars. Of course, amassing data is one thing, making scientific sense of it is an entirely different endeavor. We are reminded here of the analogous roles that TYCHO's planetary data played for KEPLER's solution of the problem of the planets. To crunch so much data, PICKERING enlisted the help of a dozen female assistants, dubbed the *Harvard Computers*, because they were basically doing what advanced machines would do today: Sift through the data and look for recognizable patterns. While hired principally as cheap labor, several of these women made significant contributions to astronomy and science. In some sense, they were the first professional female astronomers.[126]

The **Harvard classification** of stellar spectra developed in steps. At first pass, the spectra were classified by the strength of their hydrogen lines, Figure 3.31: *A type stars* had the strongest lines, followed by *B type stars*, and so forth. The work was started by ANTONIA MAURY[127] (1866–1952) and others in the 1880s. In 1895 ANNIE JUMP CANNON (1863–1941) reclassified the stars so that the development of spectral lines seemed more orderly over all types. Some spectral classes were found to be redundant, and so the official **classification scheme** has the strange lexicographical order O B A F G K M.[128] It was realized that the order is progressing from blue *O type stars* to red *M type stars*, with yellow *G type stars* like the sun somewhere in the middle. Since this is also the color appearance of a hot piece of iron, it was speculated that the spectral types really specify the **temperature** of a stars. This turned out to be correct, and so three very different pieces of information are redundant: Color, spectral type and temperature are directly related for stars. Red or *M type stars* are

---

[126] The official number one here is CAROLINE HERSCHEL, who received a stipend of £50 p.a. from the British monarch for her observations.

[127] Incidentally, she was the niece of HENRY DRAPER.

[128] A mnemonic to remember this is: *Oh! Be A Fine Girl/Guy, Kiss Me!*

3.10. SUMMARY AND APPLICATION

about 3000 K at the surface, blue or *O type stars* about 30 000 K, so there is a variation in temperature by a factor of 10 or one order of magnitude, Table. 3.2. At the time it was already known that some stars are much more luminous than others, ranging from 10,000 solar luminosities to 1/10,000 of a solar luminosity, which spans 8 orders of magnitude, a factor of 100 million. This strange pattern needed explanation that was not forthcoming for another three decades. Apparently, the energy output of a star is a very sensitive function of its surface temperature! By 1901 the untiring work of MAURY, CANNON, FLEMING, and others had led to the classification of about 1,000 stars. By the time the *Henry Draper Catalogue* of stellar spectra was published in the 1910s, an astonishing quarter million stars had been classified. By then, the classification had been refined to include subgroups with an index ranging from zero to nine. In this way a **spectral type**, say G, is subdivided into ten subtypes: $G0, G1, ..., G9$. The smaller the number, the hotter the star. So $F9$ and $G0$ stars have almost the same temperature. The sun is a $G2$ type star at 5800 K surface temperature. Note that this classification was purely descriptive. It was not known at the time what physical processes produced these spectra—they were simply ordered by appearance. First attempts were made at the time to see a evolutionary or developmental path in the data, as we have seen in Section 3.9.3. The hot O and B stars are sometimes referred to as **early spectral types**, and cool M stars as **late spectral types**. A fine and far-reaching observation was made by ANTONIA MAURY. She noticed that the spectral lines of stars in of the *same* spectral type could vary. Some stars showed well-defined, thin spectral lines, while others had spread-out, fuzzy lines. This was the first hint at the modern, two-dimensional classification scheme known as **Hertzsprung Russell diagram** of stars, which we'll encounter shortly.

Concept Practice

1. Stars are in *hydrostatic equilibrium*. This means that
    (a) ... there is not force acting on them.
    (b) ... they are made out of water (Greek: *hydro = water*).
    (c) ... that all the forces on parts of the star add up to zero.
    (d) None of the above

2. Compare an A *type* with a F *type* star. Which statement is correct?
    (a) The A *type* star is hotter.
    (b) The F *type* star is hotter.
    (c) The A *type* star is dimmer.
    (d) The F *type* star is bluer.
    (e) None of the above

3.10 Summary and Application

3.10.1 Timetable of the Emerging Expanding Universe (II)

The main steps toward an understanding of the expanding universe discussed in this chapter are described below. As always, schedules like this should be taken with a grain of salt.

- 1700–1740 Consolidation of the Newtonian universe

- c. 1740 Falsification of DESCARTES' *vortex theory* by measuring the shape of the Earth as *oblate* (predicted by NEWTON) not *prolate* (predicted by DESCARTES) by two French expeditions; one in the polar, the other in the equatorial region.
- c. 1740 The highly perturbed moon orbit is modeled with NEWTON's theory to an accuracy that allows for nautical longitude navigation (CLAIRAUT, MAYER).
- 1759 Return of Halley's comet exactly as predicted by NEWTON's theory (HALLEY, CLAIRAUT)

- 1728 Discovery of stellar aberration (BRADLEY). It is due to the finiteness of the speed of light (RØMER) and **direct evidence** for the motion of the Earth around the sun (COPERNICUS).
- 1748 Discovery of the **nutation** of Earth's axis (BRADLEY) leads to modeling the Earth as a spinning top (EULER 1760).
- 1761/69 The **astronomical unit** (AU) aka the distance to the sun is measured reliably using Venus transits observed by teams at two remote locations; CAPT. JAMES COOK oversees the 1769 measurement overseas (*Tahiti*) and discovers *Australia* while he is at it in 1770.
- 1770–1780 ARISTOTLE's teaching of four elements is replaced by their modern chemical/physical equivalents (LAVOISIER, PRIESTLEY); *air* is a mixture of "airs" (**gases**), *water* can be decomposed into hydrogen ("water former") and oxygen ("acid former"), *fire* is reinterpreted as **heat**, which is realized to be different from **temperature**.
- 1781 Uranus is discovered (W. HERSCHEL); the universe is extended beyond Saturn.
- 1755/1796 The formation of the solar system is described in terms of the **nebular hypothesis** (KANT, LAPLACE), which posits that the sun and planets were formed as a large, spinning gas cloud flattened and contracted; a universe "before the Earth" becomes conceivable.
- 1798 Measurement of the gravitational constant G in NEWTON's law of universal gravitation (CAVENDISH) allows for a determination of the Earth's mass ($M_E \approx 10^{24} kg$).
- 1801 Light is found to be a wave (YOUNG); two beams of light show interference patterns like **waves** rather than colliding like **particles** would do.
- 1812 Dark lines are found in sunlight sent through a prism (WOLLASTON/FRAUNHOFER). Later lines are identified in starlight.
- 1834–1838 The southern sky is cataloged by J. HERSCHEL, extending the universe to far below the celestial equator.
- 1838 The distance to the stars is measured with the **stellar parallax** (BESSEL).
- 1842 The Doppler effect is discovered; it pertains to both acoustic and electromagnetic waves such as light.
- c. 1850 **Energy** is found to be **conserved** (MAYER, HELMHOLTZ). It is convertible into many forms, but cannot be created or destroyed. The total amount of energy of the universe is thus a fixed number.
- 1854 **Entropy** is defined by CLAUSIUS. Its value will inevitably and irrevocably increase as time progresses, leading to an eventual **heat death** of the universe in the distant future (Kelvin).

3.10. SUMMARY AND APPLICATION

- 1848 The *meteorite hypothesis* (MAYER) is the first theory of energy production of the sun; it is falsified 1854 (KELVIN) because it predicts a decrease of the length of the year by a few seconds per year.

- 1854 The *Helmholtz-Kelvin contraction hypothesis* predicts an age of sun and therefore the universe of up to 20 million years in stark contrast to common and religious sense and sensibility.

- 1860 *Spectral analysis* is borne out of the realization that dark lines identifying chemical elements can be observed far from the signal's source (KIRCHHOFF, BUNSEN); it is realized that the dark lines in solar and stellar spectra are of the same wavelength as the bright lines of heated chemical elements.

- 1864/1888 The theories of electricity and magnetism are unified (MAXWELL); the universal theory of *electromagnetism* predicts that wave solutions are possible and travel at the speed of light; visible light is but a small wavelength interval of the vast electromagnetic spectrum; long-wavelength electromagnetic waves are produced and received in the lab (H. HERTZ).

- 1840-1872 Beginnings of *astrophotography*: the moon (1840), the sun's corona (1851), the star Vega (1850) and its spectrum (1872) are photographed.

- 1879 The amount of radiation given off by an idealized body (blackbody radiation) is found to be proportional to its temperature to the fourth power, T^4 (STEFAN, BOLTZMANN); the surface temperature of the sun is thus determined to be $T_{Sun} \approx 5,800K$.

- from 1886 Mass production of stellar spectra by invention of the *objective prism* and using *astrophotography* (PICKERING ET AL.).

- 1890–1901 Classification of stellar spectra (MAURY, CANNON).

- 1870–1907 First phase of an emerging understanding of stars; stars are hot gas balls, that is, thermodynamic, self-sustaining, energy-radiating "machines." (LANE, RITTER, EMDEN)

3.10.2 Concept Application

1. Calculate the power radiated by a star "A" of 3000 K and half the sun's radius and a star "B" of 6000 K and a quarter of the sun's radius.

    (a) In *SI units*, that is, in Watts.

    (b) As a fraction of the solar power or luminosity.

    (c) Comment on the results in light of the fact that star B is twice as hot and half the size of star A.

3.10.3 Activity: Electromagnetic Waves

Light is an electromagnetic wave. As such, it shares properties with familiar types of waves such as the wave on a vibrating string, sound in air and waves on water. The main equation states that the speed of a wave is the product of its wavelength and its frequency: $c = \lambda \times f$. The speed of light is $c = 300.000$ km/s.

Part I: Fundamental relations

1. Calculate the frequency of red light of wavelength 600 nm. If you use scientific notation ($c = 3 \times 10^8$ m/s, $\lambda = 6 \times 10^7$ m) you should be able to do this without a calculator.
2. How would the frequency change if the wavelength would be half as long?
3. How would it change if the wavelength would be twice a long?
4. Rank the speeds of the electromagnetic waves in 1–3 which have different wavelengths, and therefore different frequencies.
5. Formulate a rule (in words) that states the relationship between frequency and wavelength of an electromagnetic wave.
6. Compare red light with a radio wave. Which one, if any, has the higher frequency, the longer wavelength, higher speed, and higher energy?

Part II: Brightness and distance

The brightness of an object depends on how much light it sends out (luminosity L) and the distance between the object or light source and the observer (radial distance r). Brightness is a measure of how bright, say, a star *appears* in the sky, while luminosity is a measure of how much light or energy a source *sends out*—regardless of its distance. Luminosity is measured in Watts (1W = 1 Joule/sec), while brightness is measured in Watts per square meter (W/m²) [129]. The fundamental relation between brightness, luminosity and distance is

$$B = \frac{L}{4\pi r^2}.$$

7. You are in a big, dark room and view a 400 W light bulb from a distance of 10 meters. At this point, which of the following do you know? Brightness, luminosity, distance.
8. A second light bulb is placed next to 400 W light bulb. It appears to be four times dimmer than the 400 W light bulb. What can you conclude?
9. A third light bulb is placed somewhere in the room. It also appears four times dimmer than the 400 W light bulb. What can you say about its distance and luminosity?
10. List two possible luminosity/distance combinations that would save the appearances, that is, reproduce the brightness pattern described in 9.

Part III: Doppler Effect

11. Light is sent out from a source moving towards an observer in Figure 3.28. Four circular wave fronts have been drawn. Copy and enlarge the figure on a separate sheet. Draw in the positions of the source at the four instances when the individual wave fronts were sent out (Hint: they are at the center of the circles.)

---

[129] Astronomers often use the so-called magnitude scale to measure brightnesses of stars, in which a brighter star has a smaller magnitude number, but we'll use brightness B instead.

3.10. SUMMARY AND APPLICATION

12. Determine the "true" wavelength, that is, the wavelength sent out by the source.
13. Explain why the observer measures a different wavelength than the one the source sends out.
14. An observer left of the source would measure a longer wavelength. How can two observers disagree on the result of a scientific experiment?
15. For which observer is the wave red-shifted?
16. What can you say about the velocity of the source? How does it relate to the shift in wavelength?
17. What happens when the speed of the source is as high as the velocity of the wave? Can it be even higher?

3.10.4 Activity: Spectra and Blackbody Radiation

Everything radiates! Literally every *body* sends out electromagnetic waves of many different wavelengths. There are several laws of physics connected to this phenomenon of **blackbody radiation**, since there are several interesting questions we can ask. The amazing thing is that the amount and exact distribution of waves sent out depends on just two things: the temperature T of the body and its surface area A. Nothing else matters, and so objects as different as a hot star, the human body, and an ultra-cold gas cloud in outer space send out blackbody radiation obeying the same laws. **Kirchhoff's laws** tells us that every solid or liquid body sends out a continuous spectrum, that is, waves of all kinds of wavelengths. The **Stefan-Boltzmann law** tells us *how much* power or energy per time the object radiates. **Wien's law** tells us at which wavelength the object shines brightest. Let's look at these laws in detail.

Part I: Kirchhoff's Laws: These three laws listed in Sec. 3.7.3 and depicted in Figure 3.40. Describe the different spectra encountered in Nature.

1. Describe what we mean be the term "spectrum".
2. What does the spectrum of a hot, solid object look like? How do we call such a spectrum?
3. What does the spectrum of a hot dilute gas look like?
4. What does the spectrum of a hot solid object look like when it is viewed though a cold, dilute gas cloud?

Part II: Stefan-Boltzmann law: This law states that the power radiated by an object is $P = \sigma A T^4$. For our purposes, we can identify the power radiated with the **luminosity** L of the object. The value of the Stefan-Boltzmann constant σ is of no interest to us since we only need to understand *relative* luminosities.

5. Compare two stars of same size, star A being three times hotter than star B. Which star is more luminous and by which factor?
6. Compare two stars C and D of same temperature, with star C being half the size of star D. Which star is more luminous and by which factor?
7. Can two stars of different temperature have the same luminosity? How or why not?

Part III: Wien's law: The distribution of wavelengths an object sends out has a universal shape which is called the Planck curve, see Figure 3.35(a). The special features of the curve are not important to us, but the general features are rather obvious. Indeed, GUSTAV KIRCHHOFF figured them out in 1859 by pure speculation, without any specific data. An object will not send out very short nor very long wavelengths, so the curve must start and end at zero. This means it must rise up to a maximum, and then decrease. That's all we need. There is a peak, and since wavelengths can be arbitrarily long, the peak cannot be in the middle; it sits off center. **Wien's law** tells us where it is. It states that the peak wavelength λ_{peak} (the wavelength at which the object sends out most light) multiplied by the object's temperature T is the same for any object. This product is equal to Wien's constant $k_{Wien} = 0.0029 \text{m} \cdot \text{K}$ (meter times Kelvin): $T \times \lambda_{peak} = 0.0029 \text{m} \cdot \text{K}$.

8. Figure out *your* peak wavelength.
9. What do we call electromagnetic waves with this wavelength?
10. The sun is 20 times hotter than you at its surface. What is its peak wavelength?
11. What do we call electromagnetic waves with this wavelength?

Image Credits

Fig. 3.2: Source: https://commons.wikimedia.org/wiki/File:Problem_of_longitude.svg.

Fig. 3.6a: Copyright © 2004 by User Herbye, (CC BY-SA 3.0) at https://commons.wikimedia.org/wiki/File:Praezession.svg.

Fig. 3.10a: Source: https://commons.wikimedia.org/wiki/File:Orion_Nebula_-_Hubble_2006_mosaic_18000.jpg.

Fig. 3.10b: Copyright © 2011 by Ken Crawford, (CC BY-SA 3.0) at https://commons.wikimedia.org/wiki/File:Barnard_33.jpg.

Fig. 3.11a: Source: https://commons.wikimedia.org/wiki/File:Pleiades_large.jpg.

Fig. 3.11b: Copyright © 2016 by KuriousGeorge, (CC BY-SA 4.0) at https://commons.wikimedia.org/wiki/File:M13_from_an_8%22_SCT.jpg.

Fig. 3.12a: Source: https://commons.wikimedia.org/wiki/File:M57_The_Ring_Nebula.JPG.

Fig. 3.12b: Source: https://commons.wikimedia.org/wiki/File:Crab_Nebula.jpg.

Fig. 3.13a: Copyright © 2010 by Adam Evans, (CC BY 2.0) at https://commons.wikimedia.org/wiki/File:Andromeda_Galaxy_(with_h-alpha).jpg.

Fig. 3.13b: Source: https://commons.wikimedia.org/wiki/File:Messier_87_Hubble_WikiSky.jpg.

Fig. 3.15a: Source: https://commons.wikimedia.org/wiki/File:40_foot_telescope_120_cm_48_inch_reflecting_telescope_William_Herschel.png.

Fig. 3.15b: Source: https://commons.wikimedia.org/wiki/File:BirrCastle_72in.jpg.

Fig. 3.16a: Copyright © 2005 by Anton commonswiki, (CC BY-SA 3.0) at https://commons.wikimedia.org/wiki/File:Radiantrp.jpg.

Fig. 3.16b: Source: https://commons.wikimedia.org/wiki/File:Herschel-Galaxy.png#/media/File:Milky_way.jpg.

Fig. 3.17: Adapted from: http://coolcosmos.ipac.caltech.edu/cosmic_classroom/classroom_activities/herschel_experiment2.html

Fig. 3.18a: Source: https://commons.wikimedia.org/wiki/File:Whirlpool_by_lord_rosse.jpg.

Fig. 3.18b: Source: https://commons.wikimedia.org/wiki/File:Messier51_sRGB.jpg.

Fig. 3.19a: Copyright © 2006 by Bob Mellish, (CC BY-SA 3.0) at https://commons.wikimedia.org/wiki/File:Chromatic_aberration_lens_diagram.svg.

Fig. 3.19b: Copyright © 2006 by Stan Zurek, (CC BY-SA 3.0) at https://commons.wikimedia.org/wiki/File:Chromatic_aberration_(comparison).jpg.

Fig. 3.19c: Copyright © 2010 by DrBob, (CC BY-SA 3.0) at https://commons.wikimedia.org/wiki/File:Lens6b-en.svg.

Fig. 3.20a: Source: https://commons.wikimedia.org/wiki/File:Ole_R%C3%B8mer_at_work.jpg.

Fig. 3.20b: Copyright © 2006 by Heinz-Josef Lcking, (CC BY-SA 3.0) at https://commons.wikimedia.org/wiki/File:Mural_Quadrant_-_by_John_Bird_-_London_1773.jpg.

Fig. 3.20c: Source: https://commons.wikimedia.org/wiki/File:Groombridge_transit_circle.jpg.

Fig. 3.20d: Oliver Lodge, "Heliometer," Pioneers of Science, pp. 312, 1893.

Fig. 3.21a: Copyright © 2014 by Krishnavedala, (CC BY-SA 4.0) at https://commons.wikimedia.org/wiki/File:Gregorian_telescope.svg.

Fig. 3.21b: Copyright © 2014 by Krishnavedala, (CC BY-SA 4.0) at https://commons.wikimedia.org/wiki/File:Cassegrain_Telescope.svg.

Fig. 3.22: Source: http://www.atnf.csiro.au/outreach/education/senior/astrophysics/binary_types.html.

Fig. 3.23: Copyright © 2012 by Stannered, (CC BY-SA 3.0) at https://commons.wikimedia.org/wiki/File:Ebohr1_IP.svg.

Fig. 3.25: Source: http://images.slideplayer.com/27/9251952/slides/slide_6.jpg

Fig. 3.27a: Copyright © 2007 by ploufandsplash, (CC BY-SA 3.0) at https://commons.wikimedia.org/wiki/File:Onde_electromagn%C3%A9tique.png.

Fig. 3.27b: Copyright © 2007 by Inductiveload,NASA, (CC BY-SA 3.0) at https://commons.wikimedia.org/wiki/File:EM_Spectrum_Properties_edit.svg.

Fig. 3.28: Copyright © by IB Physics Stuff, (CC BY-SA 3.0) at http://ibphysicsstuff.wikidot.com/doppler-effect.

Fig. 3.29: Source: https://commons.wikimedia.org/wiki/File:Fraunhofer_lines.svg.

Fig. 3.30: Source: https://commons.wikimedia.org/wiki/File:Kirchhoffs_first_spectroscope.jpg.

Fig. 3.31: Source: http://quantumchemistryinnovative.blogspot.com/2011/01/emission-spectra-of-some-elements.html.

%noindent Fig. 3.32a: Source: https://commons.wikimedia.org/wiki/File:Carnot_he

Fig. 3.32b: Source: https://commons.wikimedia.org/wiki/File:Watt_steam_pumping_engine.JPG.

Fig. 3.35a: Source: https://commons.wikimedia.org/wiki/File:Black_body.svg.

Fig. 3.35b: Copyright © 2006 by Sch, (CC BY-SA 3.0) at https://commons.wikimedia.org/wiki/File:EffectiveTemperature_300dpi_e.png.

Fig. 3.36: Source: https://commons.wikimedia.org/wiki/File:Atmospheric_electromagnetic_opacity.svg.

Fig. 3.37a: Source: https://commons.wikimedia.org/wiki/File:Human-Visible.jpg.

Fig. 3.37b: Source: https://commons.wikimedia.org/wiki/File:Human-Infrared.jpg.

Fig. 3.38: Adapted from: https://freestarcharts.com/algol-eclipse-dates-times-march-2017.

Fig. 3.39a: Source: http://www.eyes-on-the-skies.org/shs/spec-rot-gb.htm.

Fig. 3.39b: Source: http://www.eyes-on-the-skies.org/shs/spec-rot-gb.htm.

Fig. 3.40: Copyright © 2005 by Pearson Prentice Hall.

Chapter 4

Interlude: Discovering the Solar System

This chapter is an *interlude* in at least two senses. For one, we interrupt our chronological tale of science and go back about a century; from the emergence of astrophysics in the late nineteenth century we regress to the late eighteenth century, when additional objects in the solar system were being discovered. Secondly, the material presented in this chapter is not tightly linked to the book's main storyline, namely the construction of the expanding universe. Nonetheless, there are some loose ends we better tie up before moving entirely beyond the solar system and deeper into the cosmos.

Guiding Questions of this Chapter

4.1 How did astronomers discover different types of objects in the solar system, and how can they determine their properties?

4.2 What can we learn about the Earth and the moon that teaches us about solar system objects much farther away?

4.3 What are the common properties of the four inner or *terrestrial* planets? How do they differ?

4.4 What are the common properties of the four outer planets? How do they differ from the terrestrial planets?

4.5 How did the solar system form? Is there a way to test the formation hypothesis?

4.6 What are the properties of the sun? What do they tell us about the stars which are more than 100,000 times farther out?

4.1 Introducing the Solar System

4.1.1 Timeline of Discoveries

For the longest time, the solar system consisted of eight objects: Earth, moon, sun, and five planets. This notion goes so far back that the latter seven became the namesakes of the days of the week, as we discovered in Section 2.1. Starting with GALILEO in 1610, moons of planets were discovered regularly, but it came as quite a surprise when W. Herschel serendipitously found Uranus in 1781. Arguably, Johann Daniel Titius (1729–1796) and Johann Elert Bode[1] (1747–1826) anticipated the discovery of further planets when they found a **numerological law**[2] in 1766/1772 for the semi-major axes a of planets from the sun

$$a_n = (0.4 + 0.3 \times 2^n)\,\text{AU},$$

where $n = -\infty, 0, 1, 2, 3, 4, 5\ldots$. This reproduces the distances of Mercury, Venus, Earth, Mars, Jupiter, and Saturn well, but $a_3 = 2.8\,\text{AU}$ is an unfilled spot. Scientific thought had emancipated from the Pythagorean dreams of Kepler's *Mysterium Cosmographicum* and did not take the numerology of Titius and Bode too serious, but the law provided some motivation to look for further planets. Bode went so far to form the so-called "celestial police" in 1800 to search for the missing planet(s). Unaware of this, Giuseppe Piazzi on New Year's 1801 found an object, *Ceres*, that fit the bill. Orbiting the sun at a distance of 2.77 AU once in 4.6 years, that is, between Mars and Jupiter, Ceres had been overlooked for several reasons. For one, it is so dim that it cannot be seen with the unaided eye. This already tells us that it must be much smaller than the other planets. Also, its orbit is not in line with the other planets', in that it is tilted over 10° with respect to the *ecliptic*. All other planets except Mercury (7°) do not deviate more than 3.5° from Earth's orbital plane. Nonetheless, it was regarded as a planet, because it did not orbit any known planet—as moons would. Astronomers were therefore somewhat startled when Heinrich Olbers (1758–1840) discovered another planet in 1802, *Pallas*, with virtually the same orbit ($a_{Pallas} = 2.77\,\text{AU}$) and yet another one in 1807, *Vesta*, with a similar orbit ($a_{Vesta} = 2.36\,\text{AU}$). In between, *Juno* was found in 1804 with $a_{Juno} = 2.67\,\text{AU}$. No other discovery was made for 38 years, and it became widely accepted that there were exactly four planets between Mars and Jupiter. Then in December 1845, *Astraea* was found with $a_{Astraea} = 2.57\,\text{AU}$, starting an avalanche of new discoveries. Indeed, ten years later no less than 37 of these objects were known, and it became clear that they were not planets, but a new group of objects. Hence, they were given the new label **asteroid**.[3]

Even a new planet, *Neptune*, had been predicted and observed within a year of *Astraea*'s discovery. This was a triumph of **celestial mechanics**, as the new planet's existence was **deduced** from the Newtonian universal **theory of gravity**. Indeed, *Uranus* had been deviating further from its predicted orbit than could be explained by anything but the gravitational attraction of another, as of yet undetected, large planet beyond Uranus' orbit. What's more, the accurately measured deviation allowed two astronomers, Urbain Le Verrier (1811–1877) and John Couch Adams (1819–1892) to predict the unseen planet's position accurately. We have to cut short the exciting story of the eventual discovery,[4] but the Frenchman Le Verrier succeeded convincing the director of the Berlin observatory in Germany, Johann

---

[1] Bode determined the orbit of Uranus following Herschel's observations and gave the planet its name. Herschel's name *Georgius Sidus* didn't stick—Britain's King George III had just lost the American War of Independence.

[2] That is, a rule which is based on nothing but numbers, characteristically lacking physical insight.

[3] The rate of discovery kept increasing. For instance, at the Heidelberg Observatory under Max Wolf around the *fin de siècle*, an astounding 800 of them were discovered. This number is dwarfed, however, by the approximately 600,000 asteroids known today.

[4] The story is telling for the climate for astronomical research in the three countries involved, namely Britain, France, and Germany.

4.1. INTRODUCING THE SOLAR SYSTEM

Figure 4.1: **Martian rotation.** Mars rotates as evident by the changing surface features visible in these photos of the 2005 opposition taken with a digital camera through a modest 8″ telescope.

GOTTFRIED GALLE (1812–1910), to look out for it by sending him a letter with his predictions. Receiving the letter in the morning, GALLE and his assistant HEINRICH LOUIS D'ARREST[5] (1822–1875) found *Neptune* less than one degree or two lunar diameters off the predicted position that night.

Also *Neptune* deviated from its prescribed orbit, and the hunt was on for yet another planet, even farther from the sun. The story of the discovery of *Pluto* is far less glamorous and certainly not a showpiece for celestial mechanics or science. When *Pluto* was finally found in 1930, it was more the incredible dedication and endurance of CLYDE W. TOMBAUGH (1907–1997) than the vague prediction of its whereabouts that secured the last discovery of a planet to date.[6]

4.1.2 Investigating the Solar System

Most of what we know about the planets today comes from space probes sent to all major planets plus Pluto. The amount of data and details these space missions have given us is overwhelming. Moreover, much of it must seem like an endless list of factoids for the novice unfamiliar with the principles of the physical sciences that tie these measurements and observations together. Our task here is to understand how scientific results are obtained, rather than focusing on the results themselves. For these reasons, we will highlight what can be said about the planets from afar, and refer the reader interested in a detailed account of the planets to the literature, for example [13, 36].

It is amazing how much we can find out out the planets by analyzing the light they send to us which is, after all, reflected sunlight. Here is a list of properties that are accessible via careful observation. We are assuming here that the orbits of the planets are known. Recall that the determination of the planetary orbits by KEPLER and NEWTON was the solution of the ***problem of the planets***—and the key result of Chapter 2.

- *Mass.* As we saw in Section 2.10, the mass of an object can be deduced from the orbits of its satellites via Newton's gravity law. For instance, we had found that the mass of the Earth is over 300 times smaller than that of Jupiter, and that Jupiter is 1,000 times lighter than the sun: $M_{Jupiter} \approx \frac{1}{1,000} M_{sun} = 0.001 M_{sun}$. Incidentally,

---

[5]Of French Huguenot descendancy.

[6]The new 2006 criterion for "planethood" is more stringent and led to Pluto's reclassification as a ***dwarf planet***.

this is why we were justified in the last section to claim that the sun incorporates over $99.99\% = 100\% - \frac{1}{1,000}$ of the mass of the solar system.

- *Size.* The diameter of a planet can be obtained from its apparent or **angular size** once the distance to the planet is known, see Section 1.2.1 and Figure 1.8(b). For instance, Jupiter at opposition is $5.2\,\text{AU} - 1.0\,\text{AU} = 4.2\,\text{AU} = 4.2 \times 149.6$ million km $=$ 630 million km away, and its angular diameter in the night sky is $42''$. Therefore, its actual diameter is 630 million km $\times \sin\left(\frac{42°}{60\times 60}\right) \approx 142\,000$ km $\approx 11 R_{Earth}$. Since Jupiter is the prime example for a **Jovian planet**, and Earth is the prototypical **terrestrial planet**, we have to conclude that Jovians are a whole order of magnitude larger, and therefore have a volume that is $10^3 = 1,000\times$ bigger than a typical terrestrial planet's.

- *Density.* The average mass per unit volume is the average density or simply *the* density of a planet. We suspect that the planets are not homogeneous; they are probably denser at their center than they are near their surface. However, it is hard to assess this **internal structure** from a distance. Still, even the average density tells us a lot. For instance, by dividing the mass of a Jovian planet by its volume we find its density ρ which is much lower than for a terrestrial planet. To wit,

$$\rho_{Jupiter} = \frac{M_{Jupiter}}{V_{Jupiter}} = \frac{2.0 \times 10^{27}\,\text{kg}}{\frac{4\pi}{3}(142\,000\,\text{km})^3} = 1300\,\frac{\text{kg}}{\text{m}^3} = \frac{318 M_{Earth}}{1,000 V_{Earth}} \approx \frac{1}{3}\rho_{Earth}.$$

That is, the Earth is made out of material that is three times denser than the material Jupiter consists of; Jupiter must be made of very light material. Comparing to the sun, $\rho_{sun} = \frac{2.0\times 10^{30}\,kg}{\frac{4\pi}{3}(700\,000\,\text{km})^3} = SI1400\frac{kg}{m^3}$, we come to suspect that Jupiter and the sun might be made of the same, light material. This is puzzling, because Jupiter is a planet, and the sun is not. On the bright side, patterns which appear counterintuitive carry the promise of learning a great deal about the universe.

- *Rotational period.* The length of a "day" on a planet or the sun, aka their **sidereal rotation period**, can be determined in several ways. If there are **surface features** then we can measure the time it takes for them to cross the face of the celestial object. We described how GALILEO used sunspots to measure the rotation period of the sun in Section 2.8; see also Figure 4.1. Another method to determine the rotation period is via the **Doppler shift of spectral lines**. Since one limb of a rotating object comes toward us and the other recesses, we can see a shift of the spectral lines if we take a spectrum of just the eastern or just the western limb, for example of the sun, Figure 3.39. The latter method is much more accurate, because lineshifts can be measured easily and accurately. Determining the rotation of the objects in the solar system yields the following pattern.

    1. *Mercury and Venus rotate extremely slowly.* Mercury's day is almost 59 Earth days, and therefore exactly two-thirds the length of the Mercury year! This cannot be an accident. Indeed, this 2:3 ratio of the two rotation periods can be explained by the tidal forces of the sun which over the eons have resulted in this **synchronous rotation**. Recall that the tidal interaction between Earth and moon has resulted in a 1:1 ratio of the moon's rotation period and its orbital period around the Earth. The Venus day lasts 243 Earth days and is longer than the Venus year of 225 Earth

4.1. INTRODUCING THE SOLAR SYSTEM

days. Also, Venus is the sole planet to rotate clockwise, so the sun on Venus rises in the west and sets in the east! It is neither known *why* Venus rotates **retrograde** nor why it rotates so excessively slowly.

2. *Earth and Mars have very similar rotation periods.* The day on Mars is about one hour longer than on Earth. It is tempting to consider this to be the typical rotation period of a terrestrial planet.

3. *The Jovian planets rotate much faster although they are much bigger.* Indeed, Jupiter and Saturn rotate in about 10 hours—more than twice as fast as the Earth. Uranus and Neptune rotate in less than 17 hours, much faster than the Earth.

- *Rotational axis and seasons.* We saw that the orbital planes of the planets are hardly inclined with respect to each other. From the side, the solar system looks flat; it is disklike, not spherical, see Figure 4.2. Contrariwise, the inclinations of the planets' axes are all over the place. We saw in Chapter 1 that the Earth's equator inclination of 23.5° with respect to the **ecliptic(orbital) plane** is the **reason for the seasons**, because it results in the sun being high in sky at noon in summer, and lower in the sky at noon in winter. We find that Mars, Saturn and Neptune have similar axis tilts, whereas Mercury, Venus and Jupiter have hardly any tilt,[7] and therefore no seasons. This leaves Uranus as the oddball. It's equator is tilted nearly maximally (90°), so that it exhibits extreme seasons. Its one hemisphere experiences forty years of uninterrupted sunshine, whereas the other has a forty-year night.

- *Atmospheres.* The existence of atmospheres is hard to observe from Earth, and usually required fortuitous circumstances, like the discovery of Venus's atmosphere by observing an arc of light during the Venus transit of 1761 by Mikhail Lomonosov from St. Petersburg observatory. We can glean the **composition** of planetary atmospheres by analyzing the light we receive through a prism. Of course, we have to be careful to distinguish between the absorption lines stemming from the sun, the planet, and the Earth's atmosphere, as the light produced by the sun is reflected by the planet, and travels through the Earth's air before it reaches our telescopes. In practice, the situation is often unfavorable for the detection and analysis of a planet's atmosphere. For instance, the fact that Venus's atmosphere is predominantly carbon dioxide was positively confirmed only in 1967 by a Soviet space probe (Venera 4), and as late as 2001 a new spectroscopic method to study planets' atmospheres was discovered and used for the first time during the 2004 transit of Venus in front of the sun.

- *Surface Temperature.* Also this can be tricky to figure out. For planets like Venus—always shrouded in thick cloud layers—the surface temperature was first *calculated* by taking into account atmospheric gases and the resultant **greenhouse effect** (RUPERT WILDT, 1940), and only later confirmed by space probes (*Mariner 2*, 1967).

We see that these observations yield quite a few patterns to be explained. For instance, a major effort has to go into explaining why there are exactly two types of planets with pretty much opposite properties, and why the Jovian planets are much farther from the sun.

---

[7]Technically, Venus's tilt is close to 180° because its rotation is retrograde.

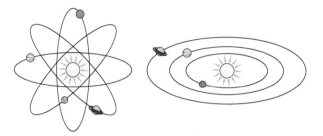

Figure 4.2: **The solar system is flat.** Two possible configurations of orbital planes. (a) Randomly oriented orbital planes. (b) Aligned orbital planes as found in our solar system.

4.1.3 The Asteroids as Clues to Solar System Formation

Between Mars and Jupiter lies the **asteroid belt**, whose existence in the form of millions of small bodies poses a question. Namely, in light of the **Kant-Laplace formation theory** of the solar system, we might come up with the hypothesis that the asteroids are a "failed planet." It seems plausible that the gravitational influence of nearby Jupiter disrupted the formation of this "planet." It is easy to falsify this hypothesis, because if true, the mass of the asteroids should add up to the mass of a planet, even if as small as Mercury. The number of asteroids is very large, but their sizes and therefore masses decline sharply with their **asteroid number**.[8] This makes sense, because by and large, bigger objects are brighter and therefore easier to spot. Judging from Figure 4.3, we can neglect the masses of asteroids with a number greater than 10 or so,

$$M_{allAsteroids} = \sum_{allAsteroids} m_i = \sum_{first\ few} m_i + \text{small rest}.$$

Even then, we have a problem: How do we determine the masses of the first 10? For the planets, we were able use the orbits of their moons, but unlike 243 *Ida* sporting moon *Dactyl*, very few asteroids have moons. As always, it is better to use a crude assumption than to do nothing, so we'll assume that the asteroids are rocky as the terrestrial planets, and of spherical shape. As a lower limit, we use the Martian density, lowest among the terrestrials, of about $3900\ kg/m^3$ to calculate the asteroid masses. Now, this is still a lot of work, so we use the fact that 1 *Ceres* is about twice as big, and therefore $2^3 = 8$ times more voluminous and therefore massive, as the next largest, 2 *Pallas* and 4 *Vesta*, to conclude that the first 10 asteroids will not have much more than twice Ceres' mass, which is according to our assumptions is

$$M_{Ceres} = (Volume) \times (Density) = \frac{4\pi}{3}(450,000\,m)^3 \times 3,900\frac{kg}{m^3} = 1.5 \times 10^{18} = \frac{M_{Mercury}}{222,000}.$$

Hence, we would need over 200,000 Ceres-sized asteroids to come up with a mass comparable to Mercury's—theory falsified!

What, then, is the significance of the hundreds of thousands of asteroids known today? What do they tell us? Clearly, the solar system contains more matter than just the planets. Probably the asteroids formed in a way similar to the planets, and they almost certainly formed at the same time as the planets. Therefore, studying asteroid material gives us a

---

[8] The asteroid number reflects the order of discovery. Hence, *1 Ceres* was the first to be seen, and *25143 Itokawa* was the 25143th to be detected.

4.1. INTRODUCING THE SOLAR SYSTEM

Figure 4.3: **Asteroids.** Portraits of asteroids taken by space probes. Note the extreme difference in size.

clue about the formation processes and the age of the solar system. If you think about this statement, you might have two major objections. First, how can we study **asteroid material** at all? After all, we are "down here", and the asteroids are "up there." Relatedly, wouldn't it be much easier to study "Earth material" to find out about the age of the solar system? After all, the Earth formed at the same time as the solar system. The problem with the latter suggestion is that while the Earth as a planet did form concurrently with the rest of the solar system, its surface is newer. As we will see below, the Earth is a dynamic planet and *geologically* quite active. One has to be thus very careful when assessing the Earth's age by dating its material only accessible at its surface. In fact, it is easier to study **asteroid material**, because it constantly "rains down" on us. You see, the Earth in its orbit travels through space thinly filled with material, and therefore "sweeps up" this (mostly dust sized) matter. From time to time a bigger object gets in Earth's way; we might even say, it *collides* with Earth. Now, most of these objects burn up in the atmosphere as they travel toward the surface at high speeds. We see them as *shooting stars* in the night sky, and as *fireballs* if they are bigger. These phenomena of light display[9] are called *meteors*. This is a slightly confusing word choice, because *meteorology* is the science of weather phenomena. When the word was coined by the ancient Greeks, however, it had an extended meaning as *all phenomena in the sky*, so the labeling makes sense. Seldom an object is big enough to make it to Earth's surface. When the object is found on the surface, it is called a *meteorite*, Figure 4.4(a) is an example. It is very unlikely but not impossible that a very large object collides with Earth. When this happens, a substantial crater may be produced, like the *Barringer crater* in Arizona, Figure 4.4(b). The two most recent encounters with "space matter" both happened in Siberia.[10] The so-called *Tunguska event* of 1908 flattened over 2,000 square kilometers of forest and produced an explosion that could be heard in London, UK, several thousand kilometers away. Strangely enough, no remains or crater were found, so the object must have exploded in mid-air. The most recent event was the *Chelyabinsk meteor* of 2013.

---

[9]Rarely is sound associated with these *meteors*.

[10]Why did they both happen in Siberia? Because Siberia is so big, and the probability of getting hit by an asteroid scales with the *area* of the "target." In fact, there are almost two thousand kilometers between Tunguska and Cheyabinsk.

Figure 4.4: **Meteorite And crater.** (a) The Murnpeowie meteorite, a 2520 pound *iron meteorite* found in the Australian Outback in 1908. (b) The Barringer crater in Arizona.

Its motional (kinetic) energy was about 29 times the energy released by the nuclear bomb detonated over Hiroshima. Fortunately, the object exploded about 30 km above ground, but damages on buildings occurred and 1,500 people were hospitalized. Nobody died, but this was the first documented incident in which humans were injured by space matter. The objects that caused these events were likely very small asteroids, less than 100 m across, sometimes known as **meteoroids**.[11] To sum up the confusing labeling, a **meteoroid** entering the Earth's atmosphere produces a **meteor** phenomenon, and falls to the ground as a **meteorite**; the latter is basically a rock. This begs the question: How do you find and identify them? You certainly do not want to look in a landscape full of (Earth) rocks, which leaves two prime locations: sand desserts and snow-covered regions. Indeed, many meteoroids are found in Antarctica and the Sahara Desert. Overall, 30,000 meteorites have been found on Earth, which confirms our assertion that the Earth picks up space matter regularly.

Meteorites can be analyzed in the laboratory. Investigating the space matter with sophisticated analyzing techniques is one of the rare occasions for astronomers to do more than just observing. In particular, the rocks can be **radioactively dated**. Their age yields an **upper bound** for the age of the Earth and the solar system, because the structures within meteorites are believed to be the earliest objects to form in the solar system. The oldest known meteorites have an age of 4,567 million years. Note that the use of four significant figures suggests that scientists are confident to date age of the solar system to within 1 million years, with a uncertainty of only 1 part in 4,567! As a cross-check, we can determine the age of surface rocks on the Earth, which have to be younger than the upper limit. Indeed, the oldest rock on Earth is about 4.4 billion years old, so about 160 million years younger than the solar system itself.

Concept Practice

1. Where in the solar system do we find most of the *asteroids*?
   (a) Closer to the sun than Mercury
   (b) Farther from the sun than Neptune
   (c) Around the Earth's orbit
   (d) Between Mars and Jupiter

---

[11] In 2010 **meteoroids** were classified as being between 10 μm and 1 m in size. The general idea is that an asteroid is so big that it can be detected with a telescope, while a **meteoroid** cannot.

4.2 The Earth-Moon System

The **Earth-moon system** is peculiar in the solar system, as the mass ratio between the planet and its moon is by far the largest,[12] and the Earth is the closest planet to the sun sporting a moon. In our brief survey of the system we will focus on the properties which are useful for expanding our knowledge of the universe. In particular, we are looking for hints to understand how planets and their solar systems form. Maybe we can get a sense as to whether planetary systems are likely to exist, and how probable it is to find a planet of similar characteristics as the Earth out there—which might harbor life.

4.2.1 The Earth

From an astronomical point of view, we are most interested in the **bulk properties** of Earth, that is, its **interior structure**. Of course, we can only access the **Earth's surface** directly. However, we can probe the interior by analyzing the seismic waves produced by geological activities such as earthquakes and volcanic eruptions. On the other hand, we are constantly reminded in observational work that the Earth is surrounded by an atmosphere,[13] which has, of course, important other consequences, not the least to make life on the planet possible at all. We thus should study the atmosphere in greater detail than we have so far.

The Interior Structure of the Earth

Several things are clear from the outset. The density of surface rocks is rather small; **granite** is about two and a half times denser than water ($\rho_{Granite} \approx 2600 \, \text{kg/m}^3$). On the other hand, we know from above that the Earth's average density is more than twice that ($\rho_{Earth} \approx 3\rho_{Jupiter} \approx 5500 \, \text{kg/m}^3$). Therefore, the Earth must contain material in its interior that is much denser than the surface rocks. What could it be? Naturally, we think of heavy metals such as lead ($\rho_{lead} \approx 11.000 \, \text{kg/m}^3$), gold ($\rho_{gold} \approx 19\,000 \, \text{kg/m}^3$) or iron ($\rho_{Iron} \approx 7900 \, \text{kg/m}^3$). When we look at the abundance of chemical elements in the Earth's crust as guidance, Figure 4.5(b), the choice is obvious. The most abundant elements in order are oxygen, silicon, aluminum, and iron,[14] with iron being the densest by far. Hence, it is likely that the Earth has a core that is mostly iron. This theory is corroborated by the fact that the Earth has a sizable magnetic field and that iron as a **ferromagnetic** material[15] displays strong magnetic effects. The story here is not as straightforward as you might think. In particular, the Earth is *not* a giant permanent magnet. Rather, Earth's magnetic field is generated by electric currents in the large mass of iron in its interior. The currents are generated by forces due to the rotation of the Earth. Another important ingredient in the mechanism is the existence **convection currents** which transport hot material upward and cooler material downward, creating vortices like in Figure 3.41. This means that the iron-rich interior of the Earth must at least partially be liquid, that is, molten. The Earth thus must be very hot inside, as the melting temperature of iron is about 1200 K at sea level pressure, and much higher at the

---

[12]The Pluto-Charon mass ratio is even bigger, but Pluto is not a planet anymore...

[13]For instance, the atmosphere absorbs electromagnetic waves of most wavelengths except visible light, radio waves, and some infrared light.

[14]Lead comes in at number 37, and gold at 72—too bad.

[15]In fact, iron (Latin: *ferrum*) is the stereotypical *ferromagnetic* material, which is no *ironic* statement, since *irony* comes from the Greek word *eironeia* meaning *simulated ignorance*.

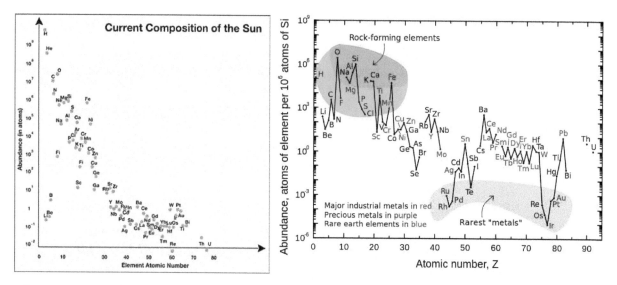

Figure 4.5: **Chemical abundances.** (a) Sun (b) Earth. Note the logarithmic scales. For instance, on Earth we find one gold (Au) atom for every billion ($10^9 = \frac{10^6}{10^{-3}}$) silicon (Si) atoms.

immense pressures inside the Earth. The existence of a planetary magnetic field can therefore tell us a lot about the planet's interior and invites follow-up questions such as *Why is the Earth so hot inside?* and *How long does it take for the Earth's interior to cool down?*

By analyzing seismic activity, such as the propagation of density waves caused by earthquakes geologists have been able to piece together a consistent picture of the Earth's structure, Figure 4.6. It is a science showcase—how different sciences came together to unravel the mystery of Earth's internal structure. The main parts of the puzzle are the **chemical** composition of the different layers, a **physical** model of pressure and temperature as a function of depth or distance from center, and the melting points of the materials under the given circumstances (pressure and temperature). Clearly, both the pressure and the temperature increase monotonously toward the center. Since one part of the model affects the other,[16] a lot of different plausible configurations were tried before settling on a **self-consistent** configuration. The internal structure obeying all boundary conditions, such as size of the core determined by seismic waves, looks as follows. As suspected, there is a solid **inner core** surrounded by a liquid **outer core**, both made mostly of iron. The solid **mantle** surrounding the core is made from lighter materials, mostly silicon-rich rock. The upper layer of the mantle called the **asthenosphere** is plastic, that is, deformable, because pressure isn't high enough to solidify this layer.

The **surface** of the Earth is unique in the solar system, because it is mostly (about 70%) covered with liquid water. The existence of large amounts of liquid water is probably the most important fact about the Earth in terms of biological life. **Astrobiologists** are virtually certain that without liquid water—chemically an almost universal solvent—life will not emerge. As we will see in Section 4.3.3, liquid water stabilizes a planet's **climate** and its presence is a sign and criterion for a stable climate. The existence of liquid water tells us two things immediately. Firstly, the atmosphere of a planet with liquid water is substantial, since liquid water can only exist if air pressure is higher than about 600 Pa, which is less than 1%

---

[16] For instance, lighter mantle material leads to less pressure on the core.

4.2. THE EARTH-MOON SYSTEM

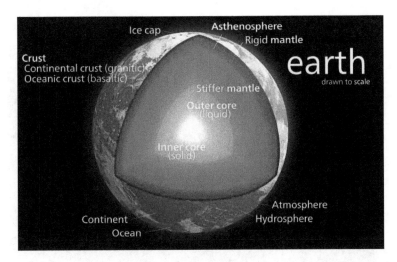

Figure 4.6: **Earth's interior.** The main features are a thin *crust*, a solid *mantle*, a liquid *outer core* and a solid *inner core*.

of Earth's atmospheric pressure. Secondly, the average temperature of the planet must be between the melting and boiling point of water, which at sea level on Earth is between 273 K and 373 K (0 °C–100 °C). In fact, water is so important for life that it is used to define the **habitable zone** around a star as the region where it can exist in liquid form on a planet, see Figure 4.7.

Clearly, the oceans are the "lowlands" on Earth covered with water, whereas the "highlands" rising above sea level are known as **continents**. The difference in altitude of the highest elevation on Earth (Mt. Everest at 8848 m ≈ 9 km above sea level) and the deepest depression (Mariana Trench 11.000 m = 11 km below sea level) is a measure of the surface gravity. Among the terrestrial planet, Earth has the largest surface gravity, so we would expect its **altitude span** of 9 km + 11 km = 20 km to be the *smallest* among those planets. We will see that similar **topological patterns** occur on other terrestrial planets and the moon. Most continents are separated by large bodies of water, which is why a good part of the surface area of the Earth was uncharted territory until about 500 years ago. As the outlines of the continents were explored in greater detail, starting with explorers like CHRISTOPHER COLUMBUS (c. 1451–1506) and FERDINAND MAGELLAN (c. 1480–1521), it dawned on scientists that they looked rather like pieces of a gigantic puzzle. In particular, the South American continent fits so well into the bay demarked by the southwest coast of Africa that this is unlikely to be an accident. In 1915 it was the meteorologist ALFRED WEGENER (1880–1930) who hypothesized that the position of the continents today is the result of a **continental drift**. That is, over millions of years the continents on both sides of the Atlantic ocean drifted apart from each other. In the distant past, therefore, the continents must all have been joint together. Indeed, there is good evidence that about 240 million years ago a single **supercontinent** existed—a continuous landmass reaching from the North- to the South Pole. WEGENER called it *Pangaea*.[17] This large-scale motion of landmasses is possible due to **plate tectonics**. Namely, the continents sit on large plates that "float" on the **asthenosphere**, the soft upper layer of the Earth's mantle. Its **convection currents** (transporting energy from the hot interior to the cold exterior, see Section 3.8) drive the movement of the **tectonic plates** on which the

---

[17]In Greek *pan* is *all* and *gaea* is *Earth*.

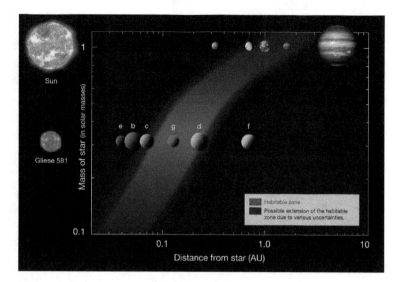

Figure 4.7: **Habitable zones.** As different stars have different luminosities, that is, power outputs, the location and extent of the zone around a star where liquid water can exist on a planet varies. The hotter the (main sequence) star, the wider and farther out is its habitable zone. Shown are two examples: our solar system and the exoplanetary system around the star Gliese 581.

continents sit. WEGENER's bold hypothesis of course needed extraordinary evidence to be accepted. Here is some.

- Identical rock formations are found on different continents thousands of kilometers apart today. This is evidence that the formations were side by side hundreds of millions of years ago.

- Where tectonic plates collide there is **seismic activity**, such as volcanoes, increased probability of earthquakes, and **geysers**. The pattern of occurrence of such phenomena coincides with the shape of the tectonic plates, Figure 4.8.

To figure out the internal structure of the Earth from these varied pieces of information is a marvelous achievement of the physical sciences. Of course, the successes of this model are not the whole story. We have to keep in mind that on the way to the accepted theory, there were many questions that had to be answered and detail problems that had to be solved. Here are two obvious questions: *Where* does all the heat trapped in the interior of the Earth come from, that is, *why* is the core temperature so high? *Why* is the Earth's interior **stratified** or **differentiated**, that is, why is the mantle made out of distinctly different material than the core? That is, why aren't the materials randomly distributed, that is, mixed? Clearly, these are questions about the formation of the Earth, and possibly the solar system, so we will postpone them until Section 4.5.

The Earth's Atmosphere

The Earth's atmosphere profoundly affects how we perceive the universe. From **atmospheric refraction** to **atmospheric absorption**, we have to take its effects into account to **reduce** our observations, that is, systematically eliminate the atmospheric distortions from our data. Much can be dealt with pragmatically. As we remarked in Section 1.1.2, KEPLER purged

4.2. THE EARTH-MOON SYSTEM

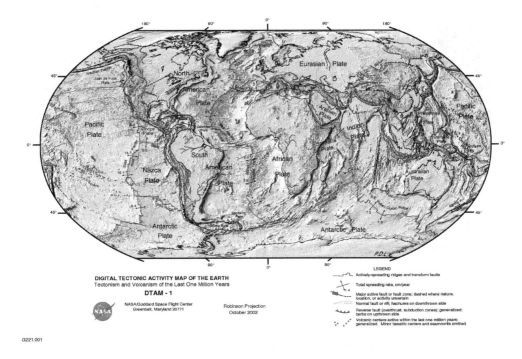

Figure 4.8: **The Earth's tectonic plates.** Note that earthquakes and other seismic activities are concentrated around the edges of the tectonic plates.

TYCHO's data of the refraction due to air almost two hundred years before its main ingredients oxygen and nitrogen were even discovered. As we have seen in Section 3.4.2, there were good reasons why these discoveries came so late. The bulk composition of the air turned out to be 78% nitrogen, 21% oxygen and 1% other gases, notably Argon. This pattern poses two problems. First, Earth's atmospheric composition is very different from other planets', even the ones in the same (terrestrial) group of planets. The latter have atmospheres dominated by carbon dioxide, which is but a trace element in Earth's atmosphere.[18] A related enigma is the existence of large amounts of oxygen. You may know that oxygen is chemically active, and, in fact, very corrosive. Oxygen in the atmosphere therefore gets "used up" by many processes, not the least to sustain animal and human life ("breathing"). To maintain the oxygen levels of the atmosphere, the element hence has to be produced continuously, or rather—*recycled.* Indeed, there is an **oxygen cycle** by which oxygen gets produced in plants via **photosynthesis** from carbon dioxide, whereas animal **respiration** in a reverse process produces carbon dioxide from oxygen, loosely speaking. In fact, there are many *cycles* on Earth that ensure a stable environment. Some, like the water cycle (**evaporation** and **precipitation**) are "inorganic," but many are due to life and biological activity (carbon and nitrogen cycles). In sum, the **composition of the atmosphere has been altered by life.** This is a unique feature of Earth—unmatched in the entire solar system.

As we saw before, the atmosphere is a self-sustaining system. Layers of gas are held and squeezed together by the planet's gravitation. Temperature differences produced by the sun's radiative energy drive air movement, **weather** and **climate.** We saw in Section 2.9.4 that

---

[18]Albeit, of utmost importance in the climate debate because carbon dioxide is a powerful **greenhouse gas**, and released by human activity such as combustion of fossil fuels.

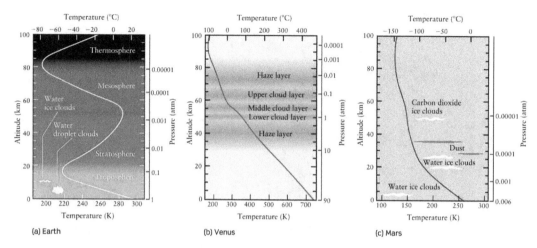

Figure 4.9: **Temperature profiles of terrestrial atmospheres.** (a) Earth; (b) Venus; (c) Mars.

TORRICELLI around 1650 realized that air pressure is but the weight of a column of air pressing down on every square inch near the Earth's surface. Two hundred years later KELVIN, Section 3.9.3, using the newly invented theory of **thermodynamics**, modeled the atmosphere as an ideal gas in **convective equilibrium**. We already touched on the importance of understanding the thermodynamics of gas-balls squeezed together by their own gravity in Section 3.9.3. In fact, the latter construct is the first model of a star invented by LANE and RITTER. Even without these sophistications, it is clear that the density of air, and certainly air pressure, should decrease as we leave the Earth's surface to climb to higher and higher altitude, for example in a balloon experiment. On the other hand, energy from the sun hits the upper layers of the atmosphere first, and so the temperature profile of the atmosphere might be more complicated. Indeed, comparing the Earth's atmospheric temperature profile, Figure 4.9(a), with that of the other terrestrial planets, Figure 4.9(b)and(c), we conclude that the Earth's atmosphere is more "sophisticated." Its temperature rises and falls dramatically driven by the distinct properties of the its four main layers, the **troposphere**, the **stratosphere**, the **mesosphere**, and the **thermosphere**. Whenever we see a compilation of data like in Figure 4.9(a)-(c), we should try to make sense of it, rather than just accepting it as a bunch of facts. Here are some examples of adequate, appreciative thoughts on the graphs. First, we might note that on all three planets the temperature rises toward the surface. In fact, the "air" is warmest right at the surface. This makes sense if we take the **greenhouse effect** into account, by which the sunlight that hits and warms the surface cannot easily leave the atmosphere, and thus "reheats" air and surface. On the other hand, we might reason that the upper layer of the atmosphere should be warmer than lower layers because it gets hit by the sun's rays first, before other layers get their share of energy. Indeed, this is a dramatic effect in the Earth's **thermosphere** whose outer layer is more than 60 Kelvin warmer than the air at its lower boundary. Strangely, the effect is barely perceptible in Venus's and Mars's atmospheres. Noticing this unexpected difference can lead to further investigation, but we are not pursuing it here. Another pattern we can appreciate is the sharp drop of temperature in Earth's **troposphere** by almost 80 Kelvin. This is reflected in the fact that low-hanging clouds are made of water droplets, whereas clouds at about 10 km altitude are mostly water ice clouds.

4.2. THE EARTH-MOON SYSTEM

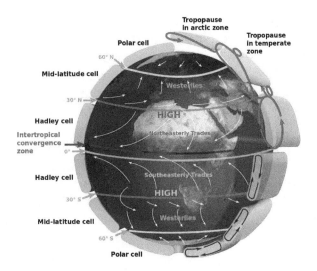

Figure 4.10: **Circulation pattern of the Earth's atmosphere.** The air moves due to **convection** and the Earth's **rotation**.

Our weather is driven by temperature differences, both vertically ("colder on top") and horizontally ("warmer at the equator"). There is air movement, aka **convection**, driven by temperature differences, for example between the equatorial and the polar regions of the Earth. Naively, we would therefore expect a constant movement of hot air from the equator to the poles, and cold air drifting the opposite direction. This behavior is upset by the rotation of the Earth, which mixes things up in a perpendicular direction. This is more than we need here, so we'll stop and conclude that air movements, Figure 4.10, just as interior temperature and pressure profiles, Figure 4.6, can be complex but are *explainable* by fairly straightforward physical processes such as **energy transfer**, Figure 3.41.

To see the importance of the atmosphere for the planet, we ask the following question: What would the average surface temperature of the Earth be if it *didn't* have an atmosphere? This is an exercise in **thermodynamics**. Basically, energy from the sun warms the Earth's surface, and due to its temperature, the Earth will radiate energy. In other words, we are dealing with an **equilibrium** due to the two processes working in opposite direction: **Absorption** of energy makes the surface warmer, **emission** of energy makes the surface colder. In Section 3.8.1 we saw that the Earth receives a power of about 1,400 W per square meter from the sun. Keeping in mind that the sun is below the horizon for 50% of locations, and that the Earth *reflects* about 30% of the incoming sunlight, we are left with about 240 Watt per square meter that are absorbed on average by the surface, and need to be re-emitted to maintain **thermal equilibrium** (lest the Earth heats up dramatically). Assuming that the Earth is emitting energy like a blackbody, we can invoke the **Stefan-Boltzmann law**, Equation (3.12) to calculate the Earth's surface temperature T_E:

$$\sigma A T_E^4 = 240 \frac{W}{m^2} A.$$

The surface area A of the Earth drops out, and we can solve for $T_E = (240 \frac{W}{m^2}/\sigma)^{\frac{1}{4}} = 255\,\text{K}$. This result is surprising, since the actual surface temperature of the Earth is about $287\,\text{K} \approx 57\,°\text{F}$, that is, much warmer than without an atmosphere. The reason for this discrepancy is the existence of an atmosphere. The process responsible for warming the surface is known as the *greenhouse effect* of the atmosphere, named after the analogous effect of the glass ceiling in a gardener's greenhouse. We will get back to this important effect in Section 4.3.3. Here, we simply note the effect that the elevated temperatures and the existence of large quantities of liquid water have on the Earth's surface. Namely, its topography is shaped by weather and its *erosive forces*. Wind, rain, ocean and river currents, as well as freezing-thawing action break down surface rocks with dramatic consequences. In particular, entire mountain ranges created over tens of millions of years by *tectonic plate movements* disappear over tens of millions of years due to *erosion*.[19] Also, the record of bombardment of the Earth's surface with space debris, aka *cratering*, is basically wiped out by the combined action of erosion and geological activity. This record is pretty much intact on the moon (as well as Mercury and Mars) which gives us important insight in the early history of the solar system.

4.2.2 The Moon

The moon is by far the nearest celestial object,[20] so it makes a lot of sense to study it thoroughly. Largely, it is the blueprint for many other bodies in the solar system. Already the old Greeks around ARISTARCHUS, Section 2.3.2, knew the distance to and size of the moon. Its surface features were studied in great detail starting with Galileo, Section 2.8. Its mass, however, was not known for quite some time because *the moon doesn't have a moon*.[21] Since the moon is about one quarter the size of Earth, we can estimate its mass by assuming the same density as Earth. Then the mass ratio is the same as the volume ratio, namely

$$\frac{M_{moon}}{M_{Earth}} = \left(\frac{1}{4}\right)^3 = \frac{1}{64}.$$

This is only an upper bound, since the moon's density is bound to be smaller as the Earth is the densest object in the solar system.

The most striking feature of the face of the moon, see Figure 2.41, is the existence of large dark areas on a background of much lighter areas. This pattern is visible to the unaided eye and gives rise to the *pareidolic images* known as the **Man in the Moon**. In reality, our brain tricks us into interpreting the pattern formed by the dark *maria* (oceans) and the bright *terrae* (highlands) as a human face or body. Telescopic observations immediately reveal that the highlands are heavily cratered areas, whereas in the *maria* regions we find much fewer of them. What is the reason for this dramatically different *topology*? This question cannot be answered by observation alone. Only after the **Apollo moon missions** had brought moon rocks back to Earth could geologists determine that the rate of impacts on the moon has changed dramatically in the history of our solar system. If we ponder the *nebular hypothesis*, Secs. 3.5.2 and 4.5, we expect the amount of "flying debris" to reduce dramatically over time. In other words, we expect the early solar system to be the era of most violent impact activity and crater creation. The analysis of the moon rocks, however, while

---

[19]For a sense of scale, compare the elevations of the peaks in the *young* Rocky Mountains with those in the *old* Appalachians: Pikes Peak (CO) 14,115 ft; Mt. Washington (NH) 6,288 ft.

[20]The *Apollo 11* missions took four days to reach the moon in 1969; future manned Mars missions will take more than half a year to reach the neighboring planet, that is, almost 50 times as long.

[21]This simply means that the usual method of determining the mass via Newton's law of gravity from the orbit of a satellite isn't applicable.

4.2. THE EARTH-MOON SYSTEM

corroborating our general expectations, revealed a second peak in impact activity dubbed the **late heavy bombardment**. This happened about 500 million years after the **early heavy bombardment** which had peppered the lunar terrae or highlands with impact craters. Scientists believe that the lunar maria were created by massive impacts during the more recent *late heavy bombardment*. Evidence is as follows. Chemical analysis shows that the floor of the *maria* consists of iron-rich **basalt**, that is, **igneous volcanic rock**. This explains the *maria*'s darker color compared to the **highlands** made of aluminum-rich rock. It should be said, though, that the moon as a whole is very dark. Namely, it reflects only 12% of the incoming sunlight, comparable with the **albedo**[22] of worn asphalt. The timing of these massive late impacts about 4 billion years ago was such that the interior of the moon was still molten, and thus the lava streams oozing out of the moon's surface cracked by the impacts flooded the lunar plains—thus creating the *maria*. **Late heavy bombardment** lasted for about 200 million years. The much reduced bombardment after this period is responsible for the light cratering of the *maria* themselves. The interplay between observations and experimental analysis thus resolved the mystery of the two distinct lunar topologies.[23] We can go even further. Namely, the scenario suggests that the lunar highlands represent the moon's crust, whereas the magma forming the *maria* came from the moon's interior, and therefore likely is **mantle** material. The difference in density ($2900 \, \text{kg/m}^3$ for highland rocks versus $3300 \, \text{kg/m}^3$ maria basalt) tells us that the **differentiation** of the lunar interior must have gone analogous to the processes that stratified the Earth's interior. If nothing else, it tells us that both bodies must have been completely molten so that the heavier materials could sink toward their centers.

It was again space travel that made possible another lunar surprise. When the first photo was taken by the Russian space probe *Luna 3* in October 1959, the world realized that the *far side* of the moon looks strikingly different.[24] Forever hidden from Earth-bound observation by the **synchronous rotation** of the moon, the *far side* consists almost entirely of highlands. Why are there no sizable *maria* on the *far side*? An amazingly simple explanation is the hypothesis that the moon's crust is thicker of on the far side, thus preventing lava from the mantle to well up to the surface. This **asymmetry** of the moon's interior, Figure 4.11, doubles as an explanation of the **synchronous rotation** itself. Due to the gravitational forces of the Earth, the moon's heavier elements ended up closer to the Earth, and this in turn puts a **torque** on the moon that eventually resulted in the lunar spin and orbital periods to become identical.

There are no indications that the moon has been geologically active in the recent past. This is consistent with its gross physical properties. Namely, the moon as a much smaller body (1/64th of Earth's volume) of significantly smaller density must have cooled much faster than the Earth. Therefore, geological activities such a volcanism or plate tectonics are unlikely after the first billion years or so of its existence.

It is pretty obvious that the moon has no atmosphere. We see no clouds, and stars occulted or eclipses by the moon show no smooth but a sharp disappearance as the moon moves in front

---

[22]The term **albedo** means "whiteness" (from Latin *albus=white*) and is the **reflection coefficient** of an object's surface. The moon's **albedo** is accordingly 0.12, that is, 12%.

[23]It will not have escaped your attention that the solution to this problem created another; why *late heavy bombardment* happened or what caused it is not known.

[24]Incidentally, nobody prior to 1959 knew what the far side looked like. It could have hosted an *alien space-port* for all we knew up to then.

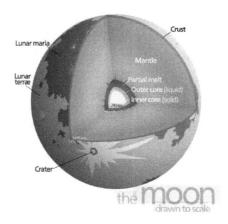

Figure 4.11: **The moon's interior.** Visible are the thin *crust*, the large *mantle*, and the very small *core*. The moon's average density is much smaller than Earth's.

of other celestial objects. The lack of atmosphere is the reason why the sky in astronautic footage appears pitch black. With the lack of weather comes the lack of **erosion** on the moon. The moon in turn keeps almost perfect record of its impact history. Indeed, only *fresh* impacts will erode old craters, and this goes very slowly in our "post-bombardment" era. Indeed, a new 10 km diameter lunar crater is created only every 10 million years or so. Keep in mind, though, that the smaller the crater, the more frequently is it created. For instance, 1-meter craters are created monthly, and 1-centimeter craters every few minutes. A quick word on the formation of craters. Naively, it may seem that they form when an object impacts on the lunar surface, thus pushing away the material and creating a hole or depression. If you think about it, and then pair your thoughts with the observation that all craters are perfectly round with a well-developed rim and often a central mountain, it seems highly unlikely that **impactors** coming from all directions, that is, along a path with random inclination toward the surface, should produce a circular crater. Indeed, the formation of impact craters happens differently than anticipated, Figure 4.12. Regardless of the initial direction of the impactor, its large motional energy deposits a lot of energy—akin to an explosion—and enough to melt the some of the materials. The impact sends shockwaves into the surface and ejects material at high speeds, both of which form a *circular* depression. As some of the pulverized and molten material settles back, it creates a rather smooth crater floor.

4.2.3 The Formation of the Earth-Moon System

Earth and moon form a close system that is unique in the solar system, since the moon is not only large compared to its host planet (the Earth), the system is also relatively close to the sun, and influenced by the large planets, in particular Jupiter. This makes the moon a **highly perturbed** object, and understanding its orbit is very demanding and frustrating, as we saw in Section 3.1.2. The moon has quite an influence on us: The **tides** are in large part due to the moon; it influences and modulates temporal patterns in living nature such as the human menstrual[25] cycle; the moon has a stabilizing effect on the orbit and axis orientation of the Earth. As we have seen in Section 2.10, the motion of the moon was a crucial ingredient

---

[25]Latin: *mensis = month*.

4.2. THE EARTH-MOON SYSTEM

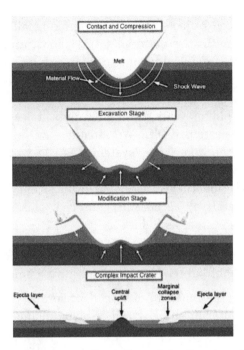

Figure 4.12: **Formation of an impact crater.** Regardless of the direction of the *impactor*, a compression of the surface and the large amount of energy depositied will lead to a melting of surface material. This ensures the symmetry of the formation ("all craters are circles"). Depending on the nature of the impact, a *complex crater* with features such as a central mountain and a partially collapsed crater walls is formed.

in Newton's solution of the *problem of the planets*. Newton went on to explain the tides by the gravitational influence of the moon on the Earth and its oceans. We saw further, Section 3.2.1 that in the eighteenth century the moon's position could be measured with such precision that minute effects such as the *nutation* of Earth's axis could be discerned. For related reasons, the moon became interesting as a *universal clock* to aid navigation at sea because it enabled a reliable determination of geographic longitude; see Section 3.1.2.

For all of these reasons, the *genesis* of such a close, "symbiotic" system of celestial objects is perplexing. Scientists have tried for a long time to understand how the Earth-moon system formed, and it is an interesting example of scientific discourse and *theory building*. To be sure, it is not clear that we will be able to explain anything. The Earth-moon system could be just a historical accident; for instance, the moon could have come so close to the Earth that it was captured into a stable orbit. However, the probability for this to happen is exceedingly small—but not zero. One of the first to speculate about the origin of the moon was GEORGE DARWIN[26] (1845–1912), who hypothesized that the moon spun off of the Earth due to the centrifugal forces associated with fast rotation about its axis. Indeed, this theory predicted the later confirmed drifting of the moon away from the Earth and became the accepted formation theory for three quarters of a century. Not until the mid-1970s did another hypothesis emerge. Inspired by the details of the theory of planetary formation, astronomers proposed that the *collision* of the Earth with a large, Mars-sized object could have resulted in the present-day Earth-moon system. If the impact was direct, it would

---

[26]He was the fifth child of British naturalist CHARLES DARWIN (1809–1882), one of the inventors of the *science of evolution*.

have completely melted both objects, thoroughly mixing all the materials and shattering both objects. In the aftermath, the material settled down under the influence of their own gravity. Some fraction of the material would subsequently form the moon, while other parts escaped and yet others—probably the majority and the densest—were incorporated in the post-collision Earth. From the age of moon rocks we know that this cataclysmic event must have happened about 4.5 billion years ago. This puts it less than 100 million years after the formation of the solar system itself. We already saw that this period of **early heavy bombardment** was when the inner solar system was rife with smaller objects, so it seems plausible and not too improbable that such a massive collision could have occurred. This **giant-impact hypothesis** nonetheless is a bold proposition, so we need to look for strong evidence to secure its (tentative) acceptance. Indeed, this scenario explains a lot of the features of the Earth-moon system as follows.

- If the heavier elements sink into the molten post-collision Earth, then the emerging moon should be made of lighter elements and have a small core. This is what we find.
- In a central collision, the material is spread preferably in the equatorial plane, so the emerging moon should orbit the Earth in a plane inclined only slightly with respect to the equator. This is the case.
- The entire surface of the moon should have been molten at some point, and we find form the Apollo rock samples that this was the case. This need not be true of the moon had formed in some other way.
- The surface rocks of Earth and moon are *chemically* and *isotopically* so similar that the explanation is almost inextricably that they have a common origin.

Concept Practice

1. The Earth's mantle is ... (a) solid (b) liquid (c) gaseous (d) unknown.
2. The moon
    (a) ... is denser and smaller than the Earth.
    (b) ... is older than the Earth.
    (c) ... has more craters than the Earth.
    (d) None of the above

4.3 The Terrestrial Planets

4.3.1 Mercury and Venus

The closest planet to the sun, Mercury, is also the smallest of the planets. For our purposes, it is well described as a *bigger version of the moon*, as we cannot learn much that we didn't already know by studying the moon. Mercury has no atmosphere, is heavily cratered, and fairly void of seismic activity. Its density is comparable with Earth's, that is, much higher than the moon's, and so it must contain a large iron-rich core.

The closest planet to Earth, Venus, is its twin—almost identical in mass and diameter. Its atmosphere is vastly different, with a grueling pressure about ninety times that of Earth's atmosphere. It consists mostly of carbon dioxide, as we can tell from the spectrum of Venus, Figure 4.13(a), which led to a **runaway greenhouse effect** resulting in incredibly high

4.3. THE TERRESTRIAL PLANETS

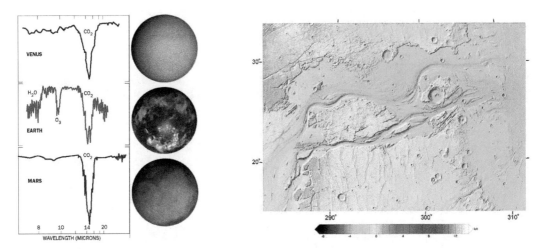

Figure 4.13: **Terrestrial spectra and evidence for water flow on Mars.** (a) Shown on the left are the spectra of the three terrestrial planets with an atmosphere. All three show absorption due to carbon dioxide, but only Earth displays water and ozone (O_3) lines. (b) Shown on the right is a large outflow channel named Kasei Valles on Mars in false colors to enhance perception of altitude differences.

temperatures of about 737 K (864 °F) at the surface. This temperature is higher than the melting point of lead. We will discuss Venus's atmosphere and how it and the Venusian climate developed in Section 4.3.3. A dramatic consequence of this dense atmosphere with heavy clouds made of *sulfuric acid* is that the surface of Venus can never be seen from space. Before the touchdown of the first Russian space probe in 1975, nobody knew what it looked liked. Today, the topography of Venus is well mapped due to radar-ranging mission of the *Magellan* space probe[27] in the early 1990s. The Magellan probe did not touch down, but it sent out microwaves which are not absorbed by Venus's thick cloud layer. When these were reflected by the rocks and mountains on Venus's surface, an accurate image of the Venusian topography could be reconstructed. It turned out that Venus's surface is not clearly divided into high- and lowlands like the Earth, the moon (and Mars, as we will see), Figure 4.14(a). Rather, the isolated, smallish highlands seem randomly distributed. To find such a different surface configuration on a planet so similar to Earth is a pattern in nature that needs to be explained. It is probably the result of a geological process which is very different from Earth's **plate tectonics**. But why should the surface-shaping processes be so different on the twin planets? As a sanity check, we compare the **altitude span** on Venus and find that it is comparable to Earth's. This is a relief—because it is what we expect from the almost identical masses of the two planets.

4.3.2 Mars

Mars, the red planet—named after the Roman god of war—as always inspired the imagination of humans. This intensified after telescopes revealed surface structures on Mars. Indeed, there was an outright Mars "frenzy" around the *fin de siècle*, when fantastic tales of life on Mars seemed to be corroborated by observational evidence. This story of the Mars canals is a interesting and cautionary tale of how scientific ambition and obsessions can go awry—but

---
[27]Named after the Portuguese explorer we mentioned earlier.

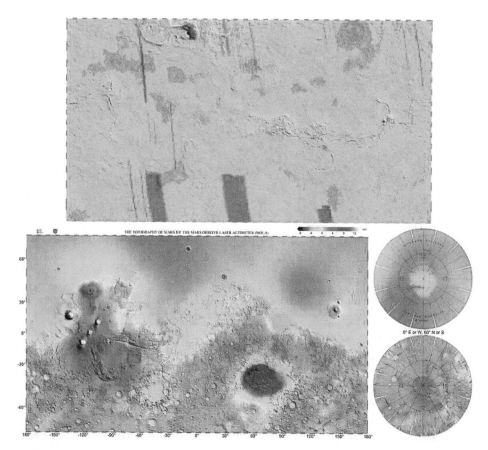

Figure 4.14: **Planet maps.** (a) Venus (b) Mars. Both maps are rendered in false colors to enhance elevation differences. Highlands appear red, lowlands blue.

beyond the scope of this introductory text, see [42] for a contemporary refutation. Even so, the ambition to find life on Mars has motivated Mars research to this day. A large number of space probes have visited the red planet in the last two decades, and they have provided convincing evidence for the existence of liquid water on Mars. Alas, most of this water has flown in the very distant past—billions of years ago. Nonetheless, the mere existence of liquid water has fueled the hope of scientists that life had a chance to develop on Mars, even if just in the form of microorganisms and fossilized today. There is compelling evidence that life is not thriving on Mars today. The atmosphere of Mars is too thin for liquid water to exist on its surface. Also, it consists almost entirely of carbon dioxide. We said before that atmospheric oxygen is a sign for and produced by life. On Mars, the existing oxygen has all been used up to produce *iron oxide*: a layer of rust covers the surface. Put bluntly, Mars is the red planet, because it *rusted out*. Part of the sustained optimism to find life relates to the similarity of Mars and Earth. Both planets have an atmosphere, similar seasons due to similar axis inclination, polar icecaps which are seasonally shrinking and growing, and days of almost identical lengths. Mars today lies at the outer edge of the sun's **habitable zone**.

The Martian surface features can be seen with the help of telescopes. They were interpreted (before the advent of space probes) as a sign that the Martian topography resembles Earth's. In truth, Mars is a much smaller planet than the Earth. At half the size, one expects $\left(\frac{1}{2}\right)^3 = \frac{1}{8}$ the mass, but its lower density results in a mass which is less than $\frac{1}{10} = 10\%$ of the

4.3. THE TERRESTRIAL PLANETS

Earth's. The lower surface gravity means that structures on Mars (like mountains and volcanoes) are larger than on Earth. Also, sand storms can engulf the entire planet, because dust settles much slower—and no ocean prevents a global sprawl. The smaller size spells doom for ongoing seismic activity, because the Martian interior likely has cooled and solidified due to its much smaller volume. Relatedly, no sizable magnetic field is expected with a heavy iron core unlikely to exist due to the rather small average density of Mars. Indeed, we find the largest volcanoes[28] of the solar system on Mars, but they are all *extinct*. They are the result of millennia of uninterrupted eruptions, because Mars does not have **plate tectonics** that could shift a magma-spewing hotspot to another position.[29] Also the largest valley in the solar system is on Mars. This super-sized version of the Grand Canyon is called *Valles Marineris* and is 3000 km long, which is more than the distance from Columbus, Ohio, to Las Vegas, Nevada. Is is likely also the result of volcanic activity. Mars has its fair share of impact craters—less than the moon and Mercury, but many more than Venus and Earth—which tells us that the Martian surface is very old, and that erosion is pretty weak. The aforementioned sandstorms are global, but they are weak,[30] causing very little erosion.

Much insight about the Martian surface features comes from studying the analogous lunar features. As we saw, a great deal about the lunar topography and geology is known owing to the moon's proximity and the fact that we can analyze actual lunar rocks in our laboratories. No Martian rocks were ever brought back to Earth, though several Martian landers were equipped to analyze rocks *on site*. Therefore, the evidence for our theories on the state and history of the Martian surface is much more indirect. That said, one striking similarity of the lunar and Martian topographies cannot be overlooked. Like the moon, Mars has heavily cratered *highlands* and relatively little scarred *lowlands*.[31] In fact, almost the entire northern hemisphere of Mars is one gigantic lowland, only interrupted by large **shield volcanoes** and the *Tharsis rise*; see Figure 4.14(b). On the same token, the southern hemisphere consistently lies several kilometers higher, save for two large circular depressions called *Hellas Planitia* and *Argyre Planitia*. It is tempting to speculate that this pattern, namely the **dichotomy** of high- and lowlands, is explainable by the same process that formed the lunar topography. Indeed, planetary scientists think that the northern Martian lowlands were formed by a lava flood possibly caused by substantial **impactors**. The circularity of the southern depressions certainly speak for this hypothesis. We can even estimate the age of the northern martian surface. Its lack of heavy cratering must mean that it formed during or after the *late heavy bombardment*, as we argued in explaining the genesis of the lunar **maria**. Detailed **radioactive dating** of the surface rocks has not been possible so far. The reason is due to the specifics of **space travel**. Namely, it is easier to land a spaceship near the equator of a planet or moon, so we have little direct data from Martian rocks at high latitudes, that is, toward the poles.

For all its similarity to the moon, Mars shows some unique and exciting features. Namely, we see structures such as shown in Figure 4.13(b) which are basically impossible to explain unless there once was flowing water on the Martian surface! This in turn must mean that Mars—in the distant past—was a very different place. We mentioned that the (past) existence of liquid water has sustained hopes to find life on Mars. So far, searches have been inconclusive at best.

---

[28] Mars's volcano *Olympus Mons* is three times the size of *Mt. Everest*.
[29] The Hawaiian Island chain of volcanic origin is evidence that this hotspot shift occurs on Earth.
[30] The 2015 movie *The Martian* grossly exaggerates the effects of such storms.
[31] We called those *oceans* or *maria* on the moon due to their darker color; on Mars they are actually lighter.

4.3.3 Atmospheres and the Greenhouse Effect

The Greenhouse Effect

We saw in Section 4.2 that the surface temperature on Earth is significantly higher than expected from a naive application of the **Stefan-Boltzmann law** based on the radiation received from the sun. There, we alluded to the fact that some gases in the Earth's atmosphere have an effect similar to the glass ceiling of a gardener's greenhouse. Namely, they hinder energy from escaping, effectively raising the temperature. This is a general effect, at work on all terrestrial planets with an atmosphere. At first, it may seem strange that a thin layer[32] of gas can have such an impact. Just to set the record straight, its impact is crucial but not large. Indeed, just receiving solar radiation an Earth without atmosphere would warm up to 255 K. The greenhouse effect gets us to 288 K, so the **relative increase** is $\frac{288\,\text{K} - 255\,\text{K}}{255\,\text{K}} = 12.9\%$, that is pretty modest. How does it work? The idea was conceived in 1824 by JEAN BAPTISTE JOSEPH FOURIER (1768–1830), who was interested in **heat conduction** and the emerging theory of **thermodynamics**. His theory was corroborated by CLAUDE POUILLET who determined the power the Earth receives from the sun (the solar constant) in 1838, as we saw in Section 3.34. At the most basic level, the idea is that specific molecules like carbon dioxide and water vapor are good absorbers in the **infrared band** of the electromagnetic spectrum. We know that visible light comes to us from sun and stars, so the atmosphere is **transparent** for electromagnetic waves in the **visible band.** Since the sun radiates mostly visible light due to its temperature,[33] this radiation will readily pass through the atmosphere to deposit its energy in the Earth's surface. The surface thus heats up and radiates according to *its* temperature, of about 288 K. Since this temperature is 20 times lower than the sun's, the Earth radiates dominantly at a wavelength twenty times longer than the sun's peak wavelength of roughly 500 nm. So $\lambda_{peak,Earth} = 20 \times 500\,\text{nm} = 10\,000\,\text{nm} = 10\,\mu\text{m}$, which is squarely in the infrared band. As this infrared radiation is trying to leave the Earth's atmosphere, it will get absorbed and re-emitted by the greenhouse molecules many times, which significantly prolongs the "leaving time." During this time, however, more radiation is received from the sun, so the net effect of the delay is that Earth and atmosphere heats up due to the extra energy deposited. It is important to realize that energy does not get "gobbled up" or stored. Just like an electromagnetic wave gets absorbed and almost immediately re-emitted by an atom or molecule (if in a different direction), so does all the energy the Earth receives get re-emitted into outer space. This may sound strange, but keep in mind that the Earth—and virtually all other bodies—are in **thermal equilibrium.** The rate at which it receives energy is the same at which it gives off energy—otherwise the temperature of the Earth would not be as constant as it is.[34] When you get the basic idea, it's easy to see that there are important caveats. For instance the energy the Earth's surface receives is not the same amount that the upper atmosphere receives. Clouds and other reflective surfaces prevent about a third of the sun's radiation to ever get deposited. Indeed, the Earth's *albedo* is about 0.30, which

---

[32]The Earth's atmosphere reaches to about 100 km. This means its extend is only $100\,\text{km}/6400\,\text{km} = 1/64 \approx 1.6\%$ of Earth's radius.

[33]Recall that the peak wavelength and the temperature of a blackbody are related by **Wien's law**, $\lambda_{peak} \times T = 0.0029\,m \cdot K$.

[34]If you put food in the fridge, it cools down before reaching thermal equilibrium after a while. During the cool-down time the food radiates more energy than it receives from its environment since it has a higher temperature than its environment; temperature differences drive heat transfer.

4.3. THE TERRESTRIAL PLANETS 285

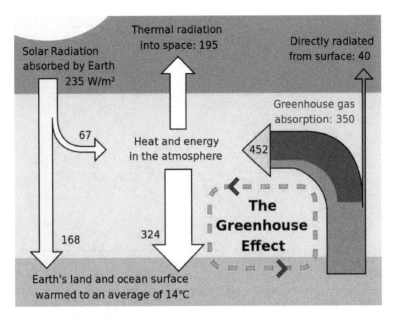

Figure 4.15: **Earth's energy balance.** The Earth emits as much energy as it receives from the sun. Still, the *greenhouse gasses* in the atmosphere ensure that the Earth's surface is warmer with than without an atmosphere.

means that the Earth reflects 30% of all incoming sunlight. This calls for a careful analysis of all energy flows to and from the Earth's surface and atmosphere, as is detailed in Figure 4.15. Be mindful that the sum of all incoming energy is exactly the same as the sum of all outgoing energy. Yet, this does not in the least contradict the fact that the "Earth with an atmosphere" has a higher temperature than the "Earth without an atmosphere." These two **physical systems** are different and therefore have a different equilibrium temperature. It might help to keep in mind that as an object heats up, it radiates away more energy due to its new higher temperature according to the **Stefan-Boltzmann law**. This prevents a runaway process, and will result in a stable temperature when the amount of energy radiated away is exactly matched—and indeed supplied by—the amount of energy received.

The other caveat is the effect of a *changing* atmosphere. If the abundances of the greenhouse gases should change, so would the temperature of the planet. This is known on Earth as **global warming**, that is, the very slow but systematic rise of the average surface temperature due to the changing carbon dioxide content of the Earth's atmosphere. Note that the pattern, predicted by Swedish Nobelist SVANTE ARRHENIUS (1859–1927) as early as 1896, is hard to measure, because the temperatures on our planet fluctuate widely due to day-night effects, seasonal change, and are very different for different latitudes. ARRHENIUS sought to explain the pattern of *ice ages*, that is, of large-scale periodic fluctuations of Earth's average temperature. He constructed the first **climate model** based on the assumption that carbon dioxide was likely the agent of such change. While water vapor is a more powerful infrared absorber and thus greenhouse gas, its abundance varies mostly in reaction to temperature itself.[35] It thus does not initiate temperature changes, although it greatly enhances them as a potent **greenhouse gas**.[36] On the other hand, carbon dioxide is produced by a wide

---

[35] In short, the hotter it is, the more humid it gets.

[36] If you've noticed the temperature drop on a clear night compared to the relatively constant nightly tem-

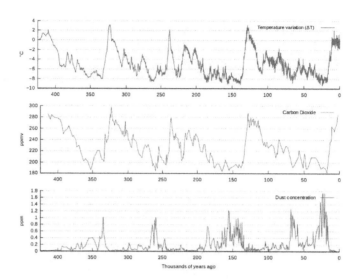

Figure 4.16: **Atmospheric CO_2 content and temperature correlation over the last 400,000 years.** The data was gleaned by studying the *Vostok ice core*.

variety of natural and and man-made processes. They range from respiration to the combustion of fossil fuels. ARRHENIUS estimated the these effects, and his prediction of the rise of global temperature due to industrial activity earned him the reputation of having been the first **global warming** proponent. The correlation between global carbon dioxide levels and average temperature on the planet is exceedingly convincing, Figure 4.16, since the geological record of the two quantities goes back hundreds of thousands of years.[37]

We see that climate can change on a planetary scale. Since a significant fraction of this change on Earth comes from human activity, some scientists have proposed to name the present era when humans for the first time have influenced nature on such a grand scale, the **anthropocene**.[38] The consequences of rising global temperatures, even if seemingly minute on the order of a few degrees, have the potential to be devastating. Massive melting of glaciers worldwide as in Figure 4.17, migration of species toward the poles, intensified droughts and floods, and many other patterns observed today unequivocally warn us of the dire consequences if we don't mitigate our actions on this fragile planet.

Terrestrial Atmospheres

In light of a changing climate on Earth, we might be interested in studying how other terrestrial planets have fared in the history of our solar system. Below we highlight the major differences between the **atmospheric histories** of Venus, Earth, and Mars. This term is used because planetary atmospheres do change and evolve. The atmosphere we find around a planet today is the product of physical processes that have influenced it over *billions* of

---

perature under a cloud (that is, water vapor) covered sky, you have produced direct evidence for its potency as a greenhouse gas.

[37]Air bubbles in Antarctic ice cores give away historic carbon dioxide levels. The temperature record is based on a measurement of the relative abundance of heavier isotopes of hydrogen and oxygen in water molecules. If it is hotter, more of the heavier-isotope water ends up in the ice because they evaporate in greater numbers from the ocean surface in warmer compared to colder times.

[38]From the Greek *anthropos* = human.

4.3. THE TERRESTRIAL PLANETS

Figure 4.17: **Melting glaciers.** Shown is the Grinnell Glacier in Glacier National Park in (a) 1981; (b) 2009.

years. No wonder that atmospheres and climates change! In fact, it is outright surprising how stable and enduring these fragile gas shrouds are. It should be noted that the *atmospheric histories* are not histories in the ordinary sense. Namely, there does not exist a historic record on the properties of atmospheres in the past. Rather, given what we know and can predict (or rather *postdict*) about the conditions in the early solar system will inform scientists' reconstructions of a planet's past. Indirect evidence for these histories is sometimes found—like the amount of cratering, or the existence of *outflow channels*, Figure 4.13(b), on Mars, suggesting the existence of liquid water at some point in the past.

Since Venus, Earth, and Mars are all terrestrial planets, they should be fairly similar in their chemical composition, and their formation process should have been virtually identical. The dissimilar climates must thus be a consequence of the position of the planets within the solar system. Naively we would expect Venus, Earth, and Mars to have successively colder climates due to their increasing distance from the sun. While this is factually correct, the reasons for this fact are much subtler than expected. For starters, Venus is much hotter and Mars much colder than expected from solar distance alone. If we take Earth as the one planet where the climate is stabilized by several intricate feedback loops and material cycles which work near perfectly, we can get a sense of what went "wrong" on Venus and Mars.

All three planets should have started out with a *primary atmosphere*. Namely, hydrogen and helium gases were accreted during the formation of the planets themselves from an interstellar gas and dust cloud rich in those elements. However, these light elements are much too volatile and the gravitational pull of the terrestrial planets is too weak to hold onto them so close to the heat of the sun. While the *primary atmosphere* thus swiftly boiled away, a *secondary atmosphere* gassed out of the interior[39] and possibly by material from comet

---

[39] It may sound strange, but rocks and minerals can contain gases or can be subject to natural processes

impact.[40] They are much thinner than the primary atmospheres (especially those of the Jovian planets) and made mostly of water vapor (58%), carbon dioxide (23%), sulfur dioxide (13%) and nitrogen (5%). On the three terrestrial planets, these atmospheres have evolved very differently mostly due to the varying amounts of solar energy ("sunlight") available.

On Earth, much of the water vapor condensed to form the oceans of liquid water that cover 70% of Earth's surface. This was possible due to the low enough (compared to Venus) and high enough (compared to Mars) surface temperature. This existence of liquid water has the additional effect that carbon dioxide could be dissolved in water, forming **carbonacious rocks**. With water vapor and carbon dioxide largely gone, nitrogen became the dominant constituent in Earth's atmosphere. The latter development was aided by the advent of biological life in the form of bacteria which extract oxygen from *nitrates*, a process which releases nitrogen gas and energy. Also, today's 20% oxygen content of the atmosphere is due to biological life. In this sense, the Earth's present atmosphere is **tertiary**, and the Earth is the only planet with life and a **tertiary atmosphere**. As we mentioned before, on Earth climate is stabilized by several cycles—effectively recycling material. There is a **water cycle** due to precipitation (rain and snow), evaporation from oceans, and runoff into oceans, which keeps water in liquid and gaseous form continuously present. There is also a **carbon cycle**, which is possible by the gaseous state of carbon dioxide which forms carbonate sedimentary rocks. These rocks get submerged by tectonic activity and spewed out by volcanic eruptions as gaseous carbon dioxide to restart the cycle. A similar cycle exists for nitrogen. The common theme here is the cyclic or periodic nature of these processes, which tends to stabilize and sustain them for a long time.

Venus is thought to have started out with liquid water on its surface, albeit at much higher temperature due to the proximity to the sun. Liquid water existed possible for several hundreds of millions of years on Venus, but early in the solar systems history, hence billions of years ago, when the solar luminosity was as much as 25% lower than today. Also keep in mind that the liquid water's temperature could have been above the 373 K nominal boiling point, because of the high Venusian atmospheric pressure. The Venusian climate back then must have been exceedingly hot and humid, and this excessive abundance of a potent greenhouse gas (water vapor) is probably what led to the development of a runaway **greenhouse effect** on Venus. As the sun's luminosity increased slowly but steadily, it pushed the inner edge of the **habitable zone** past Venus's orbit. Due to the increased energy deposit, the oceans on Venus evaporated as they heated up to above 647 K.[41] The evaporating water intensified the greenhouse effect, which led to even more evaporation. This runaway process did not stop until all liquid water was gone. What's more, the intense UV radiation from the sun broke up the water molecule, and H_2O became H_2 (Hydrogen) and Oxygen with the former light molecule vanishing into outer space. The rest of the water vapor reacted with sulfur dioxide to form *sulfuric acid*, the substance that the Venusian clouds consist of. This left Venus with a thick atmosphere made of 96.5% carbon dioxide and some nitrogen. The associated intense greenhouse effect stabilized Venus's surface temperature at about 737 K, hot enough to melt lead.

Mars's atmosphere developed due to a runaway greenhouse effect in the other direction.

---

where gases are produced in physical or chemical processes. For example, **carbonaceous rocks** like sandstone will give up their gases at high temperature.

[40] It is suspected that comets rich in icy materials have contributed significantly to Earth's water supply.

[41] This is the temperature beyond which water is gaseous, regardless of outside pressure.

Also on Mars large liquid water oceans existed probably 4 billion years ago. At that time, Mars's atmosphere was much thicker than it is today. Being 50% farther from the sun than the Earth, Mars has much lower surface temperatures, which resulted in less water vapor (a potent greenhouse gas) in the atmosphere. Moreover, increased rain and snowfall washed carbon dioxide out of the "air" and dissolved it in the ocean water, thus decreasing the warming greenhouse effect. The associated temperature drop resulted in the freezing of the water, which took the carbon dioxide permanently out of use as a greenhouse gas, and led to further temperature decrease. This runaway greenhouse effect toward lower temperatures, or **runaway icehouse effect** was exacerbated by a decrease in atmospheric pressure. Mars lost its magnetic field early in its history due to a much faster cooling of the smaller planet's interior. With the protective magnetic shield gone, Mars's atmosphere was diminished over the hundreds of millions of years of onslaught by the solar wind of charged particles which kicked out atmospheric molecules one by one. This leaves Mars today with an atmosphere that is too thin to sustain liquid water. Still, there is weather and even seasons on Mars, as the melting and freezing of the polar icecaps (made mostly of carbon dioxide ice) attest.

Concept Practice

1. All terrestrial planets
   (a) ... have moons.
   (b) ... have an atmosphere.
   (c) ... are large.
   (d) ... are dense and rocky.
   (e) None of the above

2. The *greenhouse effect*
   (a) ... is the same as *global warming*.
   (b) ... means that the Earth is getting hotter and hotter.
   (c) ... started with the *Industrial Revolution*.
   (d) ... would not exist if the Earth did not have an atmosphere.
   (e) Two of the above
   (f) None of the above

4.4 The Jovian Giants

4.4.1 Jupiter

Jupiter is the namesake of the **Jovian planets** and the largest member of this group. What is true for Jupiter is, to a large extent, also true for the rest of group, most notably Saturn.

Jupiter is truly a remarkable planet; in volume and mass it is greater than all other planets combined.[42] Jupiter has a "stripey" appearance even in small telescopes, Figure 4.18(c). Jupiter seems divided into **bands** running parallel to its equator, so it is plausible that this pattern is caused by Jupiter's fast rotation in under ten hours, as described above. The dark bands are called **belts** and the light ones are called **zones**. Telescopes show the dark North and South Equatorial Belts (NEB, SEB) separated by the broader, bright Equatorial

---

[42] Jupiter even carries the majority of the **angular momentum** of the entire solar system, including the sun, although it has only about one-thousandth of the latter's mass.

Figure 4.18: **Jupiter compared to Earth, sun and moon.** Note the belts and zones of Jupiter's atmosphere. They are formed by convectional motion, just like the analogous pattern of air circulation in Earth's atmosphere, Figure 4.10(a).

Zone (EZ). The famous Great Red Spot (GRS) is an *anticyclonic storm* in the SEB. It is about twice the size of Earth and arguably persisting since its discovery in the seventeenth century.[43] It is the result of shearing wind action in two adjacent bands of Jupiter.

What we see of Jupiter is the cloud layers in its upper atmosphere. In fact, Jupiter as a *gas planet* does not have a solid surface. It may sound strange, but you cannot land on Jupiter. What ends up happening when you get closer and closer to Jupiter is that you delve into the atmosphere which gets thicker and thicker until the immense pressure eventually crushes your vehicle.[44] Jupiter's gaseous nature is evident in its *differential rotation*. We are used to the rotation of *solid objects*, in which all the parts of the object complete a full 360° rotation in exactly the same time, because not part of a rigid object can "lag" behind other parts. This, however, is perfectly possibly for a gaseous object like the outer planets and the sun. For Jupiter this is not a large effect, but it is noticeable, and the Jovian polar regions take about 5 minutes longer to complete a full rotation. Jupiter's fast rotation has another effect which is discernible in small telescopes. Namely, the planet's disk is notably *oblate*, that is, the Jovian diameter is smaller when measured from pole to pole than when measured along the equator, the difference being almost 9300 km, or 6.5%. We saw the analogous pattern when discussing the Earth's oblate shape in Section 3.1.2. Of course, the effect is much larger for the gaseous Jovian than for the rocky Earth, where the difference is only 43 kilometers, or 3.3%.

From spectral analysis we learn that the *atmosphere* of Jupiter is mostly hydrogen and

---

[43] ROBERT HOOKE in 1664 noticed a spot on the planet, GIOVANNI D. CASSINI in 1665 a "permanent spot". Modern observation begins with the 1830 description of the spot.

[44] This is no science fiction, but actually what happen to the *Galileo probe* in 1995 as it was sent into Jupiter's clouds to study them. It survived the harsh conditions for 58 minutes traveling about 100 miles "into" Jupiter. No solid surface was detected.

4.4. THE JOVIAN GIANTS

some helium, very similar to the composition of the planet as a whole—and the sun. Indeed, 85.2% of Jupiter's atmosphere is in the form of hydrogen molecules (H_2) and 13.6% are helium atoms, leaving only 0.2% for other gases such as methane, ammonia and water vapor. This breakdown is somewhat misleading, since the atoms and molecules of different substances have different masses. *By mass*, the atmosphere is 75% hydrogen, 24% helium and 1% other substances. This begs the question why Jupiter, more than 300 times more massive than Earth and of the same composition of the sun, did not become a star itself. If it were, then the sun and Jupiter would form a **binary star system**, and since the vast majority of stars form binary or multiple star systems, this is an eminent possibility. We will see in the next chapter, however, that *mass* is the most important predictor of the life of stars, and Jupiter's $0.001 M_{sun}$ don't cut it. There simply isn't enough gravitational pressure to allow **hydrogen fusion** in Jupiter's core. On the other hand, Jupiter slightly contracts gravitationally, which produces energy; Jupiter radiates about twice as much energy as it receives from the sun.

Jupiter has a strong magnetic field. The *Voyager* spaceprobes were able to confirm its gigantic size. Indeed, the effects of Jupiter's magnetic field to redirect the wind of charged particles emanating form the sun extends beyond Saturn's orbit. This calls our theory of **planetary magnetic fields** as developed in Section 4.2.1 into question. There, we concluded that the magnetic fields are generated by moving heavy metals like iron in the Earth's core. However, Jupiter consists mostly of hydrogen, which creates a contradiction. The way out of this dilemma is the strange property of hydrogen to "turn metallic", that is, become conducting at high pressure and temperature. We are therefore forced to hypothesize that the conditions in Jupiter's interior are such that **metallic hydrogen** exists and generates a gigantic magnetic field. Here again, observing the exterior properties of an object gives us clues about its interior structure otherwise inaccessible. We will later see that these preliminary thoughts on the Jovian interior is agrees with the planetary formation theory, Section 4.5.

Due to its large mass, Jupiter's gravitational attraction influences many small objects in its vicinity. Unsurprisingly, Jupiter has many moons, of which we already know the four largest, the Galilean moons *Io, Europa, Ganymede* and *Callisto*. These moons reach planet size (Ganymede is only slightly smaller than Mercury), and are interesting to study in their own right. For instance, the inner moon Io displays active volcanism, due to the heat generated in its interior by Jupiter's immense tidal forces. Europa is a moon mostly made of ice, see Figure 4.19, and it is speculated that under a kilometer-thick ice layer there could be an ocean of liquid water. These conditions could be favorable for the development of life. Jupiter also has a (fairly small) **ring**. Namely, a large number of very small objects orbit around Jupiter's equatorial plane, just like what we observe on a much grander scale at Saturn. We will talk more about rings in the next section.

4.4.2 Saturn, Rings and Outer Planets

Saturn is about twice as far from the sun than Jupiter. It takes almost three times as long to revolve around the sun, in accordance with **Kepler's third law**. Here, we are mostly interested in the properties of the *planet*, not of the *orbit*. Therefore, we must ask what the effect of a larger orbital radius is on a gas planet. All else being equal, we expect Saturn to be a replica of Jupiter. About the same size, mass, composition, atmospheric properties, large number of moons, and so forth. This is largely what we find. Although less than half

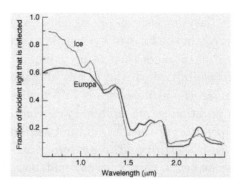

Figure 4.19: **Spectra of Europa and Ice.** A plot of the spectrum of Jupiter's moon Europa together with the spectrum of water ice. The similarity suggests that Europa is made of water ice—at least at its surface.

the mass of Jupiter, Saturn dwarfs the rest of the planets, and lives up to our expectations. It displays atmospheric bands to to its speedy and differential rotation, and the only surprises are its much more impressive **ring system** and the lower than expected helium content of its atmosphere.

The former invites us to think a little more about planetary **rings** in general. Naively, one may think of them as a solid disk that extends around a gas planet's equator, but this hypothesis is immediately refuted by applying **Kepler's third law**. The parts of the ring closer to the planet have to orbit much faster, since Saturn's gravitational force is weaker on the parts farther from the planet due to the **inverse square law**. The resulting **shear forces** would rip any solid disk to shreds, since the **ring system**'s width is substantial— about 70 000 km. So what is a ring? Another observation delivers an important clue. During the 29 years of Saturn's' journey around the sun, an observer on Earth sometimes sees the rings wide open, and sometimes edge on, as in Figure 4.20(a). In the latter position, the rings are virtually invisible. Hence, they must be razor thin. This immediately tells us that it must be made from very small objects. Since it *appears* solid, that is, without holes[45] and of large width, there must be billions of small objects orbiting around the planet. But these many objects in such a limited space must lead to many collisions, which will disrupt the ring system by depleting it of its constituents. Estimates of the life expectancy of ring systems are on the order of 100,000 years, which is but a fleeting moment in the 4,600,000,000-year history of the solar system. To explain this pattern in nature, namely the existence of ring systems, we need to find a mechanism by which rings are *formed*. Since all four Jovians sport a ring system, the mechanism must be a common occurrence, so some **catastrophe theory** will not do. It is plausible to start with the immense gravitational fields of the Jovians. The strong tidal forces we encountered as the reason for Io's volcanism in the Jupiter system, might be able to do much more destructive things. Indeed, it was shown in the 1848 by ÉDUARD ROCHE (1820-1883), that an object held together by gravitational forces only[46] can be deformed and eventually disrupted by tidal forces of a large planet, if that object gets too close to the planet. The critical distance beyond which destruction is imminent is called the **Roche limit**. In other words, we are explaining the ring systems as the result of cannibalism.

---

[45] We can discern several major **gaps** between individual rings. The most prominent are the **Cassini division** and **Encke gap**, discovered by D.G. CASSINI in 1675 and JOHANN FRANZ ENCKE in 1837, respectively.

[46] A cosmic pile of rubble, so to speak. Certainly not a piece of rock which is held together by very strong chemical bonds.

4.5. EXPLAINING THE FORMATION OF THE SOLAR SYSTEM

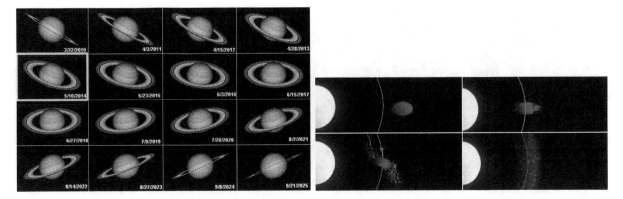

Figure 4.20: **Rings and Roche limit.** (a) Orientation of Saturn's rings relative to a terrestrial observer during Saturn's 29-year orbit. (b) Illustration of the *Roche limit* at which objects held together by their own gravity get disrupted or destroyed.

Small objects such as icy moons are disassembled by the gas giants and the debris "feeds" the rings, Figure 4.20(b).

Concept Practice

1. *Jovian planets*
   (a) ... are very large and have rings.
   (b) ... have few or no moons.
   (c) ... rotate slowly around their axes.
   (d) ... consist mostly of water ice because they are so far away from the sun.
   (e) None of the above

4.5 Explaining the Formation of the Solar System

We saw in Section 3.5.2 that already in the eighteenth century, the Kant-Laplace *nebular hypothesis* evolved which described the main features of our solar system. By positing that a gas and dust nebula—contracted by gravity and flattened by rotation—gave birth to sun and planets, the theory explained why all planets rotate around the sun, most rotate on their axes in the same, counterclockwise sense, and why all orbits lie in roughly the same plane. Here, we describe the modern refinement of the theory which is corroborated both by advanced computer simulations and observational evidence. The latter involves detailed studies of stellar birthplaces such as the Orion nebula, Figure 4.21(a), or stars very early in their life cycle, such as *T Tauri* stars, Figure 4.21(b), which show **mass ejections**, strong *stellar winds*, and a *protoplanetary disk*.

4.5.1 Two Groups of Planets Emerge Due to Temperature Differences in the Early Solar System

The most obvious pattern to be explained is the existence of two groups of planets so opposite in all their characteristics that it cannot be an accident. The question is why *terrestrial planets* are light, small and rocky, while *Jovians* are massive, big and gaseous. From our survey of the solar system in Section 4.1, we suspect that the answer has to do with the

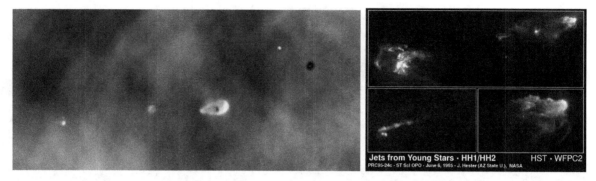

Figure 4.21: **Formation evidence.** (a) A zoom into the Great Orion nebula (M42), Figure 3.10(a), reveals protoplanetary disks, that is, nascent solar systems. (b) Mass ejections of a very young, *T Tauri* type star (Herbig Haro object).

fact that all **Terrestrials** are much closer to the sun than all **Jovians**. To say anything definitive, we have to look at the conditions as they existed close to and far from the sun *early* in the solar system's history. Keep in mind that the **primeval gas and dust cloud** was very dilute, even when substantial collapsing had taken place. In other words, the gas pressure was very much lower than required for substances to exist in their liquid state. All substances were therefore either gaseous or tiny specs of dust, comparable to snowflakes. The quantity that controls which of the two states a substance is in at low pressures is the **condensation temperature**. If the ambient temperature is above the condensation temperature of the substance, it is gaseous, if below, it "freezes out" as a solid. We really only need to distinguish two classes of substances here. The **light substances** such as water, ammonia or methane have a low condensation temperatures between 100 K and 300 K. In the other class we find substances such as iron, nickel and others which are the main ingredients of rocks. This class of **rock-forming** substances have a high condensation temperature between 1300 K and 1600 K. On the other hand, hydrogen and helium, the most abundant substances in the universe, have condensation temperatures so close to absolute zero (0 K) that they are virtually always in their gaseous state—except under extremely high pressure.

What did the primeval gas cloud look like when the solar system started to form? It had a temperature of about 50 K, so all substances except hydrogen and helium were in their solid state, frozen out as tiny dust particles. According to the nebular hypothesis, the gas and dust cloud contracted and flattened, forming a **protosun**, that is, a large mass that would turn into the sun, at its center, and a protoplanetary disk, that is, a disk of material that would turn into planets, around it. This phase of contraction and flattening lasted about 100,000 years. Scientists can estimate this by taking into account the gravitational forces between the particles with a total mass on the order of the mass of the present day sun. They can simulate on a computer how a cloud of many tiny particles develops as time goes by. Contraction of material normally leads to higher temperature, as the particles are closer together and moving faster, that is, with higher motional energies. As the protosun in the center shrank, it turned gravitational energy into thermal energy. Radiating this energy, it heated up the developing inner regions of the solar system. A temperature profile of the early solar system is shown in Figure 4.22. Note that the temperatures at this time where much higher than they are today. The reason is that the protosun had a much larger surface than the sun today, but about the same surface temperature, so, according to the **Stefan-Boltzmann law**, Equation

4.5. EXPLAINING THE FORMATION OF THE SOLAR SYSTEM

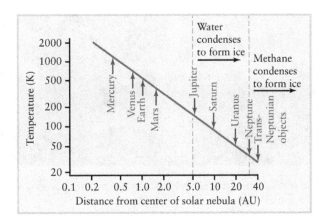

Figure 4.22: **Temperature distribution in the early solar system.**

(3.12), the protosun radiated much more energy, that is, its luminosity was much higher than the sun's today. This may seem strange, because, after all, the protosun is not a star (yet), because its central temperature was not high enough to start **nuclear fusion**.[47] As we alluded to, the energy source of the protosun was **gravitational energy**.

The important information you should gather from Figure 4.22 is that the temperature drops sharply with the distance from the protosun, such that beyond a distance of about two Mars orbital radii, it fell below the condensation temperature of water. In other words, at a distance about the Jupiter's orbital radius, water condenses or freezes out to form ice. Therefore, we have two distinct regions in the **protoplanetary disk** where planets started to form. Up to about 5 AU, the only solid particles are of the **rock-forming substances**, whereas past 5 AU both the rock-forming and the **volatile substances** (most importantly water) are in their solid state. Understandably, this resulted in two different ways in which planets formed in the two regions. Close to the sun, the dust particles collided and formed chunks, mostly held together by **chemical bonds**—not gravity. Over a few million years, these chunks further combined, gradually building up on the order of a billion objects astronomers call **planetesimals**. Think of them as asteroids with a diameter of one kilometer or so. Once this amount of mass is concentrated in objects, they can attract each other gravitationally more effectively than via **chemical bonds**. Through gravity, planetesimals coalesced into bigger objects, probably the size of Earth's moon, called **protoplanets**. These protoplanets then collided to form the **terrestrial planets** we know today. The collision of protoplanets are truly energetic events, and scientists believe that much of the heat trapped in terrestrial interiors stems from the motional energy released in these collisions. Typically, the entire planet would melt. At this time **planetary differentiation** occurred. That is, the different materials of which the planets consist—notable heavy, iron-rich and lighter, silicon-rich materials—separated and "sorted themselves." Helped by gravity, the heavier elements sank to the center of the molten planet, while the lighter materials floated toward the surface. This explains the density profile of the **terrestrial planets** very well, Figure 4.23.

This, in a nutshell, is how the **terrestrial planets** formed. What went differently for the **Jovians**? Recall that the temperatures in the realm of the proto-Jovians were low enough that also water and other volatile materials remained in their solid state. In short,

---

[47]More about **fusion** as the energy source of the sun and the stars in Section 5.4.

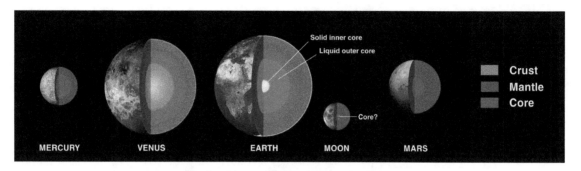

Figure 4.23: **Interiors of terrestrial planets.** Visible is the interior structure and the relative sizes of the four terrestrial planets and Earth's moon.

then, there was more "solid stuff" around, and the same process of **accretion** that formed the proto-terrestrials led to the formation of objects which would end up as the core of a Jovian planet—only much larger than in the terrestrial case, because more solid material was available in the outer solar system. This lead to a second stage of formation. Namely, the gravitational field of the large "cores" was strong enough, and the gaseous elements (hydrogen and helium) were cold, that is, slow enough, so that the core could capture them to form a gaseous envelope—becoming even more massive. This runaway process went on until most material around the nascent planet was used up. Scientists working with this **core accretion model** estimate that large planets with a core about four or five times the mass of the Earth and a thick hydrogen-helium mantle should result—in accordance with the structure of the outer planets.

Around that time the temperatures at the center of the protosun increased enough to start **nuclear fusion**. The sun thus became a *bona fide* star. Young stars go through an unstable phase known as **T Tauri stage**, in which they eject a lot of mass and put up a strong **stellar wind**. This basically clears the rest of the volatile materials out of the solar system, and the formation of planets comes to an end. The completely formed, but very young solar system had many small solid objects still floating around, and over the next hundreds of millions of years some of them would collide with planets and moons to form the impact craters we see on the moon, Mars, and other members of our solar system—the **heavy bombardment era**. Others would be "catapulted" out of the (inner) solar system by gravitational interaction with the large planets. Eventually, the solar system became the highly organized, orbitally stable, and largely empty system we know today.

4.5.2 The Nebular Hypothesis Is Challenged by Finding Exoplanets

The **nebular hypothesis**, in its modern refinement known as the **solar nebular disk model**, is thus an amazingly successful theory, and really surprisingly simple. It can coherently explain many of the patterns we find in the solar system. It makes so much sense that nobody seriously doubted it once the details had been worked out. And still: No theory is above experimental or observational testing—and this theory could not be put to a test. It was constructed to reproduce the known patterns, which it did very elegantly, but this is not the same as *predicting* other patterns that then are observed *a posteriori*. The reason for the lack of an independent test was simply that there was only one solar system. Until

4.5. EXPLAINING THE FORMATION OF THE SOLAR SYSTEM

very recently, it was a matter of pure speculation to say that there exist other planetary systems out there, where so-called *exoplanets* rotate about fixed stars other than the sun. I still remember how impossible it seemed to ever find exoplanets when I grew up. After all, planets are merely reflecting their stars' light, and are hundreds of times smaller, which means billions of times dimmer than their host star. One almost wants to coin a planetary version of the statement about the impossibility of knowing what stars are from Section 3.9: "Whether exoplanets exist and what their characteristics are we do no know and will never know." But just as spectroscopy was around the corner to enable astrophysicist to know what stars are, so were the methods to discover exoplanets being developed quickly. As recently as 1995 the first exoplanet was detected—now we know thousands of them! This new data amounted to a test of the **nebular hypothesis**—and about almost falsified it! The problem was that astronomers started to discover lots of **hot Jupiters**, that is, large gas planets much closer to their star than Mercury is to the sun. This was in direct contradiction to the theory of planetary formation that we just so plausibly developed. How could such a convincing and beautiful idea cave in on the first occasion? While this may be disappointing, it is not at all uncommon in science. There have been hundreds of elegant, simple, compelling theories that were later found not to be realized in nature, that is: wrong.[48] Being the **pragmatists** they are, scientists picked up the pieces and "rebooted" the theory. First, they realized that **hot Jupiters** are exactly what you would *expect* to find with the discovery methods they'd been using. This is called **detection bias**. If your apparatus is sensitive to big planets close to their host stars—that's what you will find. The result gives you the false impression that exoplanets *are* mostly hot Jupiters. Indeed, that is not the case. The fact is that the *only* exoplanets you can find (with your method) are hot Jupiters. Using *other* methods and, eventually, a dedicated **satellite mission** called *KEPLER* launched in 2009, several thousand exoplanets were found, giving astronomers a chance to run some reliable statistics. It was found that hot Jupiters are fairly rare, and—excitingly—that earth-sized planets are fairly common[49]. Still, the skeptic in us rears his head and insists that *"hot Jupiters* should not exist, but they do, so the theory is dead." Planetologists were swiftly finding a *patch* for the theory. According to a commonly accepted hypothesis, *hot Jupiters* are forming far away from their host star, but are **migrating** close to it. This idea seems as creative as it seems ridiculous. How can a planet migrate? It is not impossible. For instance, our Jupiter throws minor objects like asteroids and comets out of the (inner) solar system. By a close encounter with Jupiter, the orbits of these minor objects are changed dramatically by getting a gravitational "kick" which sends them far away from the sun. This is known as a **gravitational slingshot** and has been used successfully to accelerate space probes heading for the outer solar system.[50] Of course, by Newton's third law ("action equals reaction"), these minor objects exert a force in the opposite direction on Jupiter, pushing it a tiny bit *closer* to the sun. If these encounters are common, as they must have been in the early solar system, then millions of these occurring during millions of years can substantially alter the orbit of a large planet—it will migrate

---

[48]For instance, the cosmologist GEORGE GAMOV (1904–1968) had an clever theory about the genetic code in the 1950s that was rescinded because nature "chose" to be redundant instead of clever. Right now, **supersymmetry**—an incredibly versatile generalization of symmetry in particle physics *and* the underpinning of **superstring theory**—is on the chopping block, because experimentalists just can't find its predicted particles. Its demise will be a big blow to all of theoretical physics. Beauty and mathematical sophistication are a strong guiding principles, but no guarantee for finding "truth" in the sciences.

[49]The excitement, of course, comes from the hope to find life on Earth-like planets.

[50]The Pioneer and *Voyager* space probes in the 1970s and 1980s come to mind.

inward. This still sounds pretty weak as a suggestion, but there is evidence that Uranus and Neptune have migrated substantially, if in the opposite direction. These planets are too big to have formed at their present location 19 AU and 30 AU from the sun. The **protoplanetary disk** was just too dilute at these distances to form planets several times the size of Earth; they must have migrated after formation. With the new data pouring in from the newfound exoplanets, the theory of planetary formation was refined to explain the new patterns. In the normal course of the scientific method this will probably happen several times over, resulting each time in a description of the universe that is slightly better than the last one.

Concept Practice

1. There are two major groups of planets because
    (a) ... this is a historic accident.
    (b) ... Jupiter almost became a star.
    (c) ... the original gas cloud had heavier materials close to its center and lighter substances farther from its center.
    (d) ... the temperature close to the sun prevented the terrestrials from holding onto volatile substances.
    (e) None of the above

4.6 Observing the Nearest Star

Clearly, the sun is special in the solar system, because it is the only object that shines by its own light. Recall that we can conclude this from two separate observations. Firstly, we saw that the moon, Venus, and other planets show **phases**. In other words, they reveal at least part of their dark, unilluminated side. Secondly, the advent of **spectroscopy** around 1860 enabled astronomers to conclude that the planets *reflect* sunlight by seeing the same dark lines in the spectrum of the planets that they saw in the sun's spectrum—plus some additional lines stemming from the planet's themselves. The sun is thus a true star. Clearly, its proximity makes it a great study if we want to understand the other fixed stars. In Chapter 5 we will reap the benefits of our solar investigations when the **stellar energy production mechanism** is revealed. Our present interest is in the properties of the sun as they can be deciphered from telescopic observations alone. We already described in Secs. 3.8 and 3.9 some of the early ideas astronomers had about how the sun and the stars produce their energy. These hypotheses were based on one important measurement, namely the determination of the sun's luminosity, that is, the total amount of energy the sun radiates each second. This 1838 measurement by POUILLET and later JOHN HERSCHEL yielded the astonishing value $L_{Sun} = 3.8 \times 10^{26} W$, see Equation (3.8). We also saw in Section 3.8 that this enabled JOSEF STEFAN of *Stefan-Boltzmann* fame to determine the sun's surface temperature for the first time reliably and accurately in 1879 to be about $T_{Sun} = 5,800\,K$. You have to understand that this is an **effective temperature**, that is, the temperature that a perfect blackbody of the same surface area must have to radiate exactly as much power as the sun. The sun itself emits an *approximate* black-body spectrum, Figure 3.35(b). Clearly, it cannot be perfect, because it is marred by the dark absorption lines.[51]

---

[51] Note that the spectrum, Figure 3.35(b), is an **anachronism**, in that it implies the determination of the sun's energy output at many wavelengths, including ones we cannot observe from the Earth's surface due to atmospheric absorption.

4.6. OBSERVING THE NEAREST STAR

What else can we tell from afar? Other than the physical properties such as mass, radius and the surprisingly low density (mass per volume), all we can see is the *outside* of the sun. This is not quite as obvious as it seems. Today we think of the sun is a giant gas ball, and so you might wonder why we cannot look through the material of the sun, much as we look through the Earth's gaseous atmosphere to observe the universe. The sun's outer layer that defines its surface is called the ***photosphere***. Virtually all visible light we see coming from the sun stems from this very thin layer (400 km), which explains its name ("sphere of light"). The reason is that the sun's material, a hot ***plasma*** of atomic nuclei and electrons, is transparent for visible light only down to a depth of about 400 km. This distance, on the other hand, is about two thousand times smaller than the sun's radius, which is why the sun *appears* to have a well-defined surface and a sharp edge. In actuality, the gas that makes up the sun gets denser *continuously*. The sun hence doesn't have a surface; it is merely *defined* as the position of the ***photosphere***. If this sounds strange, recall that the same is true for the Earth's atmosphere. It thins out farther from the surface. Technically, it never ends, but for practical purposes we define its extent to be 100 km above the terrestrial surface, where the atmosphere is so thin that airplane flight becomes impossible. This altitude is known as the ***Karman line*** and used as the dividing line between Earth's atmosphere and ***outer space***.[52] The ***photosphere*** of the sun is extremely dilute, only 1/10,000 the density of Earth's atmosphere at sea level. Yet, it is here that the energy rising up from the interior of the sun gets converted and sent off as visible light into outer space. We can see evidence for the transport of this energy. In a telescope fitted with a ***solar filter*** one can see a fine pattern of blotches covering the sun's surface, Figure 4.24. This pattern was noticed by GALILEO and SCHEINER in the early seventeenth century, described in detail by JAMES NASMYTH (1808–1890) in 1860, and is called ***solar granulation***. It gives the sun the looks of a boiling surface. This is right on, as these structures are caused by the ***convection*** of gases in the sun's outer layer, in close analogy the rising and sinking of hotter and cooler material in the boiling process of water, Figure 3.41. Invoking the ***Stefan-Boltzmann law***, we understand that the hotter material radiates more and therefore appears brighter.[53] Careful observation reveals that the ***granules*** last up to eight hours and are about 1,000 km across. The study of patterns like this can lead to important insight into the inner workings of the sun in particular, and the stars in general.

Solar eclipses afford a more detailed view of the outer layers of the sun. When the moon obscures the brightest parts of the sun, we can see the other two of the sun's three major outer layers. Visual observations during an eclipse reveal the ***solar corona***, Figure 4.25(a), which is aptly named for its ragged appearance due to streamers of gas, reminding us of a *crown*. This outermost layer of the sun is astonishingly thin (a million times thinner than the ***photosphere***), while having an extremely high temperature, on the order of millions of Kelvins. Why can't we see the corona all the time? After all, at these high temperatures, it should radiate a lot of energy. The reason is that the corona is not a dense blackbody. Rather, it is so dilute that there is not enough material that can radiate. Much like the air inside a kitchen oven in which your hand does not immediately get burned if you put it in there for a brief moment, it does not carry a lot of energy. The high temperature of the corona means that its atoms and ions are moving at very high speeds. So high is there speed that they can

---

[52]This definition became important when the ***Ansari X prize*** was awarded in 2004 for the first reusable spacecraft to reach *outer space* twice within two weeks.

[53]The material along the edges of a ***granule*** is only 300 K or 5% cooler, but recall that the power radiated by a blackbody is a function that changes dramatically with the temperature (like T^4).

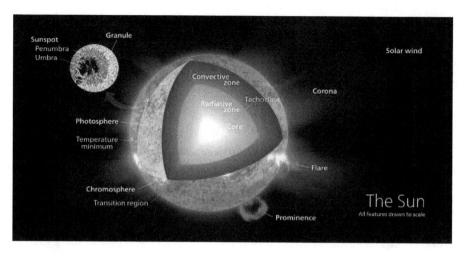

Figure 4.24: **The sun.** Visible are *solar granulation*, *sunspots*, the *photosphere*, the *chromosphere*, *flares*, and *prominences*.

escape the sun's gravity altogether. The resulting stream of (charged) particles is known as the ***solar wind***. Its interaction with the Earth's atmosphere and magnetic field gives us the ***aurorae*** or ***northern/southern lights***. Major mass ejections from the sun can even lead to failure of satellite communication systems on Earth.

The second major outer layer visible during eclipses, the pinkish glowing ***chromosphere***, Figure 4.25(b), which sits between the ***photosphere*** and the ***corona***. It is 10,000 times thinner than the photosphere, which explains two things. For one, it is much dimmer than the photosphere, and it produces an ***emission spectrum*** rather than the photosphere's ***absorption spectrum***. This in turn means that it radiates only specific wavelengths, notably the red H_α line at 656.3 nm, which explains its pinkish-red color. It is 3000 km to 5000 km deep, and features rising jets of gas known as ***spicules***. The latter are influenced heavily by the structure of the sun's magnetic field.

Indeed, the structure of the ***solar magnetic field*** is the explanation of many of its surface features. Most notably, it is the reason for the ***sunspots***, Figure 4.24. Recall that the sunspots are dark, irregularly shaped regions on the face of the sun first seen in the early seventeenth century by the likes of FABRICIUS, HARRIOT, GALILEO and SCHEINER. We saw that they can be used to determine the rotational period of the sun. In fact, careful observation reveals that the sun does not rotate like an rigid object, but displays ***differential rotation***. Namely, sunspots on the solar equator rotate faster (about 25 days) than sunspots at higher latitudes (about 33 days at 75° latitude). We saw that there was a scientific debate about what sunspots are between GALILEO and SCHEINER. GALILEO was largely correct in pointing out that sunspots are patterns found directly at the surface of the sun. They appear dark, because they are cooler than their vicinity, owing to the ever-useful ***Stefan-Boltzmann law***. Their formation and existence was long a mystery, and to ameliorate the situation, their appearance, abundance and locations were recorded in great detail to shed more light on their genesis. It is not hard to realize that in some years we find a great number of sunspots on the face of the sun, whereas in other years, we might not find but a few. After 17 years of intense solar observing, SAMUEL HEINRICH SCHWABE (1789–1875) realized in 1843, that there is a systematic, periodic ebb-and-flow of sunspot numbers. The time between two

4.6. OBSERVING THE NEAREST STAR

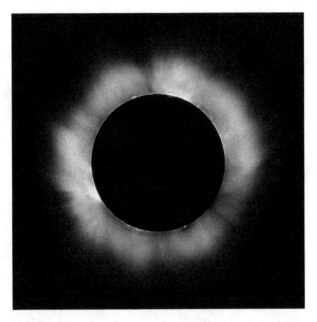

Figure 4.25: **The sun's outer layers.** The *corona* is named after its crown-like appearance. The *chromosphere* is very hot and dilute. Its pinkish appearance comes from its strong red hydrogen (Hα) emission lines.

maxima, Figure 4.26(a,top), is 11 years *on average*. The latter qualification means that the length of the cycle fluctuates somewhat, as you can tell from the **statistical noise** in Figure 4.26(a,top). Later, the Swiss astronomer RUDOLF WOLF (1816–1893) was able to reconstruct the sunspot number from historic records all the way back to GALILEO's times. And another pattern connected to sunspots was recorded. Namely, it was noticed that sunspots form at higher solar latitudes early in the **sunspot cycle**, and nearer to the solar equator late in the cycle, Figure 4.26(a,bottom). These intricate patterns of nature awaited explanation for a long time. Finally, in 1908 GEORGE ELLERY HALE (1868–1938), of large-telescope fame, discovered that sunspots are associated with the solar magnetic field. How did he do that? For starters, it was noticed that sunspots typically appear in pairs, as do the poles of a magnet. This could have been a coincidence, but HALE took the spectrum of a sunspot[54] and found that some absorption lines were split into three separate lines. This is a general behavior of **atomic energy levels**, whose energies will change in strong magnetic (and electric) fields. In fact, the influence of magnetic fields on atomic spectra had been discovered just about a decade earlier by PIETER ZEEMAN (1865–1943), after whom the effect is named. As you might guess, the splitting is larger the stronger the field is. This allowed astronomers to determine that the fields around sunspots are about 5,000 times stronger than the Earth's magnetic field. We already saw in Section 4.2.1 that magnetic fields dramatically influence the motion of charged particles. Recall that the charged particles from the sun are guided by magnetic field lines toward the polar regions, where they bump into the molecules of the atmosphere to cause **aurorae**. On the sun, the charged particles of the hot plasma rising to the surface are pushed around in such a way by the strong solar magnetic fields that a region

---

[54]It is not hard to *zoom in*, that is, increase the magnification, and take a spectrum just of a small portion of an object.

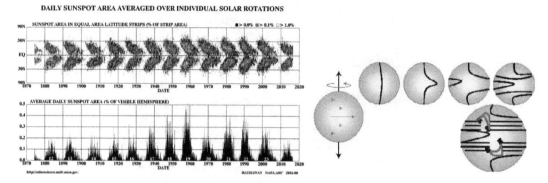

Figure 4.26: **Sunspot cycle And formation.** (a) Top: Area of sunspots (proportional to the number of sunspots) as a function of time during the last 150 years. Bottom: Appearance of sunspots as a function of solar latitude and time, showcasing *Spörer's law* ("Butterfly diagram"). (b) The Babcock model of sunspot formation: The sun's magnetic field lines get tangled up due to *differential rotation*.

of relatively cool material materializes—a sunspot is created.

There are more patterns to be discovered. Usually sunspots form in groups. They move with the sun's rotation which creates a sense of direction. There are sunspots which are in front ("preceding spots") and some that are trailing ("following spots") as the sunspot group moves. As it turns out, these two groups of sunspots always have opposite magnetic polarity. What's more, the preceding spots all have the same polarity *and* the same polarity as the hemisphere of the sun. For instance, when the magnetic north pole of the solar field is north of the sun's equator, all preceding spots in the northern hemisphere have north magnetic polarity, and all preceding spots in the southern hemisphere have south magnetic polarity. Finally, the sun's magnetic field *reverses* its polarity every 11 years, so that the magnetic north pole become the magnetic south pole and vice versa. This means that only after 22 years we are back to exactly the same sunspot situation. Astronomers therefore call this a *22-year solar cycle*. As you see, there is a lot of data we can gather by careful observation. These intricate patterns of nature displayed on the face of the sun need an explanation.

Eventually, American astronomer HORACE BABCOCK (1912–2003) came up with a very elegant and compelling theory to make a 22-year cycle plausible. If you think about it, then 22 years seem a ridiculously short amount of time for a star which has existed in stable condition, that is, in **hydrostatic** and **thermal equilibrium**,[55] for *billions* of years. The only thing moving in a comparable time as the 22-year-cycle is the sun itself as it rotates around its axis in about a month. Therefore, it is not too surprising that Babcock's **magnetic-dynamo model** has to do with the sun's rotation—and, of course, its magnetic field. The cornerstone of the model is the **differential rotation** of the sun. The magnetic fields are associated with the moving charged particles of the sun's hot plasma, and therefore the fields will deform as the solar surface gets morphed by the differential rotation, Figure 4.26(b). The field lines become so stretched as they wrap around the solar surface that **convective** motion in the photosphere can tangle up the field lines. As they develop kinks analogous to a tangled garden hose, the magnetic fields leave the surface and enter it some thousand kilometers away. Such a kink of the magnetic field thus creates a sunspot pair. The intricate, kinky field pattern can even build up to massive proportions and will push plasma material away from the sun.

---

[55]In other words, the star does not change its size nor its temperature.

We can observe these "eruptions" as violent **solar flares** lasting minutes, or more massive **prominences** which can last for weeks; see Figure 4.24. While it seems plausible then that magnetic fields could be the source for such phenomena, this is not enough evidence to support the theory. Fortunately, the **magnetic-dynamo model** allows us to explain much more. For instance, the greatest kink in the field lines after a few rotations in Figure 4.26 appears safely away from the equator, in mid-northern and southern latitudes. Therefore, the model predicts sunspots to appear away from the equator early in the solar cycle, in accordance with the observations. Later in the cycle, after many rotations, the greatest entanglement is found close to the equator—along with the associated sunspots. Even more convincing is the model's explanation of the *end* of the cycle. We can observe that the *preceding sunspots* move toward the equator while the *following sunspots* move toward the poles. However, the *preceding sunspots* of the two hemispheres have opposite polarity, so when they meet at the solar equator, their fields will cancel. This is a similar fate as meets the *following sunspots* as they drift toward the solar poles of opposite polarity. However, the effect of the *following sunspots* is such they they *reverse* the sun's magnetic field. You can see that Babcock's **magnetic-dynamo model** effortlessly explains all gross features of the 22-year solar cycle, which is why it became the (tentatively) accepted theory of solar activity.

Concept Practice

1. Most of the sun's light that reaches us comes from
   - (a) ... the corona.
   - (b) ... the photosphere.
   - (c) ... the radiative zone.
   - (d) ... the core.
   - (e) None of the above

4.7 Summary and Application

4.7.1 Content of the Solar System

The solar system contains the following celestial objects.

- *The sun.* A star which arguably *is* the solar system, since it accounts for 99.99% of its mass, even more of its energy output, and pretty much of anything else you can think of—except for angular momentum, which is largely carried by Jupiter.

- *The major planets.* There are two groups of planets: the inner, *terrestrial planets* and the outer, *Jovian planets*.

- *The moons.* All planets except Mercury and Venus have moons, that is, natural objects that orbit them.

- *Ring systems.* Only Jovians have them.

- *The asteroids.* Also known as *minor planets*, most of them orbit the sun between Mars and Jupiter in the *asteroid belt*, but a second asteroid belt lies beyond Neptune's orbit.

- *The comets.* Discovered since at least a thousand years, their status changed from *bad omen* to *dirty snowball* over the centuries. We saw the major breakthrough by HALLEY, who discovered the *periodicity* and therefore *predictability* of the short-period comets, Section 3.1.2.

- *Interplanetary matter.* Very dilute gas and dust as well as charged particles of the *solar wind* between the planets.

Concept Application

1. How do we know that the solar system is flat, that is, disklike?
2. Show that all the other planets of the solar system would fit into Jupiter if it were hollow.
3. A sunspot has a temperature of about 3,800 K. Use the *Stefan-Boltzmann law* to figure out how much power they radiate per unit area compared to the rest of the sun's surface. Does this explain why they appear dark?

4.7.2 Activity: The Greenhouse Effect

The **Earth system** (Earth plus atmosphere) gives off the same amount of energy that it absorbs from the Sun. If our planet absorbed more energy than it gives off, the Earth would continuously become warmer and warmer—which is not the case.

By setting the incoming solar radiation absorbed by Earth equal to the outgoing radiation emitted by Earth, we can estimate that the effective temperature of Earth would be a frigid 255 K (0 °F). Note that this estimate assumes that Earth has no atmosphere and reflects 30% of the incoming sunlight.

1. Judging from Figure 4.15, what is the total average amount of energy that every square meter of the Earth receives from the sun every second? (We will (over-)simplify our conversation by referring to this energy per second per square meter simply by energy.)
2. How much energy does the Earth (plus atmosphere) radiate back into outer space?
3. From which sources does the Earth's surface receive energy?
4. How much energy does the Earth's surface receive from these sources total?
5. How much energy does the Earth's surface emit in total?
6. Describe the resulting apparent paradox, keeping in mind that the **greenhouse effect** raises the surface temperature of the Earth by 40 degrees.
7. What kinds of radiation do we primarily receive from the sun? (Hint: Recall the position of the sun's blackbody curve.)

The Earth's atmosphere plays a crucial role in the **greenhouse effect**, because some forms of radiation can travel relatively unhindered in air, while others get frequently absorbed and re-emitted by **greenhouse gases**, thus being effectively slowed down. As very crude categories, use ultraviolet (UV, 0.1 to 0.2 micrometers), visible light (0.4 to 0.7 micrometers), infrared (IR, 1 to 20 micrometers).

8. Consider Figure 3.36 showing atmospheric abortion. Which types of electromagnetic radiation travel unhindered, which others do not?

4.7. SUMMARY AND APPLICATION

9. Which gases are primarily responsible for atmospheric absorption?
10. What happens to the radiation absorbed by the Earth's atmosphere?
11. Due to the radiation emitted by the Earth's atmosphere and absorbed by the Earth's surface, is the Earth's temperature near its surface going to be warmer or cooler than it would be without this radiation?
12. Describe the atmospheric *greenhouse effect* in your own words.
13. Compare the two concepts *greenhouse effect* and *global warming*.
14. What are the issues in global warming and climate change (scientific, political or others)?

4.7.3 Activity: Formation of the Solar System

According to the *nebular hypothesis*, the solar system formed from a primordial gas and dust cloud. For a schematic sketch of the process see Figure 3.14.

1. Describe the different stages of the formation of the solar system.
2. Which of the features of our solar system are explained by this theory? Name and explain at least three.

Assessing the content of the solar system, we found the following patterns:

- The sun is ten times bigger than the Jovian planets, which are ten times bigger than the terrestrial planets
- The sun and the Jovians have densities (mass per volume) that is only one third of the density of most of the terrestrial planets.

How can we explain these patterns? The crucial fact is that the Jovians are much farther from the sun. Indeed, there is a large gap between the outmost terrestrial planet (Mars) and the innermost Jovian (Jupiter). To understand how this impacts the formation of planets, we look at the temperature distribution in the early solar system as depicted in Figure 4.22. Note that the temperatures during the formation of the solar system were much higher than they are today. Also keep in mind that water freezes at 273 K and boils at 373 K.

3. What was the temperature range at the location of the terrestrial planets?
4. What was the temperature range at the location of the Jovian planets?

At low temperatures light materials such as hydrogen and helium can condense to form large planets, but when temperatures are too hot (warmer than the freezing point of water), these substances are too volatile to stick together.

5. Develop a consistent theory as to how the formation location of a planet determines its size, mass and density.
6. Does this theory *save the appearances*, that is, does it explain the features and patterns of the solar system? Explain.

4.7.4 Activity: The Sun's Properties

In Section 4.6 we learned about the Sun's properties and how scientists were able to discover them. To help you internalize and retain this information, revisit the main issues by answering the following questions.

1. How do we know how big the sun is?
2. How do we know how much mass the sun has?
3. What is the sun made of?
4. How do we know that?
5. What is the temperature of the sun on its surface and how do we know?
6. What is the temperature of the sun at its center and how do we know?
7. Is the gas pressure higher at the center of the sun or its surface? Explain how you know. Use the water pressure experienced by a scuba diver as an analogy.
8. Based on your answers above would you say that the sun's core is denser, less dense or has the same density compared to regions close to its surface?
9. How do we know that the sun produces energy?
10. How can we measure or compute how much energy the sun produces?
11. What are the outer layers of the sun?
12. How can we observe them?

4.7. SUMMARY AND APPLICATION

Image Credits

Fig. 4.3: Source: https://solarsystem.nasa.gov/galleries/vesta-sizes-up.
Fig. 4.4a: Copyright © 2011 by James St. John, (CC BY 2.0) at
https://commons.wikimedia.org/wiki/File:Murnpeowie_meteorite.jpg.
Fig. 4.4b: Copyright © 2010 by Shane.torgerson, (CC BY-SA 3.0) at
https://commons.wikimedia.org/wiki/File:Meteorcrater.jpg.
Fig. 4.5a: Copyright © by American Museum of Natural History.
Fig. 4.5b: Source: https://commons.wikimedia.org/wiki/File:Elemental_abundances.svg.
Fig. 4.6: Copyright © 2013 by Kelvinsong, (CC BY-SA 3.0) at
https://commons.wikimedia.org/wiki/File:Earth_poster.svg.
Fig. 4.7: Copyright © 2010 by ESO; adapted by Henrykus, (CC BY-SA 3.0) at
https://commons.wikimedia.org/wiki/File:Gliese_581_-_2010.jpg.
Fig. 4.8: Source: https://commons.wikimedia.org/wiki/File:Plate_tectonics_map.gif.
Fig. 4.9: Copyright © 2014 by T. Slater and R.Freedman.
Fig. 4.10: Copyright © 2013 by Kaidor, (CC BY-SA 3.0) at
https://commons.wikimedia.org/wiki/File:Earth_Global_Circulation_-_en.svg.
Fig. 4.11: Copyright © 2012 by Kelvinsong, (CC BY 3.0) at
https://commons.wikimedia.org/wiki/File:Moon_diagram.svg.
Fig. 4.12: Copyright © 2014 by Eblamble, (CC BY-SA 3.0) at
https://commons.wikimedia.org/wiki/File:Complex_Impact_Crater_Formation.png.
Fig. 4.13a: Copyright © by Mark Elowitz.
Fig. 4.13b: Copyright © 2009 by Areong, (CC BY-SA 4.0) at
https://commons.wikimedia.org/wiki/File:Kasei_Valles_topo.jpg.
Fig. 4.14a: Source: https://nssdc.gsfc.nasa.gov/photo_gallery/photogallery-venus.html.
Fig. 4.14b: Source: https://commons.wikimedia.org/wiki/
File:Mars_topography_(MOLA_dataset)_with_poles_HiRes.jpg.
Fig. 4.14c: Copyright © 2009 by Areong, (CC BY-SA 4.0) at
https://commons.wikimedia.org/wiki/File:Kasei_Valles_topo.jpg.
Fig. 4.15: Copyright © (2007) Robert E. Rohde, (GNU V1.2) at:
https://commons.wikimedia.org/wiki/File:Greenhouse_Effect.svg. A copy of the license can be found here:
https://commons.wikimedia.org/wiki/Commons:GNU_Free_Documentation_License,_version_1.2.
Fig. 4.16: Copyright © 2010 by NOAA; adapted by Autopilot, (CC BY-SA 3.0) at
https://commons.wikimedia.org/wiki/File:Vostok_Petit_data.svg.
Fig. 4.17a: Source: https://commons.wikimedia.org/wiki/File:Grinnell_Glacier_1981.jpg.
Fig. 4.17b: Source: https://commons.wikimedia.org/wiki/File:Grinnell_Glacier_2009.jpg.
Fig. 4.18: Copyright © 2012 by Tdadamemd, (CC BY-SA 3.0) at
https://commons.wikimedia.org/wiki/File:SolarSystem_OrdersOfMagnitude_Sun-Jupiter-Earth-Moon.jpg.
Fig. 4.17: Source: https://wisp.physics.wisc.edu/astro104/lecture8/F07_06.jpg.
Fig. 4.20a: Source: https://canadianastronomy.files.wordpress.com/2014/05/saturnhighlightcrop.jpg.
Fig. 4.20b: Copyright © 2014 by Theresa Knott, (CC BY-SA 3.0) at
https://commons.wikimedia.org/wiki/File:Roche_limit_(tidal_sphere).PNG.
Fig. 4.21a: Source: https://commons.wikimedia.org/wiki/File:M42proplyds.jpg.
Fig. 4.21b: Source: https://commons.wikimedia.org/wiki/File:HH1_and_HH2_imaged_by_WFPC2.jpg.
Fig. 4.22: Copyright © 2014 by T. Slater and R. Freedman.
Fig. 4.23: Source: https://commons.wikimedia.org/wiki/File:Terrestial_Planets_internal_en.jpg.
Fig. 4.24: Copyright © 2012 by Kelvinsong, (CC BY-SA 3.0) at
https://commons.wikimedia.org/wiki/File:Sun_poster.svg.
Fig. 4.25: Copyright © 1999 by Luc Viatour, (CC BY-SA 3.0) at
https://commons.wikimedia.org/wiki/File:Solar_eclipse_1999_4_NR.jpg.
Fig. 4.26a: Source: https://commons.wikimedia.org/wiki/File:Sunspot_butterfly_graph.gif.
Fig. 4.26b: Source: http://www.sott.net/image/6347/Babcock_model.jpg

Chapter 5

Extrapolating & Synthesizing: Understanding the Stars

ARCHANGEL EDDINGTON: As well we know, the sun is fated/ In polytropic spheres to shine,/ It's journey, long predestinated,/ Confirms *my theories* down the line./ Hail to Lemaître's promulgation/ (Which none of us can understand!)/ As on the morning of Creation/ The brilliant works are strange and grand.

ARCHANGEL JEANS: And ever speeding and rotating,/The double stars shine forth in flight,/The Giant's brightness alternating/ With the eclipse's total night/ Ideal fluids, hot and spinning,/By fission turn to pear-shaped forms,/ *Mine are the theories* that are winning!/ The atom cannot change the norms.

(The Blegdamswej Faust, anonymous [18])

To understand the stars, the results of astronomy and of physics had to be put together. Since the conditions inside of stars are *extreme*, the laws of nature had to be **extrapolated** to describe matter under these circumstances. Some laws had to be *modified* in the process, and new descriptions of nature cropped up. **Relativity** and **quantum mechanics** are two theories that emerged and changed our view of the universe for good. Astronomers were eventually able to **synthesize** a theory of stars from astronomical observations of stellar properties and the emerging discoveries of **subatomic physics**.

Guiding Questions of this Chapter

5.1 What is the impact of **modern physics** on astronomy?

5.2 How are the many properties of stars related?

5.3 Are stars hot gas balls? How does a hot gas ball work?

5.4 How do stars shine, that is, what is their energy production mechanism?

5.5 How does the energy production and transport determine the properties of a star? Why do stars become unstable at the end of their lives?

5.6 How do stars die? Why do they die differently depending on their mass?

5.1 Death by Extrapolation: The Birth of Modern Physics

In the late 1800s it seemed as if most laws of astronomy and physics had been found.[1] Indeed, Newtonian mechanics successfully explained all the patterns in nature from the cannon shot and the tides to the orbits of the planets and binary stars, as we have seen. In 1864 the new *theory of electricity and magnetism* was formalized; MAXWELL explained light as an electromagnetic wave, and HERTZ produced and detected other electromagnetic waves in an experiment in 1888. Finally, *thermodynamics* explained all the bulk properties of matter like temperature and pressure. We have seen that these insights into the workings of the cosmos led to the development of *astrophysics*. We repeat the underlying assumption: *The laws of nature are universal*, that is, the same laws apply up there as they do down here on Earth. This assumption allowed scientists to apply all that they had learned in the laboratory to the stars and other celestial objects. The laws of physics and chemistry, and the rules of mathematics could be used together to synthesize new insights and to predict new phenomena.

There is, however, a problem with this assumption: How do we know that we correctly induced the laws from our observations in the labs? After all, the cosmos is vast, and there likely exist conditions in the universe which we cannot (re-)create here on Earth. It is possible that the laws of nature break down, or need to be modified under extreme circumstances. What works at 300 Kelvin might not work at 300 million Kelvin. When we study objects going 100 or 1,000 miles per hour, what does this teach us about objects going 671 million miles per hour?

Indeed, the breakdown of the laws of *classical physics*, that is, the physics until about 1900, led to the construction of two revolutionary theories, *relativity* and *quantum mechanics*, that forever changed the universe by making us look at the world with different paradigms. Sometimes the new theories and their offspring are collectively called *modern physics*, and we have to be careful not to confuse this label the similar term we have used to set *modern science* starting around 1600 apart from medieval or antique science.

Of course, it is beyond the scope of the book to even attempt a description of these new theories. Rather, this section has two modest goals. Firstly, we have the duty to report these important milestones on the way to constructing the expanding universe. Our modern view of the cosmos would not be possible without these insights—even though they are outside of the field of astronomy. Secondly, the reader needs to engage with the few modern concepts relevant for astronomy. Therefore, we attempt to demystify them as much as possible.

5.1.1 Relativity

Probably the most demystifying statement one can make about this theory invented by ALBERT EINSTEIN (1879–1955) from 1905 to 1915 is that it simply deals with the influence of *relative motion* between the observer and the observed object. In short, the world looks

---

[1] This false conviction can be illustrated by the following anecdote about MAX PLANCK (1858–1947), one of the founders of *quantum mechanics*. A physics professor in Munich (PHILIPP VON JOLLY) advised PLANCK against going into physics, because "in dieser Wissenschaft schon fast alles erforscht sei, und es gelte, nur noch einige unbedeutende Lücken zu schließen (in this field, almost everything is already discovered, and all that remains is to fill a few insignificant gaps). [31]" PLANCK assured the professor that he did not wish to discover anything new, and started his physics career under the tutelage of JOLLY, although he soon switched to theoretical physics.

5.1. DEATH BY EXTRAPOLATION: THE BIRTH OF MODERN PHYSICS

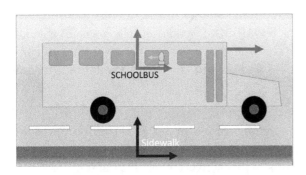

Figure 5.1: **Relativity**. (a) Students on the bus describe the world with the red reference frame, which is moving at constant velocity *relative* to the blue sidewalk reference frame. (b) An observer cannot tell whether he is on Earth or in an *accelerating* rocket. The two situations are indistinguishable. Therefore, a gravitational field is equivalent to an accelerated observer. This is an idea associated with EINSTEIN's theory of **general relativity**.

different when we move with respect to it. Imagine standing on a sidewalk and seeing a schoolbus going by on the street. The students on the bus it will describe the same scene as a stationary schoolbus with the sidewalk going by (in the opposite direction), see Figure 5.1. While different observers can disagree about what they see (here: the speed of the schoolbus and the sidewalk), they are nonetheless describing the same world, and so their different descriptions have to be consistent. The formulation of the laws that *transform* descriptions of the universe made by observers moving at different speeds is the essence of relativity. This humble agenda is reflected in the stunningly simple two *postulates of relativity*:

1. The laws of physics are the same for all observers which travel at *constant* velocity.
2. For all these observers the speed of light in vacuum has the same value $c = 300\,000$ km/s.

In other words, the world looks (or actually functions) the same regardless of the *reference frame*, as long as it is not accelerated.

These postulates lead to dramatic changes in the way we interpret the world around us. Recall that in Newton's theory the universe is merely a three-dimensional stage on (or in) which things happen, while time ticks away regardless of what happens. In *Einstein's universe*, the stage is different for different observers:

1. A moving observer measures the length of an object as being shorter compared to an observer that is at rest with respect to the object. This is known as **length contraction**.
2. An observer moving with respect to a clock sees (measures) this clock ticking slower than an observer at rest with respect to the clock. This is known as **time dilation**.

It takes quite a while for scientists to understand the full meaning of these facts, so let's home in on the punch-line: *Space and time are not absolute but relative; they depend on the relative motion of observer and object observed.* An important **corollary** is: *Quantities such as energy and momentum are not absolute but relative; they depend on the relative motion of observer and object observed.*

In particular, the **energy** of a physical system will be different for different observers. This is easy to see for motional energy $E_{kin} = \frac{1}{2}mv^2$ which depends on the mass m of the object and its speed v relative to a reference frame. If we take up the schoolbus example, the schoolbus moves and therefore has a sizable motional energy in the reference frame of the sidewalk. It's velocity and motional energy are zero for the people on the bus. Eventually, this line or arguments led EINSTEIN to conclude that there is a relation between the **energy** of a physical system (such as a particle or schoolbus) and its **mass**. This is the famous formula

$$E = mc^2 \tag{5.1}$$

which appeared as $m = E/c^2$ in EINSTEIN's original article of 1905. From our everyday experience, the connotations seem preposterous. We are so used to the permanence of objects, that is, of mass, that we cannot see how it is possible to convert mass into energy or energy into mass—yet this is precisely what is done in particle accelerators, nuclear power plants and the sun's core, as we will see.

All of the results so far are from EINSTEIN's simpler theory called **special relativity** (1905). From EINSTEIN's other theory, **general relativity**, we will only need two facts:

1. *General relativity* is the theory of a non-inertial or accelerated observer. Since it can deal with any observer (her acceleration could be zero which is the *special* case) it is more *general*, hence the name.

2. *General relativity* postulates that an observer in a windowless room cannot perform an experiment to distinguish whether the room is accelerated (as in a rocket ship) or simply close to a large mass that will result in an acceleration due to gravitational attraction; see Figure 5.1(b). Therefore, general relativity can describe the action of a gravitating object on other objects nearby. It thus replaces Newton's gravity theory with a theory of how a large mass modifies space and time around it.[2]

5.1.2 Quantum Mechanics

If the consequences of the **postulates of relativity** seem strange, then the consequences of the **postulates of quantum mechanics** are outrightly absurd.[3] However, while I encourage you to read up on this fascinating theory, we will pragmatically view **quantum mechanics** as a theory that imposes certain rules by which a consistent description of nature is constructed. In short, nature on the smallest scales is weird, but scientists found mathematical tools to describe this weirdness in a way that allows them to make predictions about the outcomes of experiments, just like with other theories. In this respect, **quantum mechanics** is not weird at all. The scaffold of the **scientific method** can—of course!—be applied to it. Put boldly, if nature is weird, we'll describe its weirdness.

It is interesting to note that **quantum mechanics**, just like **relativity** is concerned with the relation between observer and observed. Both clean up an old assumption that was always tacitly made but never fully justified: That there is a clear distinction between the world we observe (which just exists) and the independent observer which, if for a tiny moment, is "out of this world" while she makes a measurement. In other words, it was assumed that a measurement can, in principle, be made arbitrarily **non-invasive**, that is, that the interaction of the observer with the observed can be made as small as one desires. This is an idealization which becomes unrealistic if the observed objects are tiny or if the observer moves very fast with respect to the observed objects. By making this tacit assumption, scientist had effectively

---

[2]We will come back to this in the chapter on cosmology.
[3]In fact, EINSTEIN rebelled against them until his death—unsuccessfully.

5.1. DEATH BY EXTRAPOLATION: THE BIRTH OF MODERN PHYSICS

extrapolated the laws of nature: The laws were assumed to hold under extreme circumstances. But they do not, and therefore, in another revolution of Copernican dimensions, new laws of nature had to be discovered and constructed. This happened around the 1920s and was arguably the most exciting era to do science in.

To debunk a common misconception: By discovering the laws of the **modern physics**, the classical laws of nature were not falsified. True, they are not applicable under these extreme circumstances, but they are an excellent **approximation** otherwise. Even today, the laws of classical mechanics and Newton's law of universal gravitation are used justifiably almost everywhere. In fact, one of the guiding principles in developing the new laws of nature was that they'd morph into the classical laws at macroscopic distances and small velocities. The last point raises a practical question. What are "macroscopic distances" (big) and "small velocities" (slow)? The latter is clear, since we do have a gold standard for velocities: the maximal possible speed in the universe, that is, the speed of light. "Slow" means small compared to the speed of light. For instance, the speed of sound is considerable, 343 m/s = 762 mph, but small compared to the speed of light: $\frac{c}{v_{sound}} = \frac{300\,000\,000\,\text{m/s}}{343\,\text{m/s}} \approx 0.9 \times 10^6$, so light is almost 1 million times faster than sound! What about the realm of validity of classical mechanics? When would we have to use the laws of **quantum mechanics**? What is the constant of nature that distinguishes "big" from "small"? It is the **Planck constant**, $h = 6 \times 10^{-34}$ J·s. It is numerically exceedingly small.[4]

But enough of these philosophical discussions, let us focus on our quest to understand the expanding universe. For this endeavor we will need but two of the consequences of **quantum mechanics**.

- *Discrete line spectra.* Atoms can emit and absorb light only in appropriate, discrete energy chunks called **photons**, often abbreviated with the Greek letter γ (gamma). Light is an electromagnetic wave, and there is a direct proportionality between the wavelength λ and the energy E of light

$$E = \frac{hc}{\lambda}. \qquad (5.2)$$

Since electrons in atoms can have only specific, discrete energy levels, the energy or wavelength of radiation emitted or absorbed by an atom will have a specific color. This explains the line spectra of atoms; they are the consequence of the additional, quantum rules of the new theory.

- When the rules of *quantum mechanics* are applied to thermodynamics, they result in weird *properties of quantum gases*, sometimes called a **degenerate gas** or **degenerate matter**, which behaves differently compared to an *ideal gases*. In particular, the pressure of a "quantum gas" does not depend on temperature. Matter becomes degenerate at extremely high densities or extremely low temperatures (very close to absolute zero, $0\,\text{K} = -273.15\,°\text{C}$.)

---

[4]The **Planck constant** carries units of energy (Joules) times time (seconds), which is called an *action*, so one has to work a little harder to see when to apply quantum rules. For instance, an energy exchange ΔE during a small time Δt is permissible if their product is on the order of the **Planck constant**: $\Delta E \times \Delta t \approx h$.

5.1.3 The Elementary Building Blocks of the Universe

The exploration of nature at its smallest scales, that is, of the **microcosm**, starting around the *fin de siècle*, yielded discoveries just like the exploration of the night sky, the **macrocosm**. And just like astronomers find new celestial objects, such as planets and moons with specific properties like size and rotational periods, and chemists were isolating chemical elements with specific properties like mass and **chemical affinity**, physicists starting finding sub-atomic objects or particles which have specific properties.

In hindsight, what this amounted to was the search for the most fundamental building blocks of the universe. We have already seen how Aristotle's view that the universe is made from four[5] elements, was replaced in the eighteenth century by the chemists' theory that all substances can be broken down into their alleged constituents, namely the **chemical elements**. During the nineteenth century the conviction slowly gained traction that these elements are not *ideal* but *real*, and indeed the atomic, fundamental substances of nature.

This view started to change when particles smaller than the atom, that is, **subatomic particles**, were discovered. It started with Becquerel's serendipitous discovery that uranium atoms emit so-called **alpha particles** that blacken a photographic plate. This showed that atoms can **decay radioactively** and **transmute** into other other elements. They are thus not fundamental, because they consist of smaller parts. Shortly thereafter, the **electron**, an electrically negatively charged and very light particle[6] was discovered by J.J. Thompson (1856–1940) while studying cathode rays in 1897. He realized that the new particle has a specific charge (charge per mass) that was 2,000 times larger than that of the lightest particle then known, the positive hydrogen ion (the **hydrogen ion** (H^+) or **proton**, as we would say today). Since he could produce a ray of electrons from a cathode made of pretty much any material (aluminum, copper, and so forth), he concluded that electrons are constituents of *any* matter. This led him to construct a **model of the atom**. Since atoms are electrically neutral, but contain negatively charged electrons as he had just discovered, the rest of the atom, modeled as an amorphous blob, must be positively charged. This model was soon falsified by the experiments of Thompson's student Ernest Rutherford (1871–1837). Rutherford showed by shooting **alpha particles** through thin gold foils in 1911, that atoms must be largely empty—otherwise the particles would collide with them. Indeed, Rutherford determined that the **atomic nucleus** is fantastically small: 10,000 times smaller than the atom itself. The atom was subsequently modeled after the solar system: a tiny, heavy, positively charged atomic nucleus is orbited by point-like, negatively charged electrons. In this model, the nucleus plays the role of the sun, the electrons are the planets, and the gravitational force is represented by the **electrostatic Coulomb force**, Equation (3.1), between electric charges.

It was Rutherford's student Niels Bohr (1885–1962) in turn in 1913 who imposed new rules, namely that only special or **quantized** orbits were allowed around the nucleus to finally explain the absorption and emission **line spectra** of the chemical elements. Note the Keplerian nature of what Bohr did: He postulated rules (akin to **Kepler's laws**) that explained the observed patterns in nature (line spectrum vs. motion of planets) without explaining where these rules come from. Their success was their sole justification. The second

---

[5]Or five, if you count quintessence which was supposed to fill up space.

[6]We note that elementary particles a much to small to be "seen." When we speak of an electron, therefore, we have certain patterns in nature in mind that we explained by a (rather abstract) concept we label "electron."

5.1. DEATH BY EXTRAPOLATION: THE BIRTH OF MODERN PHYSICS

Figure 5.2: **The periodic table of elements.** Elements in the same *group* or column have similar chemical properties. For instance, elements in groups 1, 2, 17, and 18 are called *alkali metals, alkaline Earth metals, halogens,* and *noble gases,* respectively.

step after this phenomenological feat is to find the *general* laws of nature of which the new rules are just a *consequence*. In Kepler's case this was Newton's **law of universal gravitation**, in Bohr's case it was Werner Heisenberg (1901–1976) and Erwin Schrödinger's (1887–1961) theory of **quantum mechanics** (1925/6). With the new theory it became possible to completely understand the structure of the **periodic table** of chemical elements, Figure 5.2. Atoms are sorted by their chemical properties in the table. Columns or **groups** have similar properties, which are repeated in the rows or **periods**. The chemical properties of atoms are completely determined by the configuration of their electrons. In other words, the periodic table is a list of different atomic "species," where each species is distinguished from the next by the number of **electrons** (and therefore **protons**) it has. Hydrogen is hydrogen because it has one electron, uranium is uranium because it has 92 electrons—and 92 protons.

It soon became evident that the nucleus of heavier atoms cannot consist just of protons, because the masses of the atoms rise faster than their number of protons. For instance, the atomic weight of hydrogen with its one-proton nucleus is about 1 atomic unit, whereas helium has a nucleus containing two protons yet its weight is about 4 units. The missing particle clearly had to be electrically neutral, was discovered in 1932 and labeled **neutron**. We can thus classify a chemical element by two numbers: its **atomic number** Z, which is the number of its protons (and electrons) and the mass number A, which is the sum of the number of protons Z and neutrons N, and thus, since the electrons are 2,000x lighter, determines the mass of this atom. The general notation is

$$^A_Z \text{Symbol}$$

Here are a few examples: $^1_1 H$ is hydrogen with $Z = 1$ proton, and a mass number of $A = 1$, therefore it has no neutrons: $N = A - Z = 0$. An atom with a specific number of protons and neutrons is called an **isotope**, or when focusing on its **nuclear** properties, a **nuclide**. **Isotopes** have identical chemical properties, but differ in mass. The heaviest

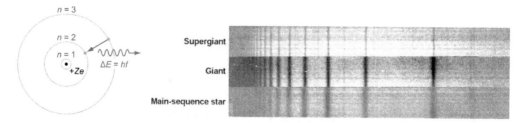

Figure 5.3: **Spectral lines.**(a) **The Bohr model.** The energy levels are modeled using electron orbits around the heavy atom nucleus in the center. In this model, spectral lines emerge as electrons jump from one atomic energy level to another. The most important red H_α line is created when an electron in the $n = 3$ level jumps into the $n = 2$ level. The energy of the $n = 3$ level is higher, and the excess energy is emitted in the form of an electromagnetic wave with energy $\Delta E = E_3 - E_2 = \frac{hc}{\lambda_{H_\alpha}}$. (b) *The effect of star size on spectral lines.* Due to the lower gas pressure and density on the surface of giant stars, their spectral lines are much sharper.

naturally occurring element is uranium. Its most abundant isotopes are $^{238}_{92}U$ and $^{235}_{92}U$. Both have $Z = 92$ protons, but the former has $N = 238 - 92 = 146$ neutrons and therefore three more than the latter. This difference changes its **nuclear properties**. For instance, the ^{238}U half-life against radioactive decay is about six times longer.

In sum, at very small distances our world looks fundamentally different. A new description of nature is necessary that is valid in this extreme regime. Rest assured, our everyday world emerges if the laws of quantum mechanics are *extrapolated* to actions large compared to **Planck's constant** h.

Concept Practice

1. What characterizes a chemical element uniquely?

    (a) The number of nucleons $N + Z$
    (b) The mass number A
    (c) The number of protons Z
    (d) None of the above

5.2 Making Sense of Stellar Classifications

5.2.1 The HR Diagram Is Constructed to Make Sense of Stellar Properties

In the last chapter we saw that **stellar spectra** were "mass produced" and classified around the *fin de siècle*. Classification is only the first step in explaining patterns in nature. A theory of stellar properties was thus sorely needed, detailing relations between the observed properties such as luminosity, temperature, spectral type, mass, size, and chemical composition. But where to start? Already in 1893 scientist WILLIAM H. MONK had asked: "Is there a relation between **luminosity** and **spectral type**?" He had observed differences of spectral line shapes, see Figure 5.3(b), and to distinguish stars with blurry spectral lines from stars with thin lines, he coined the term "giant star" for the latter. Not much more progress could be made at the time, because the distances to the stars were largely unknown. As we saw, the first distance (of *61 Cygni*) was measured with the **stellar parallax** as late as 1838, and even decades later the number of stars with known distances had not risen substantially.

5.2. MAKING SENSE OF STELLAR CLASSIFICATIONS

When Danish chemical-engineer-turned-astronomer EJNAR HERTZSPRUNG (1873–1967) entered the field[7] in 1905, many more stellar distances had been made by taking into account **proper motion** of the stars, and with the so-called **statistical parallax**. Thus, HERTZSPRUNG, working with KARL SCHWARZSCHILD (1873–1916) in Germany, was in a much better position to figure out the true nature of stars. His initial discovery was that the spectral type does not completely classify a star. Another distinctive stellar feature or **discriminant** was necessary. It turned out to be *luminosity*, just as MONK had suspected.

HERTZSPRUNG noticed that the spectral line widths in ANTONIA MAURY's Harvard classification scheme were related to the (apparent) brightness of stars. Of course, brightness is not a property of the star itself. He therefore used proper motion techniques to make the crucial step to determine the luminosity of the stars in question. In doing so, HERTZSPRUNG realized that the so-called "c type" stars in MAURY's classification were much more luminous than stars of other types—even if they were of the same spectral class! This meant that there are two different versions of stars. As an example, Vega (α *Lyrae*) is a "normal" star of spectral class A ($L_{Vega} = 40 L_{sun}$), whereas Deneb (α *Cygni*) is a much more luminous star ($L_{Deneb} = 200,000 L_{sun}$), but also of spectral class A. HERTZSPRUNG used this discovery to make tables of stars, sorting stars according to their different luminosities *within* their spectral classes. This table is therefore a *two-dimensional* classification scheme. Not just the spectral class, also the luminosity of a star is necessary to describe and understand it. HERTZSPRUNG went on in 1906 to make an exhaustive "two-dimensional" list of stars contained in the famous open cluster M45 (Pleiades). This choice was both scientifically sound and unfortunate. It made sense, because it could be justifiably assumed that all the stars in a cluster share the same characteristics. Because they are so close, they probably formed out of the same interstellar gas cloud, hence their composition and age are identical. The choice was *unfortunate* because the members of the Pleiades cluster are young, hot, blue stars, and as such, not a **representative sample** of stars.

Be that as it may, from HERTZSPRUNG's two-dimensional lists it was a short way to visualize this data in what we now call the **Hertzsprung-Russell diagram**, a plot of stars' *luminosity* versus their *spectral class*, Figure 5.4. This step was taken by HENRY NORRIS RUSSELL (1877-1955). In 1913 he presented the first **Hertzsprung-Russell diagram** at a meeting of the Royal Astronomical Society, and later published it in an English journal [33]. RUSSELL was able to show that the **giants** of MONK and HERTZSPRUNG are indeed of much larger size than "normal" stars like the sun, and coined the term **dwarf stars** for the latter. Thus, it turned out that the large luminosity is due to the large surface area of the *giants*, in full accordance with the **Stefan-Boltzmann law**, Equation (3.12). Since the large luminosity discrepancy between giant and dwarf stars could be explained by the size difference, it came to be suspected that "same spectral class" means "same temperature." However, the direct relation between spectral class and temperature wasn't fully appreciated until the mid 1920s following work of Indian physicist MEGHNAD SAHA (1893–1956) and the first female Harvard professor CECILIA PAYNE (1900-1979) as we will see.

Note that the **spectral class**, the **temperature**, and the **color index** of a star are redundant, that is, all convey basically the same information. For instance, we can refer to the sun either as a yellow-white star, a $G2$ star or a star of surface temperature 5800 K. Note that there are stars brighter and dimmer than the sun, and stars that are bluer and redder

---

[7]Once again an outsider providing much creative impulse to another field.

318 CHAPTER 5. EXTRAPOLATING & SYNTHESIZING: UNDERSTANDING STARS

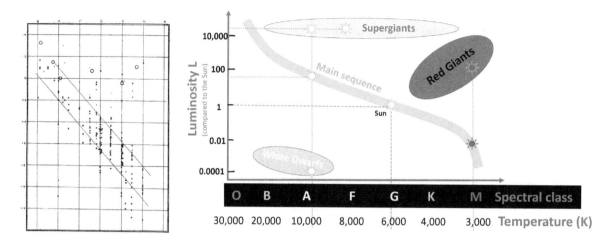

Figure 5.4: **The Hertzsprung-Russell Diagram (HRD).** (a) Russell's original plot of absolute brightness versus their spectral class of about 100 stars. (b) A modern version featuring the main goups of stars: *main sequence stars, red giants, supergiants* and *white dwarfs*.

than the sun, so the sun will appear close to the middle in a typical Hertzsprung-Russell diagram: the sun is a pretty average star.

5.2.2 Summary of Observations Encoded in HR Diagrams

The classic Hertzsprung-Russell diagram overtly displays the luminosity of a star versus its spectral type, Figure 5.4. Thus, spectral type and luminosity can be read off the diagram directly. What about other properties of stars, such as size, mass, age or composition? Can we also find out about properties of stars in general, such as the fraction of stars which are bright or young or small? Maybe it makes sense to plot several Hertzsprung-Russell diagrams, such as an **Hertzsprung-Russell diagram** of the brightest stars, a new one for the nearest stars, and maybe a diagram of all the stars in a particular star cluster, like the Pleiades in the original work of HERTZSPRUNG? All of these are good and instructive suggestions. In fact, the more information we can glean from Hertzsprung-Russell diagrams, the better is our chance of understanding how stars "work."

Here is a list of conclusions that can be drawn from Hertzsprung-Russell diagrams.

- Nine out of 10 stars (90%) lie on the narrow strip called the *main sequence* which runs roughly diagonal through the diagram from the dim, cool, red stars to the bright, hot, blue stars. Due to the fact that most stars properties fall on this line, we conclude that this must be the *normal* behavior of stars. The position of the *main sequence* is intuitive, because from *Stefan-Boltzmann* we expect hot objects to radiate much more than cooler objects, that is, hot stars should be bright, and bright stars should be hot.

- This makes it hard to fathom why we also see stars that are hot and dim (called *white dwarfs*) and which are cool and bright (called *red giants*). There are even stars which are very bright independent of their temperature or color (called *supergiants*).

- Very few stars are found outside the main sequence and red giant regions. This means that either stars with these properties cannot exist, or if they exist then they must be really short-lived or unstable.

5.2. MAKING SENSE OF STELLAR CLASSIFICATIONS 319

- There are two types of stars even at the same temperature. For instance, of red stars there is a dim, dwarf version and a bright, giant version. If and how they are related cannot be gleaned from the Hertzsprung-Russell diagram, and must come from an independent theory of stars and their properties. At this point we just state that this is a pattern in nature, and as such, it must be explained.

- In the last point we adopted RUSSELL's view that the bright, cool stars are indeed of a large size. While this cannot be directly read off the Hertzsprung-Russell diagram, it makes perfect Stefan-Boltzmann sense, if we assume that stars all function in roughly the same way. Namely, if the temperature of two objects is the same, then their luminosity (power radiated) can be different only if their surface area is different. For the red stars we find a version of $1/10,000$ L_{sun} and one that is 10,000 brighter than the sun, which is a difference of 100 million or 10^8. So the surface area must be bigger by this factor to radiate that much more at the same temperature. Since luminosity is proportional to the square of the size, we find that the red giants are 10,000 times larger than the red dwarfs. If the sun is an average star, then there are stars that are 100× larger than the sun, and stars that are 100× smaller than the sun.

- The masses of stars are not part of the data displayed in the Hertzsprung-Russell diagram, but with a suitable **theory**, we can draw lines of equal mass into the plot. It was found from other observations such as binary star periods, that stellar masses do not vary nearly as much a their luminosities. Very massive stars might have 20× the mass of the sun, and very light ones a fifth of the mass of the sun. Hence, a factor of one hundred in mass must be able to explain a factor of 100 million in luminosity! This means that the luminosity of a star depends on its mass in a very sensitive way. Our initial guess here would be $L \propto M^4$, since $100^4 = (10^2)^4 = 10^8$.

- Note that the very idea of a stellar life cycle and evolution, that is, a moving around in the Hertzsprung-Russell diagram, cannot be gleaned from the plot itself. We saw earlier that the idea goes back to the two fundamental laws of **thermodynamics**. Balancing the energy conservation equation (3.7) dictates that the energy comes from *something*, and as this "something" is used up in the process, changes of the stellar properties will occur.[8]

The new way of plotting stellar properties had astronomers excited, and many Hertzsprung-Russel diagrams were produced of different groups of stars. Soon enough, some strange new patterns in nature were discovered. In the years 1917–1920, HARLOW SHAPLEY (1885–1972) studied **globular star clusters** to decipher the structure of the Milky Way. He plotted the properties of cluster stars in Hertzsprung-Russell diagrams. Including only stars of a specific cluster in a diagram has several advantages, as was pointed out by ROBERT J. TRUMPLER (1886–1956) in 1925:

- All stars in a cluster are at the **same distance** from Earth, because the distances between stars in the cluster are very small compared to the distance of the cluster form Earth. As a corollary, stars within a cluster can be compared *even if the distance to the cluster is not known*!

---

[8]This conclusion made EDDINGTON later realize that **stellar evolution** can only be understood with a proper theory of stars, including their energy production mechanism.

320   CHAPTER 5. EXTRAPOLATING & SYNTHESIZING: UNDERSTANDING STARS

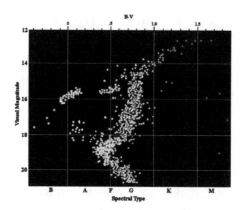

Figure 5.5: **Hertzsprung-Russell Diagram of the globular star cluster M3.**

- All stars in a cluster have the **same age**, since the whole cluster formed from an interstellar gas cloud at the same time.
- All stars in a cluster have the **same chemical composition**, since they whole cluster formed out out the same, well-mixed interstellar gas cloud.

SHAPLEY found to his surprise, that the upper part of the main sequence, namely the hot, blue stars, were *missing* in **globular clusters**, Figure 5.5. When TRUMPLER investigated further by expanding the study to **open star clusters**, he reported that the ratio of giant to main sequence stars is different for these open clusters when compared to the ratio found for sets of stars close to the sun. There are two ways to explain this pattern: **Either** some clusters have a different mass distribution of their stars than other clusters. Namely, some clusters might not have any massive stars, while others do. **Or** some clusters are older than others. The first hypothesis was falsified fairly soon. After all, it is inconsistent that some gas clouds should not be able to give birth to massive stars while others are, in particular if the process of stellar formation is **universal**, that is, the same throughout the universe. The second hypothesis opened the way to studying **stellar evolution** as we will see. Looking at star clusters of different age is like watching a **stroboscopic** time sequence of star's lives. Here is why. If we consider the Hertzsprung-Russell diagrams of star clusters of age 10 million, 100 million, 1 billion, and 10 billion years, Figure 5.6, then we can see how the individual stars change or evolve as time goes by. Of course, there is no free lunch. The show-stopper is the fact that we don't know the age of the cluster initially. In fact, this is a chicken-and-egg dilemma: We want to figure out stellar evolution from the star cluster age, but the age can only be determined if we figure out stellar evolution first. We will see later how astronomers got around this roadblock.

5.2.3 Applying New Knowledge: The Spectroscopic Parallax

In this section we will apply the insights we extracted from Hertzsprung-Russell diagrams. Our prime motivation is the construction of the expanding universe, and we have seen over and over again that the key to understanding the cosmos is the ability to measure vast—literally astronomical—distances. The stellar parallax, Section 3.6.3, was the crucial step "to get out of the solar system." It was the second rung of what can be called a **cosmic distance ladder**, after the distance to the sun (the first rung) was determined reliably by

5.2. MAKING SENSE OF STELLAR CLASSIFICATIONS

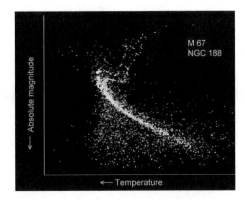

Figure 5.6: **Time sequence of Hertzsprung-Russell diagrams of star clusters.** The cluster M67 (yellow dots) has a turn-off point at higher temperature compared to NGC 188. The stars at its turn-off point are therefore more massive and must be younger, because they have not yet exhausted the hydrogen in their core. Hence, M67 is the younger cluster.

the Venus transit of 1769, Section 3.3.1. The next rung to be described here will usher in a new era, because several of the next rungs of the ladder use the same idea. Namely, the relation between brightness B, luminosity L, and distance r, illustrated earlier in Figure 2.50

$$B = \frac{L}{4\pi r^2}. \tag{5.3}$$

Since the brightness of the object is readily measured in the sky, the **luminosity** is all we need to determine the distance r to the star,

$$r = \sqrt{\frac{L}{4\pi B}}. \tag{5.4}$$

Therefore, any understanding of the celestial object that garners its luminosity, will—regardless whether the object is a star, a supernova, a galaxy or something else luminous—allow us to determine its distance.

It is fairly obvious how the Hertzsprung-Russell diagram comes in handy. After all, it is a plot of luminosity versus spectral type or temperature. For 90% of the stars—the ones on the **main sequence**—there is a strict correlation between **luminosity** and **temperature**. Therefore, we can determine one from the other. In other words, since we can determine the temperature or spectral type of a star by analyzing its light, we can translate it into a luminosity. As an example, if we determine a star to have a surface temperature 5,800 K, that is, spectral type $G2$, then we can conclude that it has one solar luminosity. This is, in essence, the method called the ***spectroscopic parallax***:

1. Observe and record the brightness of the star.
2. Measure the temperature or spectral type or color of the star.
3. Check that it's a main sequence star (broad, not sharp spectral lines).
4. Determine the star's luminosity by reading it off the Hertzsprung-Russell diagram; see Figure 5.4(b).
5. Plug brightness B and luminosity L into Equation (5.4) to find the star's distance.

Here is an example: Sirius is the brightest fixed star ($B_{Sirius} = 1.0 \times 10^{-7} \frac{W}{m^2}$) and radiates pure white light with strong and broad hydrogen lines. This tells us it is a main sequence star of **spectral class** A (more precisely **spectral type** $A1$), since A stars appear white with a surface temperature of about 10,000 K. We use the Hertzsprung-Russell diagram, Figure 5.4(b), to determine its luminosity to be roughly $L_{Sirius} \approx 10 L_{Sun}$ where $L_{sun} = 3.8 \times 10^{26} W$. Therefore,

$$r_{Sirius} = \sqrt{\frac{3.8 \times 10^{27} W}{4\pi \times 10^{-7} \frac{W}{m^2}}} = 8.6 \times 10^{1}6\,\text{m} = 9\,\text{ly}.$$

Concept Practice

1. Can you tell from the position of a star in a Hertzsprung-Russell diagram

    (a) ...how bright it is?
    (b) ...how big it is?
    (c) ... how old it is?
    (d) ...how hot it is?

5.3 Synthesizing: Putting Radiation into Star Models

5.3.1 Eddington's Standard Model of Stars

Eminent British astrophysicist ARTHUR STANLEY EDDINGTON (1882–1944) almost single-handedly drove progress in understanding stars starting in 1916 due to his ingenious **intuition**. Ideas about the energy production in stars were floating around, and since about 1910 the synthesis of macro- and microcosmos, that is, the understanding of celestial objects from their atomic composition slowly started. In 1916 these were little more than vague ideas, and EDDINGTON pragmatically decided not to wait around until these ideas solidified. Rather, he attempted a theory of stars with only minimal assumptions as to what the energy production process is. A clever decision—avoiding more than twenty years of waiting—as we will see. You have to appreciate EDDINGTON's audacity of attempting to describe stars while not knowing "what is going on!" EDDINGTON improved the meteorological gas ball models of LANE, RITTER AND EMDEN crucially by building radiation into them, that is, by assuming that energy is transported inside the star by **radiation**. It seems like a joke that star models without radiation were even considered. Of course, the **gas ball models** were radiating, but only from their surface and due to their temperature[9] that was determined by the convective process within the star. The process by which the star keeps hot inside was not part of the model; it was tacitly assumed that the star's central temperature was constant.

Recall that there are three energy transfer mechanisms: **convection, radiation**, and **conduction**. By which of these mechanisms energy is actually transported from hot to cold regions in the star depends on the circumstances. The old assumption—due to intuition gained from the process by which the Earth's atmosphere transports heat—was that **convection** is dominant in stellar interiors. EDDINGTON realized that **radiative transfer** had to be taken into account. His modification of the stellar theory was a *generalization*. Instead of excluding

---

[9]Recall that all objects radiate proportional to their temperature to the fourth power (*Stefan-Boltzmann law*).

radiative energy transport, EDDINGTON admitted it as a possibility by writing the pressure within the star as the sum of two contributions

$$p = p_t + p_{rad},$$

where p_t is the usual **thermodynamic gas pressure**, and p_{rad} the "new" **radiative pressure**.[10] As all generalizations, this led to a more complex situation. Here, the additional complexity comes from the fact that radiative pressure builds up depending on how the medium reacts to radiation. Clearly, if radiation can flow freely, no pressure builds up, whereas if the medium resists the radiation flow, it will "feel" more of a pressure. Think of a boat with sails with many substantial holes in them. As the wind flows freely through the holes, it will fail to exert pressure on the sail. The property of matter to impede radiation flow, a "radiation resistance" so to speak, is called **opacity**, and denoted by the Greek letter κ ("kappa"). Opacity of stellar matter will be different close to the center of the star than it is close to its surface—or so we may speculate. Therefore, it is safest to make it a function of the distance r from the center, $\kappa = \kappa(r)$. We don't know this function before solving the equations describing our model star, so the opacity of the stellar material becomes an additional unknown. Hence, EDDINGTON had to add another equation to the old gas ball equations to be able to solve for opacity. In a fortuitous twist, it turned out that a gas ball with radiation is still a gas ball described and solved in Emden's 1907 book [17]. Therefore, EDDINGTON could copy Emden's solution once he had realized that his new model was simply a **polytrope**[11] of index $n = 3$, that is, a special solution to the general **Lane-Emden equation** (3.15). EDDINGTON modified his model slightly due to the constructive criticism of his colleagues, in particular JAMES JEANS (1877–1946). When EDDINGTONpublished his theory of stars in the famous book *The Internal Constitution of Stars* (1926), it became known as the **standard model** of stars and did much to advance our understanding of stars and made further progress possible. This was sorely needed, because the celebrated model still lacked the core information. The energy production mechanism was effectively omitted by introducing an unspecified function $\epsilon(r)$ which is to tell us how much energy a parcel of the star produces each second. However, in the standard model it was simply assumed that this function is a constant, without attempting to explain where the energy comes from in the first place. This ability ϵ to produce energy, it became clear, depends strongly on the conditions found in the central part of the star, namely the temperature, density and chemical composition of the stellar interior. The latter was unknown. Actually, the situation was worse. Scientists at the time strongly subscribed to the wrong paradigm that the sun and stars are "hot Earths", and thus have heavy elements at their core. It seems hard to fathom today why the sun would be made out of heavy material when the universe consists mainly of the lightest elements hydrogen and helium, but this is reversing the historic development. In the early 1920s it was not known that "the universe is hydrogen," sloppily speaking. This property of the cosmos is not obvious and had to be discovered. The associated paradigm shift was met with significant resistance by the established scientists, as we will see in the Section 5.3.2. To summarize, EDDINGTON's **standard model** was based on the following assumptions:

- The star is **homogeneous**, that is, its composition is the same everywhere within the star.

---

[10]Radiation exerts a pressure (force per area) because electromagnetic waves carry momentum; the rate change in momentum is equal to a force according to *Newton's second law*.

[11]See Archangel EDDINGTON's speech at the beginning of the chapter.

- The star is in **thermal equilibrium** throughout, that is, each parcel of the star emits as much energy as it receives and produces. For instance, the sun as a whole produces $3.8 \times 10^{26} J$ of energy per second, so it emits or radiates $3.8 \times 10^{26} J$ per second into outer space to be in thermal equilibrium. The star is therefore describable by a temperature that depends only on the distance to the center $T(r)$.

- The star is in **hydrostatic equilibrium** throughout, that is, the net force on each of its parts is *zero*, see Figure 3.42. The intuition is that each part of the star "floats," that is, neither sinks toward the center, nor rises to the surface. Note that **convection currents**, see Figure 3.41, are perfectly consistent with this assumption, since material rises only to sink again; on average, it doesn't go anywhere. The stellar material is thus describable by a single density function $\rho(r)$.

- The pressure inside the star is due to **gas** and **radiation pressure**. The **energy transfer** is solely due to **radiation**. Therefore, the stellar material is ascribed an additional property called **opacity** which is a separate function of the distance to the center, $\kappa(r)$.

As you can imagine, the set of equations for the unknown functions $T(r)$, $\rho(r)$, $p(r)$, $\kappa(r)$, $\epsilon(r)$ (temperature, density, pressure, opacity and energy production rate) is hard to solve in general, so EDDINGTON made the following *simplifying assumptions*.

1. The opacity κ is constant throughout the star.
2. The energy production rate ϵ is constant throughout the star.

The equations above plus this set of assumptions defined what was called the **standard model** of stars. With it, EDDINGTON was able to explain the following patterns.

- Realizing that the pressure due to radiation grows substantially with the mass of the model star, EDDINGTON concluded that stars with masses larger than about 100 solar masses cannot exits. They simply would be blown to pieces because the pressure due to radiation is bigger than the force of gravity holding the star together. At high masses, no **hydrostatic equilibrium** is possible.

- For giant stars, presumably of large mass, EDDINGTON found that their luminosity was virtually determined by their mass only. Get this: Temperature, size, and all other stellar parameters are simply a consequence of the mass of the star. This is a powerful simplification and unification: The behavior of many different stars can be described by a formula with only one parameter. It got even better. When EDDINGTON compared his prediction with observations, he found that the formula worked even for low-mass stars. In fact, it works for all stars on the main sequence. As he found in 1924, the **mass-luminosity relation** is
$$L = kM^{3.5},$$
where k is a constant,[12] as you can see from Figure 5.7. If we express mass and luminosity in solar units $L_{sun} = 3.8 \times 10^{26} W$ and $M_{sun} = 2 \times 10^{30} kg$, then the constant drops out and we get for the luminosity of a star of two solar masses (such as the brightest fixed star Sirius)
$$L_{Sirius} = 2^{3.5} L_{sun} = 2^3 \sqrt{2} L_{sun} \approx 11.3 L_{sun},$$

---
[12]This relation can be derived mathematically from the Stefan-Boltzmann law and other assumptions.

5.3. SYNTHESIZING: PUTTING RADIATION INTO STAR MODELS

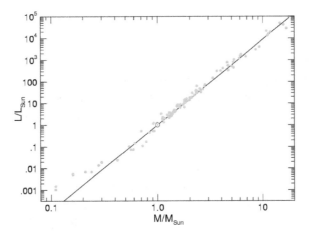

Figure 5.7: **Mass-Luminosity relation.** Shown is the luminosity of stars as a function of their mass. Both are in solar units, so the sun is the red circle at $(L = 1, M = 1)$. Note that this is a double logarithmic plot. The data is described well by a straight line, which means that the function $L(M)$ follows a power law. Since a star with mass $10 M_{Sun}$ has a luminosity between 10^3 and 10^4, the exponent is about 3.5, so $L(M) = k \times 10^{3.5}$, where k is a constant. The masses stars in the plot vary by a factor $10/0.1 = 100$, whereas the luminosities vary by $10^5/0.001 = 10^8$, consistent with the power law $((10^2)^{3.5} = 10^7)$.

so Sirius radiates more than ten times the power of the sun, even though it is only twice as massive. Luminosity is thus a very sensitive function of stellar mass.

- *Variable stars* were accessible with the **standard model**; the periodicity of δ Cepheids (see Section 5.5.4) was mathematically described by EDDINGTON in 1917.

Let's pause to appreciate the importance of the standard model. Even if you completely ignore the math involved, the picture is clear. EDDINGTON used relations from thermodynamics and radiation theory to create a viable model of a star. This is **phenomenology** at its best. Namely, he was able to quantitatively describe how stars behave, even though he didn't know the energy production mechanism.[13] Compare the situation to Kepler's three hundred years earlier. Kepler was able to *phenomenologically* describe planetary motion with a set of mathematical laws, even though he did not know the mechanism by which the planets move around the sun. In hindsight, both EDDINGTON's standard model and Kepler's laws were essential milestones on the way to understanding "what is really going on."

Not all was well, though. The mass-luminosity relation does not apply to **red giants** and **white dwarfs**. EDDINGTON's confidence in his model was so strong that he (rightly) concluded that there must be something strange going on with these stars—rather than his theory being wrong. In other words, in extreme circumstances such as ultra-dense white dwarf matter, his theory breaks down because one or more of his assumptions fails[14] and must be replaced with a different one.

---

[13] This fact was a sore point with his more mathematically-inclined competitors JAMES JEANS and EDWARD A. MILNE (1896–1950) who insisted, with some justification, that one cannot say anything definitively about stars unless one knows how they produce their energy.

[14] A good guess is that the ideal gas law does not apply to the ultra-dense **white dwarfs**, for instance.

5.3.2 Vindication: Stars Consist Mostly of Hydrogen Gas and Vary in Size

In the early 1920s several discoveries helped to support the standard model of stars and clarify some of its assumptions. This helped greatly to eventually find the holy grail of astrophysics, namely the ***stellar energy production mechanism***.

The first discovery was a direct measurement of ***stellar sizes***. As we pointed out repeatedly, stars are exceedingly far away, and even large telescopes at high magnification still show a pointlike object. Already NEWTON had estimated roughly correctly the angular size of a typical star seen from Earth to be 1/2,000 of one arc second, orders of magnitude smaller than measurable. Therefore, it seems that sizes of stars cannot directly be measured. While true for most stars, extremely big stars are amenable to size measurements with ***interferometry***. This method uses the interference of (electromagnetic) waves to measure small distances by comparing them to the exceedingly small wavelength of light. The method was championed by A.A. MICHELSON who used it first to determine the speed of light accurately in 1879, see Section 3.7, and to rule out the ***aether hypothesis*** leading to the development of ***special relativity***, before applying it to the stars. In 1920 he was able to report measurements made with his assistant at the 100-inch Hooker reflector at Mount Wilson as to the sizes of Betelgeuse (α *Orionis*) and Antares (α *Scorpionis*). The shocking result was that these stars have sizes in excess of 500 solar radii. In other words, these stars are bigger than the entire inner solar system! This and the fact that they are more than 10,000 times brighter than the sun, was such a spread in values that not a few astronomers were highly skeptical of these results.[15] On the other hand they perfectly matched and supported the assertion by HERTZSPRUNG, RUSSELL and others that giants and dwarf stars must be of vastly different size to explain the observed difference is spectral line widths.

The information contained in the invaluable ***stellar spectra*** was reassessed in another series of discoveries. With the new tools of ***quantum mechanics*** one could figure out what the lines of a line spectrum are really telling us. Recall that the spectral classification by MAURY, CANNON and others was based on the presence, absence and strength of hydrogen lines, that is, on the overall appearance of the spectrum. Naively, a star with strong hydrogen lines would seem to contain a lot of hydrogen—at least in its atmosphere or photosphere containing the cooler gas which produces the dark absorption lines. Based on the relative weakness of the hydrogen and the strength of the calcium lines in the spectrum of the sun, and the astrophysical paradigm ("Up there is the same as down here"), the sun was long considered a "hot Earth," that is, a solid or liquid body made of the same heavy elements as Earth, and surrounded by an atmosphere of light gases. EDDINGTON's standard model of stars had planted some doubts, since his assumption that stars are *gaseous* seem to work. Moreover, stellar models worked better when it was assumed that the stars, contrary to Earth, contain a sizable fraction of light elements such as hydrogen and helium.

What was missing was a direct relation between the strength of a chemical element's spectral lines and its abundance in a star or the sun. This relation was found following the work of Indian physicist MEGHNAD SAHA. SAHA showed that the strength of the lines depends crucially on the ***temperature*** of the gas that produces them. Only if the systematic temperature dependence is corrected for can the line strength be used to deduce the associated element abundance. This temperature dependence could only be deduced after it was clear

---

[15] Recall that HERSCHEL had used the assumption—plausible at the time—that all stars are roughly *equally* luminous to justify his measurement of the shape of the Milky Way.

5.3. SYNTHESIZING: PUTTING RADIATION INTO STAR MODELS

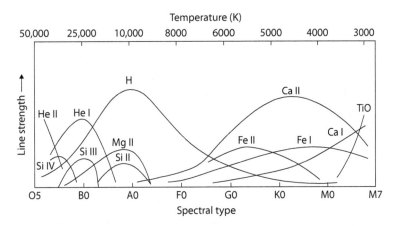

Figure 5.8: **Strength of stellar absorption lines.** Shown is the strength of absorption lines in stellar spectra as a function of the spectral type and temperature of the star. Note that the hydrogen lines are strongest for $A0$ stars with surface temperatures of about $10\,000$ K. G type stars like the sun have relatively strong calcium (Ca II) and iron (Fe II, Fe I) lines.

that the spectral lines are produced by an atomic process, that is, after the advent of **quantum mechanics**. For a rough understanding, however, one need not go further than the simple **Bohr model of the atom**, which models hydrogen as an electrically positive **proton** orbited by an electrically negative **electron**. Furthermore, there are only specific orbits allowed, which are enumerated according to their energy. The $n = 1$ orbit is the one with lowest energy, then comes $n = 2$, and so on; see Figure 5.3(a). The dark hydrogen absorption lines are produced when a hydrogen atom with an electron in its second to lowest energy state ($n = 2$) absorbs electromagnetic waves of a specific energy, see Section 5.1.2. The energy of the wave determines its wavelength λ (and therefore its color) following the Planck formula $\lambda = hc/E$. For instance, the H_α line is deeply red at a wavelength of $\lambda_\alpha = 656.28$ nm, corresponding to an energy $E = E_3 - E_2 = \frac{hc}{\lambda_\alpha} = 3 \times 10^{-19}$ J. This line is strong, when many hydrogen atoms are present which happen to have an electron precisely in the second atomic level. These conditions depend heavily on temperature. If the temperature is too high (greater than about 10,000K) electrons get kicked out of the hydrogen atoms altogether (**ionization**) by the large thermal energy available. This produces free electrons, and ionized hydrogen atoms, which cannot absorb the H_α wavelength. If the temperature is too low (lower than about $10\,000$ K), then the electron sits in the lowest level ($n = 1$) and therefore cannot absorb light the H_α wavelength. It should be obvious by now that $10\,000$ K is the sweet spot for this absorption to occur. Not surprisingly, stars of that temperature, classified as $A0$ stars, show the strongest hydrogen lines, Figure 5.8.

Analogous arguments apply to other elements, and it turns out that the sun, being of spectral type $G2$ or 5800 K surface temperature, has strong calcium lines; see Figure 5.8. We find therefore that the strength of the lines is more indicative of surface temperature than of chemical abundance. It is possible to determine the abundance nonetheless, as was discovered by CECILIA PAYNE (1900–1979) in her doctoral thesis of 1925. Her doctoral adviser was none other than HENRY NORRIS RUSSELL, and he was not pleased with the outcome. What Payne had found was that all stars have about the *same* chemical composition, and—even more shockingly—are made mostly of hydrogen (90% by numbers of atoms, 75% by mass). This was contrary to the assumptions and expectations of the astrophysics community, and consequently PAYNE had a hard time convincing fellow scientists that her conclusions were correct. Fortunately, following her work ALBRECHT UNSÖLD (1905–1995) in Germany analyzed the intensity profiles of the sun's spectral lines in 1928 to conclude without doubt that hydrogen is the most abundant element—at least in the sun's atmosphere. RUSSELL himself did a detailed spectroscopic analysis in 1929 to produce a table of the abundances of no less than fifty-six chemical elements in the solar atmosphere. Naturally, hydrogen came in on top, helium second, followed by carbon, oxygen, and nitrogen, see Figure 4.5(a). Thus, PAYNE helped to turn the chemical abundance table of the sun "upside down," and belatedly became the first female Harvard professor in 1956.

Skeptical as you are (or should be) you might say: But this is all on the outside of the stars! We cannot look *into* the stars, so (paraphrasing DOVE [48]): What the stars are made of *inside* we will never know! There are two good counter-arguments. First the simpler one. If you buy into the nebular hypothesis that the sun and stars formed from a contracting interstellar gas and dust cloud, then you'd have to admit that this cloud was initially well mixed, that is, homogeneous. Thus, if you find material on the outside of a star, the same material is—at least initially—also inside. As the star develops or ages, this may change, but likely not dramatically. The more profound argument goes back to German astrophysicist HEINRICH VOGT (1890–1968) and is known as the **Vogt-Russell theorem** (1926). It states that *the structure of a star, in hydrostatic and thermal equilibrium with all energy derived from nuclear reactions, is uniquely determined by its mass and the distribution of chemical elements throughout its interior*. Therefore, if we make a model of a star that "works," that is, explains all the properties of a star that we can observe from the *outside*, and we make specific *assumption* as to what its chemical composition is *inside*, then we can rest assured that this assumption is correct. In other words, by constructing a "good model," we have succeeded of determining the inner structure and composition of the star. In a sense, we have "theorized" our way "into" the star. We have enlarged the universe by being able to determine the **constitution of the stars**. Again the skeptic might put the finger in the astrophysical wound by stating: But we don't know the **nuclear reactions** inside the stars! Precisely that is the topic of the next section.

5.3.3 Standard Model 2.0: The Peach Star

Before nuclear physics was developed far enough to figure out the stellar energy source, other shortcomings of the standard model were discovered. Note that it was necessary to formulate the standard model first to have a frame of reference as to what was and what wasn't understood about stars. A lot of stellar patterns can be explained with the standard model, as we have seen. This is surprising, given the simplifying assumptions that had to be made in order to solve the equations. It was untenable that the star should produce its energy *everywhere* at the same rate, even though pressure, density and temperature were known to be much smaller near the surface than in the central regions of the star. Attempts to solve this paradox had to be postponed due to the lack of nuclear physics expertise at the time.

Scientists therefore focused on another blaring shortcoming. All stellar models up to this point had assumed homogeneity of the stellar material. But if the star is supposed to look the same across its entire interior, it had to be—and stay—well mixed. This is what **convection** does for you, and in that respect, the **Lane-Ritter-Emden models** where perfectly consistent. They *assumed* convection, and therefore the star remained well mixed. EDDINGTON,

5.3. SYNTHESIZING: PUTTING RADIATION INTO STAR MODELS

Figure 5.9: **The Cowling model or "peach star."** Shown are the regions of the star represented by the parts of a peach.

however, did away with convection, because he found energy transfer by *radiation* is dominant under the conditions of the stellar interior. However, a perfectly mixed, homogeneous star is impossible without a mixing mechanism. Recall that radiative transfer is by electromagnetic waves that are passing through the material from hot to cold and are not taking the material with it; the material stays where it is.

It was clear that a major upgrade of the standard model was necessary. In 1906, KARL SCHWARZSCHILD had formulated a criterion to decide whether *convection* or *radiation* transfer the energy in a gas ball. Under EDDINGTON's assumptions, Section 5.3.1, no convection was possible. Since convection was necessary to mix the star, his assumptions must be wrong. Note the beauty of deductive reasoning here. EDDINGTON's calculations were sound and therefore *sacrosanct*. Hence, his assumptions were the only thing that could be criticized and improved upon. A first step was taken by ALBRECHT UNSÖLD who showed that in cool stars EDDINGTON's assumptions can break down, since the hydrogen in the outer layers can be *ionized* (electron's are stripped off the atoms). Ionized gas behaves differently than neutral gas, and thus it was shown that the *Schwarzschild criterion* allows many stars, including our sun, to have a convective *outer layer* called the *convective zone*, see Figure 4.24. In fact, this is precisely what is observed: the *solar granulation*, Figure 4.24[Insert], was discovered 1860 and correctly interpreted as the convective motion of the sun.

What about the rest of the star? It turns out that convection in a region near the core of the star is possible if *energy production* is only happening at the stellar center. This is in contradiction to EDDINGTON's assumption that the *energy production rate* ϵ is *constant* throughout the star. In Göttingen, Germany—that is: far off the Anglo-Saxon epicenter of *stellar astrophysics* in Cambridge—LUDWIG BIERMANN (1907–1986) was able to show that if material moves ever so slightly, it would tip the balance encoded in the *Schwarzschild criterion* in favor of convection. This result was used in England by THOMAS G. COWLING (1906–1990) to construct his *point-source model* of the stars. Its name derives from the fact that he assumes energy production to be concentrated at one point, namely the center of the star, as opposed to EDDINGTON's assumption that all parts of the star produce equal amounts of energy. Note that the astrophysicist community at this point was still completely oblivious as to the energy source of the stars. However, **Cowling's model** is basically independent of *what* this source is, as long as it is located at the center of the star. This turned out to be correct, as we would expect from the fact that temperature, density and pressure are extremely high in at the center, as we saw in Section 3.9.3. The star thus looks a bit like a *peach*, Figure 5.9. The pit represents the stellar core where material is convectively mixed and homogenized. It is **decoupled** from a thick shell of material (the peach's fruit flesh) where energy is transferred by radiation. In the third, outer layer, convection again takes over to produce convection currents which manifest themselves as stellar granulation (the velvety peach skin).

Concept Practice

1. Which was not one of the assumptions of *Eddington's standard model*?
   (a) Stars are hot gas balls.
   (b) Stars produce energy by fusion.
   (c) Stars are in hydrostatic equilibrium.
   (d) Stars are in thermal equilibrium.
   (e) All were assumptions of the model.

5.4 Stars Produce Energy by Thermonuclear Fusion

The energy production processes within stars remained the lost holy grail of astrophysics until the end of the 1930s. There are several reasons why this mechanism at the crossroads of astronomy and physics could not be found earlier, as we will see now.

5.4.1 Necessary Preparatory Work

No law of science comes out of the blue. An observational or experimental discovery may come unexpectedly, but a new theory is always based on older ideas and thoughts that have circulated for awhile among a tightly-knit group of scientists—until a stroke of genius or some other shock suddenly forces the acceptance of a new paradigm: The fog clears, and suddenly the features of a new (view of the) universe are plain to see.[16] Consistent with the view of a scientific community over-saturated with "rightish" ideas which suddenly condense into a new explanation of nature is the fact that many times progress is made independently by several scientists at about the same time. The discovery of the holy astrophysical grail is no exception. WEIZSÄCKER in Germany and BETHE in the United States found it within months of each other. To see how the scientific community zeroed in on this discovery, we display the necessary preparatory work in a list.

- The laws of thermodynamics made it necessary to balance the energy book of the sun; measuring the power the sun radiates, it was recognized that only a very efficient process is able to produce the necessary enormous amount of energy it radiates.

- The **Helmholtz-Kelvin contraction theory** was ruled out because the timescale of 20 million years, Equation (3.11), is too short when compared to the **geological record** which goes back at least hundreds of millions of years.

---

[16] A similarly "nebulous" metaphor was used by WERNER HEISENBERG when he described how he discovered **quantum mechanics** [21].

5.4. STARS PRODUCE ENERGY BY THERMONUCLEAR FUSION

- The **radioactive sun** is ruled out. Excitement abounded due to the surprise discovery around the *fin de siècle* that elements like *uranium* and *radium* spontaneously decay and release energy. Yet, no coherent theory of a radioactive power generation in the sun was ever devised, and it soon became clear that there are not enough **radioactive elements** in the sun to release that much energy.

- EINSTEIN's famous formula $E = mc^2$ of 1905 puts the thought into scientist's minds that there is a lot of energy available and stored in particles' masses—if it only could be released!

- In the spirit of EINSTEIN's formula, JAMES JEANS postulates electron-proton annihilation to be the energy source of the sun. The idea is that an electrically positive particle and an electrically negative particle—together electrically neutral not to upset **charge conservation**—could "somehow" annihilate. Though "somehow" is not specified (and actually not possible as we now know), the energy released can be calculated by EINSTEIN's formula:

$$E = (m_{proton} + m_{electron})c^2 \approx (1.7 \times 10^{-27} kg)(3 \times 10^8 m/s)^2 = 1.53 \times 10^{-10} J.$$

Per reaction this is not a lot of energy, but remember that the typical number of atoms in a macroscopic object is **Avogadro's constant**, 6×10^{23}; see Section 3.4.2. Multiply by this large number, and you'll have a lot of energy at you hands!

- In 1915 WILLIAM HARPER HARKINS (1873–1951) began to study atomic nuclei and their structure and reactions.[17] By comparing the masses of hydrogen and helium, he concluded (using EINSTEIN's Equation (5.1)) that dramatic amounts of energy could be freed if hydrogen was converted to helium. Consequently, he advocated this process as the energy source of the sun.

- French Nobelist JEAN PERRIN (1870–1942) independently elaborated on the same idea. He describes not just hydrogen fusion, but essentially the formation of a main-sequence star by a contracting interstellar gas cloud already in 1919: "... the temperature gradually increases ... with the nebula contracting, encounters between light atoms capable of assembling into heavy atoms will become more and more numerous ... the formation of heavy atoms becomes more and more important, being accompanied by ultra-hard X-rays which, for the most part, do not escape from the celestial body, whose temperature becomes gigantic: the star has begun to shine."[37, p.102] A few month later PERRIN suggested more specifically that four hydrogen atoms (really: nuclei) had to be fused together to form one Helium atom (nucleus).

- EDDINGTON independently[18] came to the same conclusion in 1920, and computed that—assuming four hydrogen nuclei "somehow" could be fused to form one helium nucleus without specifying the "somehow"—one kilogram of hydrogen would yield 993 grams of helium. The seven grams mass difference would translate into $E = 0.007 \, \text{kg} \times c^2 = 63 \times 10^{13}$ J. This is the amount of energy "stored in hydrogen"—if it is released by **hydrogen-to-helium fusion**. Since the sun must produce $L_{sun} \cdot s = 3.8 \times 10^{26}$ W

---

[17] Recall that the existence and tininess of nuclei had been discovered merely four years earlier by RUTHERFORD.

[18] The time was apparently ripe for such an idea; see above.

each second,[19] it has to convert

$$\frac{\text{energy radiated}}{\text{energy stored per kg}} = \frac{3.8 \times 10^{26} \text{ J}}{63 \times 10^{13} \text{ J/kg}} \approx 6 \times 10^{11} \text{ kg}.$$

of mass each second. Since the sun's mass is 75% hydrogen, there are $0.75 \times 2 \times 10^{30}$ kg $= 1.5 \times 10^{30}$ kg of hydrogen fuel available (if all is burned), and the **thermonuclear time constant** replacing the **contraction time** is a staggering

$$\frac{\text{mass of hydrogen available}}{\text{mass of hydrogen converted per year}} = \frac{1.5 \times 10^{30} \text{ kg}}{6 \times 10^{11} \frac{\text{kg}}{\text{s}} \times 3600 \frac{\text{s}}{\text{hr}} \times 24 \frac{\text{hr}}{\text{day}} \times 365.25 \frac{\text{day}}{\text{yr}}}$$
$$= 79 \times 10^9 \text{ yr} = 79 \text{ billion years.} \quad (5.5)$$

If you didn't follow the math, just look at the last number and compare it to the geologically required hundreds of millions of years (10^9 years is 1000 million years). The point here is that *this could work!*

But wait: We had already upgraded the standard model! As good scientists, we should check that the new—allegedly improved—**Cowling model** ("the peach star") is consistent with the predictions of the old model which where, after all, confirmed by observation. A simple cross-check is a recalculation of the *nuclear timescale* above. EDDINGTON had computed it under the assumption that all of the hydrogen in the sun is available as fusion fuel, but this assumption had to be jettisoned because we needed **convection** to mix the star. In the **Cowling model** only the core of the star is convective, and thus only the hydrogen in the core is available as fusion fuel, because only it gets mixed with the helium ash. Therefore, the life expectancy of the sun may be dramatically reduced, and may become too short to comply with the *geological timescale* of hundreds of millions of years.[20] In this case we'd have to reject the new star model, or worse, even the standard model, because it cannot be improved. Before we get too anxious, let's do the calculation. We'll assume that the core of the star is a quarter of the size of the whole star. This is a far cry from a point source, but calculations have shown that this is a reasonable size for the sun's core. The core's volume can be calculated by using the *scaling* arguments of Section 1.2.6. The scaling factor is $f = \frac{1}{4}$, and thus, Equation (1.5), $V_{core} = f^3 V_{star} = (\frac{1}{4})^3 V_{star} = \frac{1}{64} V_{star}$. Hence, only 1/64th the amount of hydrogen is available for fusion, and thus the sun's life expectancy is only 1/64th of the previously calculated, or about 79 billion years/64 $\approx$ 1.2 billion years. This is getting uncomfortably close, but is still okay. Thus, until more detailed calculations have been performed, we shouldn't discard the new model.

- On to the *miracle* implied in the sentence "4 hydrogen nuclei *somehow* fuse to helium." The insurmountable problem was that the hydrogen nucleus is a single positively charged **proton**, and two like electric charges repel each other so strongly, that they cannot get close enough together to collide and fuse—even at the exceedingly hot temperatures

---

[19] Recall that the luminosity of the sun is the electromagnetic power it radiates, and that power is energy **per** unit time, whence the weird expression $L_{sun} \cdot$ seconds for the energy.

[20] In fact, by the 1930s the requirements had gotten even more constricting, because **radioactive dating** methods had put the age of the oldest rocks on Earth close to 4 billion years, that is, thousands of millions of years.

(and therefore large velocities) in the stellar core. Since **quantum mechanics** allows a lot of crazy stuff to happen that is forbidden in **classical physics** (while enforcing weird rules that are absent in classical physics), GEORGE GAMOV used the *tunneling effect* of quantum mechanics in 1928 to argue that the protons can *tunnel through* rather than *get over* the high repulsion-barrier between them.

- ROBERT E. ATKINSON (1898–1982) and FRITZ HOUTERMANS (1903–1966) calculated in 1929, based on GAMOV's work, the fusion reaction rates of several light elements into heavier ones—under the conditions in the stellar core, namely exceedingly high temperature, density and pressure [2]. In other words, they were able to figure out *quantitatively* how fast how much energy is released given the parameters describing the stellar core. These parameters in turn can be calculated using **stellar models** like EDDINGTON's.

- The "brother" of the proton, the other (electrically neutral) nucleon is found in 1932 and named **neutron**. In the same year, the electron's anti-particle is discovered and labeled **positron**, because it is positively ("anti-negatively") charged.

- Subsequently, ATKINSON suggests that the observed relative abundances of the heavy elements, Figure 4.5, can be explained by their synthesis from hydrogen (essentially a proton) and helium. As the growing nucleus captures protons, it is transmuted into the next chemical element, thus filling up the period table of elements. ATKINSON showed in 1936 that the most likely process in stellar cores is the formation of deuterium, that is, a nucleus consisting of a proton and a neutron. As it turned out later, this is the first stage of the so-called **pp chain** of **nuclear reactions** that powers the sun!

5.4.2 The Microscopic Universe: Subatomic Physics

On danger of overburdening the reader, we'll have to explain some concepts of the *new physics* that made the discovery of **stellar energy production** possible. These physical processes will become equally important when trying to understand the **early universe**. As we shall see, the nascent universe was exceedingly hot and of infinitesimal size, so it makes sense that the microcosmos of **subatomic particles** needs to be understood in order to comprehend the expanding universe from the start. Of course, we cannot hope to give a detailed account of the subatomic phenomena that take physics students many years to master. On the other hand, we need to do better than to just throw words like "fusion," "cross-section," and elementary particle names around that can only be remembered and regurgitated.

The underlying idea of subatomic physics, that is, **nuclear** and **elementary particle physics** is reductionist. Namely, it is assumed that nature at its most basic level, that is, on the smallest scales, is explainable by a few basic ingredients—the **elementary particles**—and their **interactions**. This is similar to ARISTOTLE's view that everything is a mixture of the four elements and can be understood by the properties of these elements. We have seen that as time went by, the convictions as to what is **elementary** changed. First, *water, fire, air,* and *earth* were considered elementary, then the **chemical elements** like hydrogen, helium and uranium. These atoms were subsequently found to have substructure; they are made of a tiny nucleus and pointlike electrons, as was shown in 1911 by RUTHERFORD. While the electrons are considered elementary to this day, the nucleus turned out to be a tightly bound packet of two kinds of **nucleons**, the **proton** and the **neutron**. Even the **nucleons** have not retained their elementary status, as it was found in the 1960s that they are made of so-called **quarks**. Furthermore, **quantum mechanics** posits that everything comes in smallest packages or **quanta**, a particle was ascribed to all **force fields** which are responsible for the interactions between these elementary particles.

Starting in the 1920s, subatomic physicists established rules as to what can and what cannot happen as the elementary building blocks of the cosmos interact with each other. One new feature which we can trace back to EINSTEIN's famous formula, is that mass is not conserved, only energy is. In fact, energy is **conserved** but can be transformed from one form into another. *Einstein's formula*, Equation (5.1), tells us is that mass is a *form* of energy. Therefore, other forms of energy can be transformed into mass. This is what happens in **particle accelerators**. From the huge motional energy inherent in particle collisions, new particles get created—and some destroyed, as the case may be. This happens in various forms, but such that energy, electric charge, momentum, and a bunch of other quantities are conserved. If you will, these rules define the "game of subatomic physics." If you learn them, you can predict the outcome of experiments.[21] This is more detail than we need here, but hopefully these details demystify subatomic physics a little for you. No black magic, instead the old science story: A set of rules ("a theory") is inferred from experimental data, and these are used to predict future experiments to validate the theory.

What we are interested in here are the implications for energy production in stars. Basically, nuclear physicists found that if they shoot subatomic particles at each other, sometimes they stick together in a new configuration. Often, the masses of the products of such a *nuclear reaction* were different from the masses of the particles they started with, and the difference got carried away as energy. Since there is a great many possible reactions you can try, it took nuclear experimentalists a long time to come up with quasi-complete tables listing the outcome of millions of experiments. Basically, these tables list something like this: If I shoot a hydrogen nucleus with speed v into another at rest, I have a 1.5% chance to get a deuterium nucleus plus energy plus Y if v is above a threshold speed v_0. If I shoot a helium nucleus into another one then I get .... The upshot is this: All these nuclear reactions have a certain probability to happen under specific conditions like temperature, pressure, and so forth. With a list of these reactions, we can predict what is going to happen given the conditions in the stellar interior.

5.4.3 Putting It All Together: Nuclear Fusion in Stars

After scientists had gotten this close to the holy grail, it was discovered independently by CARL F. WEIZSÄCKER (1912–2007) in Germany [41] and HANS A. BETHE (1906–2005) in Cornell [5]. As we have seen, a lot of ideas as to what process powers the stars had been thrown around. The most probable was hydrogen to helium fusion, because the energy yield is greatest. This is because the mass difference between the ingredients (4 H) and the product (1 He) is bigger than in any other reaction, say helium to carbon. Also it was known by then, see Section 5.3.2, that about 90% of a star's atoms are hydrogen, and the rest is basically helium, because all the other, heavier elements[22] together make up only about 1%. Therefore, it

---

[21] Up to a point, since quantum mechanics is stochastic at the fundamental level; only probabilities for measurements can be predicted.

[22] The elements nos. 3 to 92 are labeled by astronomers collectively as **metals**. This is a very crude classifications, the more so as some of them (carbon, oxygen, nitrogen, and so forth) are decidedly not metals. The

5.4. STARS PRODUCE ENERGY BY THERMONUCLEAR FUSION

becomes clear that whatever happened in stars, had to happen by fusing hydrogen or helium.

The devil was in the details. How do you get four hydrogen nuclei to stick together? Since they are exceedingly small, it is highly unlikely that they collide, and even unlikelier that they collide in just the right way that they stick together. Since the collision-and-stick probability—known to experts as the **cross-section** for this reaction—is so small per attempted reaction, we need a lot of attempts! This is the reason why the density (number of particles per volume) has to be so extremely high to make the thermonuclear reactions "go." Now imagine you have to have not two or three but four particles that need to collide, that is, be at the same place exactly at the same time. This is so unlikely that no realistic core density is going to suffice to get a reasonable number of reactions going every second. This is where things stood in 1938. Since no single scientist had an idea how to proceed, GEORGE GAMOW and EDWARD TELLER[23] (1908–2003) organized a conference to bring many of them together to enhance the chances that an idea would emerge. Indeed, HANS BETHE found the presentation of GAMOW's student CHARLES L. CRITCHFIELD (1910–1994) so inspiring that he decided to collaborate with him. The two devised the following circumvention of the problem of putting four hydrogen nuclei together at the same time and place. They considered a *sequence* of processes, in which only *two* particles collide at a time; such a sequence has has several *stages*. As one example for such a process (called the proton-proton chain[24] or **pp chain**), consider the following three step **hydrogen-to-helium fusion**.

1. **Two hydrogen nuclei (^1H) collide and form deuterium (^2D)**, which is a heavy version of hydrogen, namely a nucleus consisting of a proton and a neutron. This is the process ATKINSON came across in 1936; see above. If you overcome your nucleophobia and rationally analyze the reaction, it becomes clear that it is nonsense. Proton plus proton equals proton plus neutron? This would mean we have turned a proton (p) into a neutron (n) somehow, and the skeptic fears a resurgence of **alchemy**.[25] Surprisingly, we can **transmute** elements, as should be obvious from our claim that hydrogen turns into helium to power the stars. However, this has to happen according to the rules which we call **conservation laws**. The law obviously violated in the reaction $p+p \to p+n$ is the conservation of electric charge. On the left hand we have two positive charges, and on the right only one, because the neutron is neutral. We make up for it by postulating that a positively charged particle also has to emerge. Since we cannot create *another* proton, due to the **energy conservation law**, it has to be a very light particle. It turns out to be the positron or anti-electron. There is another conservation law that we need to respect, and therefore a fourth particle has to emerge. It has no charge, because we already took care of charge conservation, and has basically zero mass. It is called the **neutrino**[26]. So the first step is

$$\text{hydrogen} + \text{hydrogen} \to \text{deuterium} + \text{positron} + \text{neutrino} \tag{5.6}$$
$$^1H + {}^1H \to {}^2D + e^+ + \nu_e$$

---

idea is that **metals** is a handy, short description of the fact that in most stars they all play the same role as "non-fuel," that is, nuclei that do not fuse at the core temperatures of normal stars.

[23]Later to become the father of the thermonuclear "H-bomb"—BETHE considered himself the midwife.

[24]Actually, there are three possible **pp chains**.

[25]Alchemists tried to turn one substance into another, for example lead into gold.

[26]It was *postulated* by Nobelist WOLFGANG PAULI (1900–1958) to ensure **energy and momentum conservation** in radioactive decays.

2. **Add another proton (hydrogen nucleus) to the deuterium to form a version of helium.** Adding protons changes one element into another, which is known as **transmutation**. You might remember that the **atomic number** Z that enumerates the chemical elements (Hydrogen: $Z = 1$, helium: $Z = 2$, lithium $Z = 3$, ..., uranium $Z = 92$) is the number of protons or positive **unit charges**[27] of the nucleus of the atom. For instance, the helium nucleus has two protons in it, and is thus charged with two positive elementary units. Any element that has a nucleus which contains two protons is called helium. If it has three protons, it is no longer helium. As deuterium collides with another hydrogen nucleus and they stick together, we see that the resulting nucleus has two protons and one neutron: $(pn) + p \to (ppn)$. This is a version of *helium*, an ***isotope***[28] called ^{3}He. The second step is thus

$$\text{deuterium} + \text{hydrogen} \to \text{helium isotope} + \text{energy} \quad (5.7)$$
$$^2D + {}^1H \to {}^3He + \gamma,$$

where we have written the energy as a photon (γ), that is, the quantum of an **electromagnetic wave** which transports the energy released.

3. Finally, **two of these helium isotopes collide to produce a more stable version of helium**, the "normal" helium ^{4}He plus two hydrogen nuclei to conserve charge

$$\text{helium isotope} + \text{helium isotope} \to \text{normal helium} + 2 \text{ hydrogen} \quad (5.8)$$
$$(ppn) + (ppn) \to (ppnn) + p + p$$
$$^3He + {}^3He \to {}^4He + {}^1H + {}^1H,$$

This was quite a dose of nuclear physics *jargon*, so if you want to ignore the details, make sure you get the gist of it. Here is the shortened version:

$$4 \text{ (hydrogen nuclei)} \to 1 \text{ (helium nucleus)} + \text{(neutrino carrying energy)} \quad (5.9)$$
$$+ \text{(electromagnetic wave carrying energy)} + \text{(other particles)}$$

The production of energy (in the form of electromagnetic waves and neutrinos) is possible due to the **mass difference**, and hence **energy difference** between the initial four hydrogen nuclei and the helium nucleus produced. Let's check if this is true. If you look up the mass of hydrogen and helium, you'll find $m_H = 1.673 \times 10^{-27} kg$ and $m_{He} = 6.645 \times 10^{-27} kg = 3.972 m_H$, so helium is slightly lighter—by less than 1%—than four hydrogen nuclei. Only with this energy appearing on the right side after the reaction is the energy before the reaction equal to the energy after the reaction took place.

Meanwhile in Germany, WEIZSÄCKER thought of another way around the *"four particles at the same place and time"* problem. He postulated that all chemical elements were already present when the stars formed, effectively abandoning the *"fusion must start from hydrogen and helium"* paradigm. He was thus free to consider other—maybe more likely—nuclear reactions as long as the heavier elements were not needed in large numbers, because they

---

[27] The unit is the (absolute of the) electric charge of the electron $e = 1.602 \times 10^{-19} C$.
[28] You recall that an *isotope* is a version of an atom that has a different number of neutrons, such as the famous ^{238}U and ^{235}U. These versions of uranium have both 92 protons (otherwise they wouldn't be uranium), but one has 3 neutrons more than the other.

5.4. STARS PRODUCE ENERGY BY THERMONUCLEAR FUSION

make up only 1% of the star's atoms. The elaborate process he came up with is known as the carbon-nitrogen-oxygen (CNO) cycle, because carbon, nitrogen and oxygen nuclei act as *catalysts* in the reaction to produce helium from four hydrogen nuclei. As catalysts, carbon and nitrogen are not used up, and only small quantities need to be present in the star to make this reaction possible. The reaction involves six(!) stages, and is thus much more complicated[29] than the **pp chain**.

The question arose as to which of the two processes is heating the sun: **pp chain** or **CNO cycle**? The decisive factor turned out to be their distinct temperature dependencies. How does temperature figure into nuclear reactions? Recall that temperature is basically motional energy of particles. Typically, if the ingredients of a nuclear reaction have a larger energy, the probability to collide and stick together (cross-section of the reaction) goes up. Also, the particles are simply faster. Therefore, higher temperature means more reactions per time. And more reactions means more energy produced in total, because every reaction contributes a fixed amount of energy. Now, while the **pp chain** shows a moderate dependence on the temperature T, namely T^5, the **CNO cycle** is very sensitive to the core's temperature (T^{16}). The latter will therefore be dominant in hot, massive stars, while the former will make the cool, light stars shine. Where is the sun in this classification? It turns out that the sun is a true middle-of-the-road star, neither hot nor cool, in that both processes play a role in producing the energy in its core. The sun sun shines mostly due to the **pp chain** process, with the **CNO cycle** contributing about 7% to its luminosity. As corollaries of this discussion, we deduce the following rules:

1. *The hotter the stellar core temperature, the more energy is being produced, because more reactions happen per unit time.*

2. *Which processes happen (here: pp chain or CNO cycle) depends on the core temperature.*

In this spirit, Estonian astrophysicist ERNST ÖPIK (1893–1985) suggested in 1938 that not only main-sequence stars, buy also **red giants** shine by nuclear fusion. He hypothesized that the reactions happening inside the giants' cores are different because the central temperature of the giants is much higher than that of main sequence stars. As with the energy production process in main sequence stars, so was the way from the *idea* for a distinct red giant energy production mechanism to the formulation of a quantitative theory a long one. It didn't help that World War II intervened. Finally in 1951, the energy production mechanism of red giants was independently[30] worked out by ÖPIK and EDWIN SALPETER (1924–2008). Namely, they realized that in the cores of red giants temperatures are so high (about 400 million Kelvin versus 20 million Kelvin in stars like the sun) that helium, the "ash" of the fusion reaction where hydrogen is the "fuel," can fuse itself to release energy. Specifically, helium fuses to carbon in the so-called *triple-alpha reaction*. If you recall that *alpha* is the "radioactive name" for helium (^4He), this name makes sense, because three helium nuclei, that is, 3×2 protons plus 3×2 neutrons form a carbon nucleus. Since carbon is the sixth element in the periodic table, it has 6 protons, and since its atomic mass is 12 amu,[31] it has to have 6

---

[29]In fact, the question might be asked "How the heck did WEIZSÄCKER (and at virtually the same time BETHE) come up with *that* given that there are 92 elements?" If you are interested, hit the science history bookshelves!

[30]The time was ripe, as we have remarked before.

[31]An amu is an atomic mass unit, defined as a twelfth of the mass of the carbon-12 atom ($^{12}_{6}C$).

neutrons, so this makes sense.[32] The shortened version is

$$3 \text{ (helium nuclei)} \rightarrow 1 \text{ (carbon nucleus)} + \text{(electromagnetic wave carrying energy)}.$$

As we will see, these two fusion processes, hydrogen-to-helium and helium-to-carbon fusion, are enough to explain the energy production in most stars throughout their entire life. We suspect that other fusion reactions could happen at *higher* temperatures. In particular, we might ask why the carbon ash of the triple-alpha process shouldn't ignite to form—well, what? Doing the numbers, we might expect the production of oxygen.

$$\text{(carbon nucleus)} + \text{(helium nucleus)} \rightarrow \text{(oxygen nucleus)}$$
$$^{12}_{6}C + ^{4}_{2}He \rightarrow ^{16}_{8}O.$$

And why stop here? Pack another helium nucleus onto the oxygen to produce neon (element no. 10), and so forth. This process of building up or **synthesizing** heavier nuclei is aptly called **nucleosynthesis**, and we'll come back to it in Section 5.6.3 to prove that our bodies' atoms were synthesized in massive stars.

Concept Practice

1. The energy source of the stars is
    (a) gravitation
    (b) contraction
    (c) radioactivity
    (d) None of the above

2. To produce energy, stars must have
    (a) ... a hot enough core
    (b) ... a dense enough core
    (c) ... enough mass to build up enough pressure
    (d) All of the above
    (e) None of the above

5.5 Fusion Determines Stellar Structure and Life Cycle

All cards are on the table, so we can proceed to construct realistic stellar models. In particular, we can use the **Vogt-Russell theorem** because we know the **nuclear reactions** which feed into it. If you look at it, you'll find that the theorem is a confidence booster. It demystifies stellar astrophysics by stating that if you know the mass, composition and nuclear reactions of the star, you know everything there is to know about the star, including its past and future behavior. This last half-sentence implies that the star will change or evolve as time goes by. We already know that stars form out of an interstellar gas and dust cloud, from the **nebular hypothesis** of KANT and LAPLACE. In other words, stars are born, and so we often speak of **stellar evolution** in similar terms as we speak about biological life. Anticipating—based the

---

[32] Actually, the triple-alpha is a two-step process, where in step on a short-lived beryllium nucleus is being produced from two helium nuclei, which fuses with the third helium nucleus is the second step to form carbon. Both steps release energy.

5.5. FUSION DETERMINES STELLAR STRUCTURE AND LIFE CYCLE

second law of thermodynamics—that the star will evolve to a terminal stage when its energy sources are depleted, we might call this the death of the star, because it ceases to "function" as it did before. This in turn justifies the use of the term **stellar life cycle** for the process from the formation of a star to its death.

5.5.1 A Universal Theory of Stars

To construct realistic stellar models, we take COWLING's improved model (the "peach star" of Section 5.9), and plug in the new discovery, namely that the energy of a star is being produced at its core by hydrogen fusion. Crunching the numbers, the major features of the Hertzsprung-Russell diagram can be understood and the following patterns in nature are explained.

- **Mass-luminosity relation.** The relation between mass and luminosity, Equation (5.7), can now be *deduced* from the *general theory* of energy production by fusion. In short, more mass means higher core temperature and density, which means more fusion reactions per time, which is equivalent to more energy produced per time which is redundant with higher luminosity.

- **The main sequence.** The new standard model with fusion can be used to calculate the size of stars, and it turns out that most main sequence stars have similar sizes as the sun.[33] This in turn means according to the *Stefan-Boltzmann law* that the more massive ones of much higher luminosity, must radiate more due to their higher surface temperature, because the surface areas of main-sequence stars are similar. The new standard model therefore explains the position of the main sequence within the Hertzsprung-Russell diagram. In particular, it must run from cool, dim stars to hot, bright stars.

- **Stellar evolution.** The vindication of the *mass-luminosity relation* means that RUSSELL's theory of stellar evolution (stars keep contracting monotonously as time goes by) is wrong. Instead, the depletion of hydrogen in their cores will lead to an expansion of the stars; **main sequence stars** turn into **red giants**.

While the last point leads the way toward an understanding of giant stars, at this point it is clear that more explaining needs to be done. In particular, the new standard model does not explain the dim, hot stars known as white dwarfs. On the positive side, we can use the obvious implications of the **mass-luminosity relation** (more mass necessitates much higher luminosity) and fusion as the energy source of the stars (mass is fuel) to put together a simple formula to estimate the **life expectancy** of stars relative to the life expectancy of the sun. A star of more mass than the sun will *have* more hydrogen fuel to burn, but also needs to burn more fuel to maintain a much higher luminosity. In other words, more mass means longer life, and greater luminosity means shorter life. The life expectancy T_{star} of a star is therefore

$$T_{star} = \frac{M_{star}}{L_{star}} T_S, \tag{5.10}$$

where T_S is the life expectancy of the sun, M_{star} is the mass of the star in solar masses, and L_{star} is the luminosity of the star in solar luminosities. As an example, consider Sirius, which

---

[33] With a grain of salt: some might be three times as big, others 1/3 the size of the sun.

is twice as massive as the sun, and radiates 25× as much power as the sun, so

$$T_{Sirius} = \frac{2}{25}T_S = 0.08T_S.$$

Sirius has only 8% of the life expectancy of the sun! The sun will live over 12 times longer than Sirius even though its mass is smaller by only a factor of two. Since luminosity rises so steeply with mass according to the **mass-luminosity relation**, we can summarize the finding as follows: *Stars with higher mass have a drastically reduced life expectancy.* The **mass-luminosity relation** helped us find the most important predictor of a star's life—its **mass**.

5.5.2 Exceptions to the Rule: White Dwarfs and Neutron Stars

As we have seen in Section 3.6.3, the first **white dwarf** was discovered in the mid 1800s, although it was not fully appreciated how strange of an object it is. As a reminder, BESSEL had deduced from the wobbly **proper motion** of Sirius that it has a fairly massive companion. Sirius is thus a double star, and its bright component was dubbed Sirius A, the dim one Sirius B. The latter was discovered in 1862 visually by telescope maker ALVAN G. CLARK. He found it to be 10,000 times dimmer than Sirius A. Its luminosity is thus only one-fortieth that of the sun, $L_B \approx \frac{1}{40}L_S$. This is surprising since Sirius B is about as massive as the sun $M_B \approx M_S$, as was determined by the observed orbits of the two components, see Section 3.5.3.

After a long hiatus, the surprise turned into a shock, when in 1915 it was determined that Sirius B is much hotter than anticipated. Indeed, its white color is due to a surface that radiates at 25,000 Kelvin. We can use the luminosity and temperature of Sirius B to calculate its radius R_B in units of the solar radius $R_S \approx 700\,000$ km using the Stefan-Boltzmann law, Equation (3.12)

$$\frac{R_B}{R_S} = \sqrt{\frac{L_B}{L_S}\frac{T_B^4}{T_S^4}} = \sqrt{\frac{1}{40}\left(\frac{5800\,\text{K}}{25\,000\,\text{K}}\right)^4} \approx 0.0085,$$

so Sirius B has a radius of only $R_B = 0.0085 R_S = 0.0085 \times 700\,000$ km $= 6000$ km $< R_{Earth}$. Thus, this "star" is smaller than the Earth, and the label **white dwarf** is aptly chosen[34]. The resulting density of Sirius B, $1/0.0085^3 = 1.6$ million times higher than the sun's, was considered *absurd* at the time. However, there were a few other white dwarfs known back then, which served as independent evidence that this discovery was not a fluke.

EDDINGTON later showed that the "absurdity" goes away if one assumes that in the interiors of stars temperatures are hot enough to strip the atoms of their electrons, and thus to turn the star into a gas or plasma where atomic nuclei and their electrons independently move about at high speeds. He pointed out that this strange matter of a white dwarf must behave differently than ordinary matter. Unfortunately, EDDINGTON's calculations also led to a paradox, whereby the white dwarf can, due to its high temperature, radiate so much energy, that its remaining energy is less than that of ordinary matter at **absolute zero temperature**, where everything comes to a standstill, that is, where all motional energy is zero. Which of the laws of nature had to be jettisoned this time? RALPH FOWLER (1889–1944) found in 1926 that **classical physics** cannot be extrapolated to these extreme densities. It breaks down, and **quantum mechanical** laws have to be used to describe how matter behaves under these

---

[34]It was coined in 1922 by WILLEM LUYTEN(1899–1994).

circumstances. This was mutually beneficial for astrophysics and quantum mechanics, because a white dwarf is a macroscopic object which nonetheless exhibits the laws of the microscopic or subatomic universe. FOWLER was able to derive a new relation between temperature and pressure in such a **electron quantum gas**. Amazingly, the pressure is independent of temperature, and thus only depends on the **number density** of the electrons. This is about as non-ideal as a gas can get!

In regard to EDDINGTON's paradox, FOWLER showed using the new *quantum mechanical* laws of nature that even at absolute zero temperature, the motional energy of the electrons can be as high as in a gas of ten million Kelvin. How does this makes sense? Well, the weird behavior of electrons at extremely high densities, that is, when the electrons are really close to each other, is distinctly **non-classical**. It is the result of a **quantum rule** that makes no sense in classical physics. Specifically, the **Pauli exclusion principle** states that two electrons cannot be in the same **quantum state**. This means that states like the energy levels of the **Bohr model**, Figure 5.3(a), must be filled such that each additional electron sits at a level of higher energy, because the lower levels are already *occupied*. In a white dwarf the electrons are therefore forced to have large motional energies, because (states of) lower energies are not available. This in turn means that if there are more electrons, due to a larger mass of the white dwarf, then the velocities of the most energetic of the electrons become so large that we have to invoke the laws of **relativity**, as we saw in Section 5.1.1. What this means is that the relation between electron pressure p_e and density ρ of the gas changes

$$\begin{aligned} p_e &\propto \rho^{5/3} \text{ (non-relativistic)} \\ p_e &\propto \rho^{4/3} \text{ (relativistic)}. \end{aligned} \quad (5.11)$$

This subtle change has dramatic consequences. The relativistic pressure is too weak as a function of density, and thus gravity wins. Namely, as more mass is poured onto a white dwarf it shrinks, until pressure is not able to support the star against its own gravity. This means that the density of the white dwarf goes to infinity, as its volume shrinks to zero. This sounds like science fiction, and yet, S. CHANDRASEKHAR (1910–1985) and others were able to calculate precisely when and how this happens. As a consequence, a white dwarf will cease to exist if its mass is larger than about $1.4 M_{sun}$, which became known as the **Chandrasekhar limit**; see Figure 5.10.

The apparent absurdity led to a bitter controversy between CHANDRASEKHAR and EDDINGTON, but the result stood the test of time. Today, we do not consider the collapse of a white dwarf absurd anymore. It has been found that the imploding dwarf turns into a **neutron star** or a **black hole**. Incidentally, imploding white dwarfs play an important role in the construction of the expanding universe. You might imagine that they do not go with a whimper, and their bang is so luminous that it can be perceived from the end of the observable universe, as we will see in Section 5.6.2.

First though, we must understand stars and their evolution, and thus pragmatically forget most of these theoretical speculations. What remains is this pattern in nature: In the Hertzsprung-Russell diagram we find these hot, dim and ultra-dense stars. This is an observational fact. We have to explain it. The associated question is this: How do star-sized fluffy hot gas balls turn into planet-sized ultra-dense objects?

In this vein, it is interesting to point out that two respected observational astronomers felt bemused to speculate even further about "ultra-ultra-dense" objects hitherto not even seen.

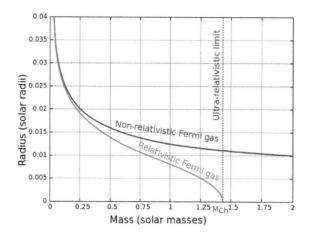

Figure 5.10: **The size of a white dwarf versus its mass.** The more mass a white dwarf has, the smaller it is. It collapses if it has too much mass, because the its gravitational force overpowers the pressure of its *electrons*. To understand the behavior of white dwarfs we need to invoke both *relativity* (because the velocities of the electrons are approaching the speed of light) and *quantum mechanics* (because the electrons are so densely packed).

WALTER BAADE (1893-1960) and FRITZ ZWICKY (1898–1974) were studying the cataclysmic events for which they coined the term **supernovae**, and were wondering what kind of process a supernova is and what it leads to. They wrote in 1934, just two years after the discovery of the neutron: "With all reserve we advance the view that a super-nova represents the transition of an ordinary star into a **neutron star**, consisting mainly of neutrons. Such a star may possess a very small radius and an extremely high density."[3] **Neutron stars** were not observed until 1967, but theoretical arguments by GAMOV in 1937 showed that neutrons can be packed even closer than the nuclei and electrons in a **white dwarf**. Under these extreme conditions, a neutron **quantum pressure** develops, exactly on the same grounds as the white dwarf electron pressure of FOWLER. However, the neutrons are about 2,000 times heavier than the electrons, and thus we expect the density and pressure of neutron stars to be much higher. Indeed, the density of matter where neutron pressure becomes effective can be computed to be about 10^{17} kg/m^3. Consequently, a one solar mass neutron star has a radius of only 10 km—it is city size[35]. You can imagine that this construction met with limited credulity, but due to its extreme and exciting properties, it became a favorite toy model for theorists. In particular, the gravitational field of neutron stars is so strong that Newton's universal law of gravity breaks down, that is, ceases to be a good approximation. Therefore, EINSTEIN's gravity theory (**general relativity**) must be used to describe neutron stars.

In summary, **white dwarfs** and **neutron stars** are ultra-dense celestial objects. Since both radiate, that is, shine by their own light, they must be stars, albeit with strange properties. In particular, they do not produce energy.[36] To explain their existence, we must find

---

[35] *Cross-check:* A neutron star is 6000 km/10 km = 600 times smaller than the white dwarf Sirius B. It is therefore, assuming equal mass, is is approximately 200 million times denser, since

$$f^3 = 600^3 = 2.16 \times 10^8 \approx 200 \text{ million}. \tag{5.12}$$

[36] Nuclear fusion has stopped in these stars, as we will see shortly.

5.5. FUSION DETERMINES STELLAR STRUCTURE AND LIFE CYCLE

a mechanism that produces such objects. Our hunch is that, since their energy production is defunct, they are the *terminal stages* of the *stellar evolution*.

5.5.3 The Construction of the Stellar Life cycle

While we have to keep the strange, ultra-dense terminal stages of stellar evolution in mind, we now use the thrust of understanding of the stellar energy production to construct a theory of a star's life cycle. Indeed, the discovery that energy in most stars is produced by hydrogen-to-helium fusion has obvious but profound consequences: The star's internal structure changes! Hydrogen is used up and helium is created. In other words, as time goes by, the star contains less and less hydrogen, and more and more helium. The other consequence of the discovery that fusion powers stars is that we have to come up with a mechanism to get the nuclear fire started.

Star Formation

The basic idea of *star formation*, follows from the laws of thermodynamics and the *nebular hypothesis*, Section 3.5.2: An interstellar gas and dust cloud made mostly of hydrogen contracts and therefore gets hotter and denser. At some point it is so small, hot and dense, that the conditions allow the fusion of hydrogen to helium at its center. It's as simple as that. Of course, the devil is in the details, and it took quite some time to refute *theoretical concerns* and gather *observational evidence* that this mechanism indeed results in the formation of stars.

To understand stellar formation, we start with the raw material. What is an *interstellar gas and dust cloud*? To put a picture in your mind, think of the *Great Orion nebula* (M42), Figure 3.10(a), that we have mentioned before. It is one of the nebulous deep sky objects cataloged by comet-hunter CHARLES MESSIER. The *Orion nebula* can be found as the middle "star" of Orion's sword dangling from the middle star of *Orion's belt* on the **celestial equator**—fairly high in the southern direction in the sky on winter nights. If you ever had the chance to look at M42 through a telescope, you'd remember the distinct sight: The nebula looks indeed like a tiny cloud in the sky, so part of the label *gas and dust cloud* is obvious. Skeptics might interject that it *looks* like a cloud, but this doesn't mean that it *is* a cloud. After all, the nebulous **Milky Way** turned out to be a multitude of stars in Galileo's telescope as we saw. So how do we decide whether it is a nebula? To develop a criterion, we'll have to first understand the distinction between gas and dust:

- **Gas** quite intuitively consists of atoms and molecules. These particles are much smaller than the wavelength of visible light. Light does scatter off the interstellar atoms and molecules like in Earth's atmosphere. Therefore, a gas cloud between a distant star and the observer, Figure 3.25, will **redden** the starlight because blue light is preferentially scattered into directions other than the line of sight. Also, the gas will **absorb** light and thus potentially put "extra" Fraunhofer lines into the spectrum of the star. The gas is incredibly dilute, about 10^{16} times thinner than air!

- **Dust** consists of grains that are much larger than atoms, but still microscopic. The dust particles have sizes comparable to the wavelength of visible light, and can therefore effectively block light of distant stars, as in the **dark nebulae** seen in Figure 3.10(b).

While very dilute, these gas and dust clouds are of gigantic proportions, and thus contain a lot of material. As good scientists, we should do an order of magnitude estimate to figure out if the hypothesis that stars are formed out of the interstellar medium has any chance of working—lest we waste a lot of time. The criterion is clearly that an interstellar cloud has to contain much more material than a typical star. This is a minimal condition, since we suspect from observing young **open star clusters** like the *Pleiades* that stars are not formed alone but in groups. The apparent size of the *Orion nebula* is very roughly half of that of the full moon, that is, $0.25° = \frac{\pi}{4 \times 180}$ rad. It is about 1000 ly away, so using Equation (1.1), we find the actual diameter $D = 1000 ly \times \frac{\pi}{4 \times 180}$ rad ≈ 4 ly. Assuming for simplicity that it is spherically shaped, the volume of the Orion nebula is

$$V_{M42} = \frac{4\pi}{3}\left(\frac{D}{2}\right)^3 = \frac{4\pi}{3}\left(2\,ly \times \frac{9.5 \times 10^{15}\,m}{ly}\right)^3 \approx 32 \times 10^{48}\,m^3,$$

where we have used the fact that $1\,ly = 9.5 \times 10^{15}\,m$, and—to avoid using a calculator—the approximations $\pi \approx 3$ and $9.5 \approx 10$. If we assume the Orion nebula is as dense as air (which is a gross overestimate!), it would contain a mass

$$\begin{aligned}M_{M42} &= \text{(volume of M42)} \times \text{(density of air)} = (3.2 \times 10^{49}\,m^3) \times \left(1.2\,\frac{kg}{m^3}\right) \\ &= 3.8 \times 10^{49}\,kg = \frac{3.8 \times 10^{49}\,kg}{2 \times 10^{30}\,kg} M_{sun} \approx 2 \times 10^{19} M_{sun},\end{aligned}$$

which is ample material to form many stars.[37]

The reason why stars form inside a gas cloud is that the latter is not in **hydrostatic equilibrium**. Since the gravitational forces between its parts is stronger than the gas pressure, the cloud shrinks until equilibrium is reached. Typically, the gas cloud breaks up into smaller pieces, each of which gives rise to a star. Thus, a gas and dust cloud like the Orion nebula will give birth to many stars. The way a star is born is outlined in the **nebular hypothesis**, Section 3.5.2. Basically, the central part of the cloud collapses, converting **gravitational energy** into **thermal energy**, thereby heating up. In this phase of stellar evolution the nascent star is called a **protostar** and still gathers material from the interstellar gas cloud. Since the collapse of the cloud is synonymous with an increase in **density** and **temperature**, at some point its central part becomes hot and dense enough to start hydrogen fusion. When this happens, a *bona fide* star has been born. However, it still sits inside a lot of material within the cloud. In other words, this phase of stellar evolution is hard to observe. Sometimes astronomers use infrared telescopes to "look into" the gas and dust cloud, since the longer-wavelength infrared waves can travel through the dust relatively freely. Infrared observations have confirmed scientists theories of star birth. To free themselves from the surrounding gas and dust, young stars typically undergo a so-called **T Tauri** stage, when they eject mass; see Fig 4.21(b). While stars are born in groups or associations, they eventually will drift apart, to become the isolated stars we see in the night sky. On the other hand, it is no surprise that we find many young stars, like the Pleiades (M45), Figure 3.11(a), in **open star clusters** and shrouded in interstellar matter.

---

[37]In fact, the mass of the whole Milky Way is only about $10^{12} M_{sun}$, so the Orion nebula must be more than a billion times thinner than air.

5.5. FUSION DETERMINES STELLAR STRUCTURE AND LIFE CYCLE

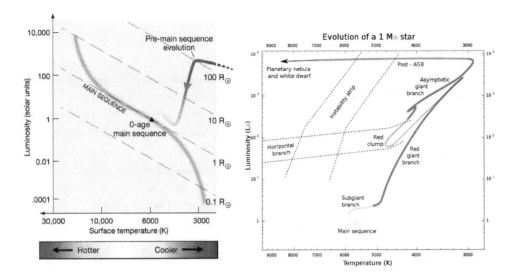

Figure 5.11: **Evolutionary tracks of a one-solar-mass star.** (a) Pre-main sequence evolution of a sun-like star. (b) Track after leaving the *main sequence*. Note the horizontal path of the star at the end of its life: Without changing its luminosity, the star becomes much hotter.

Not all the fragmented pieces of the gas cloud are of the same mass. This fact explains why some stars are more massive than others. Also note that the more massive a cloud fragment is, the faster it will evolve due to the increased gravitational forces between the particles in the cloud. This in turn means that more massive stars are born faster, since they hop onto the main sequence faster, since their nuclear fire ignites earlier. This leads to the following path in the Hertzsprung-Russell diagram, see Figure 5.11(a). Starting as a huge but cold object in the extreme upper right corner of the diagram (high luminosity, low temperature), the cloud and star get smaller and hotter, drifting diagonally down to to left in the Hertzsprung-Russell diagram, until eventually reaching the *main sequence* of normal, hydrogen-burning stars. Just how much hydrogen is burned per unit time, that is, the star's luminosity as well as its temperature, are determined via the condition of hydrostatic equilibrium by the star's mass.

Stars spend by far the longest part of their lives on the main sequence

After all these processes and changes in the protostar phase of a star's life, not much is to report in the *main-sequence stage*. Steady hydrogen-to-helium fusion provides a reliable and long-lasting energy source for the star. Consequently, a star spends by far the longest part of its life as a main-sequence star. This is evidenced by the fact that we observe 90% of stars being on the main sequence. There is, however, a slow change in the structure of the star. Namely, its chemical composition changes in its core, as hydrogen is used up and helium and energy are produced. This changing composition affects the star, and slowly increases its luminosity. The sun, for instance, is already 40% more luminous that four billion years ago. Per year, this is of course a negligible increase. Nonetheless, this change is what inevitably drives the star away from the *main sequence*.

Why do stars leave the main sequence?

At this point we have successfully explained the behavior of most stars during most of their life. In short, the **main sequence** containing roughly 90% of all stars is roughly a diagonal in the Hertzsprung-Russell diagram because of **mass-luminosity relation** and the **Stefan-Boltzmann law**, and the relative longevity of the main-sequence stage is a testament to the most efficient energy production mechanism in the universe—**hydrogen fusion**.

The most pressing question after explaining the **main sequence stage** was: What happens after it? RUSSELL's initial stellar evolution hypothesis that stars cool and move down the main sequence diagonal was discredited by the **mass-luminosity relation** of 1924, which shows that it impossible for a star just to cool and get dimmer while maintaining its mass. But how to find a new **stellar evolution paradigm**? The answer was found by Danish astrophysicist BENGT STRÖMGREN (1908–1987)—even before the theory of hydrogen fusion was worked out. He realized that fusion—whatever its details may be—necessarily causes a change of the **chemical composition of the stars**. His investigation of the consequences of a change in stellar composition within EDDINGTON's **standard model** with complete convection led him to conclude that the luminosity of a star depends on the mean molecular weight μ of its material. Put simply, as the **stellar composition** changes (helium has a different molecular weight than hydrogen) so does its **luminosity**. Taking into account the **Stefan-Boltzmann law** and the **mass-luminosity relation** between mass, radius, surface temperature and luminosity, one can draw **evolutionary tracks** into the Hertzsprung-Russell diagram. Stars move along these tracks as they age, that is, as their chemical composition changes due to fusion reactions. The tracks show, for instance, how the temperature and luminosity of a star changes as a function of its chemical composition. Keep in mind that the star's mass stays virtually the *same* throughout its entire life.[38] These tracks can be compared to the observed positions of stars in actual Hertzsprung-Russell diagrams. STRÖMGREN concluded that those of his theoretical curves with a large hydrogen content best agree with the observed main sequence. Those of his curves with less hydrogen (and more helium) lay to the right and above the main sequence. In other words, stars were drifting toward the red giant region as they used up hydrogen, Figure 5.11(b). This discovery put stellar evolution on the right track.

Of course, the details had to await the development of nuclear astrophysics, namely the discovery of the details of the fusion reactions in stars. We saw that ÖPIK and SALPETER were able to explain the existence of red giants by assuming that they burn helium in their core to form carbon by the so-called **triple-alpha process**. We also saw that the conditions for helium fusion are demanding. In particular, the central temperature of the star has to rise *twentyfold*. Assuming the **Öpik-Salpeter hypothesis** is right, we have to find a mechanism that increases the central temperature and density so dramatically, *even though the star's mass stays the same.*

The details of this process were worked out in the 1940s and 1950s by many scientists, including CHANDRASEKHAR, MARIO SCHÖNBERG (1914–1990), MARTIN SCHWARZSCHILD (1912–1997), ALLAN R. SANDAGE (1926–2010), and FRED HOYLE (1915–2001). We will describe a generalized version here, which has the additional advantage to be applicable to the most massive stars later in their lives, Section 5.6.1. Namely, the mechanism to get a sputtering stellar core "going" again involves the following steps.

---

[38] Recall that the mass converted to energy by fusion is exceedingly small.

5.5. FUSION DETERMINES STELLAR STRUCTURE AND LIFE CYCLE

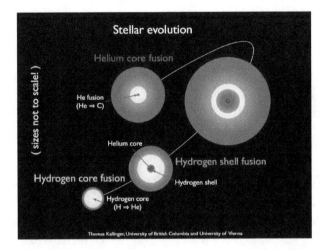

Figure 5.12: **Evolution of fusion reactions in stars.** Initially, the star is fusing hydrogen to helium in its core. As hydrogen is exhausted in the core, hydrogen burns in a shell around the core (hydrogen shell fusion). Eventually, the remaining helium core gets hot and dense enough so that helium fuses to carbon.

1. As the star is running out of fusible material in its core, energy production stalls and consequently, the pressure drops.
2. The forces on the star are not balanced anymore: Gravity is stronger than thermodynamic pressure and **hydrostatic equilibrium** is lost.
3. The star's core shrinks.
4. The core heats up due to our rule from above: "If *it* shrinks, *it* becomes hotter"—unless *it* is **quantum matter**.
5. The adjacent regions of the star also heat up and get denser because they shrink.
6. The hotter, denser material in the central region of the star produces more energy than at the beginning of the process.
7. More energy leads to a higher pressure thereby puffing up the non-burning envelope of the star. In other words the radiative region outside of the convective core or the "peach star" expands.
8. Due to the greatly expanded surface, the energy is radiated at a lower surface temperature; the star becomes *redder*.

This mechanism explains how a star drifts from the main sequence into the red giant region fairly continuously. However, there are two distinct sub-processes that heat the star in item 5, and consequently, the star traverses the red giant region twice during its evolution. In the first pass, fusion stops in the center, but due to the increased density and temperature in a shell around the center which is full of hydrogen, fusion can begin in this shell. This is known as **hydrogen shell burning**, Figure 5.12. This shell burning heats the core (and the rest of the star). It also adds helium to the core as hydrogen fuses in the shell. This added weight will shrink the core—and therefore heat it according to our "shrinks-heats" rule. At some point the core will get hot enough so that helium starts fusing because the **triple-alpha process** which produces carbon becomes possible at these high temperatures. This is known as the **helium flash**, because in lighter stars the onset of helium-to-carbon fusion happens suddenly; for more massive stars, it is a rather gradual process. At the time of the **helium flash**, the star is at the top of the first pass of its **evolutionary track** through "red giant territory," Figure 5.11.

Also the next steps in a stars life are easy to comprehend in hindsight, although it took scientists decades to figure it out. We apply the converse of our "shrinks-heats" rule to the core which is now producing energy by helium fusion. Therefore, it expands and cools. Consequently, also the shell around the core cools and thus less hydrogen fusion reactions happens. Counterintuitively, the new energy production mechanism (helium core burning) results in a *decrease* of energy production, and the luminosity of the star goes down. As a consequence, the star's envelope contracts, but if it contracts it heats up. The associated path in the Hertzsprung-Russell diagram 5.11 therefore goes toward the bluer or hotter regions. After the initial drop of luminosity after the helium flash, it stays relatively constant, and accordingly this region and stage is known as the **horizontal branch**. It is not too hard to figure out what happens next. At some point all the helium in the core is fused to carbon, and energy production ceases in the core. We can now go back to our generalized mechanism, and retread the steps that lead to shell burning. Having two shells burning (helium and hydrogen) produces a lot of energy, and the star shoots up into the red supergiant region, also known as the **asymptotic giant branch (AGB)**.

Detailed computer simulations based on the equations we have developed throughout this section show that the AGB star goes through a number of so called **thermal pulses**. The idea is that the helium in the shell is fused into carbon and oxygen in a fairly short time. This means two things. The helium fire is extinguished and the shell contracts, thereby heating itself and the hydrogen shell above it. The hydrogen shell therefore produces more energy—and helium, which rains down due to its heavier weight, restocking the (former) helium shell. Since also the carbon-oxygen core contracts and heats up, helium shell fusion begin anew. With all these energy production mechanisms firing and stopping, it is clear that the overall luminosity and other stellar properties will change rapidly. Of course, this cannot go on forever, as less and less fusible material is being produced in the pulses. What does the star in, though, is that its outer layers cool down so much that they become opaque. Energy cannot be radiated into outer space anymore. As a consequence, the mounting pressure blows the outer layers of the star into outer space. As the star loses a significant amount of mass, the formation of a **planetary nebula** has begun, Figure 5.13 and Figure 3.12. The stellar material that separates from the star moves swiftly, propelled and heated by the energy of the remainder of the star. Thus, a planetary nebula becomes very large and it glows! Indeed, the spectrum of a planetary nebula is an **emission spectrum** of bright lines.[39]

This ends the life of the star as we know it. Having blown off a good fraction of its mass, the pressure onto the remaining star (which is basically reduced to its former core) is not large enough to allow fusion. The star ceases to produce energy. However, it is exceedingly hot (as cores tend to be!) and therefore, according to the **Stefan-Boltzmann law**, radiates a lot of energy. The star is at the last turning point of its **evolutionary track** within the Hertzsprung-Russell diagram. Being as hot as 100 000 K and as luminous as 10,000 suns, it has only one way to go: colder and dimmer. Indeed, over the next billions of years, the freshly formed white dwarf cools down to 4,000 K, which is cool enough such that its strange, degenerate carbon-oxygen matter solidifies. The star becomes a crystal of sorts, and if you

---

[39]The spectrum of a (cold) planet is an absorption spectrum. While a **planetary nebula** and a **planet** might have a similar name and disklike appearance, they are two completely different objects.

5.5. FUSION DETERMINES STELLAR STRUCTURE AND LIFE CYCLE

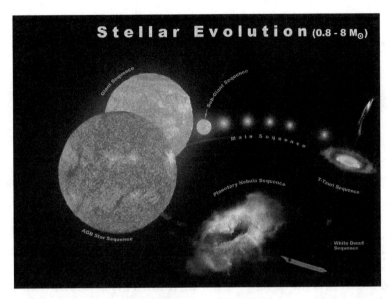

Figure 5.13: **Stellar life cycle of an average Star.** After forming from a contacting interstellar gas cloud, the star goes through a *T-Tauri stage* with mass ejections before becoming a normal *main sequence star*. Once it exhausts the hydrogen in its core, it turns goes through a *giant sequence*, ending with *helium burning* as an *AGB star*. It desintegrated by ejecting its out layers forming a *planetary nebula*. The star's carbon-rich core becomes a *white dwarf*.

consider that diamonds are carbon crystals, it becomes a diamond in the sky! However, with its declining surface temperature eventually this diamond will not shine but dim into obscurity. Its associated planetary nebula will have diluted into oblivion long before then. We can estimate based on the initial velocity of the ejected material that after roughly 50,000 years a planetary nebula fades from view.

5.5.4 Predicting: Variable Stars

We saw that the processes inside an aging star quickly become unstable. We therefore predict that the star's energy output and therefore its brightness will change quickly toward the end of its life. The horizontal branch in the Hertzsprung-Russell diagram 5.11 is consequently known as the *instability strip*. Before we jump to the (correct) solution that the internal processes of aging stars explain the existence of *variable stars* as observed in the sky, see Section 5.5.4, we have to check if this has any chance of working. The problem is one of *timescales*. So far, we have been seeing that stars are incredibly stable and long-lived. As they evolve, they change very slowly. For a sun-like star we are talking billions of years of main-sequence lifetime, and even the "short" helium-burning supergiant phase lasts tens of thousands of years.

Let's compare these timescales with what we observe in the sky. Since the late 1700s astronomers have found, by carefully comparing the brightness of a star with its neighbors, that there are more than a few so called *variable stars* which change their brightness, see Section 5.5.4. Some well-known stars like *Polaris* and *Betelgeuse* are (mildly) variable. Most variable stars will change their brightness *periodically*. They reach maximum and minimum brightness in a regular, periodic pattern known as a *light curve*, that is, the stellar brightness plotted as a function of time, Figure 5.14. Others, like the star *Mira* mentioned above, change

350 CHAPTER 5. EXTRAPOLATING & SYNTHESIZING: UNDERSTANDING STARS

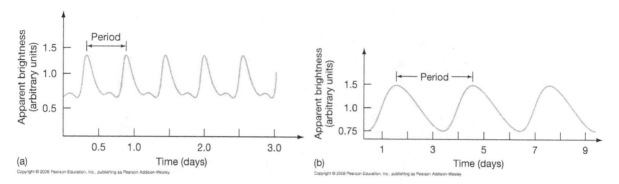

Figure 5.14: **Light curves of variable stars.** (a) *RR Lyrae* type; (b) *Cepheid* type.

their brightness so dramatically that they virtually disappear from view for several months, and their behavior is more erratic. The typical variable star will change its brightness with a period anywhere from a few days to a few months. This is an incredibly short time for an astronomical object, and we might speculate that this pattern in nature has the potential of revealing something very important about stars and the cosmos.

Indeed, the study of **variable stars** turned out to be a goldmine for astronomers, because it allowed a much improved stellar distance determination. If this revelation seems to come out of left field, think about it for a moment. If the variableness could teach us something important about a star, it is probably going to be its energy output or **luminosity**. And by our old argument, Equation (5.3), knowing luminosity and observing brightness of a star is tantamount to determining the distance to the star. This is exactly how the story played out in the early 1900s, and it led to two amazing conclusions, as we will see in the next chapter: Our galaxy, the Milky Way, is much bigger than previously estimated, and many of the other spiral nebulae are galaxies in their own right. In other words, the expanding understanding of **variable stars** expanded the cosmos!

How did this story unfold in detail? Recalling that the key ingredient is to be able to determine the luminosity of a star, it is clear that the periodicity of the brightness variation (which we can measure easily), allows us to obtain the luminosity. This can happen in two ways: Either there is a relation between period and luminosity, or the period itself directly identifies a specific type of star that has a specific luminosity. Both turn out to be viable ways. The former relation was found in a class of variables named after the star δ in *Cepheus*, the so-called *Cepheids*. In 1908 HENRIETTA SWAN LEAVITT (1868–1921) studied the behavior of these variables, and plotted their luminosities (which were known for several of them because their distances had been measured) as a function of their pulsation period, Figure 5.15. She found a strong positive **correlation** between the **period** and the **luminosity**: The longer the period P, the more luminous the Cepheid. This makes sense, because the more luminous the larger the star, and the larger an object the longer it takes for it to change. The **period-luminosity relation** plotted in Figure 5.15 is a **power law**. The formula describing how the luminosity of a Cepheid changes with its period looks something like this

$$L_{\text{Cepheids}}(P) = 3300 \times 10^{0.972(\log_{10} P - 1)} L_{Sun}, \qquad (5.13)$$

where P is measured in days. So a Cepheid with period 10 days would have a luminosity of 3300 suns. On the other hand, a Cepheid of period 100 days would be as bright as 31,000

5.5. FUSION DETERMINES STELLAR STRUCTURE AND LIFE CYCLE

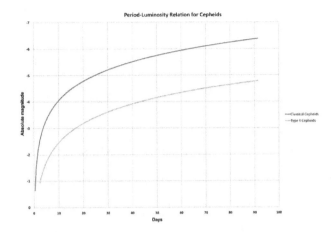

Figure 5.15: **Period luminosity relation for Cepheids.** The time between brightness maxima, that is, the *period*, is directly related to the total power output of these variable stars, that is, their *luminosity*. Since it is easy to measure the period, this is a great tool to determine the distance to the star via the *brightness-luminosity-distance relation*.

suns, so roughly ten times more luminous. This makes sense, because the luminosity goes up with the *logarithm* of the period, and $log_{10} 10 = 1$, whereas $log_{10} 100 = 2$. Here is an analogy. Consider a simple pendulum, that is, a bob at the end of a string swinging around a pivot point. The longer the string, the longer the period of the pendulum, and vice versa. If you let the string length represent the luminosity, then it can be determined from the period of the pendulum.

A few years later, another class of variables called *RR Lyrae* stars (after its prime representative in the constellation *Lyra*) became useful as a **cosmic yardstick**. HARLOW SHAPLEY (1885–1972) had determined that all stars in this group, identifiable by their small period of less than a day and a secondary maximum in their light curve, Figure 5.14(a), have a luminosity just shy of 100 solar luminosities

$$L_{\text{RRLyrae}} \approx 100 L_{Sun}.$$

This means that by identifying an RR Lyrae as such already gives away its luminosity. This is analogous to identifying a G2 main sequence star as such thereby determining that it has exactly one solar luminosity, because it sits on the main sequence for which a strict correlation between luminosity and spectral type exists.

Concept Practice

1. Which is a true statement about main sequence stars?

    (a) The more mass they have, the longer they live.
    (b) The hotter they are, the longer they live.
    (c) The dimmer they are, the longer they live.
    (d) None of the above.

2. White dwarfs
    (a) ... are star-sized. (b) ... are planet-sized. (c) ... are city-sized. (d) ... have infinite density. (e) None of the above.

5.6 Nucleosynthesis Fills the Periodic Table

> "We are stardust, we are golden
> We are billion-year-old carbon
> And we've got to get ourselves
> Back to the garden."
> *(Joni Mitchell, "Woodstock", 1969)*

5.6.1 Fusion Leads to Nucleosynthesis in Massive Stars

In the previous sections, we've come to appreciate that stars are hot, glowing gas balls. All are essentially made of the same material (mostly hydrogen, some helium, and a dash of "metals"). Consequently, we convinced ourselves that **mass** is the most important criterion to determine a star's **luminosity** and other properties—including its **life expectancy**. So far we've limited our discussion of stellar evolution to light and middle-weight stars. There is a good reason for it. These stars are by far the most common, so from a pragmatic point of view, if we understand these stars, we've understood most stars. Furthermore, a consistent theory of these "normal" stars plays the role of a **frame of reference**. With a functioning theory at hand, we can effectively point to the assumptions of that theory which would fail for other, more massive stars. On the other hand, extrapolating a theory and probing when it breaks down is a fertile approach, as we have seen when discussing the transition from classical to modern physics. You have to appreciate how much the modern theories have told us about our classical theories. We understand much better what **classical physics** is, after we've seen it breaking down at high velocities and small distances. In the same vein, we might hope that the extreme regions of the **Hertzsprung-Russell diagram** might teach us much about the true nature of stars and the universe. And indeed, the understanding of massive stars will provide us with methods to construct the expanding universe by an expanding understanding of extremely luminous processes such as **supernovae**.

The guiding question for this subsection is clearly "In what way are massive stars different from lighter stars?" On the face of it, this is a silly question. Of course, they are different in that they have much more mass. So a more refined question is: "What new effects does the additional mass spawn?" This is a crucial question, because so far we had appreciated mass as a continuous parameter. According to the **mass-luminosity relation**, Equation (5.7), more mass simply means much more luminosity, and therefore much shorter life expectancy; see Equation (5.10). So if we heap up more and more mass onto a star, at some point something dramatic and qualitatively new must happen. More mass means more gravitational pressure, and this results in higher core temperature and density. We don't have to wax too creative to see where this is going. It must be that at some point these conditions will drift into a regime where new nuclear processes are possible in the core of the most massive stars.

Where did the evolution of the light-weight stars stop—and why? Well, we've learned in Section 5.5 that a normal star fuses hydrogen to helium, then helium to carbon and oxygen, and that the end product of such a stellar life is a white dwarf, which is essentially a "planet-sized, star-massed" object consisting mostly of carbon, some oxygen, and a thin hydrogen or helium atmosphere. So the answer to our first question is: The process stopped with helium-to-carbon fusion. And we can guess the answer to the second question: It stopped there because to conditions were not right to fuse carbon to heavier elements. The same conclusions were drawn by *two* groups of scientists in the mid 1950s. Naturally, **nuclear**

5.6. NUCLEOSYNTHESIS FILLS THE PERIODIC TABLE

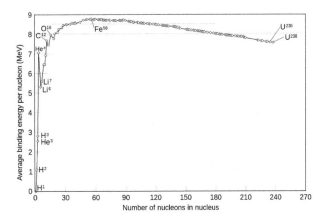

Figure 5.16: **Nuclear binding energy.** Displayed is the average binding energy per nucleon in MeV (1 MeV = 1.6×10^{-13} J) as a function of the number of nucleons, that is, for all chemically elements. The sharp rise of the curve for light nuclei (small number of nucleons) means that much (binding) energy can be gained by fusing them to form heavier nuclei. The maximum of the curve represents the most stable element in the universe, iron (^{56}Fe). The decline of the curve after iron means that energy can be freed by decomposing very heavy nuclei into lighter ones by **nuclear fission** or **radioactive decay**.

physicists needed to figure out under which conditions light nuclei can fuse into heavier ones—and what the probability for such processes is. This line of work is connected with names such as FRED HOYLE, SALPETER, ALASTAIR CAMERON (1925–2005), and Nobelist WILLIAM FOWLER (1911-1995). The second group around GEOFFREY BURBIDGE (1925–2010), MARGARET BURBIDGE (born 1919), and JESSE GREENSTEIN (1909–2002) focused on the conditions in stars, and the associated relative abundance of the chemical elements in stars. Several members of these groups worked together to write a seminal article with the title "Synthesis of the Elements in Stars" [10] in 1957. Their amazing conclusion was that the heavier elements are **made** in the most massive stars. Indeed, the universe was not "born" with all pieces in place, but it has changed—evolved—into what it is today. The universe was a different place several billion years ago. At the very least, it contained much fewer heavy elements than today. As an important corollary, we find that since we humans consist of a lot of heavy elements such as calcium, sulfur, iron, oxygen, nitrogen, and phosphorus, we couldn't have "formed" billions of years ago. Simply because our building blocks had not been produced yet by the universe, that is, its most massive stars. In other words, "we are star dust", made of heavy elements produced in stars. Of course, the last step in this line of reasoning is still missing. If massive stars produce the stuff we are made of, how is it liberated from the interior of these stars? How do the heavy elements *become* star dust? The answer to this question is in the next section 5.6.2. It has to do with the violent death of the most massive stars called **supernovae**.

Before we go there, we'll describe in more detail how the elements are produced or **synthesized** in massive stars. These **fusion processes** are collectively known as **stellar nucleosynthesis**. Recall that the generic fusion process looks like this:

(light nucleus) + (light nucleus) ⟶ (heavy nucleus) + (other particles) + (energy).

Therefore, we can use the tables produced by **nuclear experimentalists** and detailing exactly *which* light nuclei can fuse into *which* heavier nuclei, how much energy is released, and what the conditions (temperature, pressure, density) are under which these **nuclear reac-**

tions become possible. We can do even a little better than this, and *predict* which heavier elements will be produced in fusion reactions starting from carbon where the light stars quit. How? Simply look at the abundances of the chemical elements in the universe, displayed in Figure 4.5. We find that there are many more hydrogen, helium, carbon, nitrogen, oxygen, neon, silicon, iron atoms in the universe than there are other elements.[40] Therefore, we suspect that these elements are actively produced in massive stars. So we can produce a tentative, symbolic list of fusion processes in the most massive stars:[41]

$$
\begin{aligned}
(\text{carbon}) &\longrightarrow (\text{neon}) \\
(\text{neon}) &\longrightarrow (\text{oxygen}) \\
(\text{oxygen}) &\longrightarrow (\text{silicon}) \\
(\text{silicon}) &\longrightarrow (\text{iron})
\end{aligned}
$$

This hypothesis seems pretty feeble, but we can do a cross-check. We can ask which are the ***most stable elements*** in the periodic table. This is again a question for ***nuclear experimentalists***. They can produce a plot of the binding energy (per nucleon) of all the elements, Figure 5.16(a). The idea here is that more binding energy means tighter binding, which means more energy is necessary to break up the nucleus of an element. We see that precisely our list of suspects are the most stable elements in the universe, and that iron comes out on top. In other words, if we fuse iron to form heavier nuclei, the latter would be *less* stable. This cross-check supports our hypothesis, because it seems likely that nuclear reactions occur such that a stable end-product is being produced. Modern star models produced by state-of-the-art computer simulations have vindicated this chain of reactions in the most massive stars.

We can now recycle the general fusion scheme for normal, light stars of Section 5.5.3. All we have to do is to *assume* that the conditions are right so that these processes will occur. In sum, as soon as fusion of an element runs out of the element—its fuel—the core contracts, heats up, and switches to the next element, namely the one just produced. For example, if carbon fuses to oxygen, and no carbon is around anymore, temperature and density will increase until oxygen can fuse to neon, and so on. Probably the shell-burning process we discussed earlier will also take place, and thus a lot of burning shells (and a non-burning hydrogen envelope) will make the most massive stars look like an *onion*, Figure 5.17.

It seems that we might be able to produce all the naturally occurring chemical elements with this scheme, all the way up to uranium (element number 92). Accordingly, we might conclude that very massive star enjoy a long life. However, we already know that the later cannot be true, since Equation (5.10) shows that the more massive a star is, the faster it lives and the earlier it dies. There are two flaws in our reasoning:

1. Fusion of lighter into heavier nuclei produces energy only if the heavier nucleus is more

---

[40]In the solar system the astonishing dominance of the most abundant elements can be put into numbers. Out of 1 million atoms, 909,964 are hydrogen (91%), 88,715 are helium (8.9%), 477 oxygen, 326 carbon, 102 nitrogen, 100 neon, 30 silicon, 28 magnesium, and 27 iron. The rest (83 elements!) together contribute 74 atoms (0.0074%). Incidentally, oxygen, carbon, hydrogen and nitrogen make up 96% of the *human body* by mass. Only 17 elements are known to be essential for human life.

[41]We are cheating a little here by using some nuclear physics insights. Namely, the heavier neon nucleus is produced before the lighter oxygen, and nitrogen and magnesium do not fuse into heavier elements in the way the other nuclei do.

5.6. NUCLEOSYNTHESIS FILLS THE PERIODIC TABLE

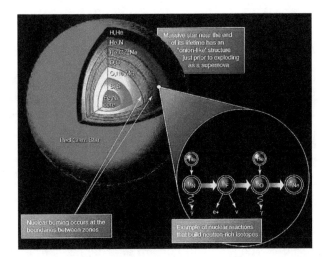

Figure 5.17: **Interior stratification of an old, massive star ("onion star")**. Very massive stars produce conditions in their interior to allow the fusion of many different elements. At the end of their lives, they end up looking like an onion, with different elements occupying shells around the stellar core: The heavier the element, the closer it sits to the core. The shells are surrounded by a (rather thick) layer of non-fusing hydrogen on the outside.

stable than the lighter one. Essentially, the extra binding energy is the mass difference, and liberated as electromagnetic energy.

2. Since the core of the star gets successively hotter and denser, nuclear reactions happen at a much higher rate, therefore the star produces much more energy than at earlier evolutionary stages, therefore the later stages become shorter and shorter; see Table 5.1.

Form the first point we conclude with the help of Figure 5.16 (the binding energy of nuclei) that iron is the ***endpoint of stellar fusion***. The reason is that fusion of iron, having the highest binding energy of all elements, costs rather than produces energy. Energy production in stars will thus terminate when iron fills up the stellar core. This has dramatic consequences, as we will see in the next section.

Before we go there, let's compare the evolution of the rest of the star, namely its non-burning hydrogen envelope with that of the lighter stars. There, we saw that the envelope is shed in several steps to form a ***planetary nebula***. Does the same thing happen in massive stars? In principle—yes. As we have said, a massive star will produce incredible amounts of energy (up to a million times that of the present-day sun), which puffs up its outer layers to gigantic dimensions. These supergiants can become as large as Jupiter's *orbit*, so about 5 AU or 700 million km in radius! This is an enormous volume ($R_S = 0.7$ million km $= \frac{1}{1000} R_{SGiant}$, so $V_{SGiant} = 1000^3 V_S = 10^9 V_S$), and even if the star has 20 times the mass of the sun, its density will be $\frac{20}{10^9} = 2 \times 10^{-8}$, or about 50 million times more dilute than the sun. This is many times thinner than air at sea level! Also, it is clear that the gravitational grip of the star onto its outer layers is very weak due to NEWTON's ***universal gravity law*** according to which the force of gravity fall off like the ***inverse square*** of the distance. This means that the supergiant exerts a force on *its* surface layers that is $\frac{M_{SG}/M_S}{(R_{SG}/R_S)^2} = \frac{20}{1000^2} = 0.00002$ that of the force that the sun exerts on its surface. This weak force is not enough to prevent dramatic mass loss. Indeed, the most massive stars lose a good fraction—up to 60%—of their

Material	Central temperature	Burning Time
Hydrogen	$4 \times 10^7 K$	7×10^6 years
Helium	$2 \times 10^8 K$	7×10^5 years
Carbon	$6 \times 10^8 K$	600 years
Neon	$1.2 \times 10^9 K$	1 year
Oxygen	$1.5 \times 10^9 K$	6 months
Silicon	$2.7 \times 10^9 K$	1 day

Table 5.1: **Burning time of fusible materials in very massive stars.**

mass in their supergiant phase. This material surrounds the dying star like a cocoon, and will sustain collateral damage when the central star disintegrates in a so-called **supernova explosion**. The material of the star will end up as a jagged **supernova remnant** like the famous *Crab nebula* in *Taurus*, Figure 3.12(b).

The process is thus very similar to the emergence of a *planetary nebula*—up to this point. It is reasonable to expect that stars—all starting out as essentially hot hydrogen gas balls—should behave similarly. Since their masses are different, we are *extrapolating* our knowledge from lighter stars like the sun to say something about massive stars. If this extrapolation goes smoothly, we would expect the life cycle of these massive stars to be somewhat (but not crucially) different from the way that the lighter stars live their lives. Sure, the more massive star might burn through its fusion fuel faster, and it may even—due to its higher core temperature and density—be able to fuse elements that a lighter star wouldn't, but, by and large, the same pattern would occur: Fuse lighter elements into heavier, more stable ones, and generate heat and pressure to counter-balance gravity. Running out of "fusibles," the outer layers of a puffed up star separate from its core because surface gravity is weak and eventually the star's ability to maintain energy production is compromised.

In the first section of this chapter, we saw how extrapolations can break down. For instance, in *relativity*, high speeds lead to modification of the laws of classical mechanics because otherwise the speed of light cannot be the same for all observers. In other words, if parameters (here: speed) get too extreme, behavior of a new quality is observed. In stellar evolution, clearly the parameter is the *mass of the star*. So what new phenomenon happens if the mass grows larger? There must be some kind of threshold, beyond which the mass of the star is "too" large. It is not too hard to figure out that "too large" means: So large that all elements can be fused until hitting the *most stable element* of the universe. When the gravitational pressure is immense enough to allow iron to be fused, an *absolute* threshold has been reached. Since iron has the highest binding energy per nucleon, see Figure 5.16, it *takes* not produces energy to fuse it into more complex or heavier elements. Taking energy out of the core of a supermassive star is deadly: The core will collapse, triggering the most violent stellar explosion in the cosmos, called a *supernova*.

5.6.2 Supernovae: The Fertile Death of Massive Stars

It may seem from what we just said that **supernovae** are the logical consequence of the processes within a very massive star, and thus a **prediction** of a general **theory of stellar evolution**. While the former assertion is correct, see the empowering **Vogt-Russell theorem**, quite the opposite of the latter is true. Namely, **supernovae** were first *observed* as a

5.6. NUCLEOSYNTHESIS FILLS THE PERIODIC TABLE

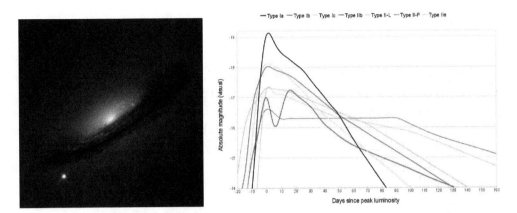

Figure 5.18: **Supernovae.** (a) The supernova SN1994D (lower left) in NGC 4526, a spiral galaxy containing about 100 billion stars. The supernova is almost as bright as the entire galaxy. (b) Schematic light curves of different types of supernovae. Plotted is the absolute magnitude, a measure for the luminosity of the supernovae, as a function of time, expressed in days since peak luminosity.

pattern in nature, and then *explained* and built into a narrative about the life of stars. Many textbooks present supernovae as the violent end of high-mass stars determined by the laws of **nuclear physics** and **thermodynamics**, but I find these fascinating accounts pedagogically questionable for two reasons. Firstly, for most readers they will be largely unintelligible, because they are full of unfamiliar words and concepts such as "neutrinos" and "supersonic shock waves" which would need scaffolding beyond the scope of introductory texts. Secondly, as amazing as it is that scientist are able to understand even these extreme events, the details will blur the focus of the book, which is, after all, the construction of the expanding universe. As it turns out we are better off for our purposes to take a much more pragmatic stance, and simply ask: What does a **supernova** look like, and what does its existence tell us about the universe?

From this point of view, we reconnect with the renaissance; namely, Tycho Brahe's 1572 discovery of a new star which was classified 350 years later as a **supernova**, testament to his superb record of its brightness as a function of time. For an observer on Earth, a supernova is nothing but the sudden appearance and eventual disappearance of a star. As always, what *appears* in the night sky is different from the *interpretation* or explanation of this pattern in nature. For instance, the notion of a **supernova** as an exploding star is an interpretation of a speck of light that is appearing and than disappearing. Note that its name aptly focuses on what had been observed. When Tycho spoke of it simply as a new star (*stella nova*), he summed up what was known about it: It's a star and it either wasn't there before, or it might have been too dim to be noticed. How did Tycho know it is a star and not something like a comet—which also appears and disappears? As we pointed out in Section 2.6.2, Tycho was unable to measure a **parallax** for the new star, which meant that it must be exceedingly far away.[42] We will later see what is "super" about it. For now, this prefix alerts us to the fact that there must be several different "suddenly appearing stars" in the sky, and thus that there must be other observations *distinguishing* the different types.

If you think about it, the only thing that Tycho could do back in the days was to

---

[42] This is a bit *anachronistic*, since Tycho would not have been able to measure the parallax of a comet either.

record how the brightness of the object changes over time.[43] There was literally nothing else to observe since the new star—a *fixed star*—did not move. But this was exactly the right thing to do to decipher this new pattern in the sky. When BAADE and ZWICKY looked at TYCHO's data 350 years after the fact, Section 5.5.2, they were able to plot the brightness of the supernova as a function of time,[44] Figure 5.18. This is known as a **light curve** for obvious reasons. How can we tell from a distance of thousands of light-years what makes this star appear and fade? If there is any chance of finding out, we must look for comparable patterns in nature. If this is a unique, once-in-universal-history event, then there is no way for us to figure it out. But we already know that Kepler in 1604 saw another (super)nova, and there were others. So, to make sense of TYCHO's "new star," BAADE and ZWICKY also plotted the light curves of other "new stars" and compared them: Some looked very similar, others differed, for example their brightness faded faster. But they could do more. In the 350 years that had passed, astronomers developed other ways to analyze starlight, as we have seen. In other words, BAADE and ZWICKY had an independent way of looking at this pattern in nature: They had an additional "eye on the sky." This was, of course, **spectroscopy**, the study of the stellar absorption lines. These two types of observations ("How does the brightness change?" and "What spectral lines do we see in the starlight?") made it clear that there are several different types of *stars-that-suddenly-appear*. Here is a list of the main types.

- Stars that "reignite," that is, can appear (or get significantly brighter) multiple times. For instance, a star in the constellation *Ophiuchus* (*RS Ophiuchi*) flared up in 1898, 1933, 1958, 1967, 1985, and 2006. Typically these stars become brighter by a factor of a few hundred and fade within one or two weeks, Figure 5.19. Stars that change their brightness in this way were labeled **novae**.

- Stars that only appear once and then irrevocable fade into oblivion. These are called **supernovae**, because they are much brighter than **novae**. Indeed, their brightness will vary by factors around 10,000 and they appropriately take much longer to fade. A typical time span here is several months. There are subtypes. From Figure 5.18 it is apparent that some supernovae are significantly brighter than others at their peak, and there are also differences in the way supernovae fade. The classification is as follows.

    - ***Type I supernovae*** are brighter (as luminous as several billion suns) and their brightness declines featureless, that is, without "hiccups."
    - ***Type II supernovae*** are about a factor 10 dimmer (as luminous as several hundred million suns) and their light curve shows another feature. Namely, after the peak brightness there is a ***plateau***, when the brightness stays relatively constant, before sharply dropping. After a while the sharp drop eases into a much more gradual decline of brightness; see Figure 5.18.

When analyzing supernova spectra, further distinctions become apparent. The main one is that the spectra of ***Type I supernovae*** do not show hydrogen lines, whereas ***Type II supernovae*** display strong hydrogen lines. This is already a strong hint toward

---

[43] Recall that well into the nineteenth century, astronomy's primary task was to record positions and brightnesses of celestial objects as accurately as possible.

[44] Remember that such plots, routinely done today, were invented by DESCARTES as part of his *analytic geometry*—half a century after TYCHO.

5.6. NUCLEOSYNTHESIS FILLS THE PERIODIC TABLE

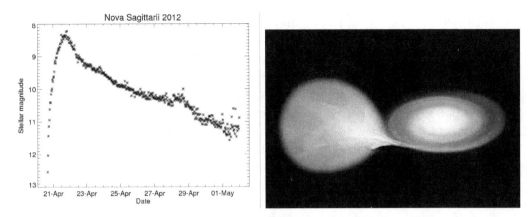

Figure 5.19: **Nova light curve and scenario.** (a) The light curve of the nova Sagitarii 2012. (b) An artist's rendering of a possible scenario or explanation of the light curve: A close binary star consisting of a white dwarf and a red giant. The white dwarf siphons off hydrogen from the red giant. When enough hydrogen has accumulated, it burns off, and the star flares up.

the mechanisms responsible for the different types. We immediately must suspect that *Type I supernovae* are "abnormal" since stars contain a lot of hydrogen which should show up in the spectrum. There a subclasses of Type I:

- *Type Ia supernovae*: Strong silicon line at 615 nm in spectrum.
- *Type Ib supernovae*: Strong helium lines in spectrum.
- *Type Ic supernovae*: Weak helium lines in spectrum.

Another observation, namely that *Type II supernovae* do not occur in galaxies without star formation (*elliptical galaxies*), suggests that the *progenitor stars* giving rise to a Type II supernova are *young* stars.

As you can see, this is quite a complicated pattern in nature! The initial observation of a new star turned into a large data set. Putting the observed light curves into different bins (categorizing) yields some order, but it doesn't explain anything. Indeed, it took decades for astronomers to come up with an acceptable theory as to what makes stars behave in these different ways. For instance, the *Type Ia supernovae* were not understood until 1973 [37]!

Let's take stock, and see what patterns of stellar evolution we still have to explain, while being on the outlook for insights that may help us in our quest to understand the expansion of the cosmos. Firstly, we did not finish our theory of massive stars. We guessed that something dramatic will happen when they run out of fusible elements, namely, when their core is rich in iron, the most stable element of the universe. We thus suspect that one of these supernova types is a dying, massive star. On the other hand, we might use the incredible luminosity of supernovae to our advantage when figuring out distances in the cosmos. What we want is a so-called *standard candle*, an object of precisely known (large) luminosity so it can be seen "from the other side of the universe." If we can understand one of the supernova types well enough to predict its luminosity, we can use it as a *cosmic yardstick*. Namely, due to the relation between luminosity L, apparent brightness B and distance r, $B = \frac{L}{4\pi r^2}$, Equation (5.3), we can figure out the distance to a celestial object by observing its brightness and determining its luminosity. We have done this before; recall the method of spectroscopic parallax, Section 5.2.3. Therefore, understanding supernovae will enable us to measure distances far out in the expanding universe—and to understand how the universe evolves!

This is an exciting roadmap, but first we have to do our homework, namely to decipher the different types of supernovae. Of course, we cannot hope to retrace all the ideas and scenarios astronomers have come up with to explain the details in the light curves of every supernova—but we don't have to! The scientists worked hard over the decades to come up with and rule out many different plausible mechanisms to make a supernova "go off." Armed with hindsight, all we have to do to is to check that their best models fit the bill, that is, explain the observations—or *save the appearances* as it was called in centuries past.

Let's try to figure this out ourselves, as there are some things which are pretty obvious.

1. *Type II supernovae show strong hydrogen lines and the progenitors are young stars.* This strongly suggests that these are normal, massive stars at the end of their lives, so this could be the missing explanation for the end of a massive star's life. Why? Firstly, because normal stars have lots of hydrogen—at least in their outer layers that produce the spectra. And secondly, because massive stars die young, because they are so luminous, that is, go through their fusible fuel fast. Fits the bill, doesn't it? Once their core fills up with iron, these massive stars cannot produce energy by fusion. Hence, their cores collapse, which sets free an incredible amount of gravitational energy that is transformed into thermal energy. This is what powers a **Type II supernova**. Of course, with the collapse of the core, the rest of the star will follow suit and rush in to fill the void. However, the core can only collapse so much. At some point, governed by the laws of quantum mechanics, the core will turn stiff and incompressible. Picture a brick wall—only a billion times more rigid. When the rest of the star rushes in, it bounces right off this "wall," and is sent back through the star as a shock wave that completely rips the star apart. What remains is the ultra compact core of the star (a **neutron star** or **black hole**, depending on the mass of the star). The rest of the star, (the "star dust" of JONI MITCHELL's song if you will) is blown and spilled all over the galactic neighborhood. This material is known as the **supernova remnant**, and enriches and fertilizes the interstellar medium, not unlike dying plants fertilizing the soil. Here, the fertility comes from the fact that the **supernova remnant** contains heavy elements such as the ones you and I are made of: oxygen, carbon, magnesium, and so forth. These heavy elements—absent in the **primordial** universe—are what the next generation of stars (so called **Population I stars**), planets, and eventually humans are made of. The creation of these heavy elements in stars is known as **stellar nucleosynthesis**, and we'll come back to it below.

2. *Type I supernovae do not contain hydrogen.* How is this possible? After all, the entire universe consists mostly of hydrogen. This is not normal. If we pause to think, there is only one place in the universe where hydrogen has been systematically "destroyed" and replaced by other elements, and that is the core of stars. If hydrogen fuses to helium, eventually there will be no more hydrogen. This strongly suggests that **Type I supernovae** are linked to the burned-out cores of stars. In other words, these must be exploding white dwarfs! This is both plausible and outrageous. Plausible because it explains the absence of hydrogen lines in the spectrum, and outrageous, because these white dwarfs are ultra-compact, ultra-stable and the potentially dangerous fusion reactions in them have stopped a long time ago. All that these ultra-hot objects do is to

5.6. NUCLEOSYNTHESIS FILLS THE PERIODIC TABLE

cool down by radiating electromagnetic waves through their tiny surfaces. Nonetheless, astrophysicists around ICKO IBEN in 1973 found a way to detonate them. The point is that a white dwarf cannot detonate alone—a *Type I supernova* involves a partner star. For these reasons the *Type I(a) supernovae* have been called **carbon detonations** (because the white dwarf is mostly carbon) or **assisted suicides** (because the partner star helps trigger the explosion). How does this work? Remember the *Achilles heel* of a white dwarf: The *Chandrasekhar limit* establishes that no white dwarf with a mass greater than 1.4 solar masses can exist. If the mass of a white dwarf becomes too large, it will collapse under its own weight. The other thing to remember is that white dwarfs are not *classical* objects. Since they are so dense and compact, the laws of classical physics cease to apply, and one must us quantum laws to describe them. In particular, the thermodynamic properties of white dwarfs are very strange. For normal matter, temperature increase leads to pressure increase which leads to expansion and eventual cooling; this is a safety valve of sorts that prevents the temperature and other parameters from rising indefinitely. However, this safety mechanism is not operable in white dwarfs. The pressure is due the rules of quantum mechanics, and completely independent of temperature. This is a dangerous thing, since no relieving expansion occurs. It is now clear how this assisted suicide will work, Figure 5.20. In close binary stars, the more massive star will live faster and develop into a white dwarfs sooner. As the smaller star enters its own giant phase, its outer layers will be siphoned off by and *accrete* on the white dwarf. As the white dwarf's mass approaches the *Chandrasekhar limit*, its internal temperature will rise due to the increased pressure. Eventually, it will get hot enough to fuse carbon and oxygen to form radioactive nickel, an element in the iron group. The problem with the onset of fusion is that the strange, degenerate matter of the white dwarf will not expand, and thus the energy production goes entirely into heating the dwarf, and therefore fanning the nuclear fire. As it gets hotter, more nuclear fusion reactions occur. These additional reactions produce even more energy, which raises the temperature even more. You can see that this is a runaway process, and runaway processes end up in catastrophes. Within seconds, the spreading reaction front eats through the entire star, and completely blows it apart, because it releases so much energy; it is aided by convection and turbulence. In that respect, the process is comparable to the shock wave that destroys the star in a type II supernova. Note, however, that the Type I detonation is fed by fusion energy, whereas the Type II explodes due to the release of gravitational energy via the collapse of the core. In both cases, the dramatic energy release rates doom the stars. Further evidence that this theory is correct comes from the fact that the oxygen-carbon fusion releases copious amounts of silicon, which is the hallmark of a Type Ia spectrum. Thus, the theory can be tentatively accepted, as it explains the observational facts. Note, however, how much this explanation relies on other theories. We would not be able to construct such a complex scenario without first understanding quantum and nuclear physics, energy transport, and so forth. Remember, observationally all we have is the appearance of a speck of light, the features of its spectrum and its light curve. The latter can be used to do another cross-check. If radioactive nickel is produced, it will decay exponentially and release energy. This should be responsible for the "afterglow" of the Type I supernova, and thus we would expect the fast and featureless decline of the light curve—which we indeed observe.

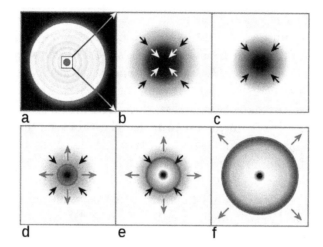

Figure 5.20: **Type II supernova.** (a) A massive star has evolved into an "onion star" with an iron core. (b) Energy production stops, and the core collapses. (c) The rest of the star collapses. (d) The core becomes *incompressible* and sends out a *shock wave*. (e) The shock wave travels through the star and blows it to pieces. (f) The stellar material is expanding and spills into outer space, ending up as a *supernova remnant*.

Even with corroborating evidence, these mechanisms are as exotic as a science fiction story. As a rule, scientists reject fantastic explanations, and so have put their fingers into every possible shortcoming of these hypotheses. Yet, they have been unable to come up with anything better. What's more, these scenarios comply with all known laws of physics, and they explain the observed patterns in nature.

We'll discuss two additional reasons why the described supernova theories have been tentatively accepted: *neutron stars* and *novae*. We have seen that the course of the stellar development of very massive stars that lead to a type II core-collapse supernova is analogous to the development that lead to the formation of a planetary nebula in lighter stars. The mass difference leads to a different, but comparable outcome. Instead of a symmetric planetary nebula like in Figure 3.12(a) due to an "orderly" blow-off of the outer layers of a light star, the turbulent asymmetric explosion of a massive star leads to a jagged supernova remnant, Figure 3.12(b). On some level, though, essentially the same thing happens: Much of the stars mass is ejected at high speeds into outer space. What's more, the core of the former star survives in both cases. The core of a light star is a carbon-oxygen white dwarf, because the last fusion reactions produced carbon and oxygen from helium. For massive stars the core will not be able to become a white dwarf because its mass is above the *Chandrasekhar limit*. As the core of a massive star collapses into something much denser than a white dwarf, it is compressed so hard that the mush of particles is not recognizable anymore as atomic nuclei. In other words, it doesn't matter what chemical elements it consisted of before (presumably iron, the end product of any energy-producing fusion cascade). Under the immense pressure the whole star becomes a gigantic nucleus, and as *protons* and *electrons* of opposite electric charge are pressed into each other, their charges neutralize, producing *neutrons* and *neutrinos* the latter of which escape, taking copious amounts of energy with them, and spawning nuclear reactions that forge many of the elements of the periodic table, see Section 5.6.3. The end product is thus an ultra-compact object, about the size of a city that consists mostly of neutrons—a *neutron star*. This final stage of a massive star is the analog of the terminal white dwarf stage of lighter stars. Now, while this is a fine

5.6. NUCLEOSYNTHESIS FILLS THE PERIODIC TABLE

extrapolation of processes established in the evolution of lighter stars, you will agree that the scientists better come up with hard evidence that this is indeed what happens. Otherwise, it sounds a like the claim that I can be invisible, but only if nobody looks. Because of their exotic properties, neutron stars are impossible to observe with conventional telescopes, and so their existence would be a matter of belief—a choice if you will.

There is an amazing punch-line to this story due to a serendipitous and seemingly unrelated discovery in radio astronomy. As we saw, radio waves are low-frequency electromagnetic waves, and they can be detected with special telescopes which are basically gigantic radio antennas. Since the Earth atmosphere is transparent for radio signals, we can open another "eye on the sky" by detecting radio waves from celestial objects. Radio astronomy basically started after WW II and by the 1960s, scientists were looking with bigger and bigger instruments at finer and finer details of the "radio sky." In 1967 JOCELYN BELL (born 1943), a doctoral student in ANTHONY HEWISH's (born 1924) research group at Cambridge University in England, discovered regular radio pulses, Figure 5.21(a), that could be traced back to a specific position in the sky. These pulses were timed with such an amazing accuracy—one pulse every 1.3373011 seconds, that they were at first considered artificial, because no natural process was know that could operate with such accuracy and so incredibly fast. We are talking about a radio source that is switched on and off about once every second with a period variation of less than one in 100 million. The "signal-of-aliens" hypothesis was soon falsified by the detection of other such sources—labeled *pulsars* for obvious reasons—in other parts of the sky. The race was on to come up with an explanation of this new pattern in nature. A natural way to create an on/off signal is the rotation of an object. However, the pulsar's period of a second is so small that all conventional objects would be destroyed by the associated *centrifugal forces* (Imagine the Earth spinning once a second instead of once a day or 86,400 seconds!). Now, if someone would say "These pulsars must be neutron stars!" two questions should immediately pop up in your mind:

1. What process can speed them up to such incredible rotation rates?

2. Why do they switch on and off? In other words, if they rotate, why does one "side" radiate and the other does not? Why are they "cosmic lighthouses?"

There are actually rather simple answers to these questions. Both involve the incredible shrinkage of the original star. Remember, we said that neutron stars are city-sized (about 10 km diameter) and we computed that they are a factor 600 smaller than the planet-sized white dwarfs, Equation (5.12). Since the Earth is about 100 times smaller than the sun (a typical star), a neutron star has an area that is $(600 \times 100)^2 = 4 \times 10^9$ or 4 billion times, and a volume that is $(600 \times 100)^3 = 2 \times 10^{14}$ or two hundred trillion times smaller than a star's! Now, the law of *angular momentum conservation* which makes figure skaters spin faster if they bring in their outstretched arms (to reduce their *moment of inertia*) tells us that an object spins faster proportional to the factor by which its moment of inertia is reduced. The moment of inertia I is proportional to the square of the size R of the object[45] $I \propto R^2$, so we expect the shrinkage of the normal to a neutron star shortening of the rotation period by the factor 4 billion computed above. The sun rotates in about 26 days, so this seems plausible since $\frac{26d \times 86,400 sec/d}{4 \times 10^9} \approx 7 \times 10^{-3}\,\text{s} = 7\,\text{ms}$.

---

[45] Sorry, you have to take this on faith, but check any intro physics text if you don't believe me.

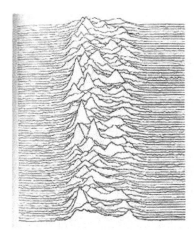 

Figure 5.21: **Neutron star signals.** (a) Radio Pulses indicative of a spinning neutron star. (b) Possible explanation or scenario: the lighthouse effect described in the text. Note that the green rotational axis is inclined with respect to the turquoise magnetic axis.

What about the lighthouse effect? Well, like the Earth, stars have a magnetic field. Usually these fields are fairly weak, but if we shrink them by a factor 4×10^9 (the number of field lines per surface area clearly constitutes an area *scaling* behavior.) then we are talking about an enormously powerful magnetic field. In fact, it is so powerful that it will prevent any radiation that is not in the direction of the magnetic axis from leaving the neutron star. If the rotation axis and the magnetic axis are not aligned, we have created a **cosmic lighthouse**, Figure 5.21(b). In other words, the beam of radiation sometimes points toward the distant observer, and most of the time does not. The signal therefore looks like: on for short time, off for long time, on for short time, and so forth. This is exactly the pattern of radio pulses shown in Figure 5.21(a). Because the neutron star is the only object able to spin so fast, and thereby explaining this new pulsation pattern in nature, it now is universally accepted as a correct theory.

The second story—concerning *novae*—is less spectacular. It addresses the following questions of skeptics: "Why does the white dwarf have to die in a **Type I supernova** explosion? Is there no safe way of burning off some accrued extra mass? Not everything has to end in a violent explosion, does it?" Indeed, it would be strange if there aren't situations where the white dwarf is fed additional mass at a rate that is low enough so that it can "deal" with it. In fact, we can do much better here than thinking of a theoretical argument—there is observational evidence that it must be possible. This is precisely what a *nova* is. You can probably guess from the name what a nova looks like: just like a *supernova* only not "super." Of course, what this sloppy statement means is that a *nova* will be much dimmer, that is, the energy released is not nearly as high (about a factor 10 million less), and a nova can occur many times, whereas a supernova occurs exactly once in the life of a star. Since this pattern in nature ("a weak supernova") exists, we have to explain it. Again it involves a white dwarf in a close binary star system. The white dwarf is fed mass (hydrogen) from a companion star, but at a low enough rate. Therefore, the white dwarf acquires a thin layer of hydrogen on its hot surface. When enough hydrogen has accumulated the increased gravitational pressure heats it up to a temperature hot enough for hydrogen fusion. This sudden flare-up is what explains recurring *novae* such as *RS Ophiuchi*. Be aware that our hand-waving argument

5.6. NUCLEOSYNTHESIS FILLS THE PERIODIC TABLE

here makes it only plausible that this could occur. A full-blown model calculation is necessary to start to convince scientists that this hypothesis indeed explains the phenomenon we call a *nova*. You can probably imagine that these calculations will be complicated, but that it is possible to *predict* how energetic such an event will be, how long it will last, and so forth. These predictions can then be compared with the data we have on recurring *novae*.

5.6.3 Extrapolating: The Elementary Evolution of the Universe

This section is a combination of a summary and an outlook into cosmological questions. First, let's compile an *executive summary* of what we've learned about stars.

- Stars can be modeled as hot gas balls made of hydrogen, some helium, and a dash of heavier elements (metals).

- Stars are highly-efficient, self-sustaining energy-production machines. They are self-sustaining because they are in hydrostatic and thermal equilibrium. The associated balancing of forces and energy flows is so stable that a typical star like the sun can radiate copious amounts of energy for about 10 billion years.

- Stars fuse light elements into heavier elements. The mass difference between the ingredients and the end products of a fusion reaction is converted into (mostly) electromagnetic energy.

- When running out of fusible material in its core, a star's core will shrink and heat up. This will either make additional fusion reactions possible to sustain the star against the pressure of gravity, or the star will disintegrate due to the gross imbalance of forces and energy flows. Since the last step in inevitable, we can state that "gravity always wins" in stellar evolution.

- The material of the star is recycled in the sense that it becomes part of the *interstellar medium* out of which a future generation of stars will form. Note that the star has been fusing lighter into heavier elements, and therefore the recycled matter is dramatically different compared to the interstellar matter out of which the star formed initially. At the very least we can conclude that there is less hydrogen and more of the heavier elements. Of course, most of the star's mass is in its envelope, not its core, so even at the end of its life the star is mostly hydrogen.

With some justification we might therefore call stars "heavy elements factories." This led scientists in the 1950s to the idea that all the 92 elements we find in the periodic table[46] were made in stars. This is a truly **cosmological question**, of the sort "Where does it all come from?" To put out the idea that we can understand where all the matter in the universe comes from was audacious—and the formulation of an ambitious goal that is still being realized today.

We'll leave it to the **nuclear astrophysicists** to figure out exactly how the different elements are produced in stars, novae, supernovae, and the like. The take-home message is that *the elements are being produced in the course of a star's life*. This could be the end of the story, weren't it for a skeptic's possible interjection: "But what about the helium that stars

---

[46]In fact, there are many more distinct nuclei, because we can add different numbers of **neutrons** to form different **nuclides**, that is, atomic nuclei with the same number of **protons**, but a different number of **neutrons** All told, there are about 1000 different nuclides in the universe.

start out with? If this is made in stars, then the first stars cannot have had any helium to begin with." This is an important point, since one can show that stars without a significant amount of helium "don't work." We therefore conclude that helium (and a dash of some other elements) were present *before* the first stars formed in the universe.

This brings us to the cosmological outlook, and, in some sense, the ultimate *extrapolation*. We postulate that some elements, notably helium, were made when the universe itself came into being. This is known as **cosmic nucleosynthesis** to set it apart from **stellar nucleosynthesis**. Since we know that **nucleosynthesis** or fusion takes place only in an incredibly hot and dense environment, we face a dilemma: At present, the universe seems like an extremely cold and dilute place, quite the opposite of what it must have been in the past when it forged the **primordial elements**.[47]

Concept Practice

1. Massive stars can fuse elements up to iron to produce energy. Why is that?
    (a) Iron is the most massive element.
    (b) Iron is the smallest element.
    (c) Iron is too loosely bound.
    (d) Iron has the most binding energy.
    (e) These stars can actually fuse *all* elements up to uranium.

2. A supernova
    (a) ... is the terminal stage of stellar evolution for all stars.
    (b) ... can only go off once.
    (c) ... produces a white dwarf, the burned-out core of a star.
    (d) None of the above

5.7 Summary and Application

5.7.1 Timetable of the Emerging Expanding Universe (III)

The main steps toward an understanding of the expanding universe discussed in this chapter are described below. Schedules like this should be taken with a **grain of salt**. The individual achievements and discoveries of the era can be grouped into the following phases of our emerging understanding of the cosmos.

- c. 1900–1939 **The birth of modern physics.** Classical physics cannot explain how objects interact and move if these objects and the distances between them are very small, or if they move very fast. To describe the very fast patterns of nature **relativity theory** is developed with the following main stages.

    – For the special case of a non-accelerated observer, **special relativity theory** describes that space and time are linked and their measurement depends on the relative speed of the observer. One consequence is the famous formula $E = mc^2$, which postulates the equivalence of mass and (a large amount of) energy. (EINSTEIN, 1905)

---

[47]Primordial elements are the elements present at the "start" of the universe.

5.7. SUMMARY AND APPLICATION

- The generalization of *special relativity* to all—even accelerated observers—called *general relativity* turns out to be a *theory of gravity*. Since a blindfolded observer cannot tell whether he or she is accelerated by rocket or simply sitting next to a large, gravitating mass, the concept of a gravitational force becomes obsolete. It is replaced by a curved four-dimensional space-time, which becomes the stage on which physics (and hence everything else) plays out. (EINSTEIN, 1915)

To describe the very small patterns of nature, the theory of *quantum mechanics* is constructed. Some stages of this development were as follows.

- 1900–1905 "Radiation is particles:" Blackbody radiation (PLANCK, 1900) and *photoelectric effect* (EINSTEIN, 1905) show that electromagnetic waves can have particle-like behavior.
- 1903–1913 "The atom is positive nucleus and negative electrons:" THOMSON's *plum pudding model* (1903), RUTHERFORD discovers the nucleus (1911), BOHR postulates quantum rules to stabilize the atomic shell, that is, the "orbits" of the electrons.
- 1925–1926 "Quantum Mechanics is complex wavefunctions:" A stringent mathematical theory is developed to "explain" all non-relativistic (slow) quantum phenomena (HEISENBERG 1925, SCHRÖDINGER 1926)
- 1927 "Nature is fundamentally indeterminate and random:" *Uncertainty principle* (HEISENBERG 1927)
- 1928–1949 "Quantum Mechanics plus Relativity is Quantum Field theory:" Starting with *Dirac's equation* (1928), a theory of matter is developed that is consistent with both relativity and quantum mechanics.
- from 1932 "The atomic nucleus is protons and neutrons:" Discovery of the neutron (CHADWICK 1932)
- from 1932 "Particles and anti-particles can annihilate:" Discovery of the positron (anti-electron) (ANDERSON 1932)
- from 1930 "Radioactivity is a weak force:" The neutrino is postulated (PAULI 1930), a theory of the weak force which describes radioactivity is developed (FERMI 1934).
- from 1938 "Lead can be turned into gold:" The *transmutation* of the elements is demonstrated (HAHN, STRASSMANN, MEITNER 1938)
- 1939 "The stars shine by Hydrogen Fusion:" A nuclear theory of the energy production process in stars is developed, see below. (WEIZSÄCKER, BETHE)

- 1909–1915 **Properties of stars are related: The Hertzsprung-Russell diagram.** Many properties of stars can be summarized by devising a graph which plots stars depending on their luminosity and temperature (HERTZSPRUNG 1909, RUSSELL 1913) There are several regions where the stars can exist: There are dim, cool stars as well as bright, hot stars as expected (normal or main sequence stars), but also bright, cool stars (*red giants*) and dim, hot stars (*white dwarfs*). The major discovery is that for a given temperature or *spectral class* two different versions of stars are possible: Giants and dwarfs, which are big and small in *size*, and hence bright and dim in *luminosity*. This is a new pattern in nature that needs to be explained.

- **1916–1926 Star models are synthesized by including radiation pressure.** (EDDINGTON 1916) Even through the mechanism for stellar energy production is unknown, many properties of the stellar interior can be inferred from these models, including the fact that energy production—whatever it is—must happen exclusively in the core of the star which is small compared to the total volume of the star. In the core the temperature and density of particles are extremely high. (EDDINGTON 1926)

- **1938–1939 Stars produce energy by hydrogen fusion.** Given the extreme conditions in the stellar core, microcosmic mechanisms, that is, nuclear fusion, can be developed to explain energy production in stars. (WEIZSÄCKER, BETHE). There are two different processes important in hot stars (carbon-nitrogen-oxygen cycle) and cooler stars (proton-proton cycle). Both happen more frequently at higher temperatures and higher densities. This puts a lower limit on a star's mass. The first cycle is more sensitive to temperature (T^{16}), so it is dominant at high temperatures and insignificant at low temperatures.

- Before 1950 **The main sequence is well understood.** Building in the energy production into the thermodynamic star models leads to a consistent theory of normal stars. It remains a mystery how stars deviate from the *main sequence* and turn into *giants*.

- Around 1950 **Hydrogen shell burning leads to red giants.** The material in the stellar core does not mix with the rest of the star. Hydrogen "fuel" gets used up in to core. This leads to fusion around the core in a shell. The shell produces more energy than the core did, and the hydrostatic equilibrium changes; the star turns into a red giant.

- from around 1955 **Other stars in the Hertzsprung-Russell diagram are reached.** The helium flash leads to supergiants, instability to the existence of variable stars, and so forth. (M. SCHWARZSCHILD)

- **1957 Synthesis of the Elements in Stars.** The article of the same name starts the field of *nucleosynthesis* by showing that the chemical elements are created in stars by subsequent fusion of lighter into heavier elements. (BURBIDGE, BURBIDGE, FOWLER, HOYLE)

- From the 1930ies **Terminal stages of stellar evolution (death of stars) are discovered:** White dwarfs (CHANDRASEKHAR), supernovae (ZWICKY, BAADE), Neutron stars (CHANDRASEKHAR, HEWISH, BELL), Black Holes.

- From 1955 **Computerized star models.** Development of digital computers affords the possibility of detailed calculations of stellar evolution. It turns out that supernovae must be simulated in full three dimensions, because to initialize the explosion *turbulence* is required.

5.7.2 Concept Application

1. Generate a plausible "theory" of stellar evolution by describing a path of your choice in a Hertzsprung-Russell diagram by stating in words what happens to the star as it matures. Use words such as "becomes hotter" or "less luminous", "contracts" or "expands."

5.7. SUMMARY AND APPLICATION

2. Draw plausible curves for the temperature $T(r)$, density $\rho(r)$ and opacity $\kappa(r)$ within a star as a function of the distance from the star's center. "Plausible" here means that you can give a good argument as to why you made this choice. For instance, a star which is hotter at its surface than at its center is *not* plausible, because heat flows from hot to cold, not the other way around (second law of thermodynamics).

3. Analyze the poem at the beginning of the chapter by writing out line by line its astrophysical content. For example, "the sun in polytropic spheres to shine" refers to EMDEN's hot gas ball model.

5.7.3 Activity: Hertzsprung-Russell Diagrams

Hertzsprung-Russell diagrams (HRD) are two-dimensional plots of stellar properties with luminosity (or equivalent) on the vertical axis and spectral type (or equivalent, like temperature) on the horizontal axis. As an example, consider the star Aldebaran (α *Tauri*). Aldebaran is spectral type K5 and 160 times more luminous than the sun. The first letter is the spectral type: K (one of the OBAFGKM sequence), the Arabic number (5) is like a second digit to the spectral type, so K0 is very close to G, K9 is very close to M. Aldebaran thus would sit to the upwards right of the sun in a HRD.

1. Put the stars of the following list into a Hertzsprung-Russell diagram: α Centauri ($1.5 L_{sun}$, G2); Sirius A ($25.4 L_{sun}$, A1); Sirius B ($0.056 L_{sun}$, A2); Betelgeuse ($120,000 L_{sun}$, M2), Regulus ($288 L_{sun}$, B8); 61 Cygni ($0.153 L_{sun}$, K5); Procyon ($6.93 L_{sun}$, F5); Deneb ($200,000 L_{sun}$, A2); Proxima Centauri ($0.002 L_{sun}$, M6).

2. Sometimes stars are classified according to their general luminosity classes. Describe the following classes, circle or indicate the region in the Hertzsprung-Russell diagram and give one example chosen from the stars above.

    (a) Supergiants (I)
    (b) Red Giants (III)
    (c) Main sequence stars (V)
    (d) White dwarfs (WD)

3. The spectral classes correspond to different surface temperatures, which are, in turn, related to different colors following **Wien's law**. Assign a temperature and color to each spectral type by adding two more horizontal axes to the Hertzsprung-Russell diagram. Possible colors: white, bluish, bluish-white, yellow, red, blue, orange. Possible temperatures: 3000 K, 30 000 K, 10 000 K, 4000 K, 6000 K, 8000 K, 20 000 K.

4. By which factor does the temperature of stars vary?

5. By which factor does the luminosity of stars vary?

6. How are the two correlated if at all?

7. What could the reason be that stars have so different luminosities, given that they are all made essentially of the same material, namely hydrogen and some helium?

8. Do all stars of the same temperature have the same luminosity? Explain with an example.

5.7.4 Activity: Stellar Models

We want to explain how stars function. That is, we want to build a model of a star that is consistent with the known laws of Nature and also explains all the properties of stars.

1. First, let's review what we actually know about the stars from observing them. List five properties of stars that we can glean from observations.

2. There are other things we know about stars from observations which are *not* properties of the star. Explain why the distance to a star and its **radial** and **transverse velocities** are not properties of the star.

3. Describe briefly which general features of stars we have to explain. Take into account that stars come in different colors, temperatures and luminosities, and are subject to the force of gravity.

4. A crude model of stars is to view them as hot glowing gas balls. Explain what we mean by this.

5. The key ingredient to a star's stability is **hydrostatic equilibrium**. Explain the concept.

6. The earliest stellar model was HOMER LANE's description of a star as a gas ball propped up by convective motion. **Convection** is one form of energy transport. Describe how it works.

7. What was EDDINGTON's contribution? How did he modify **Lane's model**?

8. One of the successes of **Eddington's standard model** was the prediction of the **mass-luminosity relation**. Explain what that is.

9. What problems were left to solve after **Eddington's standard model** of stars explained many features of stars?

5.7.5 Activity: Early Stellar Life

Here is a very short description of stellar formation. Stars form when a stellar gas and dust cloud contracts under its own gravitational force. As it contracts it becomes denser and smaller while heating up. The center of the collapsing gas cloud eventually becomes so hot and dense that light atomic nuclei (the hydrogen nucleus is but a proton) will bump into each other so forcefully that they can stick together and form more complex nuclei. The mass difference between the ingredients and the products of this **nuclear reaction** is (converted into) energy. This process is called **nuclear fusion** and the source of the energy radiated by stars.

1. Draw the main sequence into a Hertzsprung-Russell diagram.

2. Argue with the **Stefan-Boltzmann law** (luminosity depends on size and temperature) to draw the formation tracks of three stars as they develop from a large interstellar gas cloud into a Hertzsprung-Russell diagram on a separate sheet. Assume that one of the stars has a mass bigger than the sun, one has the sun's mass, and one has a smaller mass.

5.7. SUMMARY AND APPLICATION

3. How does the initial mass of the gas cloud influence the formation process and the outcome?

4. How do stars produce energy? Briefly describe hydrogen fusion.

5. The star Betelgeuse radiates 120,000 times more energy per second than the sun. What does this say about the rate at which fusion happens at its core?

6. What is the consequence for the life expectancy of Betelgeuse?

7. Betelgeuse has about 8 times the mass of the sun. How does this influence its life expectancy?

8. The sun has a life expectancy of about 10 billion years = 10,000 million years. Estimate Betelgeuse's life expectancy.

9. For main sequence stars, the **mass-luminosity relation** posits that the luminosity is proportional to the mass of the star to the 3.5th power: $L = kM^{3.5}$. For simplicity, let's assume the constant is $k = 1$. Use this information to figure out the life expectancy of a main sequence star with a mass 8 times bigger than the sun's.

10. Is Betelgeuse a main sequence star? Why or why not?

5.7.6 Activity: Cosmic Yardsticks

Variable stars can be used as cosmic yardsticks to measure the distance to celestial objects. The idea is to observe the **apparent brightness** B of the variable star, somehow find out how bright it actually is (i.e. its **luminosity** L). Then we can calculate the distance to it by comparing the two. We know that $B = \frac{L}{4\pi d^2}$, that is, the apparent brightness increases with the actual brightness or luminosity, and decreases with the square of the star's distance from us.

The three most useful variable stars are **Cepheids, RR Lyrae stars** and **supernovae**. The first two are periodically changing their brightness, supernovae flare up just once. In all cases we want to plot the apparent brightness of the star as a function of time. This so-called **light curve** will be characteristic in the sense that it can be used to identify which type of the three the variable star is. It is also use to study the properties of the star.

Cepheids: Have a light curve with a period of several days between maxima; no secondary maxima. Luminosity: between 500 and 20,000 Suns, determine with period-luminosity curve, Figure 5.15, or Eq. (5.13).

RR Lyrae: Have a light curve with a period of one or two days between maxima; small secondary maxima exist. Luminosity: about 100 Suns.

Supernovae: Have aperiodic light curves, just one peak; if plateau exists, it is Type II, otherwise Type I Luminosity: Type I: 80 million Suns, Type II: 100 million Suns.

1. Find out to which category the light curves in Figure 5.22 belong.

2. Determine the luminosity of the star described by the light curve.

3. Read off the apparent brightness of the star at the maximum of the light curve.

Luminosity	$B=1$	$B=1.3$	$B=1.5$	$B=3.65$	$B=5$	$B=10$
50	392773.4	450964.2	494472.4	1330890	2478233	24782327
100	555465.5	637759.7	699289.6	1882162	3504750	35047503
200	785546.8	901928.4	988944.9	2661780	4956465	49564655
300	962094.5	1104632	1211205	3260001	6070406	60704057
400	1110931	1275519	1398579	3764325	7009501	70095007
500	1242059	1426074	1563659	4208643	7836860	78368600
600	1360607	1562186	1712903	4610337	8584850	85848500
700	1469624	1687354	1850146	4979734	9272698	92726978
800	1571094	1803857	1977890	5323559	9912931	99129309
900	1666396	1913279	2097869	5646487	10514251	1.05×10^8
1000	1756536	2016773	2211348	5951920	11082994	1.11×10^8
2000	2484117	2852148	3127318	8417286	15673720	1.57×10^8
3000	3042410	3493154	3830167	10309028	19196308	1.92×10^8
4000	3513072	4033547	4422696	11903840	22165987	2.22×10^8
5000	3927734	4509642	4944724	13308898	24782327	2.48×10^8
10000	5554655	6377597	6992896	18821624	35047503	3.5×10^8
1×10^7	1.76×10^8	2.02×10^8	2.21×10^8	5.95×10^8	1.11×10^9	1.11×10^{10}
1×10^8	5.55×10^8	6.38×10^8	6.99×10^8	1.88×10^9	3.5×10^9	3.5×10^{10}
1×10^9	1.76×10^9	2.02×10^9	2.21×10^9	5.95×10^9	1.11×10^{10}	1.11×10^{11}
1×10^{10}	5.55×10^9	6.38×10^9	6.99×10^9	1.88×10^{10}	3.5×10^{10}	3.5×10^{11}

Table 5.2: **Distance to the star as a function of its luminosity L and its brightness B.**

4. Find the distance from Table 5.2 or use the formula $d_{star} = 1.58\sqrt{L} \times 10^{(B+21.73)/5}$, where B is the **apparent brightness** of the star in magnitudes, and L the **luminosity** of the star in solar units.

5.7.7 Activity: Late Stellar Life

Recall that **main sequence** (MS) stars convert hydrogen to helium. Eventually, they run out of hydrogen in their core.

1. In general terms, what happens if stars run out of hydrogen in their core? Argue with the concept of hydrostatic equilibrium.

2. Use the **Stefan-Boltzmann law** (luminosity depends on size and temperature) to draw the evolutionary tracks of a star of one solar mass as it develops away from the main sequence star into a Hertzsprung-Russell diagram on a separate sheet. In particular, label the following major stages:

   (a) **Subgiant stage** with hydrogen shell burning, in which the star is slightly cooler and several times more luminous than in its MS stage.

   (b) **Red giant stage** with intensive hydrogen shell burning. The star is cooler and about 100 times more luminous than in the MS stage.

5.7. SUMMARY AND APPLICATION

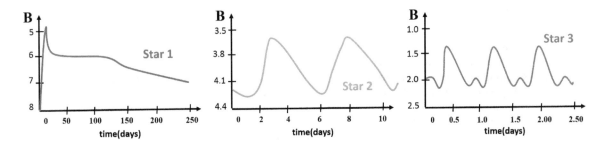

Figure 5.22: **Light curves of three different variable stars.**

(c) *Helium flash and horizontal branch* in which the star gets hotter and smaller because it starts fusing helium to carbon in its core.

(d) *Asymptotic branch giant* with helium shell burning (because the core is clogged with carbon ash). The star's outer layers are only 3000 K, but the luminosity soars to above 10,000 times the luminosity of the sun.

(e) After the *asymptotic giant phase*, the star becomes so bloated that its outer layer become opaque for radiation. The star sheds its outer layers which form a *planetary nebula*. This is an explosive event, where the star's material heats up dramatically to beyond 30 000 K, while keeping its high luminosity.

(f) Eventually, the outer layers expand and cool, exposing the burned-out core of the star as a *white dwarf*. This is a very small object with a surface temperature of about 7000 K–10 000 K.

3. How do more massive stars develop? Their lives do not end when a carbon core develops, because their greater mass can squeeze and heat up the core enough so that carbon can fuse to oxygen. Eventually oxygen will fuse to even heavier elements. With many different elements burning in different shells, the star resembles an onion. Sketch such an onion star and label its layers: non-burning hydrogen, burning hydrogen, helium, carbon, neon, oxygen, silicon, iron.

4. This build-up of more complex atomic nuclei stops at iron. Why?

5. How do the following massive stars end their lives?

   (a) Stars which are much more massive than the sun, but not excessively massive.
   (b) The most massive stars, with masses larger than about 25 solar masses.

374 CHAPTER 5. EXTRAPOLATING & SYNTHESIZING: UNDERSTANDING STARS

Image Credits

Fig. 5.1b: Copyright © 2008 by Markus Poessel; adapted by Pbroks13, (CC BY-SA 3.0) at https://commons.wikimedia.org/wiki/File:Elevator_gravity.svg.

Fig. 5.2: Copyright © 2017 by Sandbh, (CC BY-SA 4.0) at https://commons.wikimedia.org/wiki/File:18-column_medium-long_periodic_table.png.

Fig. 5.3a: Copyright © 2007 by JabberWok, (CC BY-SA 3.0) at https://commons.wikimedia.org/wiki/File:Bohr-atom-PAR.svg.

Fig. 5.3b: Source: http://slideplayer.com/slide/4495668/14/images/5/Spectral+lines+of+supergiants,+giants,+and+main-sequence+stars.jpg

Fig. 5.4a: Source: http://frigg.physastro.mnsu.edu/ eskridge/astr101/hrd_original.gif.

Fig. 5.5: Copyright © 2006 by R. J. Hall, (CC BY-SA 1.0) at https://commons.wikimedia.org/wiki/File:M3_color_magnitude_diagram.jpg.

Fig. 5.6: Copyright © by Worldtraveller, (CC BY-SA 3.0) at https://commons.wikimedia.org/wiki/File:Open_cluster_HR_diagram_ages.gif.

Fig. 5.7: Source: http://www.astronomy.ohio-state.edu/ pogge/Ast162/Unit2/structure.html.

Fig. 5.8: Source: http://www.astro.uu.se/ ulrike/Spectroscopy/PPT/spectraltype.html.

Fig. 5.9: Source: https://commons.wikimedia.org/wiki/File:Autumn_Red_peaches.jpg.

Fig. 5.10: Source: https://commons.wikimedia.org/wiki/File:ChandrasekharLimitGraph.svg.

Fig. 5.11a: Source: https://www.e-education.psu.edu/astro801/content/l5_p5.html.

Fig. 5.11b: Copyright © 2016 by Lithopsian, (CC BY-SA 4.0) at https://commons.wikimedia.org/wiki/File:Evolutionary_track_1m.svg.

Fig. 5.12: Source: https://www.nasa.gov/sites/default/files/images/532335main_slide3.jpg.

Fig. 5.13: Copyright © 2017 by Antonio Ciccolella, (CC BY-SA 4.0) at https://commons.wikimedia.org/wiki/File:Stellar_Evolution_(0.8-8_M%E2%98%89).jpg.

Fig. 5.14b: Copyright © 2008 by Pearson Education, Inc.

Fig. 5.14a: Copyright © 2008 by Pearson Education, Inc.

Fig. 5.15: Source: https://commons.wikimedia.org/wiki/File:Period-Luminosity_Relation_for_Cepheids.png.

Fig. 5.16: Source: https://commons.wikimedia.org/wiki/File:Binding_energy_curve_-_common_isotopes.svg.

Fig. 5.17: Source: https://commons.wikimedia.org/wiki/File:Nucleosynthesis_in_a_star.gif.

Fig. 5.18a: Copyright © 1999 by NASA/ESA, (CC BY 3.0) at https://commons.wikimedia.org/wiki/File:SN1994D.jpg.

Fig. 5.18b: Copyright © 2012 by Lithopsian, (CC BY-SA 3.0) at https://commons.wikimedia.org/wiki/File:Comparative_supernova_type_light_curves.png.

Fig. 5.19a: Source: https://stereo.gsfc.nasa.gov/ thompson/nova_sagittarii_2012/nova_sagittarii_2012.gif.

Fig. 5.19b: Source: https://commons.wikimedia.org/wiki/File:Making_a_Nova.jpg.

Fig. 5.20: Copyright © 2011 by R.J. Hall; adapted by Magasjukur2, (CC BY-SA 3.0) at https://commons.wikimedia.org/wiki/File:Core_collapse_scenario.svg.

Fig. 5.21a: Copyright © 2009 by ashish_soods, (CC BY-SA 3.0) at https://celestialbodies.wikispaces.com/file/detail/CP_1919.jpg.

Fig. 5.21b: Copyright © 2009 by ashish_soods, (CC BY-SA 3.0) at https://celestialbodies.wikispaces.com/file/detail/Neutron_Star.png.

Chapter 6

Refining to Infinity: Cosmology

Naturally, this (last) chapter is about astronomy on the grandest scale. We are applying and using the knowledge gained by studying the stars and their life cycle to understand the cosmos. At the same time the results of other fields, in particular particle and mathematical physics, are used to construct a consistent model of the universe. Still, in the end it is the constant refinement of our theories and descriptions informed by newfound patterns in nature that ensures progress on the way to discover the expansion of the universe.

Guiding Questions of this Chapter

6.1 What is the shape and size of the galaxy we live in?

6.2 How are systems of stars organized and distributed in space?

6.3 How can we measure distances to the edges of the universe? How does the universe look on the largest scales? Can the cosmos be considered an object?

6.4 How does the universe as a whole behave? Does it change? Does it have a beginning or an end? How can we observe the expansion of the universe we are in? What does the redshift of the light sent out by galaxies tell us? Can we determine the size or age of the universe?

6.5 What is an accelerating expansion? Why does the universe accelerate its expansion? Why is the universe flat?

6.1 The Milky Way Contains Virtually Everything

We already encountered the Milky Way several times when refining our description of the universe. In Section 1.1, we referred to the dim, milky band extending all across the sky as the Milky Way, our galaxy. Plainly visible on moonless nights in ages past, it is nowadays observable only far from our cities' light pollution. In Section 2.8 we described how GALILEO used his telescope to discover that the milky, cloudy appearance is due to a large number of stars, so close together that they cannot be resolved by the naked eye. Further on, in Section 3.5.3, we saw the first attempt by WILLIAM HERSCHEL to decipher the structure of the galaxy, and to determine the position of the sun within it.

There are many amazing discoveries that have been made since, which have given us a truly detailed and comprehensive understanding of our galaxy. For our quest to understand

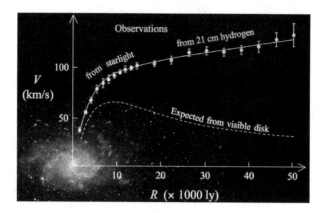

Figure 6.1: **Observing galactic rotation.** The rotation of galaxies like the Milky Way can be gleaned from observations of the motion of stars relatively close to its center, or of interstellar matter farther out. The actual rotation velocities (V) are much higher than expected from the visible mass. Instead of decreasing with the distance from the galactic center (as expected from Kepler's third law), the rotational velocity stays constant or even increases. This means that the effective mass $M_{\text{eff}}(R)$ inside an orbit of given radius R must increase with the distance from the center.

the expanding universe, however, we will focus on its broad features, which will inform and guide our exploration of the "rest" of the universe, because we will find its patterns repeated again and again. The Milky Way can therefore serve as a blueprint for the construction of the universe, much as our solar system and its gravitational dynamics served as a stepping stone to the distant stars. There is also a cautionary tale or two in the story of the exploration of the Milky Way. For one, we will again see how hard it is to decipher the shape and size of an object you are part of. Recall from Section 2.3.2 how difficult it was for the Greek astronomers to measure the size of the Earth and to show that it is a sphere—because they were standing *on it*[1]. This problem will resurface later in the chapter, when we try to measure the shape and size of the universe we live in. Secondly, it is easy to get carried away with the results of research, especially when the result is extreme and powerful (or yours). We will see how the enormous size of our galaxy misled scientists to believe it *is* the universe, and not just a small part of it.

6.1.1 The Shape and Size of Our Galaxy

What information might we get from a close inspection of the milky band of stars and matter that surrounds our sky? Two simple observations narrow down the possible shape of the Milky Way. First, we are looking at a *band*, and this means that the sun is not part of a sphere filled with stars. Since we see distant stars and nebulae above and below what we might call the ***galactic equator***, we are poised to think of the galaxy as a disk. At the very least, we can say that there is more "stuff" around the ***galactic equator*** than at higher galactic latitudes. The other observation we can make is that the galactic band is densest and brightest in the constellation *Sagittarius*. This suggests that the galactic center lies in this direction. HERSCHEL's observations of a distance at which no more stars were discernible, see Section 3.5.3, on the other hand, suggest an ***edge*** of the Milky Way. Indeed, if you think about it, there is no *a priori* reason why the Milky Way couldn't extend to infinity. Recall

---

[1] In our ***Space Age***, we can easily glean that information from satellite photos.

6.1. THE MILKY WAY CONTAINS VIRTUALLY EVERYTHING

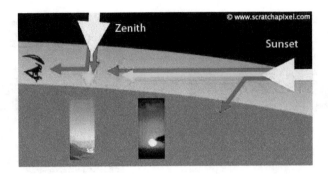

Figure 6.2: **Sun at noon and at sunset.** It is redder at sunset because light has to travel through more air to get to the observer. Blue light from the object is filtered out because it get scattered more widely.

that HERSCHEL made severe assumptions, such as equal luminosity of all stars, and therefore his conclusions of a finite, lenticular galactic shape with the sun slightly off center, Figure 3.16(b), are questionable. Indeed, our study and appreciation of the Copernican revolution and its aftermath should make us skeptical of any *special* role or place of the sun, such as at or near the galactic center. If we have learned anything from our refining description of the cosmos, it is the continued shift of Earth from a special to an average place. Instead of being at the center of the universe, the Earth rotates around a star which is average on most accounts (luminosity, size, temperature, and age). From what we've learned we would therefore expect the sun to be somewhere between the center and the edge of the galaxy. But how can we *measure* its position within the galaxy?

From our observation that the galaxy has a center in the constellation *Sagittarius* two things are clear. First, the galaxy can be modeled after the solar system: A big mass at the center drives the rotation of everything else around it. Fortunately, we already know the laws governing this rotation on a grand scale. Kepler's laws derived from Newton's universal gravity tell us that objects closer to the galactic center rotate faster than objects near the edge of the galactic disk. This behavior is altered somewhat by the fact that technically an object's rotation is driven by *all* mass sitting between it and the center of the rotation; see Figure 6.1. Contrary to the solar system, where the mass of the planets is insignificant compared to the central mass of the star, a galaxy has an quasi-continuous mass distribution, such that the effective mass M_{eff} (the mass inside an orbit of given radius r) will very considerably with the orbit's radius, $M_{eff} = M_{eff}(r)$. Second, it becomes clear that within a disklike shape of the galaxy, the only relevant information about the sun's galactic position is its distance from the center. The big obstacle for determining this distance is the fact that (unbeknownst to HERSCHEL, but clear to us from stellar formation from gas and dust) the galaxy contains a lot of interstellar matter that obscures our view of pretty much anything in the disk of the galaxy. In particular, the distant center of the galaxy, where the density of the interstellar medium seems greatest, is basically hidden from view. Incidentally, it seems curious that HERSCHEL was unaware of this matter. An analogy might help here. Why is a motorist driving in dense fog aware of the fog? You might be tempted to answer: Because he cannot see anything. However, how does he know that there is something to see? Because he has been on this road—or some road—and knows how it looks *without* fog. The astronomer does not know whether he doesn't see stars because there is a dust cloud between her and the star—or because there simply is no star in this region of space. Consequently, the notion that

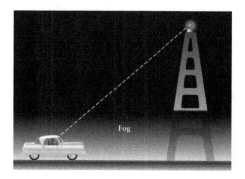 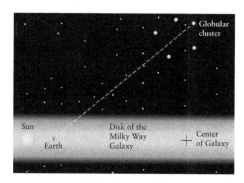

Figure 6.3: **Fog analogy.** (a) A car driving in dense fog. The driver can orient herself with the help of position lights "sticking out" of the fog. (b) An observer on Earth, stuck with the sun in the galactic disk rich in interstellar matter which absorbs light. Orientation is possible by observing globular star clusters of the galactic halo orbiting the galactic center at randomly oriented orbits, which lie outside of the galaxy's main plane.

space between stars is not empty was a hard-won discovery. Of course, we see bright emission of reflection nebulae like M42, planetary nebulae and supernova remnants, Figures 3.10 and 3.12, but we see them because they radiate! Dark clouds like the Dustlanes nebula, Figure 3.10(b), can only be discerned in front of a star-studded background. Worse, most often, matter isn't completely opaque, so we see stars, but through a veil of thin gas and dust.

Can we distinguish the effects of a foreground gas and dust cloud, when we are not even quite sure what is behind the veil? Well, we do know! We have to give ourselves some credit for the hard work we did in the last chapter, in which we came up with a fine description and understanding of stars. We do understand what their spectrum should look like if we know their temperature and their color, so if stars do not quite look as expected, then probably something is messing with the starlight. For instance, if a star with strong hydrogen lines looks yellowish, then something weird is going on, since an *A star* (identifiable by its strong hydrogen lines) should be plain white. As an analogy think about the sun seeming orange or even red at sunset. The reason that the sun, our *G2 star*, is not looking yellow—as it should—is that at sunset its light travels through a thicker layer of air than at noon, Figure 6.2. The air and aerosols (gas and dust) scatter light more strongly if its wavelength is shorter (bluer). At sunset the sun's light therefore looks redder because more of its bluer content is scattered away from the observer. Also, the foreground gas and dust cloud can add more absorption lines, much like the Earth's atmosphere adds some strong oxygen lines which superimpose the solar spectrum; see Figure 3.29.

The dust and gas is most concentrated in the equatorial plane of the Milky Way, so we need to look at objects off the plane to orient ourselves in the galaxy. This is analogous to the motorist who can still see tall buildings rising above the fog layer near the ground, Figure 6.3. Globular star clusters like in Figure 3.11(b) are playing the role of these beacons in our galaxy. Why? As was realized by HARLOW SHAPLEY, globular star clusters have randomly inclined orbits around the center of the galaxy, and many of them are found at high galactic latitudes—far away from the obscuring gas and dust of the galactic plane. If you can make a three-dimensional map of these clusters by determining their distances (their positions on the two-dimensional sky are easily gleaned) then the center of this distribution is the center of our galaxy. SHAPLEY found the distances to these globular clusters by exploiting the properties of RR Lyrae variables that are commonly found in the clusters. The reason is that the RR

6.1. THE MILKY WAY CONTAINS VIRTUALLY EVERYTHING

Lyrae variables are on the instability strip of the HRD, and thus represent a late stage in a star's life. This dovetails with the fact that globular star cluster are rather old objects, as we learned in Section 5.2.2. Since RR Lyrae stars are fairly easy to identify due to their light curve exhibiting a secondary maximum, Figure 5.14(b), SHAPLEY was able to determine the distance to all known globular star clusters (about a hundred in all). He exploited the fact that RR Lyrae are a hundred times more luminous than the sun. Comparing that luminosity L to their apparent brightness B observed in the sky, the distance is essentially the square root of their ratio, L/B, Equation (5.3). In 1920 SHAPLEY was ready to report the astounding size of our galaxy. He found that some of the globular star clusters, which form a roughly spherical *halo* around our galaxy are more than 100,000 ly away.[2] As we said, determining the three-dimensional distribution of the star clusters, which tend to hover around the galactic center, allows one to figure out the sun's position relative to the center. The modern estimate is that the sun is about 25,000 ly off center, and thus about halfway between center and edge of the Milky Way. This is what we guessed above, using our Copernican argument of mediocrity.

The size of our galaxy is enormous! Comparing the diameter of the Milky Way of about 120,000 ly, with the distances to the stars bright in our night sky, we realize that stars as distant as Betelgeuse (500 ly) are very close to us. In other words, gazing out into the night sky, we see but a tiny part of our galactic neighborhood; it's more like the view across the street. Even objects like the Great Orion nebula (6,000 ly) and the Great Hercules cluster M13 (21,000 ly) are fairly close to us compared to the size of the galaxy. In fact, *all* objects we have been talking about in this book so far (main sequence stars, solar systems, star clusters, interstellar gas clouds, neutron stars, white dwarfs, and so forth) are part of the Milky Way. In this sense, the Milky Way contains everything the universe comprises. It is therefore not surprising, that SHAPLEY merged this thought with the enormity of the galactic scale to forge the belief that the Milky Way *is* the universe. All other unclassified objects, such the Great Andromeda nebula (M31), Figure 3.13(a) were interpreted as being small and nearby (less than 60,000 ly away). This view was opposed by several astronomers, notably one CURTIS HEBER (1872–1942), and the two great astronomers faced off in April 1920 in what's known as the **Great Debate**. It was debated whether or not the unclassified nebulae are near and small, or far and large—and thus independent island universes of their own right. Interestingly, the debate did not produce a clear winner at the time. Yet, within a few years the case was settled for good by the young EDWIN HUBBLE (1889–1953) who used the **Cepheid variables** as a yardstick to establish the extra-galacticity of the Andromeda nebula, as we will see soon.

6.1.2 Galactic Structure: The Milky Way Has Three Distinct Parts

By analyzing our naked-eye observations of the Milky Way, we have identified three ***morphological structures*** of our galaxy: the plane or ***disk*** of the galaxy, the center or ***central bulge*** of the galaxy, and the ***halo*** of the galaxy, that is, region surrounding the others and containing the globular star clusters, see Figure 6.4. We found that the size of the galaxy is

---

[2] In 1930 ROBERT J. TRUMPLER (1886–1956) realized that distant star clusters were dimmer than expected from distance alone, and concluded that the interstellar medium dims starlight considerably over large distances, as detailed above. Therefore, SHAPLEY's galactic distances had to be corrected later, to about half their original value.

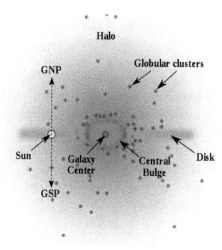

Figure 6.4: **The structure of the Milky Way.** Seen edge on, the Milky Way shows three distinct parts: the *central bulge*, the *disk*, and the *galactic halo* of *globular star clusters*. The sun is about halfway to the edge in the disk.

truly astounding. The galactic disk has a diameter of 120,000 ly while only about 2,000 ly thick, and the sun sits a full 25,000 ly off the **galactic nucleus** hidden inside the **central bulge**.

We can discern finer details using telescopes. They show us that most of the milky clouds seen are comprised of stars, but a deeper understanding of the galactic structure involves observation in wavelengths longer than visible light. The reason is the abundant galactic dust, which is largely opaque for visible light. Objects such as dust particles scatter electromagnetic waves only when their size is comparable with the wavelength of the radiation. Since dust grains have diameters up to a tenth of a micrometer, infrared radiation of a few micrometers (**near infrared**) and a few hundred micrometers (**far infrared**) can penetrate the dust to show us what is hidden in it. Moreover, we know from Wien's law, Equation (3.13), that the temperature of a blackbody determines the wavelength at which it is brightest. Typical temperatures of galactic dust range from 10 K to 100 K, which corresponds to a radiation peak at $\lambda_{peak} = 0.0029\,\text{m·K}/T_{peak} = 0.000\,29\,\text{m}(0.000\,029\,\text{m}) = 290\,\mu\text{m}(29\,\mu\text{m})$ for 10 K (100 K). This means that the galactic dust glows in the **far infrared**, whereas it practically is not radiating at all in **near infrared**. The latter, in turn, is where stars, especially the cooler *M-type stars* radiate. Observing, therefore, in the far infrared is like looking at the dust, blending out anything else. Contrariwise, taking a photo in the near infrared should show the stars very clearly, because at this wavelength we are looking *through* the dust, which cannot scatter these long wavelengths. Thus, if we are studying the galaxy with three different wavelengths, we will see different things, Figure 6.5. As advantageous as the observation in the infrared is, it was not an option before the Space Age, since the Earth's atmosphere absorbs most incoming infrared radiation, Figure 3.36. Therefore, one either has to observe at the specific wavelength "holes" of the atmosphere (around $10\mu m$), or else observe from above the atmosphere with space-based telescopes like the Spitzer Space Telescope.

While studying globular star clusters and galactic structure in the mid-1940s, WALTER BAADE discovered that there are two distinct populations of stars. **Population I** stars like the

6.1. THE MILKY WAY CONTAINS VIRTUALLY EVERYTHING

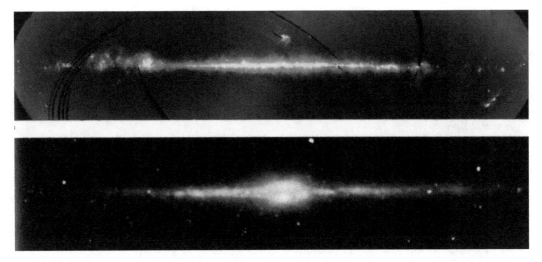

Figure 6.5: **The Milky Way in infrared light.** Upper panel: IRAS composite image at 25, 60, and 100 m (*far infrared*). Lower panel: *COBE* composite image at 1.2, 2.2, and 3.4 μ m (*near infrared*).

sun have very few **metals**[3] as we saw, but their metal content is still about thirty times that of **Population II** stars. In other words, Population I stars are "metal rich" and **Population II** stars are "metal poor." This finding has been called the second most important astronomical discovery of the twentieth century [30]. How do we know that this is the right way to interpret these stars' spectra, Figure 6.6? Maybe this is a new type of star? The idea, as often in science, is to isolate the effect from the uninteresting correlated phenomena. Practically, we would take the spectra of two stars of same spectral class and **luminosity class** (Don't compare giants with dwarfs!) of the same mass and surface temperature, but one of Population I and one of Population II. As in Figure 6.6, we would find that the metal lines in the Population I star are much stronger. We would conclude the latter star contains more metals, but otherwise the characteristics of the two stars are so similar that it makes sense to note the population as an additional property, rather than opening up a whole new classification scheme of stars. This sounds convincing, but if you'd really try to follow these steps we outlined, you would find that an important piece of the procedure is missing, which is why Baade deserves so much credit for his discovery of the two **stellar populations**. The million-dollar question is: How do you know which star is a Population I and which is a Population II star?[4] For the attentive reader, the answer is clear from our discussion of the stellar life cycle in Section 5.6: Metal-rich Population I stars must be a younger generation of stars, since they were born from the metal-enriched remnants of a past generation of stars. Therefore, we would look in old structures like globular star clusters to find old, Population II stars. This pattern of "segregation" of stellar populations is repeated on the galactic scale, and it tells us something important about the three parts of the galaxy.

- The **galactic halo** made of globular star clusters and isolated stars is an old, Population II region of the galaxy. Since virtually no interstellar matter exists here, star formation is impossible and has ceased billions of years ago as the ages of the clusters and stars—

---
[3]Elements heavier than helium, in astronomy jargon.
[4]This is reminiscent of the question BESSEL faced a hundred years earlier: How do you know which star is near and which is distant?

Figure 6.6: **The difference between Population I and II spectra.** The spectra of the old, metal-poor ***Population II*** stars (top) show much weaker metal lines, while exhibiting the same general features of the spectral classes they belong to. The younger, relatively metal-rich ***Population I*** star (below) has much stronger metal lines.

the youngest of them being billions of years old—attest.

- The **galactic bulge** consists of both stellar populations and contains gas and dust. On the other hand, its yellowish to reddish color speaks to the absence of hot, blue, young stars, and we must conclude that little star formation is happening, and a large number of old Population II stars is present.

- The **galactic disk** is the primary star-formation region of the galaxy. In fact, the bluish color of the spiral arms tells us that many hot, blue stars have just formed here. Even more, we must conclude due to their extremely short life expectancy that there is continuous star formation going on.

The Milky Way's Spiral Arms Are Star-Forming Regions

We have seen in Section 3.5.3 that astronomers starting with WILLIAM HERSCHEL began studying and classifying "nebulae"—objects which we today call galaxies. We will describe the different shapes and forms of galaxies in more detail in Section 6.2. Early observations showed that there are galaxies with and without spiral arms. The former are the stereotypical ***spiral galaxies***, the latter the so-called ***elliptical galaxies***. A natural question to ask is therefore if the Milky Way has spiral arms. This is a hard question to answer, because we get—at best—an edge-on view of our galaxy because we are in it. Therefore, we have to resort to indirect methods to figure this out. Before we do, let's ask a seemingly naive question: What *are* spiral arms? Today, this is much easier to answer, because we have magnificent, detailed, and colorful images of other galaxies. These pictures show bluish, spiraling lines that emanate from the galactic bulge. They certainly look like "arms," and were identified and studied early on by the likes of HERSCHEL and LORD ROSSE even before the mid 1800s. Their bluish color, however, which tells us so much about the stars that produce them, is *not* visible at the eyepiece, and thus the immense *luminosity* of blue, hot stars is what defines "the look" of spiral galaxies, even if the color is not discernible in visual observations at the eyepiece.

Indeed, it is a curious fact that up to the invention of color photography, astronomy was basically color blind as far as distant objects are concerned. Celestial objects other than the sun, moon, planets and some bright stars, have surface brightnesses too low to trigger our eyes' color receptors, and thus even objects like the Great Orion nebula, Figure 3.10(a), appear black-and-white to astronomers with even the largest telescopes. Only photography or modern CCD chips can gather enough color information on these faint objects to depict them in all their magnificent beauty. You might pause to think if these pictures are "real", because nature doesn't look like this—at least to us with our feeble light detectors, aka eyes. In some sense, all pictures are misleading, because the contrast or brightness has been adjusted, or they

6.1. THE MILKY WAY CONTAINS VIRTUALLY EVERYTHING

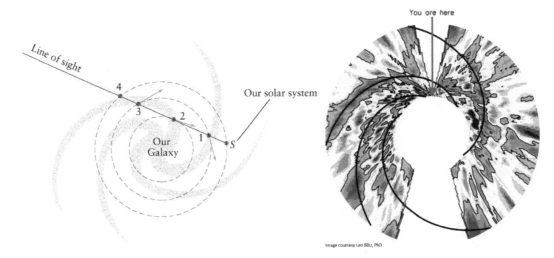

Figure 6.7: **Spiral arms.** (a) Observing the galaxy from the sun's position (S) in the disk means that the (line of sight) runs through several spiral arms. As stars at positions 1 − 4 orbit the center of the galaxy, the direction of their motion relative to the line of sight is different. They show different Doppler shifts of their spectral lines, which enables astronomers to reconstruct the spiral structure of our galaxy. (b) Map of the Milky Way in 21-cm radiation. This radiation is emitted strongly by neutral hydrogen (H I) in interstellar matter. Since interstellar gas clouds give birth to new stars, their position is indicative of the location of the spiral arms, which are shown as solid lines connecting the H I regions.

represent electromagnetic waves which we cannot detect with our eyes, like infrared or radio waves. Nonetheless, the information these pictures display is real in the sense that actual, physical processes like the emission and absorption of light has occurred. Our scruples are justified, though, because a picture alone is incomplete. We need information about exactly how it was taken. A picture is a measurement of sorts, and, as we detailed in Section 1.2.4, the "experimenter" must provide a detailed description of the measuring process, so that her results can be *reproduced* by others. Only with this additional information is the picture really useful to gain insight into the workings of the cosmos. As a direct analogy, think of a row of numbers, such as Table 1.1, listing positions of an object at different times. Without the explanation that these numbers are the distance a ball has rolled down an inclined plane at the time specified, the numbers, and hence the entire experiment are completely useless.

But back to our quest to detect the Milky Way's spiral arms. For consistency, let's assume the Milky Way *has* spiral arms. How would they look like from the sun's position? When we observe a particular section of the milky band across the sky, we should see an overlay of several spiral arms along our line of sight, as is clear from Figure 6.7(a). The line of sight hits several spiral arms which are rotating around the galactic center, so we predict different radial velocities (toward or away from us) of the different spiral arms. Note, however, that spiral arms are curved, and thus the line of sight can pierce a single spiral arm several times. Nonetheless, it should be possible for dedicated astronomers to decipher the spiral arm structure of our galaxies from the different Doppler shifts due to the different radial velocities. But wait! We have to look *through* foreground spiral arms—and interstellar matter—to see distant spiral arms. How is this possible? We already saw that long wavelength light penetrates dust clouds, and that the clouds glow in the far infrared if they have a temperature of 10 K to 100 K. To decipher the spiral structure of the Milky Way we should map the locations of **hydrogen gas**, which is mostly responsible for the formation of stars

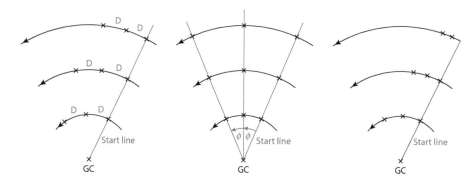

Figure 6.8: **Orbital velocity patterns.** (a) The stars are rotating around the galactic center (GC) with the same *linear velocity*, that is, they travel the same distance in a time interval. Since the stars in the outer orbits have a longer way to go, they fall behind relative to stars in inner orbits. (b) If the galaxy would rotate like a rigid disk, all stars would travel at the same *angular velocity*, that is, they would traverse the same angle ϕ in a time period. (c) If the stars travel in a Keplerian way, the stars in the outer orbits travel slower *and* have a longer way to go, so they fall behind even faster than in (a).

in the first place. Recall that the "quantum jumps" of electrons between different energy levels is the cause of the spectral lines, Section 5.1.2. In 1944 it was proposed by HENDRIK VAN DE HULST (1918–2000) that there is a pair of levels so close in energy that the energy difference amounts to an incredibly long wavelength photon, that is, electromagnetic wave. This is understandable when we combine Planck's relation between energy and frequency of an electromagnetic wave ($E = hf$, Equation (5.2)) with the relation between frequency and wavelength $c = \lambda f$. We obtain

$$E = h\frac{c}{\lambda}. \tag{6.1}$$

The energy difference of the so-called hyperfine splitting of hydrogen energy levels is 320,000 times smaller than the transition between the **Bohr levels** that causes the H_α-line of 656 nm wavelength. Therefore, the wavelength of a photon carrying away the energy difference between the hyperfine levels is 320,000 times longer, namely 0.21 m. This is the famous 21 cm line of astronomy, which enabled astronomers to discover the spiral structure of the Milky Way.[5] This was achieved after 1952 by mapping the **neutral hydrogen** content of the galaxy. A modern version of the results is depicted in Figure 6.7(b). The plot shows that neutral hydrogen is not distributed uniformly in the galactic disk, but concentrated in curved lanes that can be interpreted as the spiral arms. This evidence can be corroborated and cross-checked by another observation. The intense star formation in the spiral arms results in many hot, blue stars which produce a lot of ultraviolet radiation due to their high surface temperature. This energetic radiation can kick electrons out of hydrogen atoms, a process known as *ionization*. The ionized hydrogen, known as H II to astronomers (neutral hydrogen is H I), glows in the red light of the H_α line, because of **recombination** of electron and hydrogen nucleus with subsequent radiation as the electron cascades down to the lowest energy level. In sum, the hot, young stars make the surrounding hydrogen gas glow red. This **emission line** can be picked up easily from large distances, and therefore detecting **H II regions** is an important tool for determining the structure of galaxies, not just our own, see Figure 3.18. Once we answer the question as to what spiral arms *are*, namely regions of intense star forming in our

---

[5] The hyperfine splitting of the hydrogen atom itself was first observed in a physics experiment in 1951 at Harvard University.

6.1. THE MILKY WAY CONTAINS VIRTUALLY EVERYTHING

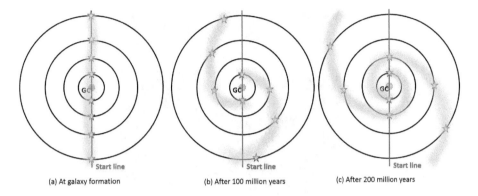

Figure 6.9: **Wrong hypothesis for spiral-arm winding.** If the spiral arms were created at the same time as the galaxy, they would wind themselves up because of different orbital periods of their constituent matter.

galaxy, the next questions pop up: Where do spiral arms come from, and how do they persist over galactic time spans? A plausible idea to start with is that spiral arms could simply be created with the galaxy, and then rotate along with the rest of the galactic disk. This theory is easy to refute. Since objects closer to the galactic center have a shorter rotation period, Figure 6.8, the spiral arms would wind themselves up as in Figure 6.9 in "only" a few hundred million years. This may not seem a short time, but the relevant time to compare to is the typical age of a galaxy, which must be at least as old as the sun, roughly 5 billion years, easily ten times as long as the "wind-up time." The conclusion is that spiral arms cannot be something material like a grouping of stars. It must be something that is continuously recreating itself. There are two leading theories explaining the sustained spiral pattern, and a final solution to the problem remains under investigation.

The ***density-wave model*** developed by BERTIL LINDBLAD in the 1940s and worked out in more detail in the 1960s by CHIA-CHIAO LIN (1916–2013) and FRANK SHU (born 1943), holds that spiral arms are ripples of compressed interstellar matter that move through the galaxy. The compression of the matter gives rise to star birth at intense rates. In this view, the spiral arms are the crests of a density wave that sweeps through the galaxy. Even when the density wave has moved past a region, it remains visible for a couple of million years until its hottest, bluest stars leave the main sequence. The cooler, redder stars of the region will still shine for a long time after that, but the bluish glow has then moved on to the next region. Indeed, overall the density of stars in the galactic disk is surprisingly uniform. The spiral arms contain only about 5% more stars than the other regions of the galactic disk. For the visual impression of an outside observer, however, it is only those ultra-bright, blue stars that matter.

If the ***density-wave*** model were right, then we should see continuous spiral arms, since density waves are continuous curves that operate on the interstellar matter of the galaxy. However, looking out into the universe we see spiral galaxies that are not that well ordered, namely ***flocculent galaxies*** with spiral arms that look patchy like little pieces of wool (hence the name), Figure 6.10. To explain this pattern, astronomers in 1976 developed the ***self-propagating star-formation*** model. According to this theory, hot stars put stresses on their environment that triggers star formation. Hot stars radiate a lot of energy, they produce stellar winds, and, not the least, eventually they explode as supernovae, all of which compresses the surrounding gas and dust clouds. As the compressed interstellar medium gives

Figure 6.10: **Flocculent galaxy (NGC4414).** The spiral arms of this galaxy are not well defined and look fuzzy. This poses a challenge for the *density wave model* of spiral arm formation.

birth to the next generation of hot stars, the cycle begins again. This way, a star-forming region *self-propagates* through the galaxy.

6.1.3 Galactic Rotation as Evidence of Dark Matter

It is natural to suspect that the Milky Way is rotating. After all, rotation is an efficient way to stabilize a dynamical system. A system of objects, each exerting a gravitational force on each other, would collapse under its own gravity. A case in point is the collapse of gas and dust clouds that give birth to stars. What prevents the collapse is a velocity of the objects perpendicular to the direction of the attractive force, see Figure 2.37, and the discussion of the moon orbit in Section 2.10. Rotation is also intuitive for another reason we already encountered when discussing the formation of the solar system in Section 3.5.2. Namely, the rotation explains the flattening of the original gas cloud, as this could explain the flattening of the galaxy into a thin disk. The real question therefore is: *How* does the galaxy rotate? There are three plausible ways; see Figure 6.8.

1. The galaxy could rotate as a **solid disk**, such that objects orbit the center of the galaxy in the same time, regardless of their distance from the center. This is unlikely, since the disk is certainly not stiff, as there is plenty of (mostly) empty space between stars even in the densest parts of the Milky Way. Also, from planetary motion we know, that objects farther from the center should be slower.

2. The galaxy could rotate according to **Kepler's third law**. Recall that the outer planets take *much* longer to orbit the sun, because they have a longer way to go,[6] and because their velocity is smaller due to the decrease of the sun's gravitational pull. This rotational pattern would occur if the mass of the galaxy is concentrated close to its center, but in a galaxy objects appear to be evenly spread out over its disk.

3. The galaxy could rotate with **constant orbital velocity**, such that the objects farther from the center take longer to orbit the center because they have longer to go, not because they move slower.

---

[6]The circumference of orbit is roughly $2\pi a$, where a is the semi-major axis of the orbit.

6.1. THE MILKY WAY CONTAINS VIRTUALLY EVERYTHING

As we said, these are three plausible possibilities, and only observation can distinguish them. How did astronomers figure out the correct one? The dilemma they were facing was that the observer herself orbits with the solar system around the galactic center. Therefore, measuring velocities via the Doppler shift of stellar spectra is not going to tell us all we need to know. We only measure the ***relative, radial velocity*** with the Doppler shift of the spectral lines. Nonetheless, analyzing the Doppler shifts of the 21 cm ***radiation*** coming to us from different parts of the galaxy was enough for astronomers to establish that all parts of the galaxy rotate in the same direction around its center. This is hardly a surprise—but it is always good to check. The more surprising insight from the 21 cm measurements is that the velocity of objects is approximately *independent* of the objects' distance from the center. This is evidence that the objects in the galaxy rotate with constant velocity (Scenario 3).

Now that we know how the parts of the galaxy move or rotate, we can use Newton's form of Kepler's third law to figure of how the mass in the galaxy must be distributed. After all, the mass distribution determines the gravitational forces on and hence the orbital velocity of the objects. Due to the spherical symmetry of the galaxy, an important corollary can be used. It was originally derived by Newton to make his theory of universal gravity watertight (see Section 2.10). Namely, for a spherically symmetric mass distribution, only the mass *inside* an orbit, that is, closer to the center than the object itself, is responsible for the force that determines the orbital motion of the object. The reason is that the mass outside the orbit attracts the objects from all angles, but such that all these forces add up to zero, aka cancel. Let's label this mass $M_{\text{eff}}(r)$, because it is the mass effective in driving the motion of an object with orbital radius r.

We can intuitively anticipate the solution—which scientist often do to cross-check their calculations. The argument is as follows. Since the orbital velocity measured to be constant as a function of distance from the center, but the gravitational pull of a fixed mass drops like the inverse square of that distance, the effective mass cannot be fixed, but must grow as a function of the distance to the center to make up for the decline of its gravitational force with the distance. More specifically, since the force *drops* like $1/r^2$, the mass must *rise* like r^2, so that the force is constant

$$F = G\frac{mM(r)}{r^2} = G\frac{m\rho r^2}{r^2} = Gm\rho,$$

where m is the mass of the orbiting object, $M(r)$ is the mass of the galactic parts within the orbit of the object, and ρ is a constant, playing the role of an *area* mass density. To achieve a constant mass density, matter in the galaxy must be uniformly spread out over the its disk. From observation, we have all reason to believe that this is true.

The next question is then: What is the ***total mass*** of the galaxy? Clearly, the answer would tell us a lot about the Milky Way. For instance, we could estimate how many stars it contains, and more generally, it would make insights into the detailed dynamics of the galaxy possible. The mass of the Milky Way follows from the old argument related to Newtonian circular motion under the influence of gravity. In short, the faster an object at a fixed distance from a center rotates, the more mass is inside the orbit. We will therefore want to measure the period of the sun's orbit around the galactic center. We already know the radius of its orbit, which is the distance to the center of about $r = 25\,000$ ly. The time it takes to travel a distance $2\pi r$ (the circumference of the orbit) with a velocity v is the period P, namely

$$P = \frac{2\pi r}{v}.$$

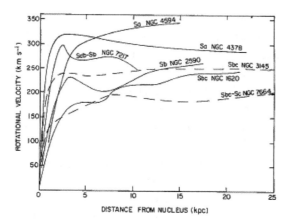

Figure 6.11: **Rotation curves of galaxies.** Shown is the velocity of objects in several galaxies as a function of their distance from the galactic center (1 kpc=3260 ly).

How can we determine the velocity of the sun around the center? We need to determine this velocity relative to a background, external to the galaxy or at least the galactic disk, since all disk objects rotate, too. Astronomers took the Doppler shifts of distant galaxies and globular star clusters[7] and averaged them. They found the velocity of the sun relative to the background to be $v_{Sun} \approx 220$ km/s. This is more than 7 times faster than the Earth rotates about the sun. On the other hand, the sun has $25\,000\,\text{ly} \times 63\,400\,\frac{\text{AU}}{\text{ly}} = 1.58$ billion AU to go. That is a 1.58 billion times longer way! Therefore, it takes the sun $\frac{1.58}{7}$ billion years ≈ 225 million years to go once around the galaxy. This is called a ***galactic year*** for obvious reasons.

To determine the mass of the galaxy, we should do a similar type of calculation for the stars at the edge of the galactic disk, because the entire mass of the galaxy lies within these largest orbits. Astronomers can do even better by measuring the velocity along the orbit or the ***orbital speed*** of objects at various distances from the center. The plot of orbital speed versus distance from the galactic center is known as the galaxy's ***rotation curve***, Figures 6.1 and 6.11.

The rotation curves defies our intuition in that we would expect the orbital speed to drop systematically at distances larger than the radius of the visible galactic disk. Beyond the visible galactic disk the effective mass should be constant, and Kepler's third law should kick in. The naive conclusion is that there there is not enough mass out there to keep up the orbital speeds, yet we see constant speeds way past the visible disk. This is known as the ***missing mass problem*** and was discovered as early as 1933 by FRITZ ZWICKY. Just how much mass is "missing?" Is this a small or a large effect? To answer these questions, astronomers took a census of the galaxy. They counted the number of stars, determined the mass of interstellar matter based on 21 cm radio observations, and estimated the number of white dwarfs, and other dim but heavy or abundant and easily overlooked objects. When they summed up all these masses, it came up far short of the mass necessary to explain the large galactic orbital speed. For instance, stars themselves represent only 10% (one-tenth) of the mass of the galaxy. The latter was determined by the rotational speeds to be in excess

---

[7]This is legitimate, since the globular star clusters are moving outside the galactic disk in random orbits, and thus their velocities should average to zero.

6.1. THE MILKY WAY CONTAINS VIRTUALLY EVERYTHING

of one trillion (10^{12}) solar masses. This in turn yields an estimate for the number of stars in our galaxy. Assuming an average star to have half the mass of the sun, we can estimate that our Milky Way contains $10^{12} \times 10\% / \frac{1}{2} = 200$ billion stars.

In conclusion, our galaxy's gravitational mass comes mostly from something that we cannot detect. This "something" acts gravitationally just like ordinary matter, but is invisible.[8] It has been labeled *dark matter*—and we do not know what it is! We can infer its presence only from the fact that it attracts other matter gravitationally. It is an exceptional claim—that something is out there which defies detection. Therefore, we need exceptional evidence to make sure that the pattern in nature we attribute to it cannot be explained by something else. We will get back to this in Section 6.3 (on cosmology), because the problem persists and is even pronounced on larger scales, for example when when we try to understand the dynamics of clusters of galaxies, that is, the gravitational attraction between galaxies.

The story of *dark matter* does not have a happy end. To this day it is not known what dark matter is.[9] For now, researchers are trying out hypotheses (usually particles with the desired properties of being hard to detect and producing a strong gravitational force). These hypotheses are tested by designing experiments to detect the postulated particles. Nothing conclusive has been found as of yet. There are other scientists who believe instead that the galactic rotation pattern is best described by a modification of Newton's laws on the galactic scale and beyond. These researchers think that the extrapolation of Newton's laws breaks down at extremely large distances and have to be replaced by a different relation between mass and acceleration. However, both of these approaches have their problems. We live in exciting times when fundamental cosmological questions are being debated! Note that we can *use* the observed pattern—even when we do not know how to explain it. Namely, there is a relation between rotational speed and luminosity of a galaxy known as the *Tully-Fisher Relation*, which can be used to determine the luminosity of a galaxy, and then be fed into the *brightness-luminosity relation*, Equation (5.3), to determine the distance to the galaxy. It is described in more detail in the next section.

Concept Practice

1. Which is not one of the major parts of the Milky Way?

    (a) Bulge (b) Ring (c) Disk (d) Halo (e) All are major parts of the Milky Way.

2. Where in the Milky Way is the sun situated?

    (a) In the halo.

    (b) Somewhere in the central bulge.

    (c) At the very center.

    (d) Somewhere in the disk.

    (e) None of the above

---

[8] "Invisible" here means not detectable by any telescope or detector. This matter is not just too cold to radiate—it does not radiate electromagnetic waves at all.

[9] In fact, it is even worse. There is yet another substance out there which dominates the universe and still is completely unknown. We will encounter it in Chapter 6.

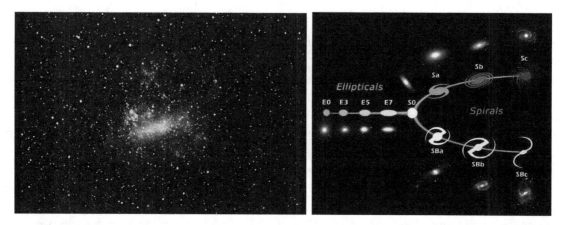

Figure 6.12: **Galaxy classification.** (a) The *Large Magellanic Cloud* of the southern sky as an example of an *irregular galaxy*. (b) Galaxy Classification according to Edwin Hubble. The types of galaxies form the so-called *tuning fork diagram*.

6.2 Other Galaxies: The Milky Way Is Not the Universe

6.2.1 Galaxies Are Island Universes

We are now returning to our main story-line of the expanding universe which we left after mentioning the **Great Debate** in Section 6.1.1. The debate was about the nature of the faint nebulosities we call *galaxies*. In 1920 it was not clear at all whether these are close and small *or* distant and large objects, since all that was known was their **apparent size**. This changed in 1924 when Edwin Hubble published his work on Cepheids in the Andromeda nebula, M31. He had been able to take photographs that showed individual stars in the spiral arms of this nebula with the 100-inch telescope on Mt. Wilson—the world's largest telescope from 1917 to 1948. His breakthrough discovery came when he realized that some of the stars were Cepheid variables. Since he was familiar with their **period-luminosity relation**, he spent the next months taking more photos to establish the Cepheid variables' light curves and determined their periods. From this he could calculate their luminosity, and the brightness-luminosity relation, Equation (5.3), enabled him to show decisively that the Andromeda nebula was far outside of our own Milky Way. Almost overnight, the Andromeda nebula turned into the Andromeda galaxy, and the universe opened up into extragalacticity and expanded manifold. At a distance of about 2.5 million light-years,[10] the small nebulosity in the constellation *Andromeda* turned out to be a galaxy of its own right, even larger than the Milky Way, having about twice its mass, and a diameter of about 220,000 ly. This means that the Andromeda and the Milky Way galaxies are about 20 galactic diameters apart. As a comparison, stars in galaxies are more than $4\,\text{ly} \approx 4 \times 63\,200\,\text{AU} \approx 253\,000\,\text{AU}$ apart. Since one astronomical unit is about $\frac{149.6\,\text{million km}}{1.4\,\text{million km}} D_{Sun} \approx 100$ solar diameters, stars are more than 25 million stellar diameters apart. From this little exercise we conclude that *it is almost impossible for stars to collide, but likely that galaxies collide*. In fact, the Milky Way and the Andromeda galaxy are predicted to collide in about 4.5 billion years. We will later learn

---

[10] Hubble underestimated the distance by a factor of one-half, due to the fact that Baade's discovery of the two stellar populations came two decades later; see Section 6.1.2. As it turns out, the metal-rich **Population I Cepheids** found in the Andromeda galaxy are about $4\times$ more luminous, and therefore two ($2 = \sqrt{4}$) times more distant than Hubble thought originally.

	Spiral Galaxies (S and SB)	Elliptical Galaxies (E)	Irregular Galaxies (Irr)
Diameter (ly)	15,000 to 800,000	3,000 to 600,000	3,000 to 30,000
Luminosity (L_{Sun})	10^8 to 10^{10}	10^5 to 10^{11}	10^7 to 10^9
Mass (M_{sun})	10^9 to 10^{12}	$10^5(!)$ to 10^{13}	10^8 to 10^{10}
Fraction of all galaxies	70%	25%	5%

Table 6.1: **Properties of the three main classes of galaxies.**

that galaxy collisions are fertile rather than fatal, and that they are a rather standard way for galaxies to change and evolve. As we have just seen, stars in distant galaxies behave like the ones in our Milky Way. Likewise, interstellar matter is present in galaxies like M31. We therefore infer that other galaxies are quite like our own, and hence **island universes** of their own.

6.2.2 Galaxies Come in Different Shapes and Sizes

From the pictures of galaxies that we've encountered, for example Figures 3.13, 3.18, 6.12(a), it is obvious that individual galaxies look different. Partly this is due to the way a galaxy is positioned relative to our line of sight. Spiral galaxies look different from the side or **edge-on**, Figure 6.4, than they do **face-on**, like the **grand design** Whirlpool galaxy (M51), Figure 3.18. But not every galactic shape is explainable by different orientation. There are clearly distinct **morphologies** (forms, shapes, and sizes). As always, these patterns in nature need explanation, and the first scientific reflex is to **classify**. HUBBLE himself came up with a **classification scheme** for galaxies that is still in wide use. Broadly, he distinguished three groups: **spiral galaxies**, **elliptical galaxies**, and **irregular galaxies**. Here, *nomen est omen*[11] as spiral galaxies (example: M51, Figure 3.18) have spiral arms, elliptical galaxies (example: M87, Figure 3.13(b)) look elliptical, and irregular galaxies look—well, irregular, see Figure 6.12(a). There are two main types of **spiral galaxies**: those with a bar ("barred spirals (SB)") and those without a bar ("spiral galaxies (S)"). Depending on whether the spiral arms are tightly wound or not, there are subtypes, for example SBa, SBb, SBc. The Milky ways is probably a SBb galaxy with a bar and moderately wound-up spiral arms. The elliptical galaxies have a numeral which represents their eccentricity. An E1 galaxy is almost circular, an E7 galaxy is very elongated. This gives rise to the so-called **tuning fork** diagram, Figure 6.12(b), with the S0, SB0 or *lenticular* galaxy type being something of a transitional type between spirals and ellipticals.

Astronomers have classified millions of galaxies, and have come up with typical values for their sizes, masses and number of stars, see Table 6.1. Some facts about galaxies are as follows. Spiral galaxies are by far the most common, and contain both Population I and II stars. They actively produce stars. Elliptical galaxies vary in size a lot. The biggest galaxies in the universe are ellipticals. However, no active star formation is going on in ellipticals. Irregular galaxies are rare, have young stars, and are small.

Note that this is only a classification. While HUBBLE devised it in anticipation of an evolutionary path as galaxies "develop," this turns out not to be the case. No process is known in which a elliptical galaxy suddenly develops spiral arms, or how a spiral galaxy can lose its arms.[12]

---

[11] Latin(loosely): The name says it all.
[12] Collisions of galaxies are somewhat of an exception to this rule.

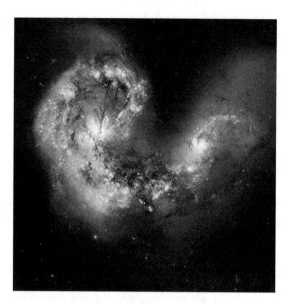

Figure 6.13: **Collision of galaxies.** The *Antennae galaxies* (NGC 4038/NGC 4039) are an example for a collision of two galaxies. Evident is the large rate of production of hot, blue young stars which make the emission nebulae glow in the red H_α light.

6.2.3 Galactic Evolution and Collisions

The evolution of galaxies is very different from the evolution of stars, as far as we know. Stars develop due to their "thermodynamic needs." Their mass and gravitational pressure necessitate an appropriate energy production mechanism. Galaxies are almost entirely driven by very weak gravitational forces due their immense sizes. On the other hand, they are likely to collide—and frequently do. We can simulate on the computer what happens in such a collision, and we see galaxies in the process of collision, Figure 6.13. Galaxy collisions take an astronomically long time to unfold even though they happen at high speeds. This is no contradiction, and is explained by their vast sizes. We mentioned that galaxy collisions are fertile rather than fatal. Why is that? As we have seen, star formation happens when gas and dust clouds contract or are compressed. This is what happens when the interstellar matter of two galaxies collides. In the collision process spiral arms can be produced or disappear, and galaxies merge—or are "gobbled up" in the case of a gross size imbalance of the involved galaxies.

These findings and plausible scenarios were over time assembled into a coherent theory backed by observational data. The initial debate raged about the formation of galaxies. Is it a "top-down" process, in which vast galaxy-sized gas and dust clouds condense into a galaxy—much like a smaller cloud would contract to form stars? Or is is a "bottom-up" process, whereby small galaxy-like objects collide and merge to build up larger galaxies? This question can only be decided by observation, since both scenarios seem plausible and logically consistent. Recall that astronomical processes usually take "astronomically" long time to play out.[13] We got around this problem when studying stellar evolution by observing *many* stars in different stages of their life cycle. For galaxies there is an amazing concept akin to a time-machine—but legal by EINSTEIN's laws of relativity. It will come in handy in the

---

[13]Exception: A supernova goes off in literally a second.

6.2. OTHER GALAXIES: THE MILKY WAY IS NOT THE UNIVERSE

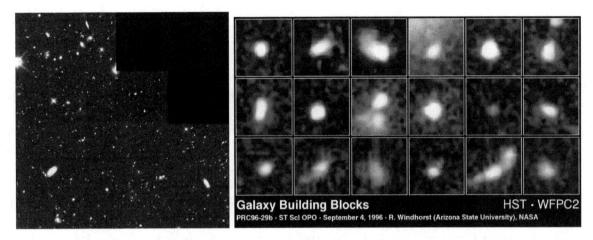

Figure 6.14: **The Hercules galaxy cluster.** (a) A Hubble Space Telescope overview photo of the cluster. (b) A close-up of eighteen galaxies in the field (a).

cosmology section, and we will look at it in detail here.

The tool is based on the finiteness of the speed of light, and known as **lookback time**. It is the simple fact that the farther you look out into the cosmos, the younger objects appear. As an example take the Andromeda galaxy which is 2.5 million ly away. Since the light traveled 2.5 million years from the Andromeda galaxy to us, we see the galaxy *as it was* 2.5 million years ago; this is the time when the light (carrying the information about the galaxy and its stars) had to leave to get to us *now*. In fact, due to the finiteness of the speed of light, we have to rethink the concept of "now." In relativity it is known as **simultaneity**, and a key concept in the modern way of thinking about nature. What happens at the same time (simultaneously) for us, does not necessarily happen at the same time for a different observer. It may seem that **physical reality** is thus "stitched" together: "Now" being *right now* in our immediate vicinity, but *20 minutes ago* on Mars (20 light minutes away), *5 hours ago* on Pluto (5 light hours away), and *4 years ago* on Alpha Centauri (4 light years away). However, for all practical purposes there is only one "now," and that is the "local now." We do not know whether a sandstorm will develop on Mars in 10 minutes from "our now" (if it does, it has happened on Mars already 10 minutes ago!), because this information has to travel to us to *affect* us. Therefore, the pragmatic thing to do is to "observe locally and interpret globally:" We take a photo of Mars and Pluto together, say, and then use our knowledge of relativity and the distances to the two objects to interpret this photo as showing how Mars looked 20 minutes ago *and* how Pluto looked 5 hours ago, because there is a 20-minute lookback time for Mars, and a 5-hour lookback time for Pluto.

What does this insight do for our quest to understand galactic evolution? Well, our knowledge of the concept of **lookback time** allows us to use the vast distances in the universe as sort of a time-machine. Sloppily speaking, if we want to see how galaxies looked 500 million years ago, we find a galaxy that is 500 million light years distant. Of course, a galaxy so far away will appear very dim and very small to us, and therefore we need a powerful telescope to look that far into the past. To do just that, the Hubble Space Telescope took an extremely long-exposed photo of a portion of sky in the constellation Hercules, Figure 6.14. We see a cluster of galaxies, and eighteen of them close up. The distance to these galaxies was determined to be 11 billion light years, so we are looking at very young galaxies, that is, the galaxies as they were 11 billion years ago.[14] Knowing the distance to the cluster gives away the actual size of the objects in the photo. The eighteen boxes each contain a very young galaxy, and are about 2,500 ly on a side. This is an amazingly small size for a galaxy—about ten times smaller in diameter than the Andromeda galaxy![15] And another remarkable fact can be gleaned from these photos which all show bluish galaxies without visible spiral arms. These are elliptical galaxies (or the central regions of spiral galaxies)— but in their infancy. This must mean that there is a burst of star formation in these elliptical galaxies early on. When they age over billions of years, their blue stars die off to give these galaxies their yellowish hue that we are used to "now." But another thing has to happen to explain the enormous size of today's largest elliptical galaxies: they have to **merge**. If all of the galaxies shown in the inset of Figure 6.14 merged, they would make up a sizable elliptical galaxy. This immediately falsifies the "top-down" galactic formation theory from above. Further corroboration of the "bottom-up" theory comes from observational evidence for the natural prediction of this theory. Namely, we should see bigger galaxies the closer we get to the present (and therefore the nearer the galaxies are).

We will suppress the details of galaxy formation to focus on the facts important for our quest to understand the construction of the expanding universe. In broad strokes, the commonly accepted theory goes like this. The rate of star formation is crucial for the evolution of a galaxy. If stars form too fast, they "use up" most of the gas in the galaxy, and an elliptical galaxy made of metal-poor Population II stars is the result—because the next stellar generation never had a chance to form. If the stellar birth rate is low enough that the left over matter can rotationally flatten into a disk, we get a spiral galaxy. The stars which formed before the disk developed are the stars and globular star clusters of the halo. You can see that this theory explains the genesis and morphology of the three main parts of a spiral galaxies. This is satisfying, but of course, far from proof. We should also acknowledge the role of "history" in the development of galaxies. Whether or not a galaxy has a close encounter or merger with another galaxy will profoundly change its morphology. Depending on the relative size of the merging or colliding galaxies, and on the exact impact parameters, very different galactic shapes and properties can result.

One thing we left out of our discussion so far is that most of the mass of the galaxies comes from **dark matter**. Since stars and interstellar matter only account for one tenth of the mass of a galaxy, dark matter has to be taken into account when explaining how galaxies and large-scale structures in the universe form and evolve. This does not invalidate the theories we have developed here. Since dark matter interacts for all practical purposes only gravitationally, it plays the role of a catalyst or accelerator of formation. Everything just happens faster when there is dark matter. We will have to keep this in mind when we talk about cosmology and **structure formation** in the universe. In particular, we will have to evaluate how much mass or dark matter is necessary (and how much is too much) to explain the structures and patterns of galaxy distributions in the cosmos.

6.2.4 Galaxy Clusters and Superclusters

If we think about the patterns in nature we have encountered so far, then there is a large degree of **order** in the universe. Planets are organized in solar systems, many solar systems

---

[14] You realize that this is more than six billion years before the Earth and the sun ever formed!

[15] Therefore, $10^2 = 100$ times smaller in disk area size.

6.2. OTHER GALAXIES: THE MILKY WAY IS NOT THE UNIVERSE

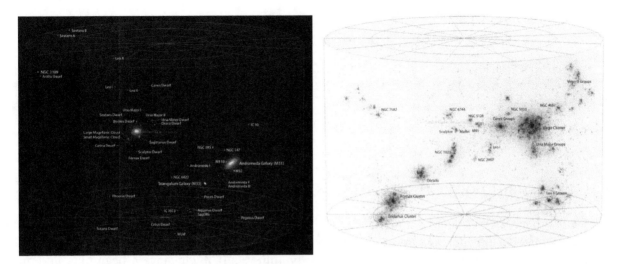

Figure 6.15: **Galaxy clusters.** (a) the *local group* containing the Milky Way and the Andromeda galaxy. (b) The *Virgo cluster* which contains the *local group* of galaxies.

and interstellar matter are organized in galaxies. Close inspection of the night sky reveals that this pattern continues. In many patches of the sky we find groups of galaxies close together. When we assemble a three-dimensional map of these galaxies, it becomes clear that several of them are organized in a galaxy cluster called the *local group*, see Figure 6.15(a), while many others are much further away, but forming other groups or galaxy clusters.

Our *local group* of galaxies has three large galaxies: the Milky Way, the Andromeda galaxy and the Triangulum galaxy M33 (the latter is much smaller than the other two. The rest of the more than fifty galaxies in the local group are small **dwarf galaxies**. The typical distance in the local group is the separation between the Andromeda and Milky Way galaxies (about 2.5 million light-years), so the size of the local group is a few times this fundamental distance, or about ten million light-years. The local group is a rather small galaxy cluster. It is associated with the much larger **Virgo cluster**. The latter is seen in the direction of the constellation *Virgo*, and occupies an area of the sky roughly $12° \times 10°$. We count more than 1,500 galaxies in this cluster, and can determine its distance with Cepheid variables to about 60 million light-years. Therefore, its apparent size translates into an actual size of about 10 million light-years. This is about the same as the *local group*, but it contains more than ten times as many galaxies. It is considered a **rich galaxy cluster**. The hierarchical structure of the universe continues with the organization of galaxy clusters into galaxy **superclusters**, consisting of dozens of individual galaxy cluster. Superclusters have a typical size of about 150 million light-years across. The **local supercluster** contains both the *Virgo cluster* and the *local group*. At these distances, mutual gravitational interaction is barely susceptible due to the inverse square law of Newton's universal gravity. Indeed, observations suggest that superclusters are not gravitationally bound; they drift apart. However, we should not forget the large, unseen parts of these superclusters: They contain mostly dark matter!

To understand the universe, we must look for the distribution of matter on an even larger scale. We are thus entering the realm of **cosmology** which asks for this distribution at the largest scales possible. The large-scale structure of the universe is the biggest pattern we find in nature. We are attempting to do nothing less than mapping the entire accessible universe!

Concept Practice

1. Which is not one of the major types of galaxies?

    (a) Circular galaxies
    (b) Elliptical galaxies
    (c) Spiral galaxies
    (d) Irregular galaxies
    (e) All are major types

6.3 Cosmology: The Cosmos as a Single Object

Cosmology is the science that deals with the origin and evolution of the universe itself. It regards the cosmos itself as an object that can be studied. This object can develop as time goes by. It might have a shape and a size—or be infinite. It might have a beginning and an end—or might always exist. We live currently in a golden age for cosmology, because many sophisticated instruments and space probes are delivering stunningly accurate measurements of quantities that directly relate to the big, cosmological questions.

6.3.1 Climbing the Cosmic Distance Ladder

Before we enter the realm of cosmology, it is wise to review the methods of distance measurement that have gotten us so far into the cosmos. Indeed, the entire *construction* of the expanding universe hinges on these methods. It is almost as if we did our investigation of the solar system, stars and their life cycle, and the dynamics of galaxies and clusters just for the purpose of finding reliable ways to measure ever increasing cosmic distances. Since the vastly different distance scales of the cosmos make different measuring tools necessary, we distinguish several "rungs" of a **cosmic distance ladder**, of methods to measure distances in the universe.

From the Earth to the Stars

As we have seen in Secs. 2.3.2 and 3.3.1, historically the determination of distances in the solar system was fairly complicated and inaccurate. It involved the following crucial steps.

1. Determine the size of the Earth (tacitly assuming its spherical shape).
2. Measure a solar system event (such as a Venus transit) which looks different from two different locations on Earth.

In other words, we determine an astronomical distance, such as the distance to the sun, from a (known) geographical distance on Earth. In the modern era starting in the 1960s, there is an easier way to determine the distance to the sun. We can bounce a radar signal (electromagnetic radio wave) off a planet like Venus, and measure how much time it takes this signal to get back to us. Since the signal has travel to Venus and back, the "double distance" to Venus can be computed by the simple formula relating distance d traveled and time t elapsed to the (constant) velocity v at which the signal travels (here, of course, the speed of light): $d = t/v$. Since we know the **relative distance** to Venus from Kepler's third

6.3. COSMOLOGY: THE COSMOS AS A SINGLE OBJECT

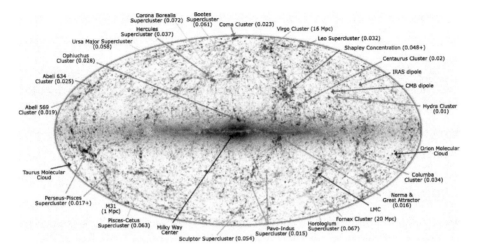

Figure 6.16: **The large-scale structure of the universe (I).** This infrared image of the *Two-Micron All-Sky Survey* shows almost 2 million galaxies. The Milky Way is clearly discernible in the center. Otherwise, the image shows that galaxies and galaxy clusters are concentrated in some areas, whereas other regions contain only a few galaxies. Overall, the impression is that of a ***filamentary*** distribution of mass in the universe. Galaxies in this image are much closer than in the 2dF Figure 6.17.

law $(1\,\text{AU} - 0.72\,\text{AU} = 0.28\,\text{AU})$, we can determine the length of the astronomical unit. Note that both the historic and the modern measurement starts with a measurement on Earth![16]

The ***second rung*** of the cosmic distance ladder takes us to the stars, as we saw in Section 3.6.3. Now that we know the size of Earth's orbit around the sun, we can use the fact that the sky looks slightly different from different points along Earth's orbit to determine the distance to a nearby star. This star's position will change in front of the background of truly distant stars, as in Figure 2.2. The more it changes, the closer the star, as is evident in the formula, Equation (2.1), $d = 3.26\,\text{ly}/p$, where d is the distance in light-years and p is the parallactic angle, that is, *half* the annual angular motion of the star.

These first two rungs of the distance ladder are arguably the most important. Having taken those two, we have freed ourselves from our earthly and solar system confines, and can press on—to explore the rest of the universe.

From the Stars to the Universe

To get to the end of the observable universe requires but one crucial insight. Namely, that the luminosity L of a celestial object, its apparent brightness B and its distance to the observer d are related by Equation (5.3), namely $d = \frac{L}{4\pi B}$. Our progress in understanding the universe (that is, its parts or objects) enables us to predict their luminosity. Therefore, the straightforward measurement of the object's brightness in the sky is enough to determine its distance to us. Here is a list of objects that can be used as rungs of the cosmic distance ladder. It should be obvious that the farther into the universe you want to look, the more luminous objects you'll need. In particular, to look to the end of the observable universe, you need the most luminous objects in the universe.

1. Use *main sequence stars* in the *spectroscopic parallax* out to 1,000 ly.

---

[16] The latter requires us to measure the *speed of light* in a lab.

2. Use *RR Lyrae* variable stars out to 300 000 ly.

3. Use the *period-luminosity relation* of *Cepheid* variables out to 30 million light-years.

4. The *Tully-Fisher relation* determines luminosities of galaxies based on their rotational speeds (accessible vie Doppler-broadening of their spectral lines) works out to 100 million light-years.

5. Supernovae, in particular Type Ia, can be used to determine distances up to billions of light years, that is, close to the edge of the observable universe.

6.3.2 The Large-Scale Structure of the Universe

We are now ready to explore the cosmos at the largest scales. That is, we are looking for the largest discernible structures in the observable universe. These are distances that are big compared to a galaxy supercluster (cluster of clusters of galaxies) which are 150 million light-years across. We need to construct a three-dimensional map of the observable universe. As we pointed out before, this will be hard due to intragalactic dust which makes observations near the galactic plane impossible, and severely hampers optical observations. As two recent examples, we exhibit the results of the *Two-Micron All-Sky Survey* project observing in the near infrared, and the *Two-degree Field Galactic Redshift Survey* which observed two two-degree slices of the sky perpendicular to the galactic plane. Both projects published their data (from observations over many years) in 2003. Figure 6.16 and Figure 6.17 present the data in a different way, but they both lead to the same conclusion. Namely, that mass (clusters and superclusters of galaxies) is not randomly distributed in the universe. There are areas with a lot of material, and also larger voids, largely empty of matter. The distribution has a **filamentary** character; the areas of large densities seem connected by strings. The distribution of mass in the universe forms a pattern that looks rather like that of soap-suds in a bathtub: large, bubbly voids surrounded by sheets of galaxies.

This is the largest pattern in nature we can discern—and we now have to explain it. How come that the matter in the universe would form such filamentary structures? It is easy to imagine a different plausible pattern, namely a random distribution, in which essentially all matter is smeared out uniformly, so that every part of the universe contains exactly the same amount of matter.[17] We need to explain these structures, as they generate a host of questions: Why should they form? Is there enough mass (gravity) to make matter clump at such large distances? Where did the initial "clumpiness" of the universe come from so that these structures can develop over time? Was everything smeared out uniformly billions of years ago?

6.3.3 Cosmological Questions

Indeed, this taps into the biggest, cosmological questions the universe poses—just by being there: Where does it all come from, see Figure 6.18(a)? Cosmological questions sound so profound and all-encompassing, that they seem hardly answerable. Hopefully, you have gained enough stamina and self-esteem from the many times we have encountered this verdict in this

---

[17]This assumption of uniform mass density is known as **cosmological principle**, and necessary to model the universe; see Section 6.4. It is a fairly good approximation on larger scales than presented here (billions of light-years).

6.3. COSMOLOGY: THE COSMOS AS A SINGLE OBJECT

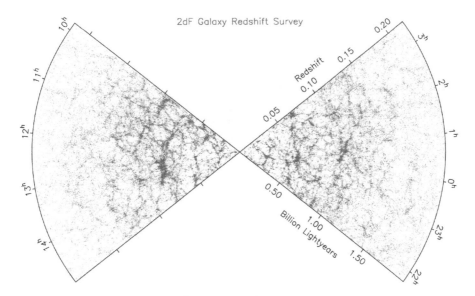

Figure 6.17: **The large-scale structure of the universe (II).** The *Two-degree Field* (2dF) redshift survey shows the distribution of about 62,000 galaxies out to five billion light-years. As in Figure 6.16, a filamentary distribution of mass in the universe is evident.

book and its dissolution at the hands of pragmatic scientists that you are confident that astronomy will march on.

Indeed, amazing things can be said about cosmological questions from trivial observations if one is willing to *think*. NEWTON did so and concluded based on the simple fact that gravity is always attractive that the universe must be infinite and filled randomly and homogeneously with stars or other matter. The reason is that any *finite* number of objects would attract each other, and—given enough time—collapse and get so close to each other as to form a big blob of matter. Only if there is a mass "behind" every mass pulling it in the opposite direction can this collapse of the universe be prohibited; see Figure 6.18(b). In Newton's view it was therefore only sensible to assume that the universe is a static, infinite assembly of masses with neither a beginning nor an end.

The German astronomer HEINRICH OLBERS (1758–1840),[18] surprised his contemporaries in 1823 with the following paradox arising in this Newtonian cosmology: Why is it dark at night if the cosmos is infinite? Indeed, if the cosmos is infinite and filled with stars, then the line of sight eventually will hit the surface of a star, and thus the whole sky should be as bright as a stellar surface. The mathematical reason is the canceling of two scaling behaviors. The apparent size of an object like the sun decreases linearly with the distance from it, as we have seen in Section 1.2.1. Hence, its apparent area and thus its brightness decreases with the square of the distance, see Figure 2.50. On the other hand, at an assumed fixed density of these objects (stars), their *number* increases quadratically with the distance from us. These two quadratic behaviors cancel, and, as can be gleaned from Figure 6.19, eventually the whole sky is filled with stars.

Consequently, at least one of Newton's assumptions must be wrong. Either the universe is not infinitely large, or it is not infinitely old, or it is not static. Newton modeled the

---

[18]He is the discoverer of the asteroids *2 Pallas* (1802) and *3 Vesta* (1807)—but probably not of *Olbers's paradox*.

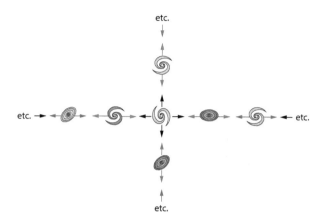

Figure 6.18: **A Newtonian universe has to be infinite.** Behind every mass is another mass pulling it in the opposite direction to stabilize the cosmos.

universe as a "flat piece" of space, extending infinitely and having no temporal boundaries. He imagined it as a *stage* on which the drama of astronomy unfolds in which the celestial objects are the players, so to speak. As OLBERS showed, this cannot be true, even though it sounds perfectly reasonable.

Concept Practice

1. Which is a correct order of the rungs of the cosmic distance ladder (starting from methods to measure the distances of close objects)?

    (a) Hubble's law, stellar parallax, spectroscopic parallax, radar ranging
    (b) Stellar parallax, spectroscopic parallax, radar ranging, Hubble's law
    (c) Radar ranging, Hubble's law, stellar parallax, spectroscopic parallax
    (d) Radar ranging, stellar parallax, spectroscopic parallax, Hubble's law
    (e) None of the above

6.4 The Expanding Universe

6.4.1 Observing the Expanding Universe

We have referenced throughout this book again and again its title, instilling the belief that the universe is expanding. But how do we know that? How is it even possible to measure the expansion—while the observer is *inside* and *part* of the universe? The expansion of the universe was inferred from observations as early as the late 1920s. The **construction of the expanding universe** was a major revolution in astronomy. Initially, the belief in a static universe was so strong that it took a long time before most scientists could bring themselves to accept the idea of an expanding cosmos. For instance, ALBERT EINSTEIN mutilated his own equations describing the space-time structure of the universe to force a static solution before accepting a dynamic, changing universe. British astrophysicist FRED HOYLE fought on until the mid 1960s—upholding hope for a static universe with his alternative cosmological theory; see Section 6.4.3.

6.4. THE EXPANDING UNIVERSE 401

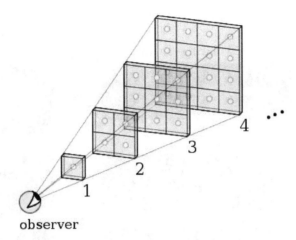

Figure 6.19: **Olbers paradox.** If the cosmos is infinite, we should see more and more stars as we look deeper into the universe. The number of stars grows with their distance squared, whereas their brightness diminishes with their distance squared, cf. Figure 2.50. Therefore, one would expect the whole sky yo be filled with stars, and the night sky as bright as the surface of the sun. This is obviously not the case and constitutes the *paradox*.

Hubble's law: Redshift Is Proportional to a Galaxy's Distance

The observations that led to this pivotal paradigm shift were made rather serendipitously. Starting in 1912, VESTO M. SLIPHER (1875–1969) noticed that many galaxies exhibited a significant redshift of their spectra.[19] This line of research was taken further by EDWIN HUBBLE in the late 1920s at Mt. Wilson. Using the powerful 100-inch telescope that had allowed him to determine the distance to the Andromeda galaxy by using the *period-luminosity relation* of isolated Cepheid variables, he extended his distance measurements to other, more distant galaxies. Seeing that the redshift of the spectra, as well as the distance of the galaxies, varied considerably, Figure 6.20, he sought to find a correlation of the two. A redshift signifies a radial velocity of the object away from the observer, a so-called *recessional velocity*. HUBBLE devised the now-famous **Hubble plot** of the recessional velocities of distant galaxies as a function of the distance to the observer, Figure 6.21. The plot shows unambiguously that recessional velocity v and distance d are linearly related[20]

$$v = H_0 d, \tag{6.2}$$

where the slope is the **Hubble constant** H_0. The current best value is about $H_0 = 70\,\text{km/s/Mpc} = 0.228\,\text{m/s/ly}$. The law holds only for galaxies that are truly distant, that is, farther than about 30 million light-years. Note the dimensions of the **Hubble constant**. Quite intuitively, it is a velocity increase per distance. For every light-year the galaxy is farther out, its velocity increases by $0.228\,\text{m/s} = 0.51\,\text{mph}$. However, the velocity itself is a distance per unit time,

---

[19] The spectrum of a galaxy is taken just like any other spectrum by analyzing light by sending it through a prism or some other *dispersive* medium. If the absorption lines of the chemical elements appear at a longer wavelength (redder color) than in the lab, the spectrum is said to be red-shifted; see the **Doppler effect** in Section 3.7.2.

[20] This relation is usually called **Hubble's law**, but was derived by Belgian priest and astronomer GEORGES LEMAÎTRE (1894-1966) in 1927 from EINSTEIN's equations of *general relativity*. LEMAÎTRE appears in the opening poem of Chapter 5.

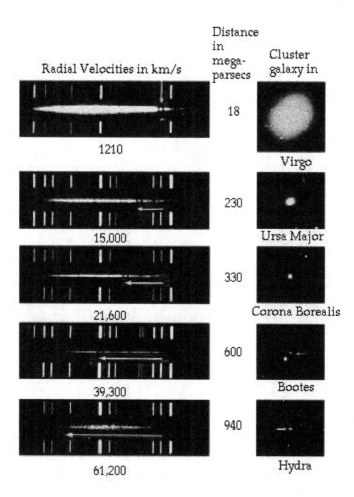

Figure 6.20: **Galaxies and their redshifts.** Shown are the spectra and distances of five galaxies. The spectral lines are red-shifted to longer wavelengths. Their distance is measured with an independent method. It is obvious that large redshift, that is, large *recessional velocity*, correlates with large distance.

and so the simplified dimension of the **Hubble constant** is that of an *inverse* time. The inverse of the **Hubble constant** is the **Hubble time**, characterizing the cosmos itself. What its interpretation is exactly, we shall see shortly. Regardless, the law has the mathematical form of a **direct proportionality**, which means practically that twice the distance yields twice the velocity, and so forth.

The discovery of **Hubble's law** is an observational feat, but its true potential is only realized when interpreted adequately. Interpretations always need a frame of reference, that is, a theory into which they can be embedded. In the case of the **Hubble law** it is EINSTEIN's **theory of general relativity** which describes how space (and time) are deformed by the presence of gravitating objects. With the malleability of the cosmos in mind, we come to a simple explanation as to why a galaxy twice as far should recede twice as fast. Namely, quite like the raisins in a bread get away from each other when the bread expands in the process of baking, Figure 6.22, so do the galaxies recede from each other as **the universe expands**.

Each raisin is now a distance d farther away from its neighbor, and that means, Figure 6.23(a), that the next-to-next galaxy is a distance $2d$ further out. It has thus moved with

6.4. THE EXPANDING UNIVERSE

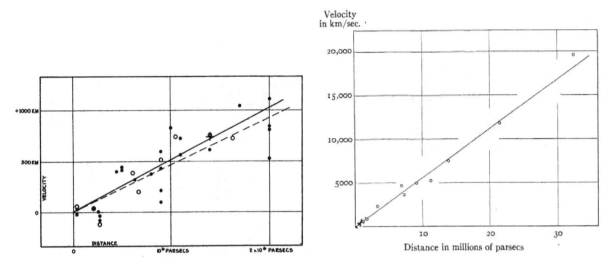

Figure 6.21: **Hubble plot.** (a) EDWIN HUBBLE's original plot from 1929. It was quite audacious at the time to describe the significantly scattered data with a straight line—but its a plausible first approximation. (b) The much-improved 1931 plot of HUBBLE and MILTON HUMASON (1891-1972) which showed impressively that HUBBLE had had the right intuition: Recessional velocity v and distance d are linearly related: $v = H_0 d$, where the slope is HUBBLE's constant H_0.

twice the velocity because it was twice as far out. It may seem that we are at the center of the universe because we observe that all galaxies recede from *us*. This is emphatically not correct. The situation is quite like that of the raisin bread: All raisins recede from all other raisins, and all could claim to be the center raisin! This is logically impossible, and there is no center of the expansion. The expansion is of space itself. That doesn't mean that *everything* is expanding. In fact, the raisins, galaxies, and the Earth are not expanding. The forces within these objects are much too strong. They keep the objects together *against* the expansion of space. In this interpretation, the universe, that is, space itself is stretching and this causes the galaxies to recede. The stretching of space is also responsible for the redshift of the spectra. Therefore, the shifted spectral lines are technically not *Doppler shifted*, which is due to a relative velocity between source and observer *in space*, but **cosmologically redshifted** due to the stretching *of space*, see Figure 6.23(b). As the electromagnetic wave travels in space, space itself stretches and therefore increases the distance between the crests of the wave. Its wavelength gets longer, therefore its color gets redder and its energy—inversely proportional to its wavelength, Equation (6.1)—gets smaller. This makes sense, because, as space expands but energy stays constant in a closed universe due to the **first law of thermodynamics**, the energy density (energy per unit volume) goes down—the content of the universe becomes diluted.

As you can see, a coherent model of the universe arises if one interprets the observed redshift of galaxies as a recessional velocity due to the expansion of the cosmos. But why should one interpret it in this way? Couldn't the universe be a static stage and galaxies just drift away from us? To understand why scientists think that the cosmos itself is expanding, we have to delve a little deeper into the theory that describes the universe itself: **general relativity**.

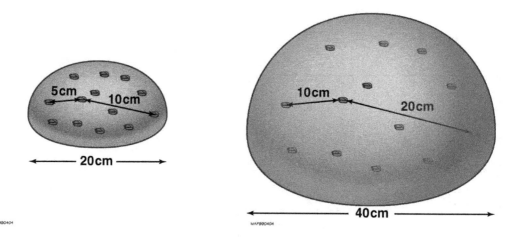

Figure 6.22: **Raisin bread as an analogy of the expanding universe.** As the bread expands in the baking process, the raisins increase their mutual distance by the same factor. Since the distance to a raisin (representing a galaxy) is proportional to the increase of its distance, the *recessional velocity* v is proportional to the distance d, that is, v grows linearly with d, to wit $v = H_0 d$, where H_0 is a constant.

6.4.2 Modeling the Expanding Universe: Standard Cosmology

General Relativity as a Cosmological Theory

It is, of course, completely beyond the scope of this book to provide a "crash course" in **general relativity**. Hopefully, the reader can glean some appreciation of the implication of EINSTEIN's theory of gravity from the following metaphors and analogies.

General relativity replaces the notion of a gravitational force by the paradigm that mass and energy *deform* space and time, as we saw in Section 5.1.1. At face value, this statement is probably incomprehensible. In particular, it is hard to accept that a warped space can have the same effect as a force. This is, however, rather easily demonstrated. Picture space as a stretchable sheet of rubber, Figure 6.24. Initially, it is completely flat, and a light object, like a small marble, will roll across it with a straight trajectory. If a heavy object, say an iron ball, is placed on the sheet, it will deform the rubber surface and make a "dent" in it. This new surface or "space shape" changes the way in which other, lighter objects move on it. The marble, for instance, will now move along a curved trajectory. If we let the iron ball represent the sun and the marble a comet, then we have a good sense of how a warped space has virtually the same effect on the comet as a gravitational force.

We have so far focused on explaining how EINSTEIN's very different view of the universe can reproduce the results of **Newton's gravitational theory** that had been so well tested. Of course, we would not replace NEWTON's theory with a different, more complex theory if the latter wouldn't be "better" in some sense than the former. Indeed, EINSTEIN's theory explains several patterns in nature that Newton's cannot. EINSTEIN convinced himself that *general relativity* was better than NEWTON's gravity theory by computing the effect of its postulates on the orbits of the planets. Naturally, the two theories differ the more the stronger gravity is, and in weak gravitational fields **Newtonian mechanics** is an excellent approximation of general relativity.[21] Mercury is thus the planet that will be affected most by general

---

[21] This is analogous to the statement that *special relativity* reduces to *Newtonian mechanics* at small velocities.

6.4. THE EXPANDING UNIVERSE

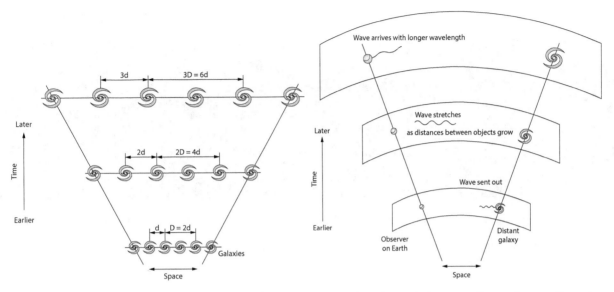

Figure 6.23: **The expanding universe.** (a) One-dimensional analogy of the expanding universe. Several galaxies are initially a distance d apart. A galaxy twice as far out has a distance $D = 2d$. After some time t, the distances have doubled due to cosmic expansion. $d \to 2d$ and $D \to 2D = 4d$. Thus, the galaxy initially twice as far out recedes with a velocity twice as large: $v_D = \frac{D}{t} = 2\frac{d}{t} = 2v_d$. (b) A wave sent out by a distant galaxy travels toward the observer for millions of years. During this time the universe expands, and therefore the wave is stretched to longer wavelength: **Cosmological redshift** of the wave has occurred.

relativity, and its orbit exhibits an effect unexplained as of 1915. Namely, the position of is **perihelion** changes systematically. Most of the perihelion motion can be understood with *Newtonian mechanics* (due to the presence of other planets), but a tiny discrepancy between observation and theory remained—a mere 43 arcseconds per century! But when EINSTEIN applied his theory to the problem, he found to his delight that it predicted exactly this amount of perihelion shift.

The effects of general relativity are exceedingly small under normal circumstances.[22] It does, however, predict new patterns in nature that were completely unexpected, which explains its fast acceptance by scientists, and its role as *the* cosmological theory. Namely, the implications of warped space are fundamental. Firstly, *everything* moving in a warped space will move on a slightly curved trajectory—including light! General relativity predicts therefore a deflection of light akin to the refraction of a lens, Figure 6.24(c). This is known as **gravitational lensing**. EINSTEIN's prediction was confirmed by EDDINGTON who mounted an expedition to take photos of stars near the sun, when the latter was obscured during a total solar eclipse in 1919. Comparing with photos taken of the same star field earlier—when the sun was not in the field—reproduced the **general relativity lensing effect** and turned EINSTEIN into a "star" overnight. Somewhat ironically, EINSTEIN was awarded the Nobel prize in 1921 for his 1905 explanation of the **photoelectric effect**—and not for special or general relativity. Secondly, a warping of space on the grandest scale, namely by the mass of the universe must be taken into account if one wishes to understand the entire universe as an single object.

---

[22] In general, general relativity needs to be taken into account when extreme accuracy is necessary. The global positioning system (GPS), for instance, does not work without taking *general relativity* effects into account.

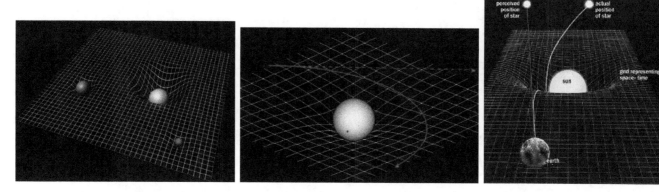

Figure 6.24: **Space-time modeled as a rubber sheet.** (a) Massive objects curve or warp space-time proportional to their mass. (b) Other objects will move on this curved space-time along **geodesics**, which are lines of shortest distance between two points. They are the **straight lines** of curved space. We perceive these paths as curved orbits of planets, and even light (c) moves on curved trajectories.

How does general relativity work in practice? Of course, it involves very advanced math (**differential geometry**), but we can catch a glimpse of its inner workings by using the rubber sheet analogy from above. What we've seen there was put into words by JOHN WHEELER (1911–2008), who did much to advance the understanding of the two relativity theories. He quipped

Mass tells space-time how to curve, and space-time tells mass how to move,

which describes succinctly the implications of general relativity for both the motion of objects and the shape of space (and time). On the symbolic level, the general relativity equation should look something like this

$$(\text{curvature of space-time}) = G \times (\text{mass distribution}), \qquad (6.3)$$

where G is the gravitational constant appearing in Newton's universal gravity law.[23] In other words, the amount of mass and its distribution in the universe determines the curvature, and therefore shape of the cosmos. In short: more mass—more curvature.

This may look like a chicken-and-egg problem, but it is not. For the initiated, this is an equation akin to the Maxwell equations of **electrodynamics**. Recall from Section 3.7 that the Maxwell equations describe how electric and magnetic fields arise due to a distribution of electric charges and currents. With WHEELER, one could say *Charges and currents tell fields how to curve, and fields tell charges how to move and currents how to flow.* This leads to the (solvable) problems of **classical electrodynamics**. In **general relativity**, one starts with a mass distribution, and calculates the ensuing curvature and shape of space (and time). But what is the distribution of masses in the universe? We have seen in Section 6.2 that the

---

[23] Just to get the elephant out of the room, we'll display the mathematical formula that drives the universe, namely the **Einstein field equation**

$$R^{\mu\nu} + g^{\mu\nu} R = 4\pi G T^{\mu\nu}.$$

where $R^{\mu\nu}$ is the curvature tensor of space, and $T^{\mu\nu}$ (the stress-energy tensor) describes how mass is distributed in space. If you just recognize the gravity constant G, don't worry. The idea is that the equation is rather easy to write down, but hard to solve.

6.4. THE EXPANDING UNIVERSE

universe is hierarchically organized on scales up to about 150 million ly. When distances are even larger, galaxies and clusters seem fairly evenly distributed. Driven by the need to solve the Einstein Equation (6.3), we choose the simplest possible mass distribution. Namely, we will *assume* that **the entire cosmos is filled completely uniformly and isotropically with mass.** In other words, we assume that the universe looks exactly the same everywhere and from any angle. This is known as the *cosmological principle*: There is no special place in the universe, which is in some sense the ultimate consequence of the paradigm introduced by COPERNICUS.[24]

Big Bang: The Expansion of the Universe Determines Its Age

Now that we've got the theory under control, we can see why the cosmos itself must be expanding—because the masses in it are moving and tell space how to bend, that is, expand. If the universe is expanding, then it has been smaller in the past! We can now make good on our promise to link the **Hubble time** $1/H_0$ to the cosmos. If we know the velocity with which galaxies are receding at a given distance, we can compute how long they took to get there. More specifically, we can compute the time in the past when all these galaxies that are receding *now* were *here*. This is precisely the *Hubble time*. In other words, in the past the universe must have been much smaller and denser. There must have been a time when the universe was basically infinitely dense and hot and *infinitesimally* small. We call this "start" of the universe the **big bang**.[25]

The amazing realization is that we can compute when the big bang happened, that is, we can determine the **age of the universe**. The reasoning is pretty straightforward. Consider a galaxy far, far away^{TM} at a distance d, and thus receding with a velocity $v = H_0 d$, determined by the **Hubble law**. The time T_0 it took this galaxy to travel this distance d is

$$T_0 = \frac{d}{v}.$$

But plugging in its velocity from the **Hubble law** yields

$$T_0 = \frac{d}{H_0 d} = \frac{1}{H_0},$$

that is, the inverse **Hubble constant** aka the **Hubble time**. The *Hubble time* is the **age of the universe**![26] All that remains is to *convert* the Hubble time into more familiar units.

$$T_0 = \frac{1}{70 \, \frac{\text{km}}{\text{s} \cdot \text{Mpc}}} = \frac{1}{0.228 \, \frac{\text{m}}{\text{s} \cdot \text{ly}}} = \frac{1 \, \text{ly}}{0.228 \, \text{m}} \text{s} = \frac{9.461 \times 10^{15} \, \text{m}}{0.228 \, \text{m}} \text{s} = 4.15 \times 10^{16} \, \text{s} = 13.1 \text{ billion years}.$$

This is an astounding result. Why? Because the universe is so young! Not in absolute terms, but compared to the age of the solar system (4.6 billion years): The cosmos is less than three times older than the Earth. This certainly settles one cosmological question: The universe has a beginning. Which in turn poses the next question: How did the universe come into existence? We will try to answer the question later. Note that the stream of science questions never seems to stop. Whether this is frustrating or exciting is a personal preference.

---

[24]That is, the continual displacement of Earth farther and farther from the center of the universe.

[25]The term was meant to be derogatory when it was coined by FRED HOYLE, inventor of an alternative cosmology, Section 6.4.3, because he didn't accept big bang cosmology.

[26]Tacitly, we have made the assumption that the expansion rate of the universe is constant. This is generally not true, but a good approximation.

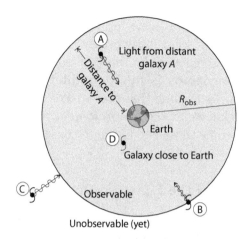

Figure 6.25: **The observable universe.** The Earth receives light from galaxies; that's how we know of their existence. The time t the light takes to get to us depends on the distance d to the galaxy: $d = vt$, where $v = c$ is the speed of light. Since the universe has a finite age, light from galaxy C has not had time to reach us yet, it remains unobservable for now. Galaxy B at a distance R_{obs} has just become observable. This means that R_{obs} is the radius of the observable universe.

The finite age of the universe has an important consequence. Namely, it defines the radius of the ***observable universe***. Since the speed of light is finite, information about celestial objects (like distant galaxies) takes time to get to us, Figure 6.25. Naively, that is, ignoring complications like cosmic curvature and expansion, the radius of the ***observable universe*** is the age of the universe times the speed of light. All we used in this calculation is the simple "distance traveled equals velocity traveled times travel time" formula, $d = vt$, here

$$R_{\text{obs}} = cT_0 = 1\frac{ly}{yr} \times 13.1 \text{ billion yr} = 13.1 \text{ billion ly}.$$

This is the dissolution of ***Olbers's paradox.*** Since the universe is so young, the light of galaxies beyond about thirteen billion light-years cannot get to us, so the night sky does appear dark. Incidentally, it is even darker than that. Light of objects close to the edge of the observable universe is extremely redshifted, and thus much less energetic. These objects thus appear much dimmer than would be expected by their distance alone.

Standard Cosmology

Standard cosmology is the name we give to the results and predictions of the standard theory of the evolving cosmos, namely ***general relativity***. The Einstein Equation (6.3) is pretty complicated, but can be solved with special assumptions. If we invoke the ***cosmological principle*** and assume a perfectly uniform mass density of the cosmos, then the solution depends on only one parameter: how much mass there is in the universe. Of course, in keeping with the mass-energy equivalence, $E = mc^2$, we have to take "energy," for instance in the form of electromagnetic waves, into account, too, when we tally up the content of the cosmos. As we already saw, scientists' attempts to determine the mass/energy content of the universe were doomed to fail until fairly recently, because it is impossible to assess its massive ***dark matter*** content directly.

Therefore, let us follow the historic path and report what scientists "did because they could." The smart thing to do in science is *not* to wait until enough expertise accumulates

6.4. THE EXPANDING UNIVERSE

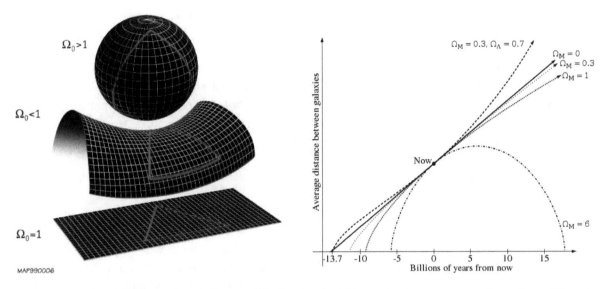

Figure 6.26: **Geometry of the universe.** (a) A universe with constant curvature can have three different geometries: A closed universe of positive curvature is "ball-shaped;" an open universe of zero curvature is flat like a tabletop; an open, negatively curved universe is "saddle-shaped." (b) The average distance between galaxies (a measure of the size of the universe) as a function of time for different density parameters Ω of the universe.

to solve a difficult problem, but to find related, less difficult problems and solve them. The hope is that by working on plausible approximations, eventually the full problem becomes accessible. This is what the scientists did. They were able to solve the EINSTEIN equations by invoking the *cosmological principle*. This way they could predict several *properties* of the cosmos which were in principle measurable, circumventing the need to directly determine the universal mass.

As it turns out, there are *three* cosmological scenarios. This makes sense, because we said the EINSTEIN equations connect mass content with curvature of space-time which can be positive, zero, or negative. The most familiar case is the zero curvature universe. This is also known as a ***flat universe*** with ***Euclidian geometry***. This is the geometry you learn in school, where you use constructions with straight lines on a flat sheet of paper. Most of the results of ***Euclidian geometry*** are so familiar that we don't think about them and simply take them for laws of nature. They are not. For one, they are results of mathematics, and math has nothing to do with nature. What is meant by this provocative statement is that mathematical theorems and their proofs follow directly from its axioms. There is no measurement necessary—and not even allowed nor possible, because the definitions of mathematics are ***abstract***. For instance, if math makes a statement about triangles, it holds for *all* triangles, and not just the few that humans are able to draw or nature is producing as a pattern. In fact, if you draw a triangle, it is bound to be flawed (ruler not absolutely straight, pencil not infinitely thin), whereas a mathematical, abstract triangle is as perfect as an actually existing object can never be. In ***Euclidian geometry***, Figure 6.26(a)[bottom], two parallel straight lines never cross, all other straight lines cross exactly once, and the sum of the angles in a triangle is 180°. What could be more obvious? Yet, none of these three facts are true in a ***curved*** space.

In a ***positively curved space***, Figure 6.26(a)[top], two parallel light beams (representing two straight lines) will ***converge***, that is, cross eventually. The sum of the three angles of a triangle is always greater than 180°. A sphere represents a space of constant positive curvature. In a ***negatively curved space***, Figure 6.26(a)[middle], two parallel light beams (representing two straight lines) will ***diverge***. The sum of the three angles of a triangle is always smaller than 180°. A saddle (***hyperbolic paraboloid***) represents a space of constant negative curvature. Together with the flat space of of zero curvature, these are the three possible shapes of the universe. Since, by definition, the universe *is* all space, it makes no sense to ask about the space outside the universe, or the space which the universe is "curved into." The universes of zero and negative curvatures are clearly without bounds. They are open universes with infinite radius. Contrariwise, the positively curved universe is finite.

Note that there is no standstill: regardless of the shape of the universe, its size will always change. The open universes will eternally expand. The flat universe is the critical case, and its initial expansion will become slow and slower, and eventually it will stop expanding. The closed, positively-curved universe will reach a maximal radius due to its expansion, but after reaching it, the gravitational force remains strong enough to make the matter fall in on itself. This will result in an "inverse big bang" or ***big crunch***; see Figure 6.26(b). In all three cases, the expansion rate of the universe is *not* constant. The expansion slows down because the masses inside the universe are attracting each other, resulting in a decelerating effect. Only in the closed, positively-curved universe is there enough mass to reverse the expansion.

So much for the theoretical understanding of the expanding universe. How about observational evidence? In other words, how can we distinguish the three scenarios? As we said, the amount of mass and energy in the universe is responsible for the curvature, but this is a dead end due to the difficulty of detecting dark matter. We need an independent way of measuring the shape of the universe. The idea is to use the different ways in which light beams travel. In short, the paths of light beams in a curved universe are curved, and this will lead to a discriminating criterion, as we will see in Section 6.5.2.

6.4.3 Steady State—Rise and Fall of the Alternative Cosmology

At the beginning of the twentieth century, the static, eternal universe was still firmly ingrained in astronomers' minds. Therefore, an expanding universe—implying a beginning and constant change—was a hard pill to swallow. It is no wonder that a faction of scientists sought a different explanation of the patterns of nature described by ***Hubble's law***. Around 1948, astrophysicists around FRED HOYLE came up with a psychologically rewarding alternative. The so-called ***steady state theory*** of the universe couldn't deny that the cosmos was expanding, but they propagated a fortified ***cosmological principle***, namely that the universe looks the same everywhere not just in space, but also in time. In other words, it avoided the "thinning out" of the universe, that otherwise could be used as a cosmic clock.[27] The group proposed to make up the diluting effect of the expansion by stipulating that matter is created albeit at a very slow rate. As it turns out, postulating that three hydrogen atoms are created each year in a volume of 1 ly^3 is enough, and basically undetectable on small scales. Of course, the comfort of having a quasi-unchanging cosmos came at a hefty price: energy conservation had to be given up, since each hydrogen atom of mass m_H has a ***rest energy*** of $E = m_H c^2$.

---

[27] The denser the universe, the younger it must be in standard cosmology.

6.4. THE EXPANDING UNIVERSE

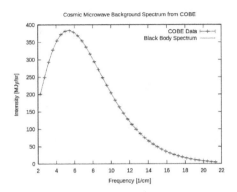

Figure 6.27: **The cosmic microwave background radiation.** Shown is the intensity of the emitted radiation as a function of the frequency. The data points are well described by the Planck curve for an ideal blackbody. The peak frequency of $5.5 \frac{1}{cm}$ corresponds according to Wien's law to a temperature of $2.7\,K$.

The controversy between the two leading cosmologies lasted for close to two decades. For science, it was quite fruitful, because it forced standard cosmologists to sharpen their arguments and look for observational evidence for their extraordinary claims of a curved, expanding universe. Moreover, the term **big bang** for the primordial fireball that started the universe, was coined by HOYLE—as a derogatory label. Already in the 1950s evidence began to mount that the universe is indeed changing. For instance, galactic radio sources are not found near to us, but certainly far from us as the emerging field of radio astronomy (developed after World War II) showed.

The final blow to the steady-state theory came via the surprise discovery of the **cosmic microwave background**. It is a background radiation in that it is getting to us from *all* directions. It was discovered in 1964 by ARNO PENZIAS (born 1933) and ROBERT WILSON (born 1936) as **background noise**. The two scientists were working on a horn antenna, Figure 6.28 (upper left corner), to send radio waves to communication satellites. They tried in vain to "debug" the antenna to get rid of some annoying noise in their signal. They ruled out terrestrial signals, the sun and extragalactic radio sources, because all of those objects are basically *point* radio sources. The two just could not seem to find a source or reason for this extremely *isotropic* radiation. Eventually it became clear that they had found what other scientists had predicted and were looking for: extremely red-shifted radiation from the "edge of the universe." In other words, they had produced evidence that the universe had been extremely hot more than thirteen billion years ago, just after the big bang. Ironically, the radiation had been *predicted* in 1948 by GAMOV, ALPHER and others, but their publications had been largely forgotten, although their ideas of a hot big bang and primordial nucleosynthesis[28] were warmed up in the early 1960s by ROBERT DICKE (1916–1997) and P.J.E. PEEBLES (born 1935). Eventually, a consistent picture of the very young cosmos arose; see Section 6.5.1.

Observationally, the cosmic microwave background has all the characteristics of a blackbody radiation (compare Figure 3.35 with Figure 6.27), with a peak wavelength of a few

---

[28]Recall that the universe contains too much helium to be explained by stars' fusion alone. Primordial nucleosynthesis solves that problem by postulating that the early universe was hot enough to mass produce helium from hydrogen in sufficient quantities. The whole cosmos was hot and dense enough to make this reaction possible. Loosely speaking, the universe was a "stellar core."

millimeters, that is, in the microwave band of electromagnetic radiation, Figure 3.27(b). This corresponds to a blackbody temperature of 2.725 K, and you can tell from the four significant figures that it is incredibly *isotropic*: The hottest spot in the sky is less than 0.0005 K warmer than the coldest spot. These extreme properties constitute a crisp pattern in nature. How can it be explained? The pivotal ideas are surprisingly easy to grasp, while we'll leave the details to Section 6.5.1. Indeed, all one has to do is to imagine an incredibly hot cosmos. Due to the intense radiation, all atoms are ionized. This hot soup of electrically positive atomic nuclei and negative electrons is called a *plasma*. The dilemma is that although there is so much electromagnetic radiation, it cannot go anywhere: No sooner is it emitted as it is reabsorbed by all the electrically charged particles. The plasma is *opaque*, acting like a dense fog. As the universe expands, it cools. In physics, we can calculate the temperature (about 3000 K) when electrons and nuclei combine to form electrically *neutral atoms*.[29] Consequently, the "universal fog" lifted suddenly when the universe hits the 3000 K mark, which corresponds to a precise age (380,000 years) and size of the universe[30] (1,100× smaller than today) according to *standard cosmology*. All of a sudden, light could travel freely—and it did. As the cosmos grew 1,100 times bigger, so did the wavelength of the 3000 K waves, and thus its associated temperature dropped by the same factor to $\frac{3000\,K}{1,100} = 2.73\,K$, precisely the temperature of the cosmic microwave background. This convincing interpretation of the observational evidence for the existence of a hot big bang ruled out the alternative *steady state theory* for good.

Concept Practice

1. Standard cosmology predicts the shape of the cosmos depending on its content. Why is this?

    (a) Due to the mathematical restrictions of the model.
    (b) Masses attract, hence they curve space-time.
    (c) The inertia of objects keeps them going after the big bang.
    (d) None of the above.

2. An closed universe

    (a) ... has a lot of mass in it.
    (b) ... has too little mass.
    (c) ... has the right (critical) amount of mass.
    (d) None of the above

6.5 The Failure of Standard Cosmology at the End of the Millennium

6.5.1 History of the Universe

Standard cosmology relates several properties of the universe, such as its size, temperature

---

[29] The reason is that temperature is a proxy for energy, see Section 3.7.3. This is a *phase transition*, similar to water *boiling* at 100 °C = 373 K. As molecules leave the *liquid phase* and enter the *gas phase*, also the reverse reaction happens, and an equilibrium is established. The ratio of the number of molecules in gas versus liquid phase depends delicately on the temperature. Under standard conditions, we have mostly liquid below 100 °C and mostly gas above 100 °C. Analogously, we have mostly neutral atoms below 3000 K and mostly plasma above 3000 K.

[30] "Size of the *observable* universe" to be precise, as the universe itself might be infinite.

6.5. THE FAILURE OF STANDARD COSMOLOGY

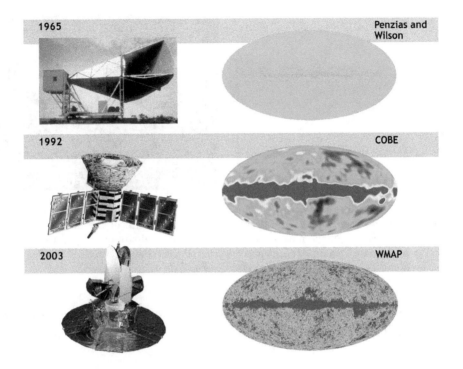

Figure 6.28: **Evolution of CMB measurements.** (a) PENZIAS and WILSON measured a completely isotropic temperature with their *horn antenna*. (b) The *COBE* satellite in 1992 found the first temperature differences. (c) the *WMAP* satellite in 2003 improved the resolution of the CMB fluctuation crucially.

and age. Therefore, we can give an account of *what* happened *when* in the cosmos. This is nothing but a ***history of the universe***. As we have seen, a pivotal date in this history occurred 380,000 years after the big bang, when the temperature of the cosmos dropped below 3,000 K, and the universe became ***transparent*** for electromagnetic radiation. In other words, the universe started to be ***observable***, although it would take several millions of years before the first stars would start to shine and galaxies would form. This formation of ***large-scale structures*** will be considered in Section 6.5.2, because it gives us important clues about the *content* of the universe.

First, we will try to get a sense of what happened *before* that time, that is, we will try to get as close to the big bang as we may. Recall that the early universe was so hot that the atoms "boil off" their electrons, and the whole cosmos is a plasma, a hot soup of electrically charged particles. Therefore, the young universe can be considered a single black body, and the peak wavelength of its radiation is inversely related to its temperature according to ***Wien's law***. In other words, the size of the cosmos and its temperature are inversely related, Figure 2.49: A universe half the size is twice as hot! Therefore, standard cosmology predicts a ***hot big bang***, because the energy contained in the universe is constant,[31] which was squeezed into a small space close to the big bang.

In other words, getting closer to the big bang means considering a hotter universe. We therefore have to understand what happens to matter at extreme temperatures to understand what the universe looked like shortly after the big bang. As high temperature is virtually

---
[31]This follows from the ***first law of thermodynamics***, because the cosmos is by definition an isolated or closed system.

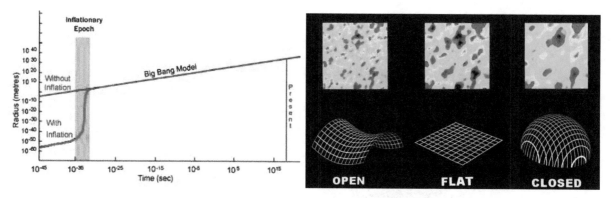

Figure 6.29: **Inflation and the curvature of the cosmos.** (a) The "radius" of the universe as a function of time. In the *inflationary epoch* shortly after the big bang, the size of the universe increased at a very large factor. (b) The curvature of the universe determines the typical size of temperature fluctuations of the cosmic microwave background. In an open, negatively curved universe they appear smaller than in an open, flat universe. The fluctuation size is biggest in a closed, positively curved universe.

synonymous with high energy, we have to look to **high energy or particle physics** for answers. Most of this is beyond the scope of this book, but the two main ideas should be understandable.

1. Close to the big bang the macrocosmos of astronomy turns into the microcosmos of particle physics.

2. As the temperatures get higher closer toward the big bang, structures are "melting," and constituents of **bound systems** emerge as free particles.

Quite like ice becoming water at higher temperatures and becoming steam when it gets even hotter, atoms "melt" into nuclei and electrons, and nuclei dissolve into *their* constituents (protons and neutrons). You can imagine that even protons can "melt" at high enough temperatures, and their constituents, called *quarks*, become free. We end up with an incredibly hot soup of **elementary particles**.

From this it should be clear that we can understand the cosmos as close to the big bang as we can understand (particle) physics at high energies. Currently, we understand physics up to energies which correspond to temperatures of 10^{27} K. Such were the temperatures when the universe was only 10^{-35} seconds old; see Figure 6.33 [bottom]. This is incredibly close to the big bang! To get even closer, physicists would have to merge **quantum physics** with **general relativity**. This hypothetical theory, sometimes called the **theory of everything**, has still to be found.

Undeterred by this intellectual brick wall, pragmatic scientists pointed out in the 1970s that there is another problem with this extrapolation of the universe to higher energy, temperature, and smaller size: The universe is too uniform! Judging from the incredible isotropy of the microwave background, the universe looked **exactly** the same **everywhere**! In other words, the universe was perfectly stirred or mixed. This is a problem, because the early universe did not yet exist long enough to allow for adequate mixing. If the "left side" of the universe has exactly the same temperature as the "right side," then "temperature information" has to be exchanged—which takes time, given the finite speed of light at which any information travels. You might think that this is not a problem, because the early universe

6.5. THE FAILURE OF STANDARD COSMOLOGY

was so small, but you would be mistaken, because it was also incredibly *young*, as scientists realized. There again is a pattern of nature that needs an explanation: Why is the universe so isotropic even though it's too big to mix? The currently accepted explanation is the so-called theory of **cosmic inflation**. It postulates that there was a very brief period in the very early universe in which the universe suddenly expanded by an incredibly large factor of 10^{50}, Figure 6.29(a). The main benefit of such a scenario is that the cosmos in this theory starts out much smaller than in non-inflationary cosmology. Therefore, mixing becomes possible, and the isotropy of the microwave background and of matter on the largest scales is explained. Inflation also predicts a **flat universe**, which was confirmed later; see Section 6.5.2. On the other hand, it is not clear what *caused* **inflation**. In this sense, the theory is in its **Keplerian phase**, that is, it describes *what* is happening (sudden increase in size), but does not explain *why* it happens.

6.5.2 The Golden Age of Precision Cosmology

It is obvious that cosmological questions are very hard to answer, because they concern the universe as a whole. Until fairly recently, measurements in observational cosmology were plagued with uncertainties in excess of 100%.[32] For instance, the 1950s were a time when cosmologists held that the age of the universe was 10 billion years (not too far off the modern value of 13.7 billion years), but other astronomers made the claim that some **globular star clusters** are 12 billion years old. Clearly, a paradoxical situation. Then, in the early 1990s astronomy entered an era of precision cosmology which is still ongoing. Mostly fueled by space-borne instruments, break-through results became almost commonplace, and the age of the universe is now known to three significant figures, that is, with an uncertainty of less than 1%. *Improving the accuracy of the cosmic microwave background map* probably is most relevant to our current understanding of cosmic expansion. The story unfolded as follows.

The results of the *COBE* space probe in 1992 provided a breakthrough almost thirty years after Penzias and Wilson's discovery. Instrument sensitivity had become high enough to map the 1 in 100,000 fluctuations of the cosmic microwave background. These tiny temperature fluctuations provide important clues about the universe, and determine its shape and content. The next step came with two competing balloon experiments starting in 1997. *BOOMERanG* and *MAXIMA* improved the resolution of the cosmic microwave background map to an extent that statistical analysis of the *size* of the fluctuations became possible. This in turn made it possible to deduce the **curvature of the universe**, as described in Figure 6.29(b). In essence, a different curvature will lead to a different size of the fluctuations: the bigger the curvature, the bigger the observed fluctuation size–the universe acts like a lens. In the late 1990s it became thus clear that we live in a flat, open universe. Of the three standard cosmologies, Section 6.4.2, this is the *critical case*, where *exactly* the right amount of matter balances cosmic expansion and gravitational contraction. Observing this delicate pattern in nature, the natural question arises: Why is the universe so finely tuned? We saw that **inflation theory** explains this pattern at least partially. The cosmic microwave background map was crucially improved upon once again in the mid 2000s with the *WMAP* space probe, Figure 6.28(c). Finally, the *PLANCK* satellite in 2013 provided the most accurate results to date.

---

[32]It is thus a factor of two by which results differed.

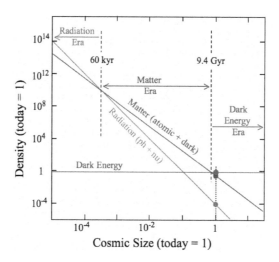

Figure 6.30: **Cosmic energy density.** Shown is the development of the energy densities of the three ingredients of the universe. **Radiation density** falls off fastest, and was dominant in the early universe. **Matter density**, understood as atomic plus dark matter, was dominant until very recently. Note that it falls off like the volume of the universe, so like (Cosmic Size)^3. Since the density of **dark energy** is constant, it is poised to become (and stay) dominant at some point in time. We live in a **dark energy**-dominated universe.

6.5.3 The Big Shock: The Universe Is Accelerating

While the previous section paints a picture of continued improvements of our understanding, an unexpected paradigm shift occurred in 1998. At this time, two independent groups of **observational cosmologists** published their findings about the distances of **Type Ia supernovae** at very large distances. They used the supernovae as **standard candles**. Namely, the supernovae's known luminosities give away their *distances* via the brightness-luminosity-distance relation, Equation (5.3). The red-shift of their spectra on the other hand determines their **recessional velocities**. Measuring these two quantities independently and plotting them, Figure 6.31, allows the astronomers to distinguish between different **cosmologies**.[33] As you can see, the data points are somewhat scattered, but they are definitely in the part of the diagram that implies that the expansion of the universe is not slowing down but speeding up. The data is consistent with the statement: We live in an accelerating universe! This is in stark contradiction to the three possible scenarios of standard cosmology. The supernovae measurements have falsified standard cosmology; the theory needs to be abandoned or modified.

Before jumping the gun, let us analyze the observations. What the researchers found was **standard candles** dimmer than expected for their distance or redshift. Indeed, they are dimmer than can be explained by a universe that is slowing down its expansion. Since any standard universe slows down its expansion due to the gravitational attraction between its objects, there must be something else in this universe, something that counteracts gravity, some kind of "anti-gravity." As of yet, scientists do not know what it is. In fact, they did not even come up with a consistent theory of what it *could be*. But they know what it does,

---

[33] As before, **cosmologies** refers to different theories explaining the cosmos as a whole, as in "ARISTOTLE's cosmology involving the four elements plus quintessence." At this point, the accepted master theory is **general relativity**, and so different cosmologies differ merely by the values of parameters such as mass density and dark matter content.

6.5. THE FAILURE OF STANDARD COSMOLOGY

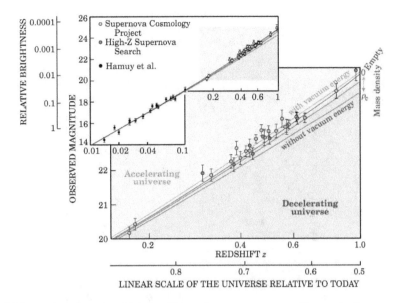

Figure 6.31: **Hubble-style plot for distant Type Ia supernovae.** Shown is the *apparent brightness* (synonymous with *observed magnitude*) of very distant supernovae as a function of their redshift z. The latter is a measure for the recessional velocity of a supernova, while its brightness is a measure of its distance via Equation (5.3) since we know its *luminosity*. The data points lie mostly in the blue region of the plot, which is evidence for an *accelerating universe*. In other words, the supernovae appear dimmer than they should for a decelerating universe. Only an acceleration of the cosmic expansion can explain why the supernovae look so dim.

and they gave it a name: **dark energy**. The name makes sense, because this "substance" is not detectable by conventional detectors (like **dark matter**), and (unlike **dark matter**) it accelerates the expansion of the universe. Mathematically, the accelerating effect of **dark energy** can be described by modifying EINSTEIN's equations. EINSTEIN himself did so in 1917 to rein in gravity to achieve a static universe. When HUBBLE later discovered that the universe is not static, he considered the introduction of a cosmological constant Λ his "biggest blunder." But in 1998 the **cosmological constant** was back! **Dark energy** acts like a cosmological constant, in that its density is a property of space itself, meaning its density stays constant, even though space expands. Since everything else *in* space dilutes over time, **dark energy** will become the dominant component of the universe. In fact, this has already happened. We live in a dark energy-dominated universe; see Figure 6.30.

How do we know? This question gives us the chance to critically reexamine the supernovae data above—which is part of the clue. Figure 6.31 shows that the data are best fit by a curve that depends on two parameters: the **matter density** (which includes dark and ordinary matter) and the **dark energy** content of the universe. Therefore, we know the relative weight of these two "substances." We can cross-check this result—as we should—by comparing to other observations. Above, we mentioned that the *BOOMERanG* and *MAXIMA* balloon experiments had established that the universe is flat, that is, has zero curvature. According to general relativity, this means that the universe contains exactly the critical mass-energy density. Often, the actual density ρ of the cosmos is expressed in units of the critical density ρ_c. This number is called $\Omega_0 = \frac{\rho}{\rho_c}$ and has three contributions: ordinary or baryonic matter plus dark matter Ω_{Matter}, dark energy Ω_Λ, and relativistic particles (light and neutrinos) Ω_{Rel}. The latter, however, is presently incredible small (suppressed by a factor of about 10,000), so

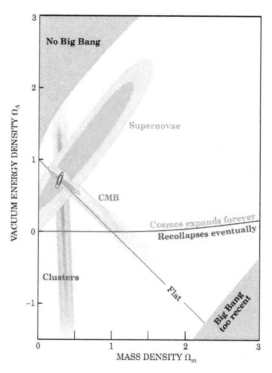

Figure 6.32: **Cosmological parameter space.** The vacuum energy density Ω_Λ and the matter density Ω_m define the properties of the cosmos. The values for these two parameters have to be determined by observations. Shown are three independent measurements: distant supernovae, galaxy cluster inventories, the cosmic microwave background. All three taken together confine the values for the cosmic densities to a very small area in parameter space around $\Omega_\Lambda = 0.7$ and $\Omega_m = 0.3$.

we can write
$$\Omega_0 = \Omega_{Matter} + \Omega_{Rel} + \Omega_\Lambda \approx \Omega_{Matter} + \Omega_\Lambda.$$

$\Omega_0 = 1$ represents a flat universe, $\Omega_0 > 1$ a closed universe ("sphere"), and $\Omega_0 < 1$ an open universe ("saddle"). In a flat cosmos, the contributions of matter and dark energy have to add up to one. This is evident in Figure 6.32, where the content of the universe is plotted (with uncertainties represented by ellipses) from three independent observational fields: properties of the cosmic microwave background, supernovae data, and large scale structure as evident in galaxy clusters. Surprisingly, all three methods to determine the content of the universe agree: We live in an accelerating universe dominated by dark energy.

This ends our quest of constructing the expanding (accelerating) universe. While we have not yet been able to explain what **dark matter** and **dark energy** are, we do know what they are "doing." We truly have come a long way in deciphering what the night sky is telling us about the universe we live in—and its history, Figure 6.33!

Concept Practice

1. The universe presently is

    (a) ... slowing down its expansion.
    (b) ... flat.
    (c) ... dominated by dark matter.

6.6. SUMMARY AND APPLICATION

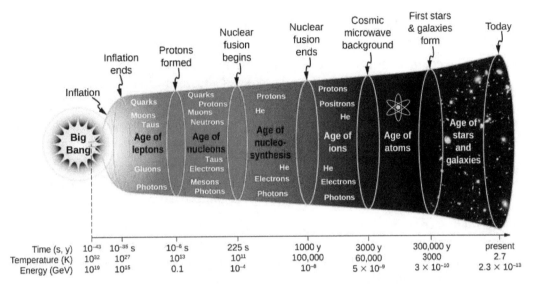

Figure 6.33: **History of the universe.** Since the big bang, the universe has expanded and cooled down. As the temperature dropped, structures have been able to form. The formation of atoms (reionization) allowed light to travel freely: The universe became transparent about 380,000 years after the big bang.

   (d) ... radiation dominated.
   (e) None of the above

2. In a universe with accelerating expansion distant supernovae

   (a) ... appear brighter.
   (b) ... appear dimmer.
   (c) ... appear bigger.
   (d) None of the above

6.6 Summary and Application

6.6.1 Timetable of the Emerging Expanding Universe (IV)

The main steps toward an understanding of the expanding universe discussed in this chapter are described below. As always, schedules like this should be taken with a grain of salt.

- 1915 EINSTEIN rolls out *general relativity*, his version of universal gravitation; there is no force, only curvature of space and time.

- 1916 Exact solutions of EINSTEIN's field equations of *general relativity* are derived (KARL SCHWARZSCHILD).

- 1919 Expedition to test general relativity with solar eclipse successful (EDDINGTON).

- 1929 A linear relation between the distance to remote galaxies and their recessional velocity is established (HUBBLE); the universe is found to change. Namely, it is expanding.

- c. 1950 The term *big bang* is coined (HOYLE) to ridicule the idea that the universe started out as an infinitely small and hot fire ball; the term stuck and is still used to vividly describe this beginning of the universe, later supported by several independent pieces of evidence.

- 1964 The *cosmic microwave background* is found (PENZIAS, WILSON); the universe has a temperature of 2.7 K; it is interpreted as leftover radiation from the **big bang** and therefore *direct evidence* for it.

- c. 1950 An alternative to the expanding and therefore diluting universe is proposed (HOYLE); the **steady state** universe does not change, because it is hypothesized that matter is generated as the universe expands such that its density stays constant.

- 1979 The concept of **inflation** is invented to explain the flatness of the cosmos and the homogeneity of the cosmic microwave background.

- 1998 The acceleration of the expansion of the universe is gleaned from the fact that very distant supernovae appear dimmer than they would in a universe whose expansion is constant or slowing down; this makes necessary an agent driving this acceleration; it is called **dark energy**.

- Presently: Nobody knows what dark energy *is*, only what it *does*.

6.6.2 Concept Application

1. Draw a *Hubble plot* showing the recessional velocities of distant galaxies as a function of their distance.

   (a) For a universe expanding at a constant rate.

   (b) For a universe expanding at a higher constant rate.

   (c) For a universe with an expansion rate that is slowing down.

   (d) For a universe with an expansion rate that is accelerating.

2. Calculate the *Hubble time* (inverse Hubble constant in units of time) for a universe where $H_0 = 100 \frac{\text{km}}{\text{s} \cdot \text{Mpc}}$.

6.6.3 Activity: The Milky Way

In this activity, we want to get a sense for the dimensions and shape of our galaxy, the Milky Way. The dimensions of the Milky Way are as follows. The diameter of the **disk** is 100,000 ly. The **central bulge** is shaped like a football with long diameter 15,000 ly parallel to the disk, and short diameter 4000 ly. The sun is in the galactic disk of thickness 1,000 ly about 26,000 ly from the center. The Milky Way has a spherical **halo of globular star clusters**. The disk contains a lot of gas and dust, whereas the central bulge contains some, and halo doesn't contain any.

1. Use ruler and pencil to draw the Milky Way on a white sheet of paper face on (bird's eye view). The scale should be 2 cm = 10,000 ly.

2. Use ruler and pencil to draw the Milky Way including some globular star clusters on a white sheet of paper edge-on (from the side).

3. Draw in the position of the sun on both sketches.

4. Indicate in the drawings how galactic latitude is measured.

6.6. SUMMARY AND APPLICATION

5. What is the galactic latitude of the sun?
6. What is the galactic latitude of most stars in the galaxy?
7. What is the galactic latitude of the globular star clusters?
8. In which parts of the galaxy (halo, disk, bulge) do stars form at the highest rate? Rank the regions and explain.
9. The bright stars we see in the night sky are very far from Earth—very far compared to the sun, that is. Recall that a lightyear is about 63,000 AU, or 63,000 times the distance to the sun. Stars in a scaled-down model can be represented by golf balls which are about 100 miles apart. Is it likely that stars collide?
10. Stars very far away by that standard are 500 ly to 1000 ly from the sun. Draw them into the sketch to scale.
11. Objects that are farther than these stars must be very luminous so we can see them. Some of these deep sky objects include M42 (Great Orion Nebula) 1500 ly, M14 (globular star cluster) 30 000 ly. Draw them in to scale.
12. How do we know that these distant objects are still part of the Milky Way?
13. The Andromeda Galaxy is about the same size as the Milky Way and 2.5 million ly away. Draw it on another piece of paper and put the two pieces of paper it in an appropriate distance to each other.
14. How likely is it that these two galaxies could collide? Compare to the collision probability of stars from above.

6.6.4 Activity: Hubble's Law

Hubble's law relates the *recessional velocity* of galaxies (how fast they move away from us) to their distance from us. It was discovered by EDWIN HUBBLE in the late 1920s by correlating the redshift of spectral lines of galaxies with their distances as determined by the Cepheid period-luminosity relation. Hubble's law states that there is a direct proportionality between recessional velocity and distance, that is, the graph of velocity versus distance is a straight line with a constant slope, the so-called Hubble constant H_0.

1. Draw a Hubble plot by putting an appropriate straight line into a two-dimensional coordinate system. Should the line go through the origin? Explain.
2. Now label the axes and commit to some scale. Typical scales are tens of thousands of km/s on the vertical and a few megaparsecs (1 Mpc = 3.26 million light-years) on the horizontal axis.
3. Calculate your Hubble constant.
4. Pick an example showing how the Hubble plot works: How can you get from the velocity of a galaxy to its distance?
5. How did Hubble get from measuring redshifts to determining the recessional velocity of galaxies?
6. Draw two other Hubble plots: one with a larger Hubble constant, and one with a smaller Hubble constant.
7. What is different about these universes?

6.6.5 Activity: Determining Hubble's Constant

Since Hubble's law is linear, that is, can be represented by a straight line, we can determine Hubble's constant (the slope of the line) by measuring the distances and recessional velocities of merely two galaxies. Galaxy A is 925 Mpc away and has a recessional velocity of 57 500 km/s. Galaxy B is 231 Mpc away and has a recessional velocity of 16 700 km/s.

1. First, let's get a sense of what the graph looks like. Plot the data in a velocity versus distance (not a distance versus velocity graph!) on a sheet of paper with pencil and ruler, and draw a straight line through the two data points. Make the graph as big as possible.

2. Determine the slope of the graph graphically. Note that the slope carries units, because it is the velocity (increase) per unit distance (increase). Do this by constructing a right triangle with a hypotenuse that is part of the straight line. The bigger your triangle is, the better you can measure the lengths of its sides with a ruler.

3. Determine the slope of the graph mathematically by using the formula for the slope, that is, Hubble constant: $H = \frac{v_2 v_1}{d_2 d_1}$, where the index 2 refers to the bigger values.

4. Does the straight line go through the origin? Should it? Explain.

5. The accepted value is 68 km/s/Mpc. What is the percent discrepancy of your result? Use
$$\text{(percent discrepancy)} = \frac{|\text{your value} - \text{accepted value}|}{\text{(accepted value)}} \times 100\%.$$

6.6.6 Activity: The Expanding Universe

One of the most important observation of cosmological relevance is the **Hubble law** discussed in the last activity. The direct proportionality between the recessional velocity and the distance of a galaxy means that a galaxy twice as distant as a reference galaxy moves away from us twice as fast as the reference galaxy. What is the interpretation of this pattern in nature? This is the subject of this activity.

1. Consider a one-dimensional universe, with five galaxies all a distance d from their neighbors. For definiteness, let's say $d = 1$ billion ly. Sketch this universe using a scale where 1 billion light-years is half an inch and label the galaxies A,B,C,D,E, with C as the middle galaxy.

2. Now imagine that this universe has expanded to twice its size, as if the galaxies where mounted on a rubber string that was stretched to twice its size. Draw this expanded universe.

3. Make a table indicating by how much each galaxy has moved from a reference point. Take the middle galaxy C as this reference point.

4. Come up with a simple law that describes the distance change (and thus the velocity) of a galaxy as a function of its distance from galaxy C.

5. Is this law compatible with Hubble's law? Why or why not?

6. Is the galaxy C special because all other galaxies are moving away from it? Redo the table of item 3—this time with galaxy B as the reference point.

6.6. SUMMARY AND APPLICATION

7. What is the Hubble constant in this universe if the time it took the universe to double its size is one billion years?

8. What is the Hubble constant in this universe if the time it took the universe to double its size is two billion years?

9. How is the Hubble constant related to the expansion rate of the universe?

10. Now on to something different. It seems curious that the universe should always expand at the same rate, that is, that the Hubble constant is a constant. To see if this is strange or natural, let's explore two scenarios. In the first scenario, early in the universe's history, galaxies were very close to each other. If they had some velocity relative to each other making their mutual distances larger and larger, what would the effect of large gravitational forces between the galaxies be? Would they speed up, slow down, or leave the velocities the same?

11. In scenario two, late in the universe's history, the galaxies are far away from each other. How do the forces and their effect differ from scenario one?

12. What is the conclusion? Is it strange or natural that the Hubble constant is constant?

13. Draw a Hubble plot in which the Hubble constant is not constant. Specifically, assume that the Hubble constant is large initially, and then gets smaller over time.

14. Let's call the distance between two very distant galaxies, say the Milky Way and a galaxy at the edge of the observable universe the *radius of the universe*. As time goes by, how does the radius of the universe change?

15. Draw the radius of the universe as a function of time. Because the expansion of the universe slows down due to mutual gravity of its galaxies, there are three possibilities. Either the universe expands forever, or it eventually stops expanding, or the universe expands up to a point and then shrinks or re-collapses. Draw these three curves.

16. The "fate of the universe" depends on which curve the universe is on. This is determined by the amount of gravitational force between galaxies. On which parameter does thus the fate of the universe depend?

6.6.7 Activity: The Expanding Ballooniverse

The expansion of the universe is not an easy concept. After all, we are in it, so how can we see the universe expanding? Here is a simple model of an expanding universe: a balloon that is being inflated.

1. With a marker, draw some galaxies on your balloon, then inflate the balloon and watch how it and everything on it gets bigger. Describe what happens.

2. Describe the model and which feature of the model corresponds to what feature of the universe, e.g. the skin of the balloon is ...

3. A model is just a model. No model correctly describes all aspects of the real thing otherwise it would not be the model but the real thing. Describe the shortcomings of your model universe.

4. Is there a center of the expansion? If yes, where is it? If not, why not?

5. If the balloon represented an actual universe, would its mass density above, below or at the critical density? How can you tell?

6.6.8 Activity: Space-Time Curvature

In this activity, you will explore the consequences of *non-Euclidian geometry*.

Task 1A: Straight Lines and Triangles—Euclidian Geometry

Take a sheet of paper and draw three straight lines with a ruler, such that you obtain a triangle. Measure the three angles of the triangle and add them up. What is the sum of the angles? Repeat this twice with different triangles. Do the results agree?

What you just performed is a task in **Euclidean geometry**, or geometry in the plane.

Task 1B: Straight Lines and Triangles on a Ball—Non-Euclidian Geometry

How can we define *straight* lines on a *curved* object? It is possible. The easiest example are those on a sphere, or on a ball. Straight lines on a sphere are those that define a plane that goes through the center of the ball. Too complicated? A rubber-band will do it for you automatically. Wrap a rubber-band around the ball. If it doesn't slip to the right nor to the left, you have a "straight line" on the sphere.

Now take another two rubber bands and form two more straight lines, so that all three rubber bands intersect. You'll get a couple of areas that look like triangles. They are called "spherical triangles". Here is the challenge: try to measure the three angles in these triangles and add them up. What do you find this time for the sum? Do this for a couple of triangles and check your results.

Task 2: Explore the world of non-Euclidean geometry

In Euclidian geometry, the angles of a triangle always add up to exactly 180°. That's what you showed in Task 1A. Also, in this geometry two non-parallel straight lines intersect only once, and parallel lines never intersect. These statements are sometimes called axioms of Euclidian geometry.

Neither of these axioms hold in non-Euclidean geometry, that is, on a surface that is not flat but curved. The amount of curvature determines what the sum of three angles α, β, γ of a triangle is. Take the ball: we say it is positively curved because the sum of the angles in a triangle on this surface is always greater than 180°. Sometimes only a bit, sometimes a lot bigger. An example for a negatively curved surface is a saddle (or hyperbolic paraboloid as the mathematicians call this kind of surface), which has $\alpha + \beta + \gamma < 180°$.

Now, go ahead and disprove the second axiom, that straight lines only intersect once. Wrap two rubber-bands around the ball and try to find a "biangle", that is, a shape with only two angles and two sides. The sides will be two straight lines that intersect twice!

6.6. SUMMARY AND APPLICATION 425

Image Credits

Fig. 6.1: Source: https://commons.wikimedia.org/wiki/File:M33_rotation_curve_HI.gif.

Fig. 6.2: Source: https://www.scratchapixel.com/lessons/procedural-generation-virtual-worlds/simulating-sky/simulating-colors-of-the-sky.

Fig. 6.3a: Copyright © 2014 by T. Slater and R. Freedman.

Fig. 6.3b: Copyright © 2014 by T. Slater and R. Freedman.

Fig. 6.4: Copyright © 2012 by RJHall, (CC BY-SA 3.0) at
https://commons.wikimedia.org/wiki/File:Milky_way_profile.svg.

Fig. 6.5: Source: https://mwmw.gsfc.nasa.gov/mmw_images.html.

Fig. 6.6: Copyright © 2014 by T. Slater and R. Freedman.

Fig. 6.7a: Copyright © 2014 by T. Slater and R. Freedman.

Fig. 6.7b: Copyright © 2014 by T. Slater and R. Freedman.

Fig. 6.10: Source: https://commons.wikimedia.org/wiki/File:NGC_4414_(NASA-med).jpg.

Fig. 6.11: Copyright © 1978 by Vera Rubin, W. Kent Ford, and Norbert Thonnard. Fig. 6.12a: Source: https://commons.wikimedia.org/wiki/File:Large.mc.arp.750pix.jpg.

Fig. 6.12b: Source: https://commons.wikimedia.org/wiki/File:HubbleTuningFork.jpg.

Fig. 6.13: Copyright © 2013 by ESA/Hubble and NASA, (CC BY 4.0) at
https://commons.wikimedia.org/wiki/File:Antennae_Galaxies_reloaded.jpg.

Fig. 6.14a: Source: http://imgsrc.hubblesite.org/hvi/uploads/image_file/image_attachment/2703/full_jpg.jpg.

Fig. 6.14b: Source: http://imgsrc.hubblesite.org/hvi/uploads/image_file/image_attachment/2723/web_print.jpg.

Fig. 6.15a: Copyright © 2011 by Andrew Z. Colvin, (CC BY-SA 3.0) at
https://commons.wikimedia.org/wiki/File:5_Local_Galactic_Group_(ELitU).png.

Fig. 6.15b: Copyright © 2011 by Andrew Z. Colvin; adapted by Uwe Trittmann, (CC BY-SA 3.0) at https://commons.wikimedia.org/wiki/File:6_Virgo_Supercluster_(ELitU).png.

Fig. 6.16: Adapted from: https://commons.wikimedia.org/wiki/File:2MASS_LSS_chart-NEW_Nasa.jpg.

Fig. 6.17: Source: http://www.2dfgrs.net/2dFzcone_big.jpg.

Fig. 6.19: Copyright © 2009 by Htkym, (CC BY-SA 3.0) at
https://commons.wikimedia.org/wiki/File:Olbers%27_Paradox.svg.

Fig. 6.20: Source: http://teacherlink.ed.usu.edu/tlnasa/reference/imaginedvd/files/imagine/YBA/M31-velocity/images/galactic_redshift.gif.

Fig. 6.21a: Source: https://inspirehep.net/record/1236611/files/FigHubblePlot1929.png.

Fig. 6.21b: Source: http://adsabs.harvard.edu/abs/1931ApJ....74...43H.

Fig. 6.22a: Source: https://commons.wikimedia.org/wiki/File:Raisinbread.gif.

Fig. 6.22b: Source: https://commons.wikimedia.org/wiki/File:Raisinbread.gif.

Fig. 6.24a: Source: http://sci.esa.int/science-e-media/img/72/ESA_LISA-Pathfinder_spacetime_curvature_above_orig.jpg.

Fig. 6.24b: Source: http://www.physicsoftheuniverse.com/photo.html?images/relativity_curved_space.jpg&Gravity%20causes%20space-time%20to%20curve%20around%20massive%20objects.

Fig. 6.24c: Source: https://scienceisbrilliant.files.wordpress.com/2013/10/space-time1.gif.

Fig. 6.26a: Source: https://commons.wikimedia.org/wiki/File:End_of_universe.jpg.

Fig. 6.26b: Source: https://commons.wikimedia.org/wiki/File:Friedmann_universes.svg.

Fig. 6.27: Source: https://commons.wikimedia.org/wiki/File:Cmbr.svg.

Fig. 6.28: Source: https://map.gsfc.nasa.gov/media/081031/081031_3000W.jpg.

Fig. 6.29a: Source: http://astronomy.swin.edu.au/cms/cpg15x/albums/userpics/bigbang2.jpg.

Fig. 6.29b: Source: http://www.kcvs.ca/martin/astro/au/unit6/153/chp15_3_files/wmap_geometry.jpg.

Fig. 6.30: Source: http://people.virginia.edu/~dmw8f/astr5630/Topic16/t16_density_evol.gif.

Fig. 6.31: Source: https://physicsforme.files.wordpress.com/2011/10/perlmutter3.jpg.

Fig. 6.32: Source: https://physicsforme.files.wordpress.com/2011/10/perlmutter5.jpg.

Fig. 6.33: Source: https://commons.wikimedia.org/wiki/File:CMB_Timeline300_no_WMAP.jpg.

Appendix A

Useful Data

A.1 Constants

Mathematical Constant(s)

These are pure, abstract constants, the same in any universe conceivable.

- $\pi = 3.1415926535897932384626433832795028841971693993751058209749445923078\ldots$ is the ratio of the circumference C to the diameter d, that is, $\pi = \frac{C}{d}$ of *any* circle in *any* universe. Note the efficient notation: one (Greek) letter for infinitely many digits[1]! This is *labeling* at its best: capture the essence of the concept while hiding the inconveniently lengthy reality.

Physical Constants

These are pure constants that define how Nature "works" in our universe; they could take on different values in other universes.

- **Gravitational constant:** $G = 6.67 \times 10^{-11} \frac{\text{Nm}^2}{\text{kg}^2}$. Tells us how strong gravity is in our universe. The numerical value is equal to the force in Newtons that an object of mass 1 kg exerts on another object of mass 1 kg at a distance 1 m. This is a tiny force; gravity is very weak!

- **Speed of light:** $c = 299\,792\,458 \frac{\text{m}}{\text{s}} \approx 3 \times 10^8 \frac{\text{m}}{\text{s}}$. The maximum speed at which particles and information can travel is the speed of light in vacuum. Light is slower in (transparent) media—which causes **refraction**. **Classical Newtonian mechanics** is a good approximation to **special relativity** if the velocity of objects are *small* compared to the speed of light.

- **Planck constant:** $h = 6.626070040(81) \times 10^{-34} Js \approx 6.6 \times 10^{-34} Js$. The numerical value is equal to the energy in Joules of an electromagnetic wave of frequency 1 Hz. **Classical Newtonian mechanics** is a good approximation to **quantum mechanics** when the product of the uncertainties in position and momentum of a particle are *large* compared to the Planck constant.

---

[1] Of course, you could argue that its one letter for one *real number*. Then again, this shows the power of the definition of real numbers.

A.1. CONSTANTS

- **Elementary charge:** $e = 1.602\,176\,620\,8(9\,8) \times 10^{-19}\,\text{C} \approx 1.6 \times 10^{-19}\,\text{C}$. Charge (in *Coulombs*) is quantized, that is, it comes in integer multiples of the **elementary charge**, which is the (absolute of) the charge of the **electron** and the **proton**.

- **Electric force constant:** $k_e = \frac{1}{4\pi\epsilon_0} \approx 9 \times 10^9 \frac{\text{N·m}^2}{\text{C}^2}$. The numerical value is equal to the mutual force (in Newtons) two particles carrying a charge of one Coulomb each exert on each other if they are 1m apart. In short, it tells us the strength of the **electrostatic force** in the universe.

- **Magnetic constant:** $\mu_0 = 4\pi \times 10^{-7} \frac{\text{m·kg}}{\text{s}^2\text{A}^2}$. The numerical value is the mutual force (in Newtons) two long straight wires carrying a current of one Ampère each exert on each other if they are $2\pi m \approx 6.28$ m apart. In short, it tells us the strength of the **magnetic force** in the universe.

- **Stefan-Boltzmann constant:** $\sigma = 5.670\,367\,(13) \times 10^{-8} \frac{\text{W}}{\text{m}^2\text{K}^4} \approx 5.7 \times 10^8 \frac{\text{W}}{\text{m}^2\text{K}^4}$. The numerical value is the power in Watts radiated by each square meter of a blackbody's surface divided by its absolute temperature to the fourth power. In short, it tells us how much light or radiation an object of a given temperature and area emits in our universe.

- **Lightyear:** $1\,\text{ly} = 9.4607 \times 10^{15}\,\text{m} = 63\,241\,\text{AU} \approx 10^{16}\,\text{m}$. The distance light travels in vacuum in one *Julian year* of 365.25 days.

Astronomical Constants

These are parameters characterizing objects in the universe. These are, if you will, historic accidents. If the universe had developed slightly differently these values would be different, even if the **physical constants** from above are unchanged.

- **Mass of the Earth:** $M_E = 5.974 \times 10^{24}$ kg.
- **Mass of the sun:** $M_S = 1.99 \times 10^{30}$ kg $= 333,000 M_E$.
- **Mass of the moon:** $M_M = 7.349 \times 10^{22}$ kg $= \frac{1}{81} M_E$.
- **Mass of Jupiter:** $M_J = 1.89 \times 10^{27}$ kg $= 318 M_E = \frac{1}{1047} M_S$.
- **Radius of the Earth:** $R_E = 6371$ km
- **Radius of the sun:** $R_S = 697\,500$ km $= 109 R_E$
- **Radius of the moon:** $R_M = 1737$ km $= \frac{1}{3.67} R_E$
- **Radius of the Jupiter:** $R_J = 71\,492$ km $= 11.2 R_E$
- **Acceleration due to gravity on Earth:** $g = \frac{GM_E}{R_E^2} = 9.806\,65 \frac{\text{m}}{\text{s}^2}$ The speed that an unsupported object near the surface of the Earth *gains* in each second it falls. Obviously, its value would be different if the Earth happened to have a different radius or mass.
- **Astronomical Unit (distance to the sun):** $1\,\text{AU} = 149.6\text{million km} = 1.496 \times 10^{11}$ m $= 8.31$ light minutes.
- **Distance to the moon:** $384\,000$ km $= 1.28$ light seconds.
- **Parsec:** $1\,\text{pc} = 3.0857 \times 10^{16}$ m $= 3.261\,56$ ly $= \frac{360 \times 60 \times 60}{\pi} = 206.260$ AU. One parsec is the distance at which one **astronomical unit** subtends an angle of one arcsecond.

A.2 Properties of Astronomical Objects

Planets

Orbital data:

Planet Name	Symbol	Semimajor axis a (million km)	(AU)	Sidereal Period P (years)	(days)	Eccentricity e	Inclination to ecliptic (°)
Mercury	☿	57.9	0.387	0.241	87.97	0.206	7.0
Venus	♀	57.9	0.723	0.615	224.7	0.007	3.4
Earth	⊕	149.6	1.00	1.000	365.256	0.017	0.0
Mars	♂	227.9	1.52	1.88	686.98	0.093	1.9
Jupiter	♃	778.3	5.20	11.86		0.048	1.3
Saturn	♄	1429	9.55	29.46		0.053	2.5
Uranus	♅	2871	19.2	84.1		0.043	0.8
Neptune	♆	4498	30.1	164.9		0.010	1.8
(Pluto)	♇	5906	39.5	248		0.249	17.1

Physical data:

Planet	Radius R (km)	(R_E)	Mass M (10^{24} kg)	(M_E)	Density ρ ($\frac{kg}{m^3}$)	Rotation (solar days)	Inclination (°) Equator to Orbit
Mercury	2,440	0.383	0.330	0.0553	5,430	58.6	0.5
Venus	6,052	0.949	4.87	0.815	5,243	243	177.4
Earth	6,372	1.000	5.97	1.000	5,515	0.997	23.43
Mars	3,397	0.533	0.642	0.107	3,934	1.026	25.19
Jupiter	71,492	11.2	1,890	318	1,326	0.414	3.12
Saturn	60,268	9.45	569	95.2	687	0.444	26.73
Uranus	25,559	4.01	86.8	14.5	1,318	0.718	97.86
Neptune	24,264	3.88	102	17.2	1,638	0.671	29.56
(Pluto)	1,187	0.183	0.013	0.0022	1,860	6.39	122.5

Largest Moons

Moon (Planet)	Semimajor axis a (mill. km)	(R_{planet})	Period (days)	Radius (km)	Mass (10^{22} kg)	Eccentr. e	Date of discovery
Moon (⊕)	0.384	60.3	27.3	1737	7.349	0.0549	—
Io (♃)	0.4216	5.90	1.769	1821	8.932	0.0041	1610
Europa (♃)	0.6709	9.39	3.551	1560	4.791	0.0094	1610
Ganymed (♃)	1.07	15.0	7.155	2634	14.82	0.0011	1610
Callisto (♃)	1.883	26.4	16.689	2400	10.77	0.0074	1610
Titan (♄)	1.2219	20.4	15.95	2575	13.4	0.0288	1655
Triton (♆)	0.548	23.4	5.877	1353	2.15	0.0000	1846

A.2. PROPERTIES OF ASTRONOMICAL OBJECTS

Important Constellations

Constellation			Sun in constell.	Best observed	Significance
Name	Abbr.	Symb.			
Ursa Major	UMa		never	Circumpolar	Stereotypical northern constell.
Ursa Minor	UMi		never	Circumpolar	Polaris (α UMi) close to CNP
Aries	Ari	♈	April	Autumn	Zodiac
Taurus	Tau	♉	May	Winter	Zodiac, Aldebaran, M45
Gemini	Gem	♊	June	Winter	Zodiac, Castor & Pollux, solstice
Cancer	Cnc	♋	July	Spring	Zodiac, M67 open star cluster
Leo	Leo	♌	August	Spring	Zodiac, Regulus
Virgo	Vir	♍	September	Spring	Zodiac, autumnal equinox
Libra	Lib	♎	October	Spring	Zodiac
Scorpio	Sco	♏	November	Summer	Zodiac, Antares
Sagittarius	Sag	♐	December	Summer	Zodiac, solstice, Milky Way Center
Capricorn	Cap	♑	January	Autumn	Zodiac
Aquarius	Aqr	♒	February	Autumn	Zodiac
Pisces	Psc	♓	March	Autumn	Zodiac, vernal equinox
Orion	Ori		never	Winter	M42, Betelgeuze, Rigel
Canis Major	CMa		never	Winter	Sirius (α CMa)
Canis Minor	CMi		never	Winter	Procyon (α CMi)
Boötes	Boo		never	Spring	Arcturus (α Boo)
Hercules	Her		never	Spring	M13 Globular Star Cluster
Lyra	Lyr		never	Summer	Vega (α Lyr), summer triangle
Cygnus	Cyg		never	Summer	Deneb (α Cyg), summer triangle
Aquila	Aql		never	Summer	Altair (α Aql), summer triangle
Pegasus	Peg		never	Autumn	Great Rectangle
Andromeda	And		never	Autumn	Andromeda galaxy is in constell.
Cepheus	Cep		never	Circumpolar	Important variable: δ *Cephei*
Ophiuchus	Oph		June	Summer	Ecliptic runs through modern const.
Centaurus	Cen		never	Southern H.	Closest star α *Centauri*
Crux	Cru		never	Southern H.	Stereotypical southern constell.

Typical Distances

1 AU is the typical distance of planets to the sun.
1 parsec is the typical distance to the (nearest) stars (α Cen = 1.3 pc).
1 meter is the typical distance from the head to the feet of a human (average height 1.7 m).

Star Maps

Star maps are readily available on the internet. For instance, `https://www.fourmilab.ch/yoursky/` has very convenient features. A good and free planetarium software package is available at `www.stellarium.org`.

Noteworthy Stars

Star Name	Bayer D.	Spectral Class	Dist. (ly)	Lumin. (L_{sun})	Brightn. (Sir.=1)	Mass (M_{sun})	Comments
Sirius A	α CMa A	A1 V	8.60	25.4	1.00	2.02	Brightest fixed star
Sirius B	α CMa B	DA2	8.60	0.056	0.0001	0.978	First White Dwarf
Arcturus	α Boo	K1.5 III	36.6	170	0.278	1.05	Giant
Vega	α Lyr	A0 V	25.3	40.1	0.261	2.135	Main Sequence
Aldebaran	α Tau	K5 III	65	518	0.122	1.5	Giant
Deneb	α Cyg	A2 Ia	2650	200,000	0.092	19	Supergiant
Betelgeuse	α Ori	M1 Iab	427	120,000	0.157	12	Red Supergiant
Rigel	β Ori A	B8 Ia	773	120,000	0.240	23	Blue Supergiant
Regulus	α Leo	B7 V	77.5	288	0.077	3.8	Main Sequence
Spica	α Vir	B1 III-IV	250	12,100	0.103	10.25	Giant
α Centauri	α Cen	G2 V	4.4	1.52	0.270	1.1	Closest star
61 Cygni	61 Cyg A	K5.0 V	11.4	0.153	0.021	0.7	First known parallax
δ Cephei	δ Cep A	F5-G1Ib	887	2000	0.0061	4.5	Variable star
Procyon	α CMi A	F5 IV-V	11.4	6.93	0.196	1.5	Main Sequence

Interesting Deep Sky Objects

Name		Const.	Type	Dist. (ly)	Diam.(ly)	Difficulty
Crab Nebula	M1	Tau	SN Remant	6500	11	hard
Great Hercules Cluster	M13	Her	Glob.Clust.	22,200	84	easy
Andromeda galaxy	M31	And	Spiral Gal.	2.5 mill.	220,000	easy
Great Orion Nebula	M42	Ori	Em.Neb.	1300	12	easy
Pleiades	M45	Tau	Open Cluster	444	14	easy
Ring Nebula	M57	Lyr	Plan.Neb.	2300	1.3	medium
Beehive	M67	Cnc	Open Custer	2700	10	medium
Virgo A	M87	Vir	Ell.Gal.	54 mill.	120,000	hard
Lg Magellanic Cloud	LMC	Dorado	Irr Gal.	163,000	14,000	obvious
Sm Magellanic Cloud	SMC	Tucana	Irr Gal.	200,000	7,000	obvious
Horsehead Nebula	Barnard 33	Ori	Dark Neb.	1500	3.5	very hard
Flocculent galaxy	NGC 4414	Coma	Gal.	62 mill.	56,000	hard

Glossary

aberration Original meaning is "a straying from a prescribed path." We encounter *stellar aberration* (an annual motion of stars with respect to a distant background due to the changing direction of orbital velocity of Earth) and *chromatic aberration* (the fact that simple lenses focus different colors at different focal points, which leads to a color distortion of telescopic images. 130, 136, 144, 146, 186, 187, 189, 190, 192, 209, 214, 218, 219, 226, 228

abscissa The horizontal axis of a two-dimensional ("xy") plot, on which the values of the *independent variable* are measured. 22

accepted value The result of a measurement that is confirmed by many different and usually varied experiments or measurements. 24, 194

accurate Something that agrees with an *accepted value* or notion. 3, 24, 27, 33, 70, 84–86, 101, 105, 106, 108, 109, 132, 137, 162, 183, 184, 194, 202, 214, 215, 242, 264, 281, 396, 415

aether In antiquity: the fifth element out of which the heavens are made. Chiefly in the nineteenth century: the alleged medium in which light travels. Also spelled *ether*. 80, 141, 148, 189, 214, 223, 326

alt-azimuth system A coordinate system for an observer (usually on Earth) which has an *altitude angle* giving angular separation of an object from the horizon, and an *azimuth angle* specifying the horizontal direction counted from north. 14, 15, 39

altitude The "vertical" angle of a celestial object with respect to the horizon . 7, 33, 58

analytic The approach of separating things and phenomena into their constituents or simplest elements. 19, 135

angle The figure formed by two rays sharing a common endpoint, called the vertex of the angle. Angles are measured in degrees (°) or radians (rad). A full circle is 360° or 2π rad. 7, 14, 15, 18, 35

angular size The angular size or apparent size is an angle describing how large an object appears from a given point of view. 18, 84, 168, 264

annular eclipse A solar eclipse in which the moon does not fully cover the sun's disk. A bright ring (Latin: *annulus*) is seen around the black disk of the new moon. 51

aphelion The point of an orbit where the planet is farthest from the sun. 119

approximation A quantity, shape or theory that is similar to a desired reference but easier to use. 18, 21, 31, 32, 97, 189, 313

asterism A group of stars forming an interesting or easy to remember pattern. 5, 12, 13

asteroid A small object in the solar system, also called a *minor planet*. 27, 218, 262

asteroid belt The region in the solar system between Mars and Jupiter where many of the *asteroids* or minor planets have their orbits. 266, 303

astronomical unit The average distance between the sun and the Earth, 1 AU = 149.6 million km. 121, 162, 168, 189, 193, 254, 390, 397, 427

astrophysics The branch of astronomy that uses results and principles from physics (and chemistry) to figure out the nature and inner workings of celestial objects. 150, 209, 213, 230, 243, 245–247, 251, 261, 310, 326, 327, 329, 330, 338, 341, 346

atmosphere The envelope of gases surrounding the earth and other celestial objects. 287, 290

aurora A natural light display in the Earth atmosphere chiefly close to the poles. It is caused by charged particles from the sun. Guided by Earth's magnetic field they bump into the molecules of the atmosphere, thereby exciting them. Upon de-excitement the molecules give off light in characteristic colors ("emission lines"), e.g. green in the case of nitrogen. Also called *polar* or *northern lights*. The terms *aurora borealis* and *australis* refer to the northern and southern lights, respectively. 300, 301

Avogadro's constant The number of molecules in a certain amount of substance (one mole). It is a fantastically large number (6×10^{23}). 26, 331

axis The axis of rotation of a celestial object, like the Earth, the moon, planets or the sun. 10, 36, 38, 50, 87, 103, 104, 183, 188, 278, 282

axis tilt The angle by which the rational axis of an object is not perfectly perpendicular to its orbit or some other plane of reference. 265

azimuth The horizontal angle of a celestial object as seen by an observer counted from the *north* point on the horizon. 7, 33, 58

belt The darker parts or stripes of the atmosphere of a Jovian planet. Oriented parallel to the planet's equator. 289

black hole End stage of very massive stars. Technically has infinite density and zero volume. The gravitational force is so strong close to the surface of the black hole that nothing—not even light—can escape, so appears black. 1, 19, 341, 360

brightness A measure of how bright an object *appears*. Also called *apparent brightness*. A very luminous object sends out a lot of light, but might appear dim if it is far away. Brightness is measured in Watts per square meter. 17, 111, 208, 417

Glossary

Celestial Equator The projection of the Earth's *equator* onto the celestial sphere. It is the circle of points equally far from the Celestial North and South Poles and divides the sky into the *northern* and *southern sky*. 38–40, 42, 58, 59, 87, 210

celestial meridian The great circle going through north, zenith, and south in the observer's sky. 8

Celestial North Pole is the projection of the *geographic north pole* onto the celestial sphere which seems to turn around it counterclockwise due to the rotation of our planet. 8, 69

celestial sphere An imaginary sphere of which the observer is the center and on which all celestial objects are considered to be situated. 37, 38, 58

chromosphere The outer layer of the sun above the photosphere. Much hotter but less dense. It sends out a pinkish glow visible during total solar eclipses. 244, 300

chronological A record of events starting with the earliest and following the timely order in which they occurred. 1, 69, 70, 108, 261

circumpolar Never setting or rising. Stars that are close to the celestial poles will rotate around the pole without sinking below the horizon. 37, 43, 59

conjunction The time when a planet is in the direction of the sun. For *inferior planets* there are two different conjunctions: the *inferior conjunction* when the planet stands between the Earth and the sun, and the *superior conjunction* when the sun stands between Earth and planet. For *superior planets*, an *inferior conjunction* is impossible. 56, 57

constellation A group of stars or *asterism* that is easy to remember and therefore has been given a name by astronomers, like *Taurus* or *Ursa Major*. 5, 12, 41

conversion factor A numerical factor that has to applied to change units on a quantity. For instance, to change units from inches to centimeters you multiply by the conversion factor $2.54 \frac{cm}{in}$. 11, 18, 29

Copernican revolution The shift from a geocentric to a heliocentric worldview, as epitomized by the switch from the Ptolemaic model to the Copernican model of the solar system. 25, 56, 377

corona Outer layer of the sun. Named after its crown-like appearance. The *corona* is extremely hot and a million times thinner than the *photosphere*. 255, 299

crescent A *phase* of moon or planets, as in *crescent moon* or *crescent Venus*, when the disk of moon or planet appears less that half illuminated. 4, 47, 49, 50, 128, 130

daily motion The apparent motion of objects in the sky due to the rotation of the Earth around its axis. Also called *diurnal motion*. 20, 38, 54, 59

dark energy A form of energy that does not dilute as the universe expands. Therefore it will become more and more dominant. Currently about 70% of the universe. As of now, scientist do not know what it *is*. 416–418, 420

dark matter A form of matter that has mass (is gravitationally interacting), but will not send out (known forms of) radiation, so it is "invisible". As of now, scientist so not know what it is, only what it does. 389, 394, 408, 417, 418

declination Celestial latitude, that is, the angular distance of a point on the celestial sphere from the *celestial equator*. 38, 39, 43, 44, 88, 187

deduction The process of getting from general premises or assumptions to a logically certain conclusion concerning a special case. 21, 55, 180

deep sky object A celestial object that is very far away; in particular: not part of our solar system. 6, 203, 204, 210, 211, 343

deferent An imaginary circle surrounding the Earth on whose circumference the center of a planet's epicycle was supposed to be carried around. 90–97, 105, 167, 168

differential rotation The rotational behavior of liquid or gaseous objects. Parts of them rotate faster than others. It is different from the familiar rotation of a stiff and solid object like the Earth or a billiard ball. 131, 290, 292, 300, 302

eccentricity A parameter e that represent by how much an ellipse deviates from a circle. A perfect circle has $e = 0$, and ellipses are confined to values of e smaller than one. The value $e = 1$ actually characterizes a *parabola*. 117–119, 391

ecliptic The apparent path of the sun among the stars. It is a circle tilted 23.5° with respect to the celestial equator. 39, 40, 59, 86, 88, 187, 188, 262

electrodynamics The theory describing the forces between charged particles, and the interactions between charges and electric and magnetic fields. 406

elongation The angular distance of the planet from the sun as seen from Earth. For superior planets all angles are possible, for the inferior planets, there is a greatest possible angle associated with the notion of a greatest (eastern or western) elongation. 57, 59, 95, 105

energy An abstract quantity telling us how much work an object can do. Possible units are *Joules* or *Newtons* times *meters*. 20, 198, 213, 221, 233, 234, 295, 311, 344

energy conservation The *physics* definition of conservation of energy has it that energy cannot be created nor destroyed; it only changes its form. For instance, kinetic energy can be transformed into thermal energy when an object slows down under the influence of friction. 20, 233, 335

epicycle A small circle whose center moves around the circumference of a larger circle called *deferent*. 91, 92

equator The collection of points on a sphere equally far off its poles; a circle. 9

equatorial coordinate system The coordinate system based on the equator, and therefore on the rotation of the Earth about its axis. 6, 14, 38, 87

Glossary

equinox The two intersecting points of ecliptic and celestial equator. The sun reaches these points on March 21 (vernal equinox) and September 23 (autumnal equinox). At these two dates day and night have the same ("equal") length of twelve hours. 39, 40, 87, 88, 120, 215

error An uncertainty in measurements. Two main classes are *random error* and *systematic error*. The former can be reduced by repeating the measurement many times over, the latter by understanding the measurement process or device better. 24, 30, 87, 110, 139

estimate The result of a quick and rough but *rational* method to measure or calculate a quantity. 15, 22

evidence Anything presented in support of an assertion, regardless of whether the support is strong or weak. 25, 73, 79, 104, 107, 112, 130, 136, 146, 153, 161, 165, 190, 247, 271, 272, 277, 280–283, 286, 287, 293, 298, 299, 303, 340, 343, 361–364, 384, 387, 389, 394, 410–412, 417, 420

exoplanet A planet orbiting a star other than our sun. 28, 297

exponent Represents the power to which a given number or expression is to be raised, for instance 3 in the expression $2^3 = 2 \times 2 \times 2 = 8$. 26, 325

extrapolation The extension of a graph, curve, range of values, or more generally the behavior of a theory, beyond the originally observed range of validity, by inferring unknown values or behaviors from trends in existing data. 21, 356, 363, 366, 389

falsified Shown to be incorrect, usually by an observation contradicting the prediction of a theory, as in "The geocentric theory was falsified by GALILEO's observation of Venus's full phases". 25, 31, 32, 48, 76, 107, 111, 112, 118, 128–130, 134, 147, 161, 195, 197, 201, 218, 221, 235, 255, 266, 297, 313, 314, 320, 363, 416

first quarter A *lunar phase*, when the disk of the *waxing* moon appears exactly half illuminated. 47, 49, 52

fit A (good) description of the relation between two or more quantities by a mathematical function. 31, 417

fixed star A speck of light that seems to be fixed to the celestial sphere, thereby rotating 360° in 23h 56m, that is, in a *sidereal day*. The sun in this sense is not a fixed star, since it moves with respect to the stars, that is, with respect to the celestial sphere, and rotates in exactly 24 hours (solar day). *Physically* the sun is a star like the others: it radiates energy gained from fusing hydrogen. 13, 14, 35, 36, 39, 41, 54, 96

force A push or pull exerted on an object by another object. Force has a direction and a strength or magnitude. The latter is measured in *Newtons* (a unit equal to $1\,\text{kg}\frac{\text{m}}{\text{s}}$). 20–22, 29, 80, 100, 117, 151

Fraunhofer lines The dark lines in a spectrum produced by atoms and molecules when absorbing light of a specific color or wavelength. 31, 228, 247, 343

full A *phase* of moon or planets, as in *full moon* or *full Venus*, when the disk of moon or planet appears fully illuminated. 47, 50–53, 58, 128, 344

galactic equator The plane of the Milky Way. The Milky Way is a disk rather than a sphere, but its rotation and in particular its rotation axis invite the description of positions within the galaxy in terms of latitude and longitude. 205, 376

galaxy An "island universe", that is, a collection of many stars, gas clouds, and other forms of matter. There are different forms like spiral, elliptical, and irregular galaxies. The Milky Way is a spiral galaxy. 6, 7, 150, 203, 204

gas cloud An interstellar assembly of gas (and often dust). 7, 127, 206, 244, 249, 254, 294, 317

geocentric model The model of the solar system in which the Earth (Greek: *ge* or *gaia*) is at the center, and is orbited by all other objects in the solar system including the sun. 25, 42, 87, 113, 128, 130

gibbous moon A *phase* of moon or planets, as in *gibbous moon* or *gibbous Venus*, when the disk of moon or planet appears more that half illuminated. 47, 50, 129, 130

graph A diagram showing the relation between (typically two) variables, also called a *plot*. 29, 143, 367

greenhouse effect The effect in which radiation from a planet's *atmosphere* warms the planet's *surface* to a temperature higher than what it would be without its atmosphere. 265, 274, 276, 280, 288, 289

heliocentric model The model of the solar system in which the sun (Greek: *helios*) is at the center, and is orbited by all the planets including Earth. 25, 42, 88, 104, 130

Hertzsprung-Russell diagram A plot of the luminosity of stars (on the vertical axis) versus their temperature or color or spectral type (on the horizontal axis). 317–322, 339, 341, 345, 346, 348, 349, 352, 368

horizon The apparent line that separates the Earth from the sky. At many locations, the *true horizon* is obscured by trees or buildings. 4, 5, 7–10, 15, 16

Hubble's law Hubble discovered the fact that recessional velocities of distant galaxies are proportional to their distances from us. 401, 402, 410

hydrostatic equilibrium If a object is in hydrostatic equilibrium, the net force on each of its parts is *zero*, hence each part "floats" in place, so to speak. 249, 250, 253, 324, 330, 344, 345, 347, 368

hypothesis A proposed explanation made on the basis of limited evidence as a starting point for further investigation. By contrast, *theories* or *models* are typically more complex combinations of several *hypotheses* (plural). 22, 23, 25, 42, 102, 108, 118, 121, 132, 145–147, 246

Glossary

idealized Represented as perfect or better or simpler than in reality. 10, 25, 38, 45, 152, 164, 255

induction The process of getting to a general conclusion or theory from a set of premises which supply evidence for the conclusion. Contrary to deductive reasoning, the truth of the conclusion is probable but not certain. 21, 55

industrial revolution A period of major industrialization that took place around the year 1800. Start date is often taken to be the crucial improvement of the *steam engine* by JAMES WATT in 1769. 199, 213, 232

inferior planet A planet closer to the sun than the Earth. 56, 57

Jovian planets Big, gaseous planets in the solar system (Jupiter, Saturn, Uranus, Neptune) or in exosolar environments. 289, 293, 303

Kepler's laws Three laws of planetary motion that describe how a planet moves around the sun. 115, 314, 377

last quarter moon A *lunar phase*, when the disk of the *waning* moon appears exactly half illuminated. Also called *third quarter moon*. 47, 71

latitude An angle describing the (angular) distance of a point off the equator on a sphere. 9, 10, 12, 15, 33, 35–38, 40, 42–44, 85, 88, 183, 187, 191, 193, 194, 210, 214, 285, 300

law Also called *physical law*; is a relation between physical quantities extracted from the results of repeated scientific experiments and observations over many years which have become accepted universally within the scientific community. 23, 26, 92

law of universal gravitation A formula that states how strong the gravitational force is between two objects. According to NEWTON, the strength of the force depends linearly on the two masses and falls off with the square of their mutual distance. 31, 117, 137, 151, 162, 195, 250, 254, 313, 315

local noon The time when the sun culminates (is highest in the sky) at the location of the observer. 7, 10, 11, 33, 44

longitude The angle on a sphere describing the angular distance of a point on a sphere from a chosen *prime meridian*. 9–12, 38–40, 44, 87, 88, 184, 185, 187, 193, 215, 279

luminosity The amount of energy (literally: light) that an object sends out per unit time. Measured in *Watts*, which is a unit of *power*. 17, 111, 226, 316, 317, 321, 346, 350, 351, 367, 382, 417

lunar eclipse When the shadow of the Earth falls onto the full moon, it appears orange and not black, since some light is scattered into the shadow region by the Earth's atmosphere. 51–53, 79, 88

lunar phases The moon's hemisphere facing the sun is always illuminated, but an observer on Earth sees different fractions of it—depending on the relative position of Earth, moon and sun. These are the lunar phases. For example, when the moon is opposite of the sun, we see the entire illuminated hemisphere of the moon and call the phase *full moon*. 47, 49, 71

main sequence Main sequence stars are "normal" stars which fuse hydrogen to helium and have an intuitive luminosity, in the sense that they are the more luminous the hotter they are. 272, 318, 320, 321, 337, 339, 345–347, 367, 368

maria The "oceans", that is, darker and sparsely cratered lowlands of the moon's surface. 50, 276, 277, 283

Milky Way Our own spiral galaxy. The sun rotates around its center every 250 million years, sometimes known as a *galactic year*. 7, 205–207, 209, 211, 343, 350

model A comprehensive way to describe patterns in nature. For example, the *standard solar model* is a well-defined theory of the sun. The term is often used anonymously with *theory*, as in the *heliocentric model* or *heliocentric theory*. 22, 23, 45

nebular hypothesis A theory to explain the formation and evolution of the Solar System and other systems. Sometimes called the *Kant-Laplace theory* after its main inventors. 206, 254, 276, 293, 297, 343, 344

neutron star Ultra-dense end stage of massive stars. Has extreme magnetic fields and rotates very fast. 341, 342, 360, 362–364, 379

new A *phase* of moon or planets, as in *new moon* or *new Venus*, when the disk of moon or planet appears dark . 47, 49–53, 59, 128, 129

node line The line between the *ascending node* and the *descending node* in an orbit. The *nodes* are the points where the orbit crosses the ecliptic plane. 46

Occam's razor A principle for *problem solving* that gives preference to solutions which are as simple as possible while still describing all known facts and relations. 76, 100, 117

operational definition A definition of a term by a set of instructions. For instance, the solar day is defined as the time from noon to noon, which implies that you would measure its length with a stop watch started at noon and stopped at noon the next day. 5, 7, 16

opposition The time when a planet appears opposite of the sun in the observer's sky. 56, 58, 106, 163

ordinate The vertical axis of a two-dimensional ("xy") plot, on which the values of the *dependent variable* are measured. 22

parsec A unit of distance equal to 3.26 lightyears. A star one parsec away will move one arcsecond due to the parallactic effect, hence the name "parallactic second". 219, 220, 427

partial lunar eclipse A *lunar eclipse* in which the full moon only *partially* enters into the shadow of the Earth. 52

partial solar eclipse A solar eclipse in which the moon does not travel centrally in front of the sun's disk. 51, 113

penumbra The half-shadow of the moon as it falls on Earth during a solar eclipse. 51

perihelion The point of closest proximity to the sun of a planet's orbit. 119, 185, 186, 405

photosphere The deepest of the outer layers of the sun. It sends out most of the light we receive from the sun. 238, 247, 248, 299, 326

planetary nebula The outer layers of a normal star are shed at the end of its life, forming an often symmetric gas cloud which looks a bit like a planet from lightyears away. 5, 205, 244, 348, 349, 355, 356, 362, 378

planetesimal Chunks of mass in the early solar system akin to asteroids with a diameter of about one kilometer. 295

precession The (very slow) rotation of the Earth's axis around the *pole of the ecliptic*. It results in a change of the location of the equinoxes. This will result in a systematic change in celestial coordinates which have the vernal equinox as their origin ("zero point"). 40, 86, 88, 92, 167, 183, 186, 191

precise Sharply defined, as in several measurements yielding (close to) the same, but not necessarily the accepted or *accurate* result. 24

prime meridian A great circle of constant longitude on a planet that serves as the origin of the longitude angle. On Earth it goes through Greenwich, UK. 9, 11

proper motion The motion of stars with respect to the background of other, more distant stars. Proper motion is very slow, on the order of arcseconds per year even for the fastest stars. 15, 20, 205, 208, 218, 317, 340

proportional Corresponding in size or amount to something else, as in "The distance traveled at constant speed is proportional to the time elapsed". 29, 31, 100, 121, 126, 132, 152, 154, 156, 157, 159–161, 187, 189, 195, 198, 215, 227, 233, 238, 250, 255, 302, 319, 363, 403, 404, 406

protoplanetary disk A rotating dense circumstellar disk of gas and dust surrounding a recently formed star. 293, 298

qualitative Qualitative information is concerned with qualities that are descriptive, subjective or difficult to measure. 22, 72, 77, 81, 90, 138, 140, 157, 179

quantitative Quantitative information is based on quantities obtained using a quantifiable measurement process. 14, 15, 17, 22, 78, 81, 90, 138, 140, 157, 179, 333

quantum mechanics The theory of fundamental interactions between particles. Needs to be used when the objects are very small, and "influences" are comparable to the Planck constant, which is on the order of 10^{-34} Js, that is, exceedingly small. 138, 153, 165, 228, 239, 244, 309, 310, 312, 313, 315, 326, 327, 333, 334, 342, 367, 426

red giant A star in the later part of its life, when the hydrogen fuel in its core has bee used up. The star puffs up ("giant") and becomes cooler and therefore redder. 318, 319, 325, 337, 339, 346–348, 359, 367, 368

relativity Einstein's modern theories of physics. *Special* relativity posits that the speed of light is the same for every observer, whence other quantities and concepts such as including time and simultaneity become *relative*, that is, depend on the relative velocity of observer and observed system. *General* relativity is Einstein's theory of gravity. Einstein does away with the notion of gravitational forces. The effects of gravity are due to the fact that each massive object will put a dent in space-time, so that other objects have to travel through curved space-time, which makes their (three-dimensional) trajectories look like orbits. 21, 25, 108, 150, 153, 189, 223, 309–312, 326, 341, 342, 356, 366, 392, 393, 402–406, 408, 414, 416, 417

retrograde motion The apparent motion of a planet around the time of its *opposition*. While the planet typically moves eastward with respect to the stars (like the sun and the moon—so called *prograde motion*), in retrograde motion it moves westward. 58, 59, 78, 93, 103

Right Ascension Celestial longitude, that is, the angular distance of a point on the celestial sphere from the *vernal equinox*. 38, 39, 88, 215

ring A collection of small objects and fragments around the Jovian planets. 291

save the appearances An expression subsuming the main task of a theory or explanation of celestial (and other) phenomena. A theory has minimally to describe "what is happening", that is, the phenomena which appear (in the sky or elsewhere). 38, 95, 102, 105, 360

scale Scientific jargon for *typical size* as in "The scale of the solar system is the astronomical unit", meaning that the typical distance between planets are on the order of 1 AU = 149.6 million km, but can range from a tenth to over ten AU. 26, 193, 194, 244

scaling The idea that properties of objects in nature and math must behave in a special way as a function of their overall size. 27–29, 84, 332, 364, 399

scientific method A method of procedure in the modern natural sciences. It consists of systematic observation, measurement, experiment, and the formulation, testing, and modification of hypotheses. However, there is no strict set of rules that need to be followed. In practice, the scientist does whatever he or she has to to advance scientific knowledge. 17, 99, 108, 124, 133, 135, 142, 161, 181, 185, 298, 312

scientific revolution The *early modern period* around the year 1600 ("from Copernicus to Newton") when crucial scientific discoveries where made, especially in physics, astronomy and mathematics (but not chemistry, which had its own revolution—delayed by a century). 70, 101, 122, 123, 135, 142, 149, 179

seasonal motion Yearly changes in the sky. For example, the sun appears to move along the *ecliptic* among the stars as months go by due to the orbital motion of the Earth. As a result, we see different constellations at night in different seasons. 20, 41, 59

seeing A term astronomers use to describe the overall quality of the observing conditions. Good seeing refers to calm, clear skies when minute details are discernible. Bad seeing means that turbulence, haze and aerosols in the atmosphere blur out details. 111, 218

semi-major axis Half the long axis of an ellipse, analogous to the *radius* of a circle. 117, 118, 121, 123, 157, 207, 386

sidereal day The time (23h 56m) it takes the Earth to make a full rotation, that is, to rotate 360° with respect to the stars or space. 39, 40, 55, 59

sidereal month The "star month", that is, the time in which the moon makes a full 360° rotation about the Earth with respect to the stars. The sidereal month of 27.3 days is shorter than the synodic month (from full moon to full moon) of 29.5 days. 47, 160

solar day The time (24 hours exactly) from noon to noon, that is, between two culminations of the average sun. 10, 37, 39, 55

solar eclipse The new moon obscuring the sun's disk by being exactly in the line of sight, so that the observer on Earth, the moon and the sun form a straight line. 41, 49, 51–53

solstice The two points of maximal angular distance from the celestial equator of the sun along its apparent path among the stars, that is, the ecliptic. The sun is 23.5° north of the celestial equator on June 21, and 23.5° south of the celestial equator on December 21. 35, 40, 41, 70, 84, 188

space-time The four-dimensional stage (three space and one time dimension) on which the action of the universe unfolds. 367, 400, 406, 409, 412

speck of light A somewhat dim point of light in the sky, a phenomenon we observe the night sky with the naked eye. We cannot tell *what* the speck of light is (a star, a planet, a star cluster, a galaxy) without further study. 4, 5, 7, 14, 46

spectroscopy The investigation and measurement of *spectra* produced when matter interacts with or emits electromagnetic radiation. For example, the study of dark lines in sunlight as it falls through a prism. 209, 213, 221, 228, 244, 245, 251, 298

star hopping A method to find a dim object (star or otherwise) in the sky by hopping from one bright and easy to find star to the next to get closer to the dim object with a telescope or binoculars. 14

Stefan-Boltzmann law A law of thermodynamics stating how much energy per unit time an object sends out—based on its temperature and surface area. 240, 244, 248, 275, 284, 285, 294, 299, 304, 317, 322, 339, 346

stellar parallax The apparent annual motion of close stars with respect to a distant background. The reason is the change in position of the observer due to the orbital motion of Earth. 103, 192, 226, 242, 243, 316

sunspot A region on the surface of the sun that is a few hundred Kelvin cooler than its surroundings. It appears black because it radiates much less due to its lower temperature. 247, 300, 303

supergiant Exceedingly luminous stars at the later stages of their lives. 318, 348, 349, 355

superior planet A planet farther from the sun than the Earth. 56

supernova The violent end stage of a very massive star. The entire star explodes. There are different types of *supernovae* (plural). 112, 134, 205, 321, 342, 356

synchronous rotation A rotation that is synchronized with some other event. For instance, the synchronous rotation of the moon, in which the moon rotates around its axis in exactly the same time as it orbits the Earth. 48, 50, 264, 277

synodic month The time from full moon to full moon (29.5 days). It is longer than the sidereal month (360° rotation with respect to the star) of 27.3 days. 50, 58, 59

synodic period The time between two successive identical configurations of sun and planet in the sky. For example, the time between two consecutive *inferior conjunctions* of Mercury is 116 days, and the time between two *oppositions* of Mars is 780 days. 57, 58, 95, 131

telescope Devices that collect light to produce a magnified image of an object. The two main types are *reflecting telescopes* (in which the main optical device focusing light is a parabolic mirror) and *refracting telescopes* (in which the main optical device focusing light is a lens). 125, 126, 151, 215

tentatively accepted Conclusions or theories in science are accepted until something better or simpler comes along. 25, 108, 212, 362

terrae The heavily cratered, lighter highlands of the moon's surface. 50, 276, 277

terrestrial planets Small, earth-sized, rocky planets in the solar system (Mercury, Venus, Earth, Mars) or in exosolar environments. 293, 295, 303

theory A well-established and wide-reaching description of nature. Examples include Newton's gravity theory, Einstein's relativity theory, Descartes's vortex theory, Newton's corpuscular theory, electromagnetic theory, thermodynamic theory, quantum theory. 22, 23, 31, 80, 108, 319

Glossary

thermal equilibrium When objects are in thermal equilibrium, the rate at which they receive energy is the same at which they gives off energy. 275, 284, 302, 324, 328, 330, 365

thermodynamics A branch of physics that deals with heat, temperature and relations between different forms of energy. 142, 179, 199, 202, 213, 221, 231, 232, 234, 274, 284, 310, 319, 403

total eclipse An eclipse of the sun or the moon in which the object is completely obscured. 51, 244

tropical year The time it takes the Earth to orbit the sun from vernal equinox to vernal equinox; since the equinoxes move, this is not the same as the *sidereal year*. The tropical or *solar year* is the basis of our calendars. This way, the date of the beginning of spring is fixed. 11, 71, 87

two-sphere model An antique model describing the cosmos as essentially two spheres: the celestial sphere with all heavenly objects, and the sphere of the Earth with the observer. If the observer changes location on Earth, this will change his view of the sky. 44, 46

umbra The shadow of the moon as it falls on Earth during a solar eclipse. 49, 51

uncertainty A parameter characterizing the range of values attributed to a measured quantity. Depending on the average value of the quantity which can be big or small, it is often better to use *relative uncertainty* or *percent uncertainty*, which compare the uncertainty with the average value of the quantity. 21, 24, 30, 33, 92, 118, 184, 219, 220, 268, 415

useful A concept or method is useful when it achieves a purpose in an efficient way. 8, 13, 53, 97, 106

Venus phases Analogous to the moon phases, planets too show phases. In particular, the *inferior planets* Venus and Mercury go through phases similar to the moon's. 130

waning A moon (or planet) phase that is "diminishing", that is, the next day the disk of the moon (or planet) will appear to be illuminated a little *less*. 47, 50

waxing A moon (or planet) phase that is "growing", that is, the next day the disk of the moon (or planet) will appear to be illuminated a little *more*. 47, 50

well-defined Unambiguous, or having a unique interpretation. 13, 19, 22, 76, 438

white dwarf End stage of normal stars like the sun. Basically their burned out cores. Have high density, high temperature ("white") and are planet-sized ("dwarf"). 216, 220, 318, 325, 339–342, 348, 349, 351, 352, 359, 361–364, 366–368, 379, 388

winter constellation A constellation that culminates around midnight during the winter months. For instance, *Orion, Gemini, Taurus*. 41

work A transfer of energy. For instance, by lifting a rock from the ground we transfer energy by giving the rock kinetic energy and potential energy we are doing *work* on the rock. 20, 234

zenith The point right above the observer. 5, 15, 34, 87

zodiac An area of the sky that extending approximately 8° north or south of the *ecliptic*, that is, the apparent path of the sun on the celestial sphere. 40

zone The lighter parts or stripes of the atmosphere of a Jovian planet. Oriented parallel to the planet's equator. 289

Bibliography

[1] ARONS, A. B. *Teaching Introductory Physics*. Wiley, 1997.

[2] ATKINSON, R. E., AND HOUTERMANS, F. G. Zur Frage der Aufbaumöglichkeit der Elemente in Sternen. *Z. Phys.* **54** (1929), 656–665.

[3] BAADE, W., AND ZWICKY, F. On Super-Novae. *Proc. Nat. Acad. Sci.* **20** (1934), 254–259.

[4] BESSEL, F. W. Bestimmung der Entfernung des 61sten Sterns des Schwans. *Astronomische Nachrichten* **365/366** (1838).

[5] BETHE, H. A. Energy Production in Stars. *Phys. Rev.* **55** (1939), 434.

[6] BIEŃKOWSKA, B. Heliocentric Theory in Polish Schools. In *The Reception Of Copernicus' Heliocentric Theory*, J. Dobrzycki, Ed. Springer, 1973.

[7] BLESS, R. *Discovering the Cosmos*. University Science Books, 1996.

[8] BOSWELL, J. *The Life of Samuel Johnson*. New York: Penguin Classics, 1986.

[9] BRIDGMAN, P. W. *Reflections of a Physicist*. Philosophical Library, 1955.

[10] BURBIDGE, M., BURBIDGE, G., FOWLER, W., AND HOYLE, F. Synthesis of the Elements in Stars. *Rev. Mod. Phys.* **29** (1957), 547–650.

[11] BURKE, J. *Discovery of the Universe*. Little, Brown, 1985.

[12] BUTTERFIELD, H. *The Origins of Modern Science*. Free Press, 1949.

[13] CHAISSON, E., AND MCMILLAN, S. *Astronomy: A Beginner's Guide to the Universe*. Pearson, 2013.

[14] DIJKSTERHUIS, E. J. *Die Mechanisierung des Weltbildes*. Springer, 1956.

[15] EDDINGTON, A. S. The Internal Constitution of the Stars. *Observatory* **43** (1920), 341–358.

[16] EINSTEIN, A., PODOLSKY, B., AND ROSEN, N. Can Quantum-Mechanical Description of Physical Reality Be Considered Complete? *Phys. Rev.* **47** (1935), 777.

[17] EMDEN, R. *Gaskugeln*. Teubner, 1907.

[18] GAMOV, G. *Thirty Years that Shook Physics*. Dover Publications, 1985.

[19] HANKINS, T. *Science and Enlightenment*. Cambridge University Press, 1985.

[20] HEISENBERG, W. Über die quantenmechanische Umdeutung kinematischer und mechanischer Beziehungen. *Z. Phys.* **33**(1) (1925), 879–893.

[21] HEISENBERG, W. *Der Teil und das Ganze*. Piper, 1969.

[22] HUNT, L. *The Making of the West: Peoples and Cultures: A Concise History: Volume II: Since 1340*. Boston: Bedford/St. Martin's, 2007.

[23] KANT, I. *Beantwortung der Frage: Was ist Aufklärung?* 1784.

[24] KEPLER, J. *Astronomia Nova*. Heidelberg, 1609.

[25] KIRCHHOFF, G. R. Ueber das Verhältniß zwischen dem Emissionsvermögen und dem Absorptionsvermögen der Körper für Wärme and Licht. *Annalen der Physik* **187** (1860), 275–301.

[26] KUHN, T. S. *The Copernican Revolution*. Harvard University Press, 1957.

[27] LINDBERG, D. C. *The Beginnings of Western Science: The European Scientific Tradition in Philosophical, Religious, and Institutional Context, Prehistory to A.D. 1450*. University of Chicago Press, 2008.

[28] LIPSCHUTZ, S., AND LIPSON, M. *Schaum's Outline of Linear Algebra*. McGraw-Hill, 2012.

[29] NORTH, J. *Cosmos: An Illustrated History of Astronomy and Cosmology Revised Edition*. University of Chicago Press, 2008.

[30] OSTERBROCK, D. E. Walter Baade's Discovery of the Two Stellar Populations. In *Stellar populations proceedings of the 164th symposium of the International Astronomical Union*, G. G. Pieter C. van der Kruit, Ed., vol. **164**. Kluwer, Dordrecht, 1994, p. 21.

[31] PLANCK, M. Vom Relativen zum Absoluten. *Naturwissenschaften* **13** (1925), 52.

[32] PURRINGTON, R. *Physics in the Nineteenth Century*. Rutgers University Press, 1997.

[33] RUSSELL, H. N. Relations Between the Spectra and Other Characteristics of the Stars. *Nature* **93** (1914), 252.

[34] SCHWARZ, O. *Zur historischen Entwicklung der Theorie des inneren Aufbaus der Sterne von 1861 bis 1926*. Veröffentlichung der Archenold-Sternwarte Berlin Treptow Nr. 22, 1992.

[35] SCHWARZ, O. Die frühe Entwicklung der Theorie des inneren Aufbaus der Sterne. In *Entwicklung der Theoretischen Astrophysik*, G. Wolfschmidt, Ed., vol. **4**. Nuncius Hamburgensis, Beiträge zur Geschichte der Naturwissenschaften, Hamburg, 2011, pp. 267–281.

[36] SLATER, T., AND FREEDMAN, R. *Investigating Astronomy*. Freeman, 2014.

[37] TASSOUL, J.-L., AND TASSOUL, M. *A Concise History of Solar and Stellar Physics*. Princeton University Press, 2014.

[38] THOMSON (LORD KELVIN), W. On the convective equilibrium of temperature in the atmosphere. *Manchester Phil. Soc. Proc.* **II** (1862), 170–176.

[39] TRUZZI, M. On Pseudo-Skepticism. *Zetetic Scholar* **12/13** (1987), 3–4.

[40] VAUCOULEURS, G. D. *Discovery of the Universe*. MacMillan, 1957.

[41] VON WEIZSÄCKER, C. F. Über Elementumwandlungen im Innern der Sterne. *Phys. Zeitschr.* **38** (1938), 633–646.

[42] WALLACE, A. R. *Is Mars Habitable? A Critical Examination of Professor Percival Lowell's Book "Mars and its Canals," with an Alternative Explanation*. Macmillan, 1907.

[43] WEBER, M. *Wissenschaft als Beruf (Science as a Vocation)*. Hackett, 2004.

[44] WEINBERG, S. *To Explain the World: The Discovery of Modern Science*. Harper, 2015.

[45] WOLF, M. Der Einfluss kosmischer Probleme auf die Entwicklung der Spektralanalyse. *Z. f. Elektrochem.* **18** (1912), 457–465.

[46] WUSSING, H. *Isaac Newton*. Teubner, 1990.

[47] ZÖLLNER, J. K. F. *Photometrische Untersuchungen mit besonderer Rücksicht auf die physische Beschaffenheit der Himmelskörper*. Engelmann, 1865.

[48] ZÖLLNER, J. K. F. *Wissenschaftliche Abhandlungen*, vol. **4**. Commissionsverlag v. L. Staackmann, 1881.

Printed in the USA
CPSIA information can be obtained
at www.ICGtesting.com
LVHW082025290824
789547LV00018B/51